Probability and Statistics

Models for Research

DANIEL E. BAILEY

University of Colorado

JOHN WILEY & SONS INC.

New York · London · Sydney · Toronto

Library of Congress Catalogue Card Number: 75-130428

ISBN 0-471-04160-2

Printed in the United States of America

10 9 8 7 6 5 4 3 2 1

To Sara

Preface

At one time quantitative methods in the social and behavioral sciences were either a specialist's domain or were restricted to a limited range of simplified calculational procedures. Fortunately, the day is approaching when the average social and behavioral scientist will not only know how to calculate a statistic but also why. The trend on which that conclusion is based is the increasing number of publications in which mathematical work is reported and the number of publications reporting work in which creative use of quantitative techniques is a basic component. This book is a presentation of probability and statistics geared to the increasing need for social and behavioral scientists to understand probability and statistics as models for research. As such, it is an interpretation of the basic state of mathematical probability and mathematical statistics in the context of applications in the social and behavioral sciences.

In this book I present a range of material in the depth and detail sufficient to prepare advanced undergraduate students and beginning graduate students (1) to read the contemporary literature in the social and behavioral sciences, (2) to read portions of the basic literature in the fields of probability theory and mathematical statistics, (3) to communicate with and consult with mathematical statisticians, and (4) to ask research questions so as to make available to themselves the methods of probability and statistical theory and practice. This is not a textbook in mathematical probability or mathematical statistics. Instead, it is an interpretative presentation of the mathematical and logical basis of probability and statistics, indulging in some mathematics, but concentrating on the logical and scientific meaning of probability and statistical reasoning. Where mathematical concepts and skills are needed they are developed in the context of their use. The mathematical training needed from the outset in approaching the book is simple algebra. In addition to a modest background in algebra, an ability to work with mathematical symbols and abstraction is the sole prerequisite that is assumed. No previous work in probability or statistics is needed.

No attempt has been made to present sophisticated or representative social and behavioral science content in the examples and problems used in the development of the probability and statistics presented. However, all of the examples and problems with content reference are couched in terminology and motivation coming from the social and behavioral sciences.

Much of the statistical application in the social and behavioral sciences is really better labeled "miss-application." The development of statistical decision theory based on probability theory provided a very powerful tool. However, this tool has not been used properly. Too much emphasis on precision in decision making in statistical applications has pervaded social and behavioral science, relative to the sophistication and precision of the research and resulting data. I do not mean the technical errors in the applications of the sort that are often referred to, such as violation of distribution assumptions in using a statistical test. Instead, I feel that the social and behavioral scientist often uses filter paper when chicken wire is the right mesh to catch the fish that are of interest. A great deal of emphasis has been put on problems of meeting assumptions in statistical tests and statistical significance. It is my strong conviction that many social and behavioral scientists have misspent talents and efforts in trying to achieve statistical significance when they would have made better use of their lives and the literature by attending first to scientific significance. Use of the powerful and sensitive tools of statistical decision heory is appropriate when substantive theory and methodology is sufficiently advanced. However, insistence by journal editors on the use of parts of statistical decision theory (often an inappropriate part) in the face of relatively primitive scientific development is not appropriate, although widespread. The use of statistical calculations and models to elucidate the meaning and structure of research findings should not always be accompanied by the appendage of decision theory and the reporting of "significance levels." Only when research is sufficiently well founded in scientific contexts is the application of sophisticated statistical decision theory justified. This book addresses the techniques and the foundations of sophisticated statistical decision theory, but the implication that it is always the method of choice is not to be drawn.

I have used the materials presented in this book for eight years in a first-year graduate level course. The text has undergone at least two rewriting stages. The initial stages were distributed to the students in my courses in the form of copies of typescript drafts. The course is a two-semester course (a total of approximately thirty-two weeks with three fifty-minute lectures a week). I find that I cover the first fifteen or sixteen chapters in one semester and the remainder in the second semester. In this way, I cover probability theory and basic statistical theory rather rapidly and spend a proportionately longer time on applications of statistical test theory. During the first semester, the laboratory (two hours a week) is devoted to instruction of computer programming (each student learns Fortran and executes numerous small projects and a "major" project) without any coordination with the lecture portion of the course. During the second semester, the laboratory is closely coordinated with the lecture. Problems in the analysis of data and the design of experiments are presented to the students in the laboratory in the second semester, coordinated with the material being covered in the lecture portions of the course. In addition, during the second semester, the students are taught to use the "canned" programs available for statistical analysis and they execute programming projects in connection with statistical calculations.

The book has been in preparation, theoretically, for sixteen years. As a graduate student at Berkeley in 1953 I felt that a need existed for the type of book that I have written. That feeling developed into a conviction when I spent two years as a postdoctoral fellow in the Department of Statistics at Berkeley. During that tenure I discovered that there was a rational basis to the statistical practice that was described in the textbooks then available to social and behavioral scientists. When I was faced with the practical problem of teaching a statistics course for the first-year graduate students in psychology at the University of Colorado in 1962, the conviction was transformed into activity.

Many persons have contributed to the production of this book in the sense that they provided the stimulation and inspiration behind its development. A listing of their names is an inadequate expression of their role, but no other vehicle for acknowledgment is available. The initial inspiration and training for this job was provided by Professor Rheem F. Jarrett of the Department of Psychology at the University of California, Berkeley. My decision to pursue my studies in mathematical statistics was largely a product of the skilled and sensitive teaching of Leo Breiman, an instructor in mathematics at the University of California, Berkeley, at the time I was a student in a "mathematics for social scientists" course he taught when I was in my third year of graduate school. The National Institutes of Health made two years of study of mathematics and statistics possible for me as a research fellow following my doctorate at Berkeley. Professor David Blackwell, then chairman of the Department of Statistics at Berkeley, sponsored my application for a postdoctoral fellowship and subsequently provided a great deal of stimulation in several courses. During my tenure as a postdoctoral student I was fortunate to have the experience of studying under Professors Edward W. Barankin, Joseph L. Hodges, Erich L. Lehman, Henry Scheffé, and David Blackwell. Whatever merit this book may have, and none of the faults, can be assigned to the benefit I gained from my work in the Department of Statistics under these distinguished and stimulating mentors.

I gratefully acknowledge the contribution of the nearly five hundred students who have taken the course using these materials. Their criticisms and suggestions were important in preparation of the final draft.

Special thanks must be given to Professor James Greeno, who read the manuscript at two stages of its development and provided comment, criticism, and advice that have proven helpful beyond measure.

Mr. Brian A. Knowles assisted in checking illustrations and examples, and participated in the sometime grueling task of proofreading galley proofs. His assistance is gratefully acknowledged.

Several persons assisted in the preparation of the manuscripts in a vitally important way. The reworking of the manuscripts required retyping of the manuscript. That rather tedious and demanding task was performed with remarkable patience, skill, and dispatch, at various times, by Judith Blessing Bowman, Barbara A. Salaman, Sylvia Wade, and Nancy Geidel. I am deeply grateful for their assistance.

DANIEL E. BAILEY

Contents

CHAPTER 1

A Conceptual Foundation

The approach of this book differs from that of many books in quantitative methods for psychology. Too often the practical and utilitarian aspects of statistics are allowed completely to dictate the material presented. This is unfortunate, because it leads to a heavy emphasis on "formulas" and on the attendant computational procedures, at the expense of a basic, honest, and meaningful development of the foundations of the formulas. Statistics can be a matter of rote memory, motivated by a fear that one might be called upon to use statistics in the uncertain future.

This book stresses the basic underpinnings of statistical theory and practice—primarily in probability theory. As a result, the usual topics dealt with in a course in statistics for psychology naturally appear. A wide range of statistical topics are covered in a meaningful and coherent fashion, trusting brute-force memory but little.

To begin with, the basic structure of psychology with respect to the role of quantitative methods is discussed. In the process, some of the concepts of probability and statistics are introduced as accomplished facts, with no effort to make them seem meaningful or to develop the rationale behind them. Next, the theory of probability and statistics is developed, step by step.

In discussing the structure of psychology in this way, and with the brevity that must be observed, many of the refinements of psychological science are glossed over. However, there are six topics of major concern.

Theories and Conceptual Systems. These are the facts and hypotheses one has about nature. If they are adequate they will correspond to nature.

Formalization. The conceptual system is formalized in more or less explicit quantitative terms. This formalization is referred to as the model, which is assumed also to be a direct, if somewhat idealized, representation of nature.

Deduction of (Empirical) Consequences of the Model. The question is asked: "What does the general model imply about particular states of nature?" This step results in the statement of "hypotheses," usually in statistical language, about the state of nature.

Operationalization. The derivations from the model are translated into operational terms. Implications of the model for the real world are of no interest until operations are specified whereby states of nature are measured, controlled, or observed. Thus, this and the preceding step are inseparable in the practical pursuit of scientific knowledge.

Observation. The results of the operations constitute the observations of the study. This step produces the "raw" data of scientific inquiry.

Description. The data are summarized, and descriptive tables, figures, graphs, and so on are used to describe them. This is the first step in which the use of probability and statistics is of self-evident or obvious importance. Although probability and statistics play an important role at each previous step, their use is generally more subtle and sophisticated in the first steps than in the description of the data.

Testing. The agreement (support value) of the data with respect to the consequences of the model is tested. The question is asked: "How likely is it that the observed data could have been observed if the hypothesis (and hence, the model) were indeed true?" When this question is answered in explicit mathematical terms, the procedures and the answer are called a statistical test.

Theories and Conceptual Systems

The character of theories and conceptual systems, their development, and their place in psychology are subjects requiring special discussion in their own right.

In spite of their importance and relationship to the topics of probability and statistics, no attempt is made here to engage in a discussion of them. Traditionally these topics are discussed in courses and texts on the philosophy and history of science.

A word is appropriate, however, concerning this use of the term "theory." By "theory" we mean a system of statements regarding the relationship of objects, the structure of objects, and the function of objects in nature. That is, a theory is a statement concerning the state or condition of nature. A theory may thus be purely speculative, or based wholly or in part on scientific data. Examples of theories under this definition might include such simple things as the statement that intelligence and manual dexterity are "related." On the other hand, an example of a complicated theory is D. O. Hebb's theory of behavior based on the concepts of nerve networks and reverberating circuits. This schema assumes that such a theory underlies the following steps even though the theory is stated very casually and may be trivial.

Theory Formalization—The Model

The development of a formal model from informal theory and conceptual systems is directly involved in this book. However, the theory of probability is necessary for an adequate discussion of model building. Discussion of the principles and methods of theory formalization must be postponed. An adequate treatment of the topic requires knowledge of more mathematics and logic than most readers initially possess. As the necessary logic and mathematics are developed, examples are introduced illustrating the translation of theoretical statements into more or less formal models.

Deduction, Operationalization, Observation

These three topics are dealt with in much the same way as is formalization of theories. They are more appropriately discussed in full detail in courses in scientific methods, mathematics, and logic. Not only do they involve the philosophic, mathematical, and logical realms, but they also are concerned with the practical features of scientific research, including instrumentation, techniques of measurement, and physical access to subjects. Only the necessary rudiments of these topics are discussed in detail in this book.

Description of Observed Data

The description of the results of observations is dealt with intensively. In fact, a large part of the next few chapters is devoted to a discussion of the probability background to methods of data description (or, in the jargon, "descriptive statistics").

The ultimate results of making observations in an empirical experiment are quantitative—that is, numerical in one form or another. This is not to say that naturalistic, nonquantitative, observations are not useful. However, these sorts of observations have their primary value in formulating quantitative models which, in turn, provide the backbone of any science. Often, in the early stages of the development of a science, nonquantitative observations are all that is possible.

Perhaps the simplest sorts of numerical data result when one number (for example, "zero") is assigned when the objects observed (such as human subjects in an experiment, rats in mazes, receptor cells) do not "possess" a specified characteristic (for example, are not married, are not schizophrenic, do not make the "correct" turn at a particular point in the maze, do not fire when stimulus energy strikes them) and another number, such as "one," when the objects possess the characteristic (married, schizophrenic, and so on). The values of the numbers assigned in this way may not have any meaning in themselves, but they lead to numerical data, for example the number of "zeros" observed in a given situation, or the proportion of "ones" observed out of the total number of observations.

When the observed data are in the form of attributions of qualitative characteristics to the experimental objects, the resulting data may be described by noting the relative frequency with which the characteristic is attributed to the objects observed. If, for example, ten patients being admitted to a hospital for mental disorders are classed as "schizophrenic" or "not schizophrenic," the observed data would be a list of patients and the "schizophrenic" or "not schizophrenic" property of each patient.

Not all the details of a list of attributes are necessary to give us a good idea of the data observed. The essence of the observations can be described by stating that, for example, four of the ten patients were observed to be schizophrenic. Or, more simply, the relative frequency of schizophrenic patients, f(schizophrenic), was .4:

$$f(\text{schizophrenic}) = \tfrac{4}{10} = .4$$

It may seem trivial, but this development is rather important, especially where large numbers of observations are involved. The entire list of data is distilled to a single number, which expresses its essence (at least one of the essential features).

A more sophisticated approach to the same descriptive principle avoids the actual counting and employs a slightly higher level of mathematical formalism. Assign the value "1" to schizophrenic patients and the value "0" to nonschizophrenic patients. Now add together all ten values assigned to the patients. The sum of the values is divided by the number of values included in the sum to calculate the average value. This average is .4 in the example, the same value found for f(schizophrenic). The two methods of describing the data are mathematically identical. The "zero-one" method is important, because counting often becomes difficult in more complex situations whereas the value-assignment method does not.

Other sorts of observations result in data that can be dealt with by counting methods only with difficulty and little usefulness. Among such data are quantitative data—for example, those data answering questions of "How much?" In this case the data are inherently quantitative (as contrasted to the qualitative observation of the presence or absence of a characteristic)—that is, numbers with more inherent meaning than the zeros and ones in the example above. The intelligence quotient of ten patients might be observed, giving the following results:

	IQ
Patient 1	104
Patient 2	96
Patient 3	142
Patient 4	110
Patient 5	90
Patient 6	76
Patient 7	84
Patient 8	90
Patient 9	82
Patient 10	100

Again, some way of describing these data is needed. Obviously it would be impractical and foolish to present all the data if there were ten thousand instead of ten observations. The relative-frequency solution of the first case has a great deal of merit in the simple example, but its drawback is obvious from Table 1.1.

Virtually nothing is gained. Instead of the list of ten IQ's, there are now nine IQ values *and* the nine relative frequencies. To be sure, thousands of observations might yield only one hundred values of IQ, and the relative-frequency table would be about ten times smaller than the table of observations.

Another method of describing "how intelligent" the patients are is to find the highest and the lowest values observed—that is, the extreme observations describe

TABLE I.I

Value Observed	Relative Frequency of Value Observed
76	.10
82	.10
84	.10
90	.20
96	.10
100	.10
104	.10
110	.10
142	.10

the values of IQ observed. Two values, and two values only (out of all the observations) are involved (in this case, 76 and 142). All other values are ignored, and the "descriptive statistic," 76–142, describes the total body of data very poorly.

The solution provided by probability theory is to average the values observed. Thus, every observation is represented in the description—the average represents the entire set of data. In the example above, the average is found by the following formula:

$$\text{Average IQ} = \frac{\text{sum of IQ's}}{\text{number of observations}} = \frac{974}{10} = 97.4$$

In addition to questions of "how many" or "how much," other questions can be asked about the data observed. For example, "How do the observations differ among themselves?" It is the rare experiment that produces observations that are all the same. And when an experiment does result in such "constant" observations, it is likely that some artificial constraint is operating in the experiment. It is one of the fundamental facts of psychology that organisms differ from one another and observations of organisms will tend to be variable. This is an important feature of data, and ways of describing the differences among observations are needed.

One method of describing the differences among observations is to list the differences between all possible pairs of observations. For example, with four observations O_1, O_2, O_3, and O_4 there are twelve differences:

$$(O_1 - O_2), (O_1 - O_3), (O_1 - O_4)$$
$$(O_2 - O_1), (O_2 - O_3), (O_2 - O_4)$$
$$(O_3 - O_1), (O_3 - O_2), (O_3 - O_4)$$
$$(O_4 - O_1), (O_4 - O_2), (O_4 - O_3)$$

Since each pair is represented twice, for example, $(O_1 - O_2)$ and $(O_2 - O_1)$, with the same size difference but with opposite sign the sum of all of these is equal to zero, regardless of the number of values of the observations. Thus the average value would always be zero. Other mathematical characteristics of these differences, which are not developed here, show that this is a poor way of dealing with the differences among observations.

Another way is to find the difference between the largest and smallest values observed—that is, the range of values observed. In the example of the intelligence quotients of hospital patients, the range is 66. However, this measure can be criticized for the same reasons the "extreme values" measure of the magnitude of the observations was criticized.

The solution generally accepted is to compare each observation with some constant reference value, in particular the average of the observations. This method gives the same number of differences (between observation and average observation) as the number of observations. To simplify this the average of the squares of the differences is calculated (the squaring is based on reasoning examined later). This average, called the variance of the observations, is a useful and descriptive measure of the differences among the observed data. The formula for the case of four observations is

$$\text{variance of } O = \frac{\text{sum of (Observation minus average Observation)}^2}{4}$$

$$= \text{var}(O) = \tfrac{1}{4} \sum_{i=1}^{4} (O_i - \overline{O})^2$$

where $\overline{O}$ is the average observation.

A third basic descriptive characteristic of major interest arises when pairs of observations for each object are studied. For example, if ten students of statistics are asked if they have had college algebra and if they found statistics easy, the results might be as shown in Table 1.2.

Interest in such an inquiry might center on how these two things go together: "Are the characteristics related?" The techniques for answering this question are too complicated to discuss before developing some technical apparatus of probability theory. However, the data can be represented in such a way as to measure the degree of relationship among observations. A table with four cells is drawn up, one cell to represent each of the combinations of yes's and no's. The number of students answering the questions with each combination of yes's and no's is counted. Four students not having had algebra did not find statistics easy, one who had not had algebra found statistics easy, and so on. The results of this analysis are summarized in Figure 1.1.

TABLE 1.2

	Had Algebra	Statistics Easy
Student 1	Yes	Yes
Student 2	No	No
Student 3	Yes	No
Student 4	No	No
Student 5	No	No
Student 6	No	No
Student 7	No	Yes
Student 8	Yes	Yes
Student 9	Yes	Yes
Student 10	Yes	Yes

A relationship between having had or not had algebra and finding statistics easy or not easy is apparent in Figure 1.1, although the relationship is not perfect (there are cases where students having had algebra find statistics hard and vice versa).

Many additional examples of descriptive methods and their bases are introduced in the material that follows. The general point is clear with these three examples: The method used to describe data depends on the sorts of questions that are asked of the data and the quantitative procedures available. There are few cut-and-dried methods applicable to experimental and observational data in general. A stereotyped approach to the description of observed data can be fruitful only where the questions are themselves stereotyped. It is unfortunate that much original and inventive thinking in psychology is spoiled by stereotyped and unimaginative approaches to data description and quantitative analysis. The examples in this book do not teach one how to be imaginative in approaching data but suggest some general principles from which the inventive research scientist can develop techniques that go to the very heart of the questions he asks in his research.

		Statistics Is Easy	
		No	Yes
Had	No	4	1
Algebra	Yes	1	4

FIGURE 1.1

Table of joint frequencies.

Testing the Support Value of the Data

This last step in investigating a problem is as important as the description of data. The foundation for a meaningful treatment of testing hypotheses is to be found in the theory of probability, just as it is for the methods of data description.

Probability theory is more evident in hypothesis testing than in data description, however. Statements made concerning the results of a statistical test are primarily statements of probability. Two aspects of the steps in the process of investigating psychological theory are involved. Specific consequences for empirical operations are derived from the general formal model. These consequences are statements of what one should expect to find if the operations involved are actually executed and if the model is true. On the other hand, there are the observations that should correspond in one or more respects to the results expected on the basis of the model. The theoretical predictions—that is, the consequences derived from the model—may involve only one of the salient aspects of the detailed observations. For example, the theory may concern the intelligence of patients entering a hospital and, in particular, state that the mental patient's intelligence, on the average, is not different from the average intelligence of nonpatients. Thus it is deduced that measured IQ of new hospital patients and of the general nonhospital population should be comparable. But how are two large sets of numbers compared?

The problem is highlighted even more by noting that the number of patients observed may be fewer (or greater) than the number of nonpatients observed. One solution to the dilemma is to let some descriptive and salient aspect of the entire set of scores, for each group, stand for the group. For example, the average IQ can stand for the entire set of data in each of the two groups. This is reflected in the formal model, and the model is stated in terms of averages instead of individual observations:

$$\text{Average patient IQ} = \text{average nonpatient IQ}$$

It is obvious that much detail and richness of detail is ignored in this formal model. However, the model does capture the substantive sense and meaning of the theoretical statement. Many models involved in psychological research are no more complex than this simple model.

The next step is to operationalize the model. This is done simply by administering a standard intelligence test such as the Wechsler Adult Intelligence Scale to a group of nonpatients and to a group of patients. The selection of subjects is more complicated than just implied, but at this point assume that biases (for example, all college graduates in the nonpatient group versus all low-educational-achievement patients) are not built into the selection of patients. Likewise, a complete operational specification would attend to details of the testing situation, past testing experience of the subjects, and so on.

The data observed, the average IQ's in the two groups will seldom, if ever, be identical. For example, average IQ's of 104.3 and 104.6 might be observed. The difference in average IQ in the two groups is only .3, whereas the difference specified by the model is .0. Just how much reliance can be placed on the .3 value? Is it really a reflection of a true difference in average IQ's?

The answers to these questions depend on many factors. The groups of subjects used may not be representative of the classes of persons with which the model is concerned. There is undoubtedly some error involved in measuring IQ. There probably are even slight fluctuations in intelligence from day to day, and the .3 may be a reflection of this; on another day the difference could be in favor of the other group. The error in one subject's measured IQ may be positive, and in another negative. The error may likewise be present in a great variety of degrees, some cases with very little error and others with large error. Many other reasons for a difference of .3 might be adduced when the "true" difference is .0.

Probability theory provides solutions to this dilemma. By transforming the formal model, as stated above, into a formal probability model and making a few additional reasonable assumptions, general and meaningful statements can be made about the data in relationship to the model. The apparatus of doing this is the topic of this text. The solution, at this point, can only be hinted at.

By applying the notions of probability theory and theory of statistics, it is possible to make statements such as the following:

The probability of observing a .3 or greater difference in mean IQ is .54 when the model is true.

That is, even if the true difference were .0, differences in mean IQ as large or larger than .3 would be observed 54 times out of 100 such experiments. Thus, one would have little confidence that the .3 represents a significant difference in intelligence of patients and nonpatients. The basis of such statements lies in the ability to take advantage of the chance factors in the experiment—the sample of subjects and error (for example, fluctuations of intelligence, changes in experimenter, changes in testing situations). Some methods discussed in this text are based on chance factors of sampling rather than error. Other methods deal with similar results where magnitude of error is the basic concept. By the methods of probability theory, a probability is assigned to every possible outcome of the experiment. These probabilities are determined prior to making the observations and act as a part of the formal model. Each possible result of the operation is specified in addition to the particular result required by the theory on which the model is based. These probabilities indicate the likelihood that, if the theory is correct, the chance factors influencing actual observations lead to specified results not in agreement with the theory. According to the probability model involved, the particular observations made would occur with a probability of such-and-such if the theory were true. If this probability is very low, there is reason to doubt that the theory is empirically adequate. If these probabilities are high, there is no reason to question the validity of the theory, at least in the context of the specific observations involved.

The next nine chapters develop in detail the formal and theoretical apparatus with which the methods and techniques of quantitative analysis in psychology are formulated. The specific utility of the content of these chapters may not be immediately apparent, but in the context of the last half of the text the purposiveness of the theoretical developments becomes eminently clear.

CHAPTER 2

Probability Models

$\mathbf{C}$hapter 1 dealt with the place of formal models in the structure of scientific psychology. Models formulated in terms of probability theory provide means of making inferences about the validity of the theories. This chapter is devoted to the study of the theory of probability and probability models.

Probability models are always abstractions and formalizations. Thus, they often are not characterized by the richness of detail found in actual empirical phenomena. However, the formalism leads to a simplicity and clarity which provides a great deal of power in the analyzing and understanding of complicated processes and conditions.

There is a direct parallel between the probability model and what is known about the real world. In fact, the structure of a probability model is largely determined by knowledge of the real world and the methods of scientific investigations. The criterion of mathematical convenience and simplicity is also a guide in the formation of probability models.

Although probability theory is fully developed and is of interest as a field of study itself, we are motivated in our interest by the use of probability theory as a model for experiments and quantitative analysis in scientific research. Since our subject is "experiments," we must be very clear about the meaning of the term "experiment" and provide a formal definition so that we can determine whether or not there is justification for adopting the theory of probability as a model for a given experiment. The form of the definition of an experiment and the consequent form of the justification are of general interest to psychologists and are fully developed here. First, the definition of an experiment is spelled out and the consequences derived. Then the parallel forms in probability theory are developed and the justification completed. Following this, some ramifications and inter-pretations of the elements in the definition and the model are drawn out. In particular, we discuss the character of the sorts of objects dealt with in experiments and probability models. The chapter is concluded with a brief excursion in graphical representation of experiments and probability models, and an illustration of a simple experiment and its model.

Experiments

An experiment consists of making an empirical observation where repetition (replication) of the experiment can lead to observation of a different result. Examples are easy to think of. Toss a penny and observe the result: The penny will turn up "heads" or "tails." On some of the tosses "heads" will be the result observed, while on other tosses "tails" will be the observation. Tossing the penny is the experiment; the showing of heads or tails are the possible results.

More interesting examples are found in experiments from psychology. If a rat is placed in a simple "T" maze, it can get to a feeding station by turning left or turning right at the choice point. The experiment consists of placing a rat in the maze and observing the direction it turns when it reaches the choice point. The results observed are "turned right" or "turned left."

The earlier example of measuring the intelligence of hospital patients is an example of an experiment: observe the IQ of a patient. The result observed is the patient's IQ. In the specific example of the IQ's from Chapter 1, nine different values were observed. Only two of the experiments produced the same result, IQ = 90. Using the same operations, a more complicated experiment might be defined where the experiment was the "simultaneous" observation of all ten patients. Thus the entire set of ten numbers is the result observed. Only when the same collection of ten numbers is observed in a repetition of the experiment (that is, observe IQ for ten patients) is the result duplicated. This is the basic concept underlying the tests of hypotheses about means. The mean of a number of values observed (as in patients' IQ's) is taken as a result in an experiment. The formulation of the experiment adopted, where there are several formulations, is determined by the availability and completeness of models for the experiment.

When an experiment is performed, some specific result is observed. Thus, results are the things that may be observed on performing an experiment. In the coin-tossing experiment there are two possibilities (barring the freak occurrence of a coin standing on its edge): heads and tails. The results r_1 and r_2 correspond to heads and tails, respectively. In the experiment with a rat in a "T" maze there are two possible results: r_1 stands for the result "turned left" and r_2 for "turned right." In an experiment where there are k different results, the total collection of possible observations makes up the list of results: $r_1, r_2, r_3, \ldots, r_k$. This would be the case in the study of the IQ of hospital patients, where r_i is the ith possible result if the results are numbered 1 through k by the subscripts of $r_1, r_2, \ldots, r_k$. The index i is, of course, no smaller than 1, but yet not greater than k: $1 \leq i \leq k$.

If an experiment is repeated a number of times, one of the possible results being observed each time, the number of times a specific result is observed is the frequency of that result. For example, if an experiment is repeated N times (N being some integer number, $N \geq 1$), the number of times the result r is observed is called the frequency of r. This is denoted by

$$\text{(frequency of result } r) = n(r)$$

If, for example, a penny is tossed ten times and "heads" is observed in four of these experiments, then $n(r_1) = 4$. Since the only possible results are heads and tails, (r_1 and r_2), then tails must have been observed on six of the ten experiments; therefore, $n(r_2) = 6$. If, in the experiment with the rat in the "T" maze, the rat is run 100 times, and observed to turn left 55 times (45 times to the right), $n(r_1) = 55$, and $n(r_2) = 45$.

These examples illustrate two basic requirements in defining experiments: the results of an experiment must be (1) exhaustive, and (2) mutually exclusive.

In order for the list of results to be exhaustive, all the possible observations in the experiment must be included. Interpreting this strictly would require that the results in the coin-tossing experiment be listed as "heads," "tails," and "edge." However, realism suggests that the result "edge" can be ignored as a freak. It occurs with such a small frequency that ignoring it will not substantially alter any experimental conclusions. On the other hand, rats can curl up and go to sleep in a maze, children can refuse to play the psychologists' games, and so on. Unless some way can be devised to make the subjects "behave," such additional categories as "did not run," and "would not cooperate" are necessary. The requirement that the results must be mutually exclusive and exhaustive implies that the sum of the frequencies of the results must be equal to the total number of experiments, that is,

$$\sum_{i=1}^{k} n(r_i) = n(r_1) + n(r_2) + \cdots + n(r_k) = N \tag{2.1}$$

This property of experiments provides a fundamental link between experiments and probability models.

The criterion of mutual exclusion is formalized by the statement that if r_i occurs on an experiment, then r_j (where $i \neq j$) cannot also occur on that experiment. This criterion is satisfied by the experiments discussed in examples up to this point. However, it is not difficult to find examples in which the list of results includes "results" that are not mutually exclusive. For example, if incoming mental hospital patients are "scored" as to whether they were suffering from mental deficiency, schizophrenia, organic brain damage, paranoia, and so on, some individuals would be scored as suffering from more than one of these maladies. Adding up the number of results (that is, the observation of each new patient and the malady suffered) gives

$$\sum_{i=1}^{k} n(r_i) > N$$

because some of the experiments would result in the occurrence of two or more "results," for example, a mentally deficient patient with organic brain damage.

The requirement for mutually exclusive results can be dropped once some simple theorems in probability theory have been developed. In fact, many of the applications we are interested in hinge on results that are nested; that is, for one result to be possible other results would have to be observed. For the present discussion, however, we will restrict our attention to the basic definition of an experiment, with mutually exclusive results exhaustively specified.

Any given experiment may be characterized by a variety of sets of results, any one or more being of interest. For example, in the "T" maze experiment the results of interest may be the number of seconds the rat took to reach the food box, the number of pellets of food he ate, whether an error was made at the choice point, the weight of the rat, and so forth. The set of results that is used in developing a model of the experiment depends on the purpose of the experiment. A good deal of attention is given in the following chapters to the question of experiments in which the results are not mutually exclusive or in which there are several different ways of defining results of interest. The entire fabric of the more complicated and interesting theory of experiments is woven from the simple definition of experiments that we have just developed.

The "T" maze experiment above was replicated 100 times to find $n(r_1) = 55$ and $n(r_2) = 45$. The frequencies are determined, in part, by the total number of experiments performed—$n(r_1) \leq N$ and $n(r_1) + n(r_2) = N$. If 200 repetitions of the experiment were observed, the frequencies might have been $n(r_1) = 110$ and $n(r_2) = 90$, quite different numbers. The relative frequency with which r_1 was observed is, however, the same in both instances—55 percent of the experiments resulted in observing r_1. This leads to a formalization of the notion of relative frequency. The relative frequency of a result is the raw frequency divided by the total number of times the result could have been observed. The relative frequency of the result r_i is denoted

$$f(r_i) = \frac{n(r_i)}{N} \tag{2.2}$$

The values, $n(r_i)$ and N that enter into $f(r_i)$ imply that $f(r_i)$ is limited by .0 and 1.0 as the smallest and largest values:

$$0 \leq f(r_i) \leq 1 \tag{2.3}$$

This follows from the fact that $n(r_i)$ is zero or greater, but never larger than N. The two extreme values of $n(r_i)$, zero and N, correspond respectively to the cases where *no* repetition of the experiment resulted in an observation of r_i and those in which every one of the N experiments resulted in r_i. Thus the smallest and largest values of $f(r_i)$ are given by

$$\frac{n(r_i)}{N} = \frac{(\text{no instance of } r_i \text{ occurred})}{N} = \frac{0}{N} = 0$$

$$\frac{n(r_i)}{N} = \frac{(\text{every repetition resulted in } r_i)}{N} = \frac{N}{N} = 1$$

The definition of $f(r_i)$ and the requirement that the results be exhaustive and mutually exclusive imply another characteristic of relative frequencies. The sum of relative frequencies of all the results must be equal to 1.0:

$$\sum_{i=1}^{k} f(r_i) = f(r_1) + f(r_2) + \cdots + f(r_k) = 1 \tag{2.4}$$

The proof of this important result is left to the reader. (Hint: Substitute the expression $n(r_i)/N$ for the corresponding expression $f(r_i)$ and apply the rules of summation.)

The properties of relative frequencies, derived from the logic of the definition of an experiment and simple algebra, provide a strong justification for using the general mathematical theory of probability as a model of experiments. The definition of experiments, the results of experiments (mutually exclusive and exhaustive), and the consequences [equations (2.3) and (2.4)] parallel the definition and structure of probability theory: The formal structure of experiments is identical with the axiomatic formal structure of probability theory. Where an identity of this sort is found, the mathematical theory is said to be a model of the empirical theory. Thus, probability theory may be taken as a mathematical model of experiments. Whereas there are specific lists of results of specific experiments in the theory of experiments, and whereas there are no mathematical tools to develop the theory of experiments, the theory of probability is a general theory and is stated in a mathematical language permitting mathematical development.

Probability Theory

Probability theory is of interest here because it provides (or is) a general model of experiments. Thus a probability model is defined as a mathematical abstraction or idealization of an experiment in terms of the definitions, axioms, and theorems of probability theory.

Events are the primary elements of probability theory. Events are directly parallel in meaning and function to results in experiments. Where results are defined by the possible observations on an experiment, events are the abstract parallels in the model. There is a one-to-one correspondence between the list of events in a probability model of an experiment and the list of results defined in an experiment. To distinguish between events and results the notation e_i is used for the ith event. To keep the correspondence between e's and r's clear it is assumed that they are paired with the same subscript. That is, the pairs will be

$$(e_1, r_1), (e_2, r_2), \ldots, (e_k, r_k)$$

All of the examples of results can be translated directly into examples of events in the models of the experiments in which the results were defined. As in the definition of experiments and results, the events of a probability model *must* be exhaustive of the logical possibilities and mutually exclusive. The emphasis on the "must" points out the correspondence between definitions in a mathematical theory and that theory itself. If some other definition of events were given, the mathematical theory would not be the same as the theory of probability. However, in the discussion of experiments, a different definition might have been proposed without changing the concern with the practical business of research. Such a definition might have led to a different model. The reader is encouraged to try to contrive such definitions and the consequent models.

Probability is defined in the mathematical theory of probability in a formal way that does not imply any specific values associated with the events. Rather, the probability of an event e_i is defined as a number $P(e_i)$ attached to (or associated with) the event e_i for $i = 1, 2, \ldots, k$ such that

$$0 \leq P(e_i) \leq 1 \tag{2.5}$$

and

$$\sum_{i=1}^{k} P(e_i) = 1 \tag{2.6}$$

These values are abstractions, not specific numbers like relative frequencies. The values of relative frequencies of results were determined by concrete and specific observations of results in a given number of repetitions of an experiment. The definition of experiments led to the general statements (2.3) and (2.4) regarding the formal properties of relative frequencies. No such considerations are made in the definition of probabilities—the properties of probabilities are defined, not derived. Expressions (2.5) and (2.6) are the axioms of probability theory when that theory is presented as an axiomatic mathematical theory. When relative frequencies are calculated from the observations made in a given number of repetitions of an experiment, they will satisfy the conditions of (2.3) and (2.4). On the other hand, before we can justify identifying numbers associated with the events in a model as probabilities, it first must be ascertained if they satisfy the conditions expressed in (2.5) and (2.6). Table 2.1 displays the parallel character of the formal structure of experiments and probability models.

The importance of this development cannot be stressed too much. The quantitative theory of experiments has not been developed to the point where derivations can be taken very far. In addition, since the list of results and the relative frequencies of each experiment are specific to that experiment, it would be

TABLE 2.1 Tabulation of the Parallel Between Experiments and Probability Models

	Experiments	Probability Models
Objects Involved: Collection	Results $r_1, r_2, \ldots, r_k$	Events $e_1, e_2, \ldots, e_k$
Properties of collection	Exhaustive and mutually exclusive	Exhaustive and mutually exclusive
Quantities	Relative frequencies	Probabilities
Formal structure	$0 \le f(r_i) \le 1$ $\sum_{i=1}^{k} f(r_i) = 1$	$0 \le P(e_i) \le 1$ $\sum_{i=1}^{k} P(e_i) = 1$
Source of formal structure	Consequence of definition of experiments	Axiomatic

difficult to know when a consequence of the formal structure of a given experiment could be generalized to another experiment. However, probability theory provides a general mathematical expression for experiments in which derivational consequences will be true for all experiments to the extent that the model is a good model for experiments.

A word of caution is needed about the meaning of derivations and consequences of probability theory. Not all the theorems and consequences of probability theory have a direct interpretation and meaning for experiments. It is too much to expect that every step and derivation in a formal theory will have an obvious interpretation in the objects and processes being modeled. A great deal of the major derivational consequences of probability theory as a model for experiments will have direct meaning for experiments and the analysis of experiments. There are many intermediate derivational steps to obtain the general theorems of probability theory. Many of these intermediate steps are without meaningful interpretation in the framework of experiments.

Probability models, as introduced here so far, are very general—applying to everything meeting the definition of experiment. In addition to studying this general model, it may be specialized so that it becomes a model for specific experiments. This is done by specifying the collection of events and assigning values as probabilities of the events in the model. A probability model can be found for any specific experiment if the experiment is sufficiently well defined. The details of probability models for some experiments are more complete than for others. If the sampling procedures or the population from which a sample is drawn are well defined, the probability model can be defined with a great deal of detail. If the necessary things are not known about the population and if the sampling techniques do not give the necessary logic to derive sampling distributions, it may be possible only to develop the probability model for the experiment to a lesser degree of detail. It is the unusual experiment that does not give rise to relatively well-developed probability models. Often an assumption regarding the number of repetitions is sufficient to formulate a probability model that is complete with respect to questions posed in behavioral science research. At other times, assumptions must be made about the bio-social character of the experimental observations.

Since a great deal is possible in the way of formulating models of experiments to provide the formal vehicle for quantitative analysis of experiments, it is important to be well acquainted with the theory of probability models. If the theory of probability models is general, that is, not tied to specific illustrations, then applications to specific experimental problems can be made easily. The following material is intended to provide the general framework in which experiments can be translated into models.

It should be clear that a knowledge of the theory of probability models is important in the formulation of experiments. If there are a number of different experiments that can be used in a study of a given phenomenon, they may well differ with respect to the ease and completeness with which they can be translated into probability models. If the scientific purpose of the experiments is satisfied equally well by two experiments, then the experiment that admits to a more adequate expression in a model is the experiment to be chosen.

It is possible to design an experiment in such a way that no adequate probability model can be found for the experiment. Behavioral science literature is not lacking in examples of such poorly defined experiments. The fault seldom is stupidity, but rather the lack of adequate knowledge of the principles of the design of experiments and the principles of the formulation of probability models for experiments. If the behavioral scientist is thoroughly versed in the theory of probability models of experiments, he can design experiments that will be suitable to answer questions he wishes to pose. Generally, it is too late to inquire about a model for an experiment after the data have been collected. Serendipity may provide a model appropriate to the scientific problem, but science is not well founded on serendipity.

This text cannot provide the reader with a complete knowledge of the theory and practice of the design and analysis of experiments. It would take several times as much space to cover even the more widely used designs. The purpose of this book is to present the general theory of probability models and a selection of the more important and customary models in the context of generalized scientific problems to which the models are applied. Current usage and practice in psychology determined the selection of problems and models discussed.

In order to develop the basic theory of probability, that is, to define the properties of probability models and the sorts of things that come out of probability models, very simple experiments are dealt with first. Once the basic and elementary conclusions or derivations in probability theory are presented, a special branch of probability theory, the theory of decisions, is presented. More complicated sorts of experiments can be modeled on the basis of these two theoretical bodies, logic and mathematics. Only a few illustrations of specific models of wide use and applicability to psychological experiments are introduced. The general methods for developing specific models are presented, up to methods involving integral calculus. Where the experiments of interest are limited to discrete and limited numbers of results or where the results can be transformed to this sort of framework, then the techniques presented in this book provide the power to develop a wide variety of probability models. The results of developments involving more advanced mathematical methods are presented with nonrigorous explanations of the basic concepts behind the developments.

The Language of Probability Theory

The mathematical nature of the theory of probability implies that the language of probability is mathematical. Any technical or reasonably complete discussion of probability theory must be to some degree mathematical. The presentation in this book cannot escape this, regardless of the customary disdain psychologists show for mathematics. A certain skill with simple algebra and a tolerance for abstractness and symbolization are necessary to understand and master the material to be presented. Very little is assumed beyond college algebra. A few other mathematical ideas are needed, and they are developed in the detail and degree of rigor proportional to the use to which they are put.

First, a rather modern branch of abstract mathematics, set theory, is introduced and developed to a level commensurate with that usually associated with chapters in survey texts for first or second year college mathematics courses. Set theory is our common vernacular throughout this book. Without the notions of sets and operations on sets, the fundamental theorems of probability and statistics must be stated in imprecise language. Chapter 3 deals with set theory as "the mathematics of events."

Types of Events and Results: Measurement

Psychologists are interested in the meaning of the numbers that come out of experiments and analyses of experiments in terms of the structure and function of the organisms they study. Probabilists and statisticians are not intrinsically interested in these problems.

This divergence of concern has led to a good deal of confusion in the psychological literature about the use and misuse of statistics. The problem has centered on the application of certain developments in the theory of measurement (a scientific, not necessarily a mathematical, concern) to statistical thinking. A good deal of this application has been misplaced and misleading. Most of the statistical methods and procedures familiar to the psychologist, until quite recently, were developed by statisticians without so much as a moment's concern with the problem of measurement. The substantive meaning and methods of interpreting statistical results constitute a scientific problem going beyond the interest and competence of the statistician as statistician.

The psychologist must play a wider role than is expected of the statistician. The psychologist must be able to assume the roles of the practical scientist *and* of the statistician, often in rapid alternation. Thus, in a book such as this, concerned with applications of probability and statistics in scientific psychology, it is natural to be concerned with both roles.

The key lies in the acceptance of statisticians' rules when statistics is at issue and scientists' rules when interpretation of statistical results is at issue.

The statistician makes a very simple distinction among models. There are, first, models that deal with nonquantitative events, such as sex, country of birth, and other events forming nominal categories. The only other sort of models recognized by the statistician are those dealing with quantitative events, such as the number of successes out of so many possible, the number of correctly answered questions in a quiz, the amount of brain tissue destroyed, and the time between a signal and a response. Thus the statistician is concerned with two classes of events: nominal and quantitative. The word "variable" (or more frequently in the context of statistics, random variable) is used in the case of quantitative events. The word "classification" is sometimes used when dealing with nominal events.

Among quantitative events the main subdivision useful from the statistical point of view is discrete valued quantities and continuous valued quantities. This subdivision is important not because of any important statistical or scientific reason but because of the different mathematical tools relevant to derivations and proofs for continuous and discrete variables. In a large part of probability theory and the theory of statistics, it does not matter what sorts of events are nominal— discrete quantities or continuous quantities. In addition to the general elementary theory of probability and statistics, there is a large body of special results and methods relevant only to quantitative events. These special bodies of knowledge are introduced in this book as they are needed in the orderly development of the subject matter. In the meantime, however, the generality of the theory of probability implies that it can be developed for the nominal case and then applied to the quantitative cases. Chapters 4, 5, and 6 spell out the theory of probability for nominal events. Chapters 7 and 8 present the special mathematical results that are obtained when the events are assumed to be quantitative and discrete. Chapters 9 and 10 treat the case of continuous variables and the special case of normally distributed random variables, respectively.

The reader who is familiar with the "popular" behavioral science approach to statistics based on classification of statistical methods appropriate for levels of measurement will possibly be asking where measurement got lost in the above scheme. The answer is straightforward and simple: Measurement was never a part of the theory of probability and random variables, it did not get lost. The theory of measurement is outside the basic realm of probability and statistics.

In order to clarify the issue, a brief discussion of measurement is appropriate. We begin with an unspecified collection of objects, such as coins being tossed, people in a survey of opinion, rats running in mazes. The objects in the collection are taken as simple objects with no relationships, no identifying characteristics, no numbers attached or any other properties except their distinctness from one another. This collection is represented in Figure 2.1 by the collection of dots in Region A. The fact that the dots are in particular positions in Region A is not to be taken as a significant attribute of the figure.

With each of the objects in this collection there is associated an attribute, such as its political persuasion, some attitude, membership in a certain behavior class. These attributes in this example are taken to be characteristics of the objects—not attributes assigned on the basis of some observational scheme. They are the "real" properties and attributes of the respective objects. If there are a number of objects that have the same attribute, they are said to form an identity class (with respect to that attribute). If there are a number of specifics in the attributes for the collection

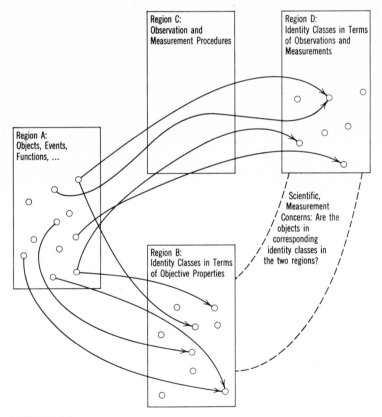

FIGURE 2.1

Structure of scientific measurement.

of objects, for example political persuasion in terms of party identification (Democrat, Republican, Socialist, Communist, Prohibitionist, . . .), there will be several identity classes. Each identity class is composed of all objects that have the same specific attribute. The segregation of the original collection of objects into identity classes is indicated in Figure 2.1 by the arrows taking each object from Region A to Region B. If the attributes involved were different from the specific example used, the collection of objects may have been partitioned into different identity classes. Assume, for example, that the original collection is of person-objects and the identity classes contain all individuals of certain ethnic origins: Mediterranean, Asiatic, Baltic, ... These two sets of identity classes possibly are quite different. Each object is in one and only one identity class for each of the two sets of attributes, and the objects in specific identity classes for one set of attributes are possibly different from objects in any identity class for the other set of attributes.

A third Region, C, in Figure 2.1 introduces the measurement or observational aspect in scientific work. Each object is assessed by some specific instrument of measurement or observational technique in an attempt to ascertain the identity class membership of the object. In the two examples above, the measurement instruments might be questions posed by political pollsters, attitude questionnaires, registration records, and so on in the first case. In the second case the instruments might be indexes of head shape, skin pigmentation, bone structure, ... The scientist lays down certain rules or criteria that are applied to the observations and measurements and that lead to the attribution to each object of membership in a category corresponding to the identity classes designated as Region B in Figure 2.1. These new categories are identity classes, but with respect to the scientific measurement or observations. These classes are represented in Region D of Figure 2.1.

The behavioral scientist, as scientist, is primarily interested in the relationship of the identity classes in Region B and the identity classes in Region D. Of particular interest to the scientist is the accuracy of the representation of the classes in Region B through the classes in Region D.

The statistician, as statistician, is not concerned with the accuracy of representation but rather with certain formal properties of the identity classes. The formal work of the probability theorist is unaffected in the main by considerations of the region of Figure 2.1 in which he is working. The next three chapters deal with objects and observations on objects of this sort where the only considerations of relationship between objects are whether or not they are in the same identity class or in different classes. The specific attributes defining an identity class are of interest only to the extent that they imply some formal property of the classes relevant to probability considerations.

Turning to quantitative events, we can see that quantitative attributes and characteristics have the same status as nonquantitative attributes and characteristics with the addition of one consideration. In addition to the identity class considerations made above with nonquantitative attributes, the quantitative nature of the attributes implies that the identity classes are ordered and correspond to numerals that represent the magnitudes of the quantities involved. When dealing with quantitative events in the behavioral sciences, one often assumes that the numerals representing identity classes of events are real numbers or some subclass of the real numbers such as the integers. Figure 2.2 illustrates this. The difference between Figures 2.1 and 2.2 is readily apparent—Regions B and D involve identity classes arrayed along a continuum in such a way that they are said to be ordered. The order implies different magnitudes of the characteristics represented by the continuum. Two identity classes at different places on the continuum represent objects with different magnitudes of the property or characteristic being represented in the identity classes. In Region B these values represent the magnitude of the objective characteristic without respect to any consideration of measurement or observational conventions or devices. In Region D the values represent the magnitude of the characteristic for each of the objects with respect to the measurement or observational conventions.

Figure 2.1 is a diagram of a classification scheme, not really measurement in the sense the term is used in the physical and mathematical sciences. However, the scheme of Figure 2.2 is that of direct measurement. A thorough and complete discussion of this would lead far beyond the main purpose of this book. However, some points call for discussion here.

Region A in Figure 2.2 is identical to the corresponding region in Figure 2.1, but Region B is a highly structured component of Figure 2.2. Three main features stand out in Region B: (1) the location of an origin or zero point, (2) the scale of the continuum (that is, which object would be mapped into the point on the continuum corresponding to the magnitude 1.0), (3) two objects mapped onto the continuum are characterized by a difference in attribute magnitude in terms of the difference in the values associated with the objects.

The existence of objects having the value zero on the continuum is not guaranteed in behavioral science. Examples are not difficult to manufacture. It seems reasonable that the nature of organisms precludes the existence of an organism with zero ability of some specific sort, or of a human being with zero personality or attitude of certain kinds, and so on. In many other instances in behavioral science, however, the magnitude zero is a real possibility, at least in theory. When there is a true zero point on the continuum, some object is selected as the representation of the unit of the scale. This object is "abstract" in terms of the continuum of Region

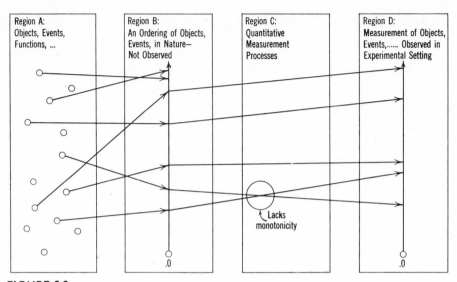

FIGURE 2.2

Structure of scientific measurement.

B but serves to complete the definition of the scale or units along the continuum.

If there is no natural zero point on the continuum, some arbitrary value (usually not zero) is associated with some object. This arbitrary value serves as an anchor for the continuum in place of the natural origin. The scale units are established in this case by selecting an object with a different magnitude of the attribute involved and assigning it a value different from the anchor value by an amount equal to 1.0. The arbitrary values thus assigned are thence taken to represent the magnitudes of the attributes involved. Once the values are assigned in this fashion, a zero point is implied. The implied zero point may not have any objective meaning, and if the arbitrary anchor or unit values are unhappily chosen, the values assigned to some objects in the collection may have negative values associated with their attribute magnitudes.

When one has defined a zero point (directly or implicitly) and a scalar unit for the continuum, it is evident that the differences in the values associated with the attribute magnitudes are an important property of this conceptualization. Two objects with different magnitudes of the attribute will have different values, for example y_i and y_j for the ith and jth objects. The difference of the magnitudes of the characteristic for the two objects is represented by the difference in the two values, that is,

$$d_{ij} = |y_i - y_j|$$

where the absolute value of the difference is involved in order to avoid negative differences. If two objects have the same magnitude of a property, then the values associated with those objects will be equal:

$$y_i = y_j$$

or, in terms of d,

$$d_{ij} = |y_i - y_j| = 0$$

Where two objects have different magnitudes of a property the values will be different. For example, if the ith object has "more of" the attribute than the jth object, then

$$d_{ij} = y_i - y_j > 0$$

This inequality implies that the objects are ordered with respect to the quantities representing the attribute magnitudes.

One advantage accrues when it is possible to establish the existence of (or the possibility of the existence of) an object, say the oth object, for which the attribute is entirely lacking and hence $y_o = .0$ for that attribute. Comparing the ith and jth objects with the oth object with values y_o, y_i, y_j in terms of the difference of the objects i and j relative to the object o is a meaningful comparison, since the values

$$d_{io} = y_i - y_o = y_i$$

and

$$d_{jo} = y_j - y_o = y_j$$

represent the magnitude of the attributes for the i and j objects. This permits direct comparisons of the attribute magnitudes of i and j objects by the ratios of y_i and y_j:

$$r_{ij} = \frac{y_i - y_0}{y_j - y_0} = \frac{y_i}{y_j}$$

Thus, if $y_i = 42$ and $y_j = 21$, then $r_{ij} = 2$ and the attribute magnitude of the ith object is said to be twice the attribute magnitude of the jth object.

This ratio comparison is not meaningful with magnitudes that are not anchored by a directly determined or natural zero point in the continuum. If, for example, an object i is selected to provide an anchor point and value y_i is assigned to object i, comparing g and h by the ratio gives

$$r_{gh} = \frac{y_g - y_i}{y_h - y_i}$$

which does not reduce as r_{ij} did above, because the values in this case are not absolute magnitudes as they were in the previous example. If the object j had been selected to represent the zero point in the continuum with $y_j = y_i + 10$,

$$r_{gh}' = \frac{y_g - y_j}{y_h - y_j} = \frac{y_g - y_i + 10}{y_h - y + 10} \neq r_{gh}$$

This result indicates that the ratio comparison is variant with the arbitrary selection of the object to represent the anchor point. Of course, ratio comparisons with the implied $y_o = .0$ are numerically invariant but have no objective meaning for attribute magnitudes.

Another assumption is made in defining the continuum. If the values representing the magnitudes of the attributes for two objects are different by a value d_{ij} while for two other objects the difference is d_{gh}, and if $d_{ij} = d_{gh}$, then the magnitudes of the characteristics are said to be equally different regardless of the values involved. This is true even if i and j come from the mid-range of value on y and g and h are represented at the extreme. This property is called the equal interval property. Equal differences (intervals) on the scale represent equal differences in the characteristic measured. For example, on an equal interval scale, a difference of ten points in the neighborhood of 150 is the same as a difference of ten points in the neighborhood of 90. The values representing magnitudes on the scale may be arbitrary.

Region C in Figure 2.2 corresponds to a measurement or psychometric procedure through which observations are made and values are assigned to the objects in Region A.

The values in Region B are not observable. The process of obtaining the actual observed values represented in Region D is intimately connected with the particular subject matter and the attributes being studied. The reader can imagine an operation by which the values on the continuum X are found. The issues germane here are the values and their characteristics relative to each other and relative to the corresponding values in the covert continuum Y, and not practical measurement.

The first requirement of all measurement is that the objects be represented on X in the same order as they are represented on Y. That is, if two objects are involved, i and j, then we require that

$$y_i > y_j \quad \text{implies that} \quad x_i > x_j$$

at least within the limits on the accuracy of evaluation of the values on X. To the extent that this order-preserving requirement is not satisfied the measurement will be a poor representation of the magnitudes of the attributes involved. If the order is badly preserved, the measures are worth little in representing the magnitudes of the attributes.

There are four special cases of measurement in terms of the relationship of the differences in the values in X compared with corresponding differences in the continuum Y: (1) ordinal measurement with an arbitrary zero point, (2) ordinal measurement with a natural zero point, (3) interval measurement (arbitrary zero point), and (4) ratio measurement (natural zero point).

In the first case, ordinal measurement with an arbitrary zero point, the X values order the objects measured in the same way that the values of Y order the objects. There is no other property preserved. For two objects i and j the following relationships between the values in Y and the values in X hold in ordinal measurement with an arbitrary zero point:

$$y_i = y_j \quad \text{implies} \quad x_i = x_j$$
$$y_i > y_j \quad \text{implies} \quad x_i > x_j$$

Stated in a mathematical language, the relationship between Y and X is single valued and monotonic; that is, for every value of Y there is one and only one value on X, and the larger-than and smaller-than relationships in Y are preserved in X for any pair of objects represented in Y and X.

In the second case one additional condition is imposed, that the zero point in Y be preserved. That is, in addition to the relationships above,

$$y_i = 0 \quad \text{implies} \quad x_i = 0$$

for the object i.

In the third case, interval measurement, order and relative magnitude differences are preserved. This means simply that for four objects g, h, i and j,

$$[(x_g - x_h) = (x_i - x_j)] \quad \text{implies} \quad [(y_g - y_h) = (y_i - y_j)]$$

in addition to the conditions for ordinal measurement with arbitrary origins. The origin is not preserved. The sorts of functions that can relate Y and X in ordinal measurement with arbitrary origins are unlimited, except for monotonicity and single-valuedness. However, in interval measurement the relationship between Y and X must be a linear function, that is,

$$X = a + bY$$

where a and b are constants associated with the measurement process.

Finally, ratio measurement is a specialization of interval measurement. Ratio measurement preserves a natural origin as well as satisfying the requirements for interval measurement. This means that the absolute magnitudes of the attributes are proportional to the values assigned in the measurements. The ratios of measurement values give the relative magnitudes of the attributes measured. This condition can be satisfied mathematically if the values of X are equal to some linear function of the values of Y where the constant a is equal to .0, that is,

$$X = bY$$

where b may be any arbitrary constant ($b \neq 0$). This constant is determined by the unit object that is assigned the value 1.0. If the object is the same in the magnitude evaluation as in the measurement, then $b = 1.0$, and the measurement values correspond numerically to the magnitude values. However, if $b \neq 1.0$, then the measurement values are not the same as the attribute magnitudes. Just as important, though, are the ratios: The measurement value ratios are the same as the attribute magnitude ratios. Take objects i and j with measurement values x_i and x_j; then

$$r_{ij} = \frac{x_i}{x_j} = \frac{by_i}{by_j} = \frac{y_i}{y_j}$$

Thus, if we are able by measurement or theoretical calculation to discover the attribute magnitude of any object (other than the object o with magnitude value $y_o = .0$), we can determine the attribute magnitude of all other objects by implication from

$$x_i = by_i$$

$$b = \frac{x_i}{y_i}$$

For any other measurement values x_j, then,

$$y_j = \frac{x_j}{b}$$

Although the relationship of measurement values and attribute magnitudes is of vital importance in interpretation of observed values, it plays no important role in probability models and statistics. If a measurement gives only an order function of attribute magnitudes, then there is no justification for comparing the measurement values of two (or more) objects except in terms of order. Thus, $x_i = 32.41$ and $x_j = 34.21$ do not reflect an attribute difference of 1.80 if the measurement is not an interval or a ratio measurement. The attribute magnitudes of i and j do not necessarily represent equally different magnitudes from those of objects g and h when $|x_i - x_j| = |x_g - x_h|$ in ordinal measurement. However, regardless of the weak meaning of X for the corresponding attribute magnitude, the statistician is concerned only with the values of X, their relative frequencies in a set of observations, and their probabilities in a probability model.

A specific example here may be useful. We begin with an attribute magnitude dimension Y having a natural origin. Imagine two measurement procedures, one preserving only the order of objects and one giving a ratio scale with $b = 1.0$. Denote the continua resulting from the measurements X and Z, respectively, for the order-preserving measures and for the ratio-preserving measures. Graphs of these measurements are given in Figure 2.3, where only some of the values are shown. These graphs make it clear that Z gives direct knowledge of the attribute magnitudes whereas knowledge of X provides little (other than order) knowledge about Y.

From a measurement point of view, scientific knowledge regarding the attribute involved in ratio measurement is superior.

From the probability and statistical point of view, no claim of superiority for Z or X can be made on the basis of the information we have in Figure 2.3. The values of Z and X are collections of real numbers and share the properties of real numbers regardless of their relationship to Y. To the statistician the important feature of Z and X is the way probabilities are associated with the values. If the probabilities (or relative frequencies) have certain properties (large in certain segments of Z and X and small in other segments, and so on, as for example a

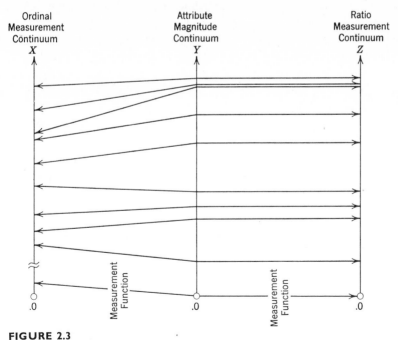

FIGURE 2.3

Comparison of ordinal and ratio measurements.

normal distribution), then certain probability and statistical theorems can be applied in order to test hypotheses about the X and Z values. In the event that Z does not have these properties and X does, then X may be the preferred measurement. This would be true where application of statistical procedures was required to obtain precision in making scientific decisions and the probability or relative frequency properties of Z were not appropriate for statistical analysis. Regardless of the measurement advantage of Z, X might be more appropriate for some scientific purposes.

The general principle is as follows. Probability and statistics are concerned only with the collection of values in a measurement process and with the attendant probabilities or relative frequencies. Where these satisfy certain conditions, regardless of measurement properties, appropriate theorems of probability and statistical theory may not be applicable. The implication is that measurement properties are irrelevant to the use of probability and statistics in the evaluation and analysis of experiments.

This general principle does not apply to the interpretation of probability and statistical findings in terms of the attribute magnitudes being measured. Clearly, if Z were such to allow application of statistical procedures, then any results of statistical analyses would also apply to the attribute magnitude continuum Y. It is not clear that results of statistical analyses on X would apply directly to the attribute magnitudes represented in X. The strongest statement possible might be limited to the order of magnitudes. The question of how broadly the statistical results can be generalized in terms of attribute magnitudes intimately concerns the measurement properties of the observed measurements. The answer depends on the scientific purpose and character of the experiments involved.

Limits of space prohibit a further elaboration of this problem. However, there are several excellent references in which one may find thorough discussions of the issues. Particularly interesting is the paper by Burke in Marx (1963).

Specifying Probabilities

Probability, as a value, has been defined in several ways. The three most prevalent ways are: (1) probability is a statement of the likelihood of results occurring in the real world, that statement being based on considerations of logic, symmetry, and identity; (2) probability is the value of the relative frequency of a result when the experiment is repeated an unlimited number of times; (3) probability is a weight (or value) attached to an event that indicates the strength of someone's expectations (or degree of belief) that the event will occur at some future time.

The importance of the method of determining the values in a probability model depends on the purpose of specifying the values. If the values are used in strictly mathematical work in probability theory and statistics, it makes no difference how they are obtained. Derivations and proofs of probability theorems do not generally involve specific values, and hence the proofs are general.

The method of determining the specific values is crucial in the assignment of probability values to events in such a way that they may match actual probabilities of results. Frequently it is possible to design experiments in such a way that logical definitions of probability values are an adequate reflection of the real world or of the theoretical concepts being investigated. This is the case where models are formulated to take advantage of sampling techniques. In some instances the relative-frequency methods of determining values of probabilities must be used. This usually occurs when little is known about the logical structure of the phenomena being studied, and when the experimental design does not specify events in a way so as to determine the probabilities. Sometimes even this method—looking at many cases in point and calculating relative frequencies—cannot be applied because of the difficulty, cost, or undesirability of making the many observations necessary. In such a case subjective probability is often used; that is, "experts" are asked to estimate probabilities. This amounts to using the "experience" or "insight" of experts as a substitute for determining the relative frequencies.

Since the first two methods listed above are used in the following chapters, examples are in order.

1. Consideration of Theory, Logic, Symmetry and Identity. A die is presumed to be a perfect cube, homogeneous in construction, each side being indistinguishable (identity) from all others while having a relation of symmetry to other sides of the die. We can argue that when a die is rolled fairly, each side has the same chance of being up as any other side. Thus the probability of any given face being up is the same as that of any other. The number of dots on the face is irrelevant, since they could be changed around in an arbitrary fashion and not change the basic nature of the die. Thus, to each face of the die, and hence to each "number of dots," we assign probabilities of equal size that satisfy the axioms of probability: $\frac{1}{6}$. If $e_1 =$ one-dot face, $e_2 =$ two-dot face, $\ldots$, $e_6 =$ six-dot face, then $P(e_1) = P(e_2) = \cdots = P(e_6) = \frac{1}{6}$.

The example of rats in a "T" maze illustrates the application of another logical principle—the principle of insufficient reason. If we knew nothing of rats and "T" mazes there is insufficient reason to assume the rat would not treat the two symmetric sides of the maze in the same fashion. The animal would turn left or right with equal probability. In accordance with the probability axioms and this reasoning, probabilities would be assigned as follows:

$$P(e_1) = \tfrac{1}{2} \qquad P(e_2) = \tfrac{1}{2}$$

where e_1 and e_2 refer to left turn and right turn, respectively.

In an early application of mathematics to modern learning theory L. L. Thurstone (1930) developed a "rational equation for the learning function." This development began with a simple probability model. Imagine that the learning situation evokes a number of acts that the learner emits in an attempt to solve a problem. Let s be the total number of acts that are relevant to the situation and that lead to successful steps in the learning. Let e be the total number of acts that are relevant to the situation and that lead to errors or loss of time, and so forth, in learning. If we assume that each relevant act is equally attractive to the learner at a given point in the learning, then

$$p = \frac{s}{s + e}$$

is the probability of a successful act being emitted, and

$$q = \frac{e}{s + e}$$

is the probability of an unsuccessful act. (The reader should prove that p and q are probabilities in the formal sense.) If in a given actualization of this model by an experiment there are 60 acts relevant and 15 of them lead to successful steps, then

$$p = \frac{15}{45 + 15} = .25$$

$$q = \frac{45}{45 + 15} = .75$$

are the probability values specifying the general model for the experiment. From this humble start and with the application of additional reasoning and mathematical arguments, Thurstone was able to derive a rather general and accurate equation for the learning curve.

A different sort of example comes from genetics. A hybrid population is of genotype Aa and the phenotype is determined by A where A is the dominant genetic element and a is recessive. When two hybrids are crossed the combinations possible at the genotypic level are AA, Aa, aA, and aa. Under the hypothesis that the genetic elements combine from parents without discrimination, that is, all combinations occur equally often, the probabilities of the four genotypic combinations are all .25. However, because of the genetic identity of Aa and aA and because of the dominance of genotype A over a with respect to producing the phenotype, the proportion of observed phenotypes corresponding to A and a will be three to one:

$$P(\text{phenotype } A \text{ from a cross of two } Aa \text{ hybrids}) = .75$$
$$P(\text{phenotype } a \text{ from a cross of two } Aa \text{ hybrids}) = .25$$

2. The Observation of Relative Frequencies. Toss a coin one thousand times and observe the number of times the result is heads, $n(r_1)$, and the number of times the result is tails, $n(r_2)$. Calculate the relative frequencies of the two results:

$$f(r_1) = \frac{n(r_1)}{1000} \qquad f(r_2) = \frac{n(r_2)}{1000}$$

For example, if $n(r_1) = 459$ and $n(r_2) = 541$, then $f(r_1) = .459$ and $f(r_2) = .541$. The rule then states $P(e_1) = .459$ and $P(e_2) = .541$ where e_1 and e_2 refer to events "heads" and "tails," respectively.

The reader should attempt to discover what $P(e_2)$ would be as dictated by the rule of insufficient reason (argue from identity, symmetry, and so forth). Which assignment is preferable, (1) from a logical point of view, (2) from an empirical (real-world) point of view?

It is interesting to note an example combining two methods of determining probabilities. Imagine loading a die in an attempt to make the "6" face show more frequently than "normal" at the expense of the "1" face. The degree of "loading" is undetermined. Define the values $P(e_1)$ and $P(e_6)$ by observing relative frequencies, and $P(e_2)$, $P(e_3)$, $P(e_4)$, and $P(e_5)$ by logical argument. To do this, roll the die, say, 1200 times and from the results construct Table 2.2.

Assume that the relative frequencies obtained in 1200 trials are stable (that is, not likely to change much by increasing the total number of trials). On the basis of the observations we set

$$P(e_1) = .083 \qquad P(e_6) = .250$$

The arguments of logic, symmetry, and identity are used in defining $P(e_2)$, ..., $P(e_5)$. In particular, $P(e_2) = P(e_3) = P(e_4) = P(e_5)$. Since

$$\sum_{i=1}^{6} P(e_i) = 1$$

TABLE 2.2 Probabilities of a Loaded Die

Number of Spots	1	2	3	4	5	6
Results	r_1	r_2	r_3	r_4	r_5	r_6
Frequency: $n(r_i)$	100	220	180	200	200	300
Relative Frequency $f(r_i)$	.083	.183	.150	.167	.167	.250

then

$$\sum_{i=2}^{5} P(e_i) = 1 - [P(e_1) + P(e_6)] = 1 - .333 = .667$$

If this sum is divided equally among e_2, e_3, e_4, and e_5,

$$P(e_2) = P(e_3) = P(e_4) = P(e_5) = \frac{.667}{4} = .167$$

Putting these probabilities together completes the probability model for the experiment "toss the crooked die." Other models may be more satisfactory, and the empirical scientist would proceed next to test the model against empirical observations. In this specific case there is little interest in rolling the die, because we have already rolled it 1200 times in determining the values used in the model. Unless a model of this sort has some usefulness in the study of phenomena that were not involved in determining the values in the model, there is little point of testing the model.

Graphical Representation of Probability Models

Often it is convenient to display the values in a probability model graphically. The basic device is to show the probability of events as a function of the events in rectangular (Cartesian) coordinates. The events generally are indicated as points on the X axis and the probabilities as points on the Y axis. Two special cases are implied by the distinction of quantitative and nominal events. Where the events are quantitative in nature there is a natural ordering of the events and the order of the points representing the events on the X axis is defined by that natural ordering. Where the events are nominal no natural ordering is defined and the order of the points on the X axis is arbitrary. The graphs are called probability graphs.

Figures 2.4 and 2.5 are examples of the graphical representation of two models developed in the sections above. In the genetics example the events are not quantitative and the graph of the probability model is presented twice with different ordering of the events on the X axis. This does not change the values of probabilities or their representation on the Y axis. The vertical lines are simply visual aids to make the location of the points of the graph stand out more readily.

The direct parallel of observed relative frequencies and probabilities provides the rationale for constructing graphical representations of experiments. If an experiment is repeated N times, each result occurs with a certain relative frequency, $f(r_i)$. If these are plotted in the same way that $P(e_i)$ is plotted in a probability graph,

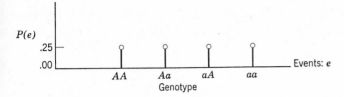

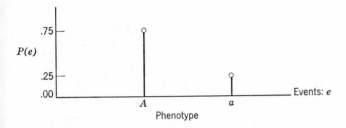

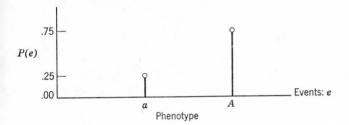

FIGURE 2.4

Graphs of a genetics model.

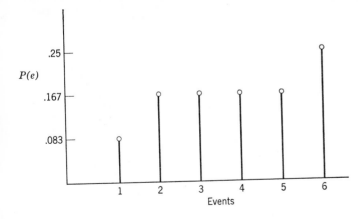

FIGURE 2.5

Model for "toss a crooked die."

the resulting graph is a graph of the experiment. This graph is called a relative-frequency graph. A direct comparison is possible between an experiment and its model by plotting the probabilities and relative frequencies on the same axes. An example of this is presented in the next section.

A frequency graph involves the relative frequencies, but the same result is obtained by plotting the frequencies themselves. Since frequency and relative frequency correspond value for value through a constant term $1/N$, the scale of the Y axis is the only thing that differentiates graphs of frequencies and graphs of relative frequencies.

An Illustration: An Experiment and Its Model

Imagine an experiment defined by the instructions "pick a number from the digits $0, 1, 2, \ldots, 9$ in such a way that each number has no more chance than any other number of being chosen." The "pick a number" part is easy, but the restriction of the fairness of the choice is more difficult to achieve. In order to evaluate the degree to which the conditions are met in the experiment, a model of the experiment must be formulated. The basic logic of the model is founded in the specific operations and conditions of the experiment.

The model is easy to formulate. There are ten events, for example, $e_1 = 0$, $e_2 = 1$, $e_3 = 2$, $\ldots$, $e_{10} = 9$, each of which has the same chance of being chosen on a given repetition of the experiment. Translating the word "chance" into the word "probability" implies that

$$P(e_1) = P(e_2) = \cdots = P(e_{10})$$

which implies in turn that

$$P(e_i) = .1 \quad \text{for} \quad i = 1, \ldots, 10$$

It is easily verified that these values are indeed probabilities and that therefore the probability model is complete.

The experiment was done in class with thirty four students participating, each providing one repetition of the experiment. The frequencies of each of the ten results are graphed in Figure 2.6 along with the graph of the model.

The lack of complete agreement between the experiment and the model of the experiment is apparent. This lack of agreement might be interpreted a number of ways. First, the conditions of the formal definition of the experiment may not have been achieved in the execution of the experiment. Second, it is possible that the

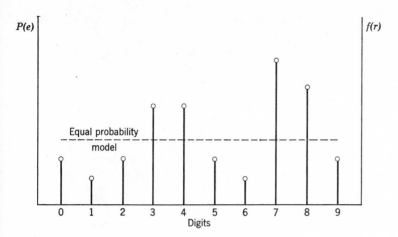

FIGURE 2.6

Model and data for "pick a digit" experiment.

model does not adequately reflect (that is, is not a model of) the experiment. Third, the sample size, the number of repetitions of the experiment, is not very large, and relative frequency is an adequate approximation to probability only with very large numbers of repetitions. Other interpretations can be thought of (for example, should the same person make all of the choices?).

The first interpretation is the most interesting from the point of view of the psychologist. If the instructions are not reflected in the data—that is, the actual results observed—this indicates something about the response bias or response preferences of human subjects in such a simple task as picking a digit. In order to have any faith in this sort of interpretation we must first rule out the other interpretations, or at least make them less plausible. Several points are worth mentioning about admissibility of the model.

The abstract definition of the experiment is not clear enough to be executed. The actual apparatus of the experiment is left unspecified. Many details are involved, any of which might make a difference in the observations. The experiments were performed in a class setting with statistics students, the choices were performed in a class setting with statistics students, the choices were written on separate pieces of paper, one choice for each student, and so on. None of these specific properties of the instrumentation are included in the considerations leading to a formulation of the model. Perhaps a social psychologist could point out reasons why some of these conditions would lead to a different formulation of the model; for instance, if each subject assumed that the most likely thing for the other subjects to do was to choose a number near the middle of the range of values and tried to correct the presumed inequality of chance by choosing one of the extreme values, then the method of doing the experiment, in a social situation, would lead to a violation of the conditions as stated in the initial formulation of the experiment. Thus there is a possible confounding of the "trivia" of the experiment and the basic nature of the experiment leading to a model. If the experiment is not what was intended and the model represents what was intended, it cannot also be a model of the actual experiment.

Other possibilities are suggested. Perhaps the word "chance" is not legitimately translated into the word "probability." Perhaps the logic of setting the probabilities to .1 is in error, or perhaps there was an error in the number of events used to determine the .1 value. The force of the arguments leading to the model of the experiment as formulated originally is impressive, however.

The third interpretation of the deviation of relative frequencies and probabilities in the example is based on statistical considerations. If we could calculate the probability of observations as extreme or more extremely deviant from the model in thirty four repetitions of the experiment, we might be able to judge the significance of the deviations observed. A detailed discussion of this interpretation and the mechanics of investigating the validity of the interpretation must be delayed until the theory of statistical decisions has been introduced. In the meantime, however, the crux of the argument can be illustrated. It is not possible at this point to judge whether the numbers were fairly chosen by the human subjects. It is, however, possible to devise a machine that is not biased in "choosing numbers." In particular, we might refer to a table of digits selected by such a machine and "choose" the numbers from the table (and hence let the machine determine the choice). Such a table is referred to as a random digits table—a table of random digits is provided at the back of the text. If the table is to be used to select fewer digits than are presented in the table and never again, selection can be started at the first of the table. If the table is to be used repeatedly, care should be taken to begin at "random" points in the table and to make different movements from one digit to another (up columns, right to left, and so on) in the table in selecting digits. The electronic computer offers a more elegant way of selecting digits. The table provided in this text was actually calculated by a computer program described in Lehman and Bailey (1968). This computer program is known to be an unbiased generator of digits (hence the term "random number generator"); that is, if a large enough number of digits were generated by this program, the relative frequencies of all the digits so generated would be .1. This implies that we can perform the experiment precisely in accord with the requirements that led to the model.

Twenty repetitions of the computerized experiment were performed. There were several computer experiments giving a relative-frequency graph less like the model than the classroom experiment. An important lesson is gleaned from this demonstration. Where the model does reflect reality, the limited number of repetitions of the experiment may produce observations not in agreement with the model.

SUGGESTED READINGS

Chaplin, J. P., and Krawiec, T. (1960), *Systems and theories of psychology*, Holt, Rinehart, and Winston, New York.

Churchman, C. (1948), *Theory of experimental inference*, Macmillan, New York.

Freudenthal, H. (1961), *The concept and the role of the model in mathematics and natural and social sciences*, D. Reidel Publishing Co., Dordrecht, Holland.

Green, B. F., Jr. (1963), *Digital computers in research*, McGraw-Hill Book Company, New York.

Hays, W. (1963), *Statistics for psychologists*, Holt, Rinehart, and Winston, New York.

Hooke, R. (1963), *Introduction to scientific inference*, Holden-Day, San Francisco.

Kaplan, A. (1964), *The conduct of inquiry, methodology for behavioral science*, Chandler, San Francisco.

Kemeny, J., Snell, J., and Thompson, G. (1957), *Introduction to finite mathematics*, Prentice-Hall, Englewood Cliffs, New Jersey.

Lehman, R., and Bailey, D. E. (1968), *Digital computing: Fortran IV and its applications in behavioral science*, John Wiley & Sons, New York.

Lewis, D. (1960), *Quantitative methods in psychology*, McGraw-Hill, New York.

Marx, M. (1963), *Theories in contemporary psychology*, Macmillan, New York.

McNemar, Q. (1962), *Psychological statistics*, John Wiley & Sons, New York.

Nagel, E. (1961), *The structure of science*, Harcourt, Brace, & World, New York.

Scott, W., and Wertheimer, M. (1962), *Introduction to psychological research*, John Wiley & Sons, New York.

GLOSSARY

Experiment.   An experiment consists of making an empirical observation where repetition of the experiment can lead to observation of a different result.

Results.   Results are the things which may be observed on performing an experiment; $r_1, r_2, \ldots, r_k$.

Frequency of result r. $n(r)$

(2.1)
$$\sum_{i=1}^{k} n(r_i) = N$$

(2.2) Relative frequency
$$f(r_i) = \frac{n(r_i)}{N}$$

(2.3)
$$0 \le f(r_i) \le 1$$

(2.4)
$$\sum_{i=1}^{k} f(r_i) = 1$$

Probability Model. A probability model is a mathematical abstraction or idealization of an experiment in terms of the axioms and theorems of probability theory.

Event. An event is the abstract element in a probability model corresponding to a result in an experiment; $e_1, e_2, \ldots, e_k$.

Probability. The probability of an event e_i is defined as a number $P(e_i)$ attached to the event where $P(e_i)$ is characterized by relationships (2.5) and (2.6):

(2.5)
$$0 \le P(e_i) \le 1$$

(2.6)
$$\sum_{i=1}^{k} P(e_i) = 1$$

PROBLEMS

1. Imagine that fifteen throws of a die resulted in the following observations:

$$1, 3, 1, 1, 6, 5, 2, 2, 6, 4, 4, 1, 3, 6, 1$$

Designate the first observation as X_1, the second as X_2, and so forth up to X_N ($N = 15$). Find

(a) $\sum_{i=1}^{15} X_i$ (b) $\sum_{i=1}^{8} X_i$

(c) $\sum_{i=9}^{15} X_i$ (d) $\sum_{i=1}^{8} X_i + \sum_{i=9}^{15} X_i$

Do your answers for (a) and (d) agree? Why?

2. In problem 1, there are six results $r_1, r_2, \ldots, r_6$.
   (a) Find the frequency $n(r_i)$ for $i = 1, \ldots, 6$.
   (b) Plot the frequencies against the values of the results.
   (c) Find $\sum_{i=1}^{N} n(r_i)r_i$. Does this agree with your answer 1(a) above? Why?

3. Two rats were taught to run a maze. The number of errors made per trial (from start to stop on a given run through the maze) is given below.

							Trial					
	1	2	3	4	5	6	7	8	9	10	11	12
Rat 1	10	8	7	9	4	4	5	1	1	0	0	0
Rat 2	11	10	6	4	4	6	7	5	4	2	0	0

(a) Make a graph showing the number of errors as a function of trial number for each rat.

(b) Regard each possible number of errors as a result. Find the frequency for each result for each rat. Plot these as a frequency graph.

(c) Let X_i be the number of errors on the ith trial for rat number 1. Let Y_i be the number of errors on the ith trial for rat number 2. Find

$$\sum_{i=1}^{12} X_i \quad \text{and} \quad \sum_{i=1}^{12} Y_i$$

Find

$$\sum_{i=1}^{12} (X_i + Y_i)$$

(d) Prove that

$$\sum_{i=1}^{N} (X_i + Y_i) = \sum_{i=1}^{N} X_i + \sum_{i=1}^{N} Y_i$$

4. Sample 100 digits from the random digits table.
(a) List the sample.
(b) Find $n(r_i)$ for $r_i = i$ where $i = 0, 1, \ldots, 9$.
(c) Find $f(r_i)$ for $i = 0, 1, \ldots, 9$.
(d) Construct a relative-frequency graph of $f(r_i)$.
(e) Construct an equal probability graph of the e_i corresponding to the r_i.
(f) Are the graphs for (d) and (e) the same? If not, why does $f(r_i) \neq P(e_i)$?
(g) Compare your $f(r_i)$ with those of a classmate. Are they the same? If not, why not?
(h) Find $\sum_{j=1}^{100} X_j$ where X_j is the jth digit sampled. Compare this with $\sum_{i=0}^{9} n(r_i)r_i$ where r_i is one of the values, say, $r_0 = 0, r_1 = 1, \ldots, r_9 = 9$.
(i) Compute the average of the observations:

$$\bar{X} = \frac{1}{N} \sum_{j=1}^{N} X_j$$

(j) An alternate way of calculating the average is given by

$$\bar{X} = \sum_{i=0}^{9} f(r_i)r_i$$

Calculate this. Go through the necessary algebra to show the identity of the two definitions of $\bar{X}$.

5. Assume that tonight the TV weather forecast says "snow by morning." From what you know of the local weather and weather forecasters in general, estimate the probability of (a) snow, (b) rain, and (c) sunny and warmer for the weather tomorrow.

6. Assume that you watch two more weather reports and they give the same forecasts. Now answer the question of problem 5. If the answer is different, explain why.

CHAPTER 3

The Mathematics of Events: Set Theory

Lists of results of experiments and events in models of experiments are composed of mutually exclusive and exhaustive elements. For this reason they are best discussed and manipulated in terms of a very elegant branch of mathematics—set theory. By applying the theory of sets to events we can more readily develop powerful theorems concerning the relationship among events, and in particular the probabilities associated with events and combinations of events.

Sets

A set is simply a collection or aggregation of "objects." The objects may be numbers, as in most mathematical applications of set theory; physical objects, such as houses, people, dinosaurs, unicorns, grains of sand, hospital patients; actions or processes such as speeches, ball games, planetary orbits by a satellite; or any number of other sorts of things.

In particular, we deal with sets that are aggregations of events. For example, the set of all events in our model of a rat in a "T" maze is the collection of objects

e_1 and e_2: $e_1 = \{$left turn$\}$, $e_2 = \{$right turn$\}$. The set of events involved in a die-tossing model is composed of $e_1 = \{$one dot showing$\}$, $e_2 = \{$two dots$\}$, ..., $e_6 = \{$six dots$\}$.

The definition of sets, and the elements of sets must obey rules similar to the definition of events or we are not justified in using the mathematics of events of a model. Events are by definition mutually exclusive and exhaustive of all possible consequences of an experiment. Correspondingly, when a set is well defined we can say clearly and with no ambiguity whether any object whatsoever belongs to the set or whether it does not belong to the set.

Defining a set is not always an easy matter. Think of the consternation of the zoologists who first discovered the platypus. A fur-bearing, billed, web-footed beast that lay eggs, incubated them, and suckled the hatched young! Was it a member of the set "fowl"? Was it a member of the set "mammal"? Clearly here was an object for which a zoological classification was unclear.

A set also must be composed of distinct objects. An object is not counted or listed twice if we count or list the objects in the set. In short, a set is a well-defined collection of distinct objects. The objects in a set are its elements, and each element is said to belong to or be a member of the set. The members, collectively and individually, of a set are contained in the set.

A set may be specified in two ways: (1) by listing all the elements contained in the set, and (2) by specifying the defining character of the elements or the members of the set. In the first case we just list the elements: The set A is composed of professors, graduate students, seniors, juniors, sophomores, freshmen, administrative personnel, clerical assistants . . . . In the second case the defining characteristics of the members of a set A is composed of all persons involved in the operation of a university.

The notation used to facilitate discussion of sets and the algebra of sets is as simple as the definition of sets themselves. The student should be careful not to look for hidden meaning or inherently difficult logic. The notation allows simple and short symbolic statement of verbal propositions and statements. Learning the basic language of set theory is like learning a very simple foreign language.

Take a collection of objects, $a_1, a_2, \ldots, a_n$. Here $A = \{$all a's$\} = \{a_1, a_2, \ldots, a_n\}$. The curly brackets are read "the set of all [things inside the brackets]."

The set of events in the "T" maze experiment is

$$T = \{e_1, e_2\}$$

where T is used as the name of the set for no better reason than to suggest that it is the set of "turns" the rat can make: $e_1 = \{$turn left$\}$, $e_2 = \{$turn right$\}$.

If e_1 is a member of T, we write $e \in T$ where the symbol $\in$ is read "belongs to" or "is a member of" or "is contained in" or "is an element of."

There are some very special sets—sets of numbers. These are the basic coin of mathematics and thus deserve special attention. The natural numbers, that is, the positive integers, 1, 2, 3, . . . , are used with great frequency in probability models in psychology. Some examples are: the number of cases of a given mental disorder admitted to a hospital; the number of correct responses made on a test of intelligence; the number of correct responses made on a test of intelligence; the number of trials required in a learning experiment; the number of ways subjects in an experiment can be allocated to groups being treated differently in the experiment. If the integers make up the set I, we have

$$I = \{1, 2, 3, \ldots\}$$

Clearly the set is well defined and the elements are distinct. Taking two numbers, say 94 and 46.3, clearly

$$94 \in I \quad \text{and} \quad 46.3 \notin I$$

That is, 94 is a member of the set I and 46.3 is not.

Another mathematical set is the collection of all real numbers: R. It is impossible to write out this set or even to start writing it out as we did for I. Thus, R, the set of real numbers, contains all of the numbers in our ordinary numbers, including the positive and negative integers and all fractions (or decimal equivalents) and the mathematical construct infinity. The two numbers mentioned explicitly above, 94 and 46.3, are both members of R:

$$94 \in R \quad \text{and} \quad 46.3 \in R$$

To allow a better representation of R we need to introduce a new notation:

$$R = \{x \mid x \text{ is a real number}\}$$

That is, a set may be specified by the defining characteristic of the elements in the set: R is the set of all x's such that x is a real number. The vertical line is interpreted to mean "such that."

Subsets

A very special sort of set is the set that is a portion of another set: a subset. In the sense that a set is a collection of objects, a subset is a part of a collection of objects.

For example, the set of all possible scores on a true-false test with one hundred questions is

$$S = \{0, 1, 2, \ldots, 100\}$$

The subset of scores (that is, subset of S) that are classified as "A" grades may be

$$A = \{91, 92, \ldots, 100\}$$

The subset of scores classifying as "F" grades may be

$$F = \{0, 1, 2, \ldots, 59\}$$

Both A and F are subsets of S. The subset for "B" grades might be defined, using the more sophisticated notation

$$B = \{x \in S \mid 81 \leq x \leq 90\}$$

Two special sets are important: the null or empty set and the universal set.

$$\text{null set} = \emptyset$$

$$\text{universal set} = \mathscr{U}$$

The null set is the set which has no members. We speak of *the* null set because there is but one null set. If two sets do not contain any elements, the one set can contain no elements not in the other. However, to be unequal, one set must contain elements not in the other. If sets are not unequal they must be equal. And, by definition, any two sets that are equal contain exactly the same elements and thus are identical. This can be argued for any pair of null sets, implying that all null sets are identically one and the same. Thus, there is but one null set—*the* null set.

At the opposite extreme is the universal set—the set that contains everything of concern in the model or experiment. This corresponds to the notions of frame of reference and universe of discourse. Where there is any question about our universe of discourse we must be very careful to specify the universal set. Otherwise, the universal set is implicit in all statements made about sets, subsets, and probabilities associated with the subsets.

Graphic Representation of Sets

The relationships among sets and subsets is not always immediately clear and simple to grasp. There are analytical, mathematical methods to assist in the study of sets, subsets, and their relationships. However, graphic devices are useful in demonstrating these sorts of things although they do not provide direct proofs. A commonly used device is the Venn diagram.

In Venn diagrams a universal set is represented as an enclosure with the other sets represented as regions of the enclosed space. The spatial character of the

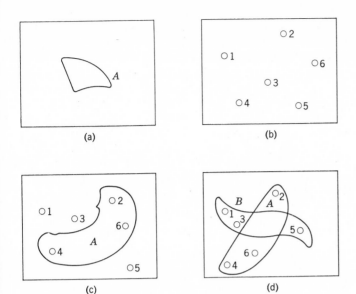

FIGURE 3.1

Illustrations of Venn diagrams.

diagrams is incidental to the graphic method—it is not intended to imply that sets are two-dimensional. The universe $\mathcal{U}$ with the set A is shown in Figure 3.1a. If we had as elements in the universe the integers 1, 2, 3, 4, 5, 6, we might represent them as dots or points in the enclosure (Figure 3.1b), the specific locations of specific points being quite arbitrary. If now we take a subset of these points, such as $A = \{x \in \mathcal{U} \mid x \text{ is even}\}$, we can draw an enclosure about $A = \{2, 4, 6\}$ to represent A (Figure 3.1c). It is important to notice that the enclosed space corresponding to A is larger than necessary to enclose just $\{2, 4, 6\}$. We may wish to draw diagrams of two sets in $\mathcal{U}$ and have to overlap our diagrams as in Figure 3.1d. This should not be taken to imply that A and $B = \{1, 3, 5\} = \{x \in \mathcal{U} \mid x \text{ is odd}\}$ have anything in common except that they belong to the same universe.

In general, we do not represent points in the diagram but deal with circles and their overlap in $\mathcal{U}$ without representing the separate elements.

Operations on Sets

There are three basic operations applicable to sets and subsets: complementation, intersection, and union. Take two sets A and B; $A = \{a \mid a \in \mathcal{U}\}$ and $B = \{b \mid b \in \mathcal{U}\}$.

Complementation. The complement of A with respect to $\mathcal{U}$ is composed of all elements in $\mathcal{U}$ not in A. The complement of A is written A' and read "A complement":

$$A' = \{x \in \mathcal{U} \mid x \notin A\} \tag{3.1}$$

If $\mathcal{U}$ is the collection of all possible grades on the 100-item true-false test and $A = \{91, 92, \ldots, 100\}$, then $A' = \{0, 1, 2, \ldots, 90\}$. In terms of the Venn diagram of Figure 3.2a, the complement of a set is shown in dark. It may be that

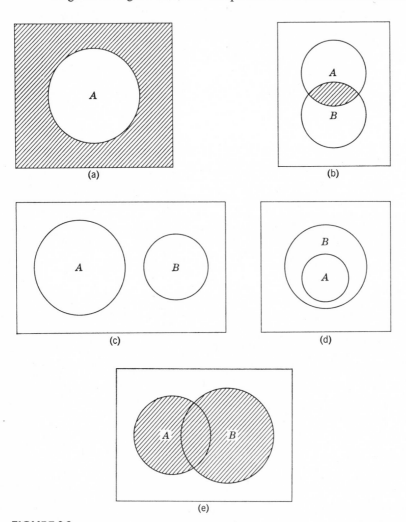

FIGURE 3.2

Venn diagrams showing set relationships.

there are no points in A'; that is, the dark portion of the Venn diagram 3.2a has no points in it. Then we conclude $A' = \emptyset$ and $A = \mathcal{U}$. This follows directly from the definitions of complementation and the null set.

Intersection. The intersection of sets A and B is the set of elements that belong to *both A and B*. This is written $A \cap B$ and read "A intersect B":

$$A \cap B = \{x \in \mathcal{U} \mid x \in A \text{ and } x \in B\} \tag{3.2}$$

The set making up $A \cap B$ is the set of points "shared" by or "in common" to A and B. Take the two sets from $\mathcal{U}$ in the example for complementation above: all scores warranting a "B" or higher define a set $B = \{81, 82, 83, \ldots, 100\}$ and all scores warranting an "A" define a set $A = \{91, 92, 93, \ldots, 100\}$. Thus, $A \cap B = \{91, 92, 93, \ldots, 100\}$. The score sets for letter grades "A" and "F" have a special sort of intersection:

$$A \cap F = \emptyset$$

That is, there are no elements in common to the two sets A and F—their intersection is the null set. Two such sets are said to be disjoint or mutually exclusive.

In the Venn diagram of Figure 3.2b the intersection of two sets is the shaded portion. This may or may not have any members, but it is common practice to represent two sets as overlapping in the diagram. At times, however, when we want to stress the fact that the sets are nonoverlapping, that is, their intersect is $\emptyset$, then the diagram is drawn like that of Figure 3.2c.

Another case of intersection is important. If, as in our letter grade example with sets $B = \{x \in \mathcal{U} \mid x \geq 81\}$ and $A = \{x \in \mathcal{U} \mid x \geq 91\}$, we have a set A composed of the members of another set B, then we say the set A is contained in the B. It is diagrammed as in Figure 3.2d.

A special symbol is used to indicate this relationship of two sets. If A is contained in B, we write

$$A \subseteq B$$

This symbol covers two possibilities: (1) If A is made up entirely of elements B but there are some elements of B not in A, then

$$A \subset B$$

(2) If B has no elements not also in A, then

$$A = B$$

In any case, if all elements in A are in B, we know the

$$A \subseteq B$$

and speak of A implying B. By implication we mean that if we know an object $a \in A$ and that $A \subseteq B$, then $a \in B$ since all elements in A are also in B by virtue of the definition of the relation $\subseteq$.

Union. The union of two sets A and B is the set of the elements belonging to one or the other (or both) of the sets A and B. That is, the union of sets is a set pooling the elements in the unioned sets. We write the union of A and B as $A \cup B$ and imply that $A \cup B$ is set C such that

$$C = \{x \in \mathcal{U} \mid x \in A \text{ or } x \in B \text{ or both}\} \qquad (3.3)$$

For example, if $\mathcal{U} = \{\text{English alphabet}\}$ and $A = \{a, e, i, o, u\}$, $B = \{a, b, c\}$, we have $A \cup B = \{a, b, c, e, i, o, u\}$. Figure 3.2e represents $A \cup B$.

The Algebra of Sets

Before we give a law of set algebra, we review briefly the corresponding law of the algebra of real numbers.

Law R1. For any x and y that are real numbers, $x + y$ and $x \times y$ are real numbers.

This says that when the operations of addition and multiplication are performed on real numbers, the results are real numbers and hence $(x + y)$ and $(x \times y)$ are of the same abstract character as x and y were.

Law S1. For any sets A and B of the universal set $\mathcal{U}$, $A \cup B$, $A \cap B$ and A' are sets.

Simply: Operations on sets produce sets. These laws are called the closure laws.

Law R2. For any x, y, and z that are real numbers

$$(x + y) + z = x + (y + z)$$
$$(x \times y) \times z = x \times (y \times z)$$

Law S2. For any sets A, B, C of $\mathcal{U}$

$$A \cup (B \cup C) = (A \cup B) \cup C$$
$$A \cap (B \cap C) = (A \cap B) \cap C$$

These two laws are known as the associative laws and tell us that the grouping of several operations of the same kind ($+$, $\times$, $\cup$, or $\cap$) are not important. The parenthesis does not have any basic meaning for two or more algebraic operations of the same kind or for the set operations of union and intersection, with respect to adjacent unions and intersections.

Law R3. For any x and y that are real numbers

$$x + y = y + x$$
$$x \times y = y \times x$$

Law S3. For any sets A and B of a universe $\mathscr{U}$

$$A \cap B = B \cap A$$
$$A \cup B = B \cup A$$

These two laws indicate that the order in which we perform the operations $\cup$ and $\cap$ to a pair of sets is of no concern. These are the commutative laws.

Law R4. For any x, y, and z which are real numbers

$$x \times (y + z) = (x \times y) + (x \times z)$$

Law S4. For any sets A, B, and C of a universe $\mathscr{U}$

$$A \cup (B \cap C) = (A \cup B) \cap (A \cup C)$$
$$A \cap (B \cup C) = (A \cap B) \cup (A \cap C)$$

These laws are the distributive laws. There are two parts to Law S4 and only one for R4. This is the first major difference in the two algebras. In the real numbers it makes no difference whether y and z are added before multiplying by x, or whether y and z are first multiplied by x and then the products added. For sets we must state two such general properties, one for unions of intersections and the other for intersections of unions. The basic interpretation is that a set unioned to an intersection distributes to form an intersection of unions. Similarly an intersection of a set with a union distributes to form a union of intersections.

Law R5. For any real number x there are two unequal numbers zero and one such that

$$x + 0 = x \qquad x \times 0 = 0$$
$$x \times 1 = x \qquad x \times \frac{1}{x} = 1$$

Law S5. For any set A of the universe $\mathcal{U}$ there exists a set $\emptyset$ such that

$$A \cup \emptyset = A \qquad A \cap \emptyset = \emptyset$$
$$A \cap \mathcal{U} = A \qquad A \cup \mathcal{U} = \mathcal{U}$$

These laws define the nature of one and zero and the nature of $\emptyset$ and $\mathcal{U}$. The similarity of the behavior of $\emptyset$ and zero, and $\mathcal{U}$ and 1 should be noted. These are called the identity laws.

Law R6. For any real number x there is another real number $-x$ such that

$$x + (-x) = 0$$

Law S6. For any set A of $\mathcal{U}$ there is another set A' such that

$$A \cup A' = \mathcal{U}$$
$$A \cap A' = \emptyset$$
$$(A')' = A$$

These laws, the complement (or inverse) laws, define the behavior of the negative of a number and the complement of a set.

Two special laws govern the behavior of sets for which we have no parallel laws in the real number system. These are called the idempotent law and De Morgan's law, respectively.

Law S7. (Idempotent) For any set A of a universe $\mathcal{U}$

$$A \cup A = A \qquad A \cap A = A$$

The set is composed of its elements, each included once and once only. Thus, $A \cup A$ and $A \cap A$ are still the collection of elements defining A.

Law S8. (De Morgan) For any sets A and B of the universe $\mathcal{U}$

$$(A \cup B)' = A' \cap B'$$
$$(A \cap B)' = A' \cup B'$$

In words, this law states that the complement of a union (intersection) is the intersection (union) of the complements.

These eight set laws govern the behavior of sets and their relationships through the three basic set operations. Some of the laws can be proved by other laws and the definition of sets and operations on sets. We cannot in this book devote the time necessary to this task. Instead, the student is urged to read more about sets from one of the sources on sets given at the end of this chapter.

In the probability proofs that we shall develop in the following parts, these laws will be used in finding sets and the probabilities of sets. For now we leave it to the student to prove that the following equalities hold. Use the set laws as you would the algebraic laws, changing each set expression into others that allow the basic truth or falsity of the relationship postulated to be proven.

1. $(A' \cap B')' = A \cup B$
2. $[A' \cap (A \cup B)]' = A \cup B'$
3. $(A \cap B) \cap (A \cap B') = \emptyset$
4. $A \cap (A \cup B) = A \cup (A \cap B) = A$
5. $[A' \cap (B \cap C)]' = A \cup B' \cup C'$
6. If $A \cap B = \emptyset$ then $[(A \cap C) \cap (B \cap C)] = \emptyset$
7. $(A \cap B) \cap (C \cap D) = [A \cap (B \cap C)] \cap D$

The following expressions can be simplified by using the set laws:

1. $(A \cap A) \cup (A \cap A)'$
2. $(A \cup \mathscr{U}) \cap (A \cup \emptyset)$
3. $(A \cup \mathscr{U}) \cap (B \cap \emptyset)$
4. $[(A \cap B) \cup C] \cap \emptyset$
5. $(A \cap \emptyset) \cup (\mathscr{U} \cap A)$
6. $A \cap (\mathscr{U} \cup \emptyset)$
7. $(A \cup A \cup A) \cap \mathscr{U}$
8. $(\mathscr{U} \cup A) \cup (\mathscr{U} \cap A) \cup A$

Numbers as Sets

In later chapters we have a special interest in numbers (variables). At that time we shall develop the idea of probabilities associated with these numbers. Our work is facilitated considerably by thinking of numbers as elements of sets. Hence, when we have worked out the theory of probability for events and sets of events, the same theory applies directly when the events and sets of events have numerical meaning.

One of the properties of numbers is that they are ordered (for example, they can represent magnitudes), and hence we may represent them in an ordered array:

$$0, 1, 2, 3, \ldots, N$$

The elements in this set of numbers stand in the relations "less than" and "greater than" the adjacent numbers, that is,

$$0 < 1 < 2 < 3 < \cdots < N$$

In addition, we may represent the set as points on a line by choosing a place for 0 and a place for 1:

Let I be the set of $N + 1$ values above; then $I = \{0, 1, 2, \ldots, N\}$. If N is very large and we took two values from I, for example a and b, $b \neq 0$, we could show that the set X is the set of all real numbers, where

$$X = M \cup P$$

$$P = \left\{ \frac{a}{b} \mid a \in I, b \in I \text{ for } N \text{ arbitrarily large} \right\}$$

$$M = \left\{ -\frac{a}{b} \mid a \in I, b \in I \text{ for } N \text{ arbitrarily large} \right\}$$

As the reader will have anticipated, X is without limit in counting measure, $n(X) = \infty$.

Some interesting properties of the elements of X are of use to us. The idea of inequality, symbolized by the relational sign, $<$, above refers to the ordering of the values of X. If the magnitudes represented by elements in X are larger as the values are represented further to the right of zero on the line representing the real numbers, then for some specific number y,

$$Z = \{x \mid x < y\}$$

is represented as the line segment falling to the left of the point representing y. Note that $y \notin Z$. From this sort of reasoning we can rationalize the following rules of inequalities.

$$Z = \{x \mid x < y\}$$

1. If the constant k is added to or subtracted from both sides of an inequality, the new inequality is in the same direction: If $x < y$ then $x + k < y + k$.

2. If the constant k is multiplied or divided into both sides of an inequality, the direction of the inequality changes if $k < 0$ and stays the same if $k > 0$: for k negative $(k < 0)$, $y < x$ means $ky > kx$; for k positive $(k > 0)$, $y < x$ means $ky < kx$.

To continue we need the idea of absolute value. The absolute value of x is denoted by $|x|$ where

$$\text{if } x \geq 0 \text{ then } |x| = x$$
$$\text{if } x < 0 \text{ then } |x| = -x$$

Hence $|3| = 3$, $|-3| = -(-3) = 3$. The combination of inequalities and absolute values to form subsets of X is illustrated by the set $T = \{x \mid |x| < 4\}$. This has a representation as a segment of the line

Hence, in general, the set defined by the inequality $|x| < y$ for some value of $y > 0$ is an interval extending equally to either side of zero.

Applying similar reasoning to the inequality $|x - k| < y$ for some value $y > 0$, the interval covered by the set

$$W = \{x \mid |x - k| < y\}$$

extends y units to either side of k. Let $k = 4$ and $y = 3$ to obtain the segment

$$W = \{x \mid |x - 4| < 3\}$$

The operation of complementation of sets has an important parallel in this context. The complement T' of $T = \{x \mid |x| < 4\}$ is given by taking all the values of x not included in T. This means that T' contains anything equal to or greater than 4 or equal to or smaller than -4 (see the diagram of T). Hence the strict inequality of T is changed in direction and equality as well as strict inequality is allowed:

$$T' = \{x \mid |x| \geq 4\}$$

The complement of W, that is,

$$W' = \{x \mid |x - 4| \geq 3\}$$

is similarly defined as extremes (or tails) on the real number line

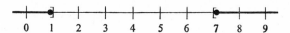

The inclusion or exclusion of a specific value (as in 4 for T and T') depends on the type of inequality, strict $<$, or not strict $\leq$, involved.

Consider the set

$$V = \{x \mid |x| < 4, x \text{ an integer}\}$$

This set does not contain any values other than integers, hence 3.5, or -2.1243, or any other "fractional valued" number of $\mathcal{U} = \{x \mid x \text{ is a real number}\}$ is not in V. The set V can be enumerated

$$V = \{-3, -2, -1, 0, 1, 2, 3\}$$

(zero is considered here an integer). Sets of this nature can be represented as a collection of points, enclosed in square brackets:

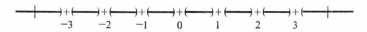

Then V' would have a graph like

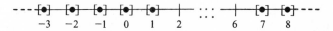

As the final example we graph the set

$$Y = \{x \mid |x - 4| \geq 3, x \text{ an integer}\}$$

This has a graph like

Counting Functions. If we simply count the number of elements in a set or subset and attach that number to the set or subset as its function value, we have a counting function. The counting function values of the set $E = \{x \mid x \text{ is a vowel}\}$ is $n(E) = 5$. The value of the set $A = \{a_1, a_2, \ldots, a_{1000}\}$ is $n(A) = 1000$. In general, the counting function value of a set or subset with m elements, $S = \{e_1, e_2, \ldots, e_m\}$, is $n(S) = m$. As a special case we see that the counting function value of an event is 1. The value of a counting function of a set is sometimes called a measure. If we denote the counting function value of a set E by $n(E)$, then $n(E)$ is said to be the counting measure of E.

Two special cases are important. The counting measure of the null set $\emptyset$ is zero: $n(\emptyset) = 0$. The counting measure of the universal set, $\mathcal{U}$, is the total number of possible elements, for example m. Thus $n(\mathcal{U}) = m$.

The Counting Measure of a Union of Sets. The unions of sets and their counting measures have a special place in the discussion to follow. Take two sets, A and B, that have no elements in common. If their measures respectively are $n(A)$ and $n(B)$, then it is clear that, since $A \cup B$ is the set including all elements of A and all elements of B and each element is in one set only, $n(A \cup B) = n(A) + n(B)$. However, if $(A \cap B) \neq \emptyset$, that is, if they have elements in common, then $n(A \cup B) \neq n(A) + n(B)$. If we count the elements in A, some of the elements counted are elements shared by A and B. Then if we count the elements in B, we are counting again those elements shared by A and B. Adding those two counts, $n(A)$ and $n(B)$, "counts in" twice the elements shared by A and B. The elements of $A \cap B$ are represented twice in the sum. However, the set $A \cap B$ is defined to include the elements of the set $A \cup B$ once and once only—any element is represented only once in a set. Thus, $n(A) + n(B) \geq n(A \cup B)$ with $n(A) + n(B) > n(A \cap B)$ when $(A \cap B) \neq \emptyset$ (that is, when $n(A \cap B) > 0$). And, in general $n(A \cup B) = n(A) + n(B) - n(A \cap B)$. If $n(A) + n(B)$ is larger than $n(A \cup B)$, it is larger by exactly $n(A \cap B)$ because the elements of $A \cap B$ were counted in both $n(A)$ and $n(B)$ and hence twice in $n(A) + n(B)$. We can prove this using the algebra of sets.

THEOREM. $n(A \cup B) = n(A) + n(B) - n(A \cap B)$ $\hspace{2cm}$ (3.4)

Proof. Let $(A \cap B) = D$ and $(A \cap B') = E$. Then $(E \cap D) = \emptyset$. That is, E and D split A into two nonoverlapping parts: $A = D \cup E$. (Draw Venn diagrams of A, D, and E to demonstrate this relationship.)

(1) By the definitions of complementation and intersections:

$$n(A \cap B') + n(A \cap B) = n(A)$$

Similarly: $n(A' \cap B) + n(A \cap B) = n(B)$.

(2) Adding, $n(A) + n(B) = n(A \cap B) + n(A \cap B) + n(A' \cap B) + n(A \cap B')$, and thus $n(A) + n(B) - n(A \cap B) = n(A \cap B) + n(A' \cap B) + n(A \cap B')$.

(3) The three terms in the right of the last expression are for nonoverlapping segments whose union covers all points of the union $A \cup B$.

(4) Thus, $n(A) + n(B) - n(A \cap B) = n(A \cup B)$. As a special case of this relationship, note that if $(A \cap B) = \emptyset$ then $n(A \cap B) = 0$ and $n(A) + n(B) = n(A \cup B)$.

Functions on Sets

The concepts of set, subset, and operations on sets are interesting for their own sake, but our primary motivation in discussing them is to be able to develop simple and powerful methods of dealing with probabilities associated with events. We have dealt with events, sets of events, and subsets of events as abstract objects, well defined, exhaustive, and mutually exclusive. The notion of sets and subsets is extended by attaching to each element, set, and subset certain numbers. The numbers are of interest themselves. If for every element of the set $A = \{a_1, \ldots, a_k\}$ we define a quantity $X(a)$ (read "X of a" not "X times a"), then the collection of pairs

$$(a_i, X(a_i)) \quad \text{for} \quad i = 1, 2, \ldots, k$$

is a set function.

This definition immediately reminds one of the process of measuring things in the everyday sense: the length of a line, the weight of an object, the intelligence of a person, the strength of a response, and so on.

Two simple sorts of functions are important here: (1) counting functions, and (2) probability functions. Later we talk about functions in general and deal with them as "variables." In particular, since we are studying random experiments, we treat some functions as "random variables."

Probability Functions. A probability function's value of an event, set of events, or subset of events is the probability attached to that event, set, or subset. Since the values are probabilities, they are quantities having the characteristics of probabilities discussed above.

SUGGESTED READINGS

Kemeny, J., Snell, J., and Thompson, G. (1956), *An introduction to finite mathematics*, Prentice-Hall, Englewood Cliffs, New Jersey.

Goldberg, S. (1960), *Probability theory, an introduction*, Prentice-Hall, Englewood Cliffs, New Jersey.

Nahikian, H. M. (1964), *A modern algebra for biologists*, University of Chicago Press.

Feller, W. (1950), *An introduction to probability theory and its applications*, John Wiley & Sons, New York.

Halmos, P. (1960), *Naive set theory*, Van Nostrand, New Jersey.

GLOSSARY

Set. Collection or aggregation of things, acts, abstractions, and so on, particularly events and results. A set A with elements $a_1, a_2, \ldots, a_k$, for example, is denoted,

$$A = \{a_1, a_2, \ldots, a_k\}$$

Subset. Part of a set.

Null Set. $\emptyset$, the set containing no members.

Universal Set. $\mathcal{U}$, the set defining the universe of discourse.

(3.1) $\qquad A' = \{x \in \mathcal{U} \mid x \notin A\}$ complementation

(3.2) $\qquad A \cap B = \{x \in \mathcal{U} \mid x \in A \text{ and } x \in B\}$ intersection

Implication. $A \subseteq B$ means that if $a \in A$ then $a \in B$.

(3.3) $\qquad A \cup B = \{x \in \mathcal{U} \mid x \in A \text{ or } x \in B \text{ or both}\}$ union

Functions on Sets. For each element in $A = \{a_1, a_2, \ldots, a_k\}$ a quantity $X(a)$ is defined to give the set function

$$(a_i, X(a_i)) \qquad i = 1, \ldots, k$$

Counting Functions (Measures) of Sets. The number of elements in a set, $n(E)$.

(3.4) $\qquad n(A \cup B) = n(A) + n(B) - n(A \cap B)$

Laws of Set Algebra. For sets A, B, and C,

Closure: $\quad A \cup B, A \cap B, \text{ and } A' \text{ are sets}$

Associative: $\quad A \cup (B \cup C) = (A \cup B) \cup C$
$\qquad\qquad\quad A \cap (B \cap C) = (A \cap B) \cap C$

Commutative: $\quad (A \cap B) = (B \cap A)$
$\qquad\qquad\qquad (A \cup B) = (B \cup A)$

Distributive: $\quad A \cup (B \cap C) = (A \cup B) \cap (A \cup C)$
$\qquad\qquad\qquad A \cap (B \cup C) = (A \cap B) \cup (A \cap C)$

Identity: $\qquad\quad A \cup \emptyset = A$
$\qquad\qquad\qquad\quad A \cap \emptyset = \emptyset$
$\qquad\qquad\qquad\quad A \cup \mathcal{U} = \mathcal{U}$
$\qquad\qquad\qquad\quad A \cap \mathcal{U} = A$

Complement: $\qquad A \cup A' = \mathcal{U}$
$\qquad\qquad\qquad\quad A \cap A' = \emptyset$
$\qquad\qquad\qquad\quad (A')' = A$

Idempotency: $\qquad A \cup A = A$
$\qquad\qquad\qquad\quad A \cap A = A$

De Morgan: $\qquad (A \cup B)' = A' \cap B'$
$\qquad\qquad\qquad\quad (A \cap B)' = A' \cup B'$

PROBLEMS

1. Prove all the equalities on page 55.

2. Perform all the simplifications on page 55.

3. Draw two samples of random digits of five digits each, using the table of random numbers at the back of the book.

Let
$$\mathcal{U} = \{0, 1, \ldots, 9\}$$
$$E = \{\text{digits observed in first sample}\}$$
$$F = \{\text{digits observed in second sample}\}$$

(a) Define $\mathcal{U}$, E, and F for your samples. Is $4 \in E$? Is $6 \in F$? Is $E \subseteq F$? Is $F \subseteq E$?

(b) Draw Venn diagrams of $\mathcal{U}$, E, and F, showing each element as a distinct point.

(c) Write out the sets that define the following in your samples:

(1) E', (2) F', (3) $E \cap F$, (4) $F \cap \mathcal{U}$, (5) $E \cup F$, (6) $E \cup \mathcal{U}$, (7) $E \cap \mathcal{U}$, (8) $F \cup \mathcal{U}$, (9) $E' \cup F'$, (10) $E' \cap F'$, (11) $E' \cup F$, (12) $E \cup F'$, (13) $E' \cap F$, (14) $E \cap F'$, (15) $[E' \cap (E \cup F)]'$, (16) $(E \cap F) \cap (E \cap F')$, (17) $E \cap (E \cup F)$, (18) $E \cup (E \cap F)$

(d) Find counting functions on all sets in (c).

4. For x a real number write a set definition of the following. Represent each of the sets on a graph.

(a) x is numerically equal to or less than 1.

(b) x is between -1 and 1 inclusively.

(c) x is numerically greater than 4.

(d) x is numerically smaller than -4.

(e) x is numerically greater than 4 and numerically less than -4.

(f) x differs from 0 by 1.96 or less.

(g) x differs from 10 by 2.56 or less.

5. For the graphs of sets on the real number line specify the set in the notation of sets, inequalities, and absolute values.

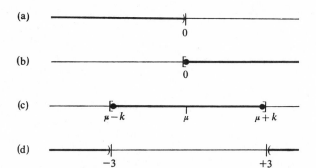

(e)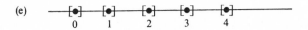

(f)

6. List all of the nonnull subsets that can be formed from the set

$$\mathscr{U} = \{e_1, e_2, e_3, e_4\}$$

How many subsets may be formed from sets with n elements (one subset is the null set)?

7. Find the counting function value of the following sets if

$$n(\mathscr{U}) = 100 \qquad n(B) = 40$$
$$n(A) = 30 \qquad n(A \cap B) = 15$$

(a) $A' \cap B$ (d) $A' \cup B'$

(b) $A \cap B'$ (e) $A' \cup B$

(c) $A \cap B$ (f) $A \cup B'$

8. Draw Venn diagrams for the sets of problem 7, showing the counting function value of each region by writing in the numbers in the disjoint parts of $\mathscr{U}$, A, B, $A \cap B$, and so on.

CHAPTER 4

Sample Spaces and Probability

In the previous two chapters the concepts of events, probability of events, sets of events, and subsets of events were introduced. In this chapter we draw all of these together into an integrated theory of probability of sets and subsets in general and show its relation to real and conceptual experiments.

Sample Space

We always begin with the collection of possible observations of a random experiment conceptualized as the set of events in a probability model. The events are defined in accordance with the requirements for elements of sets, that is, events are exhaustive and mutually exclusive elements represented once only in the list of events. This set of events is called a sample space for the corresponding random experiment. For example, the sample space for the random experiment "toss a coin and observe which side is up" would be $S = \{H, T\}$. Or "toss a nickel and a penny and record the number of heads." This would give a sample space $S = \{0, 1, 2\}$.

These examples anticipate our formal definition of a sample space. A sample space S associated with an experiment is a set such that each element in S denotes a possible observation in the experiment, and any actual experiment must result in an observation that corresponds to one and only one element of S.

Other examples may help clarify this concept. In the example above, the observations used to define the sample space S corresponded to the number of heads. The same experiment (toss a nickel and a penny) defines other sample spaces. A "mirror image" of the space would indicate the number of tails observed. These two spaces are mutually reciprocal, and one determines precisely the other. Both of these sample spaces are defined by the numbers of more elementary events. Thus, we may define a "larger" sample space that preserves all the details of the experiment. Let H_n be a head on the nickel, T_p a tail on the penny, and so forth. If we write H_nT_p as a head on the nickel and a tail on the penny, and so on for all four possible combinations, the sample space S_1 (as distinguished from S above) is

$$S_1 = \{H_nT_p, H_nH_p, T_nH_p, T_nT_p\}$$

This sample space has four events, that is, $n(S_1) = 4$, compared with three in the first sample space in the same experiment.

The principle is general. For each experiment there may be more than one sample space. The sample space appropriate in an experiment depends on the purposes of the experiment and on the analysis performed on the results of the experiment. However, it is generally better to define the sample in as great detail as possible. If this turns out to be unnecessary (after the fact), nothing is lost except the minor inconvenience of being careful. Experience will show safe ways of simplifying sample spaces when shortcuts are desirable or necessary.

Imagine that we are interested in the relationship of introductory psychology students' success on sensory physiology questions to their success on personality dynamics questions. Six true-false or multiple-choice questions of comparative difficulty are chosen for each topic. One student is chosen and he takes the two short exams. The different types of possible observations are quite numerous. Our subject might have the first sensory question right or wrong, which we shall symbolize (that is, assign a name) $S1+$ and $S1-$, respectively. Suppose that he has $S1+$; then he may have the first personality question right or wrong: $P1+$ and $P1-$. The four possibilities on the first questions alone are

$$(S1+, P1+) \quad (S1+, P1-)$$
$$(S1-, P1+) \quad (S1-, P1-)$$

Considering, in addition, only the possible observation on the second sensory question, we get the following list:

$$(S1+, P1+, S2+) \quad (S1+, P1-, S2+)$$
$$(S1-, P1+, S2+) \quad (S1-, P1-, S2+)$$
$$(S1+, P1+, S2-) \quad (S1+, P1-, S2-)$$
$$(S1-, P1+, S2-) \quad (S1-, P1-, S2-)$$

Adding the second personality question gives

$$(S1+, P1+, S2+, P2+) \quad (S1+, P1-, S2+, P2+)$$
$$(S1-, P1+, S2+, P2+) \quad (S1-, P1-, S2+, P2+)$$
$$(S1+, P1+, S2-, P2+) \quad (S1+, P1-, S2-, P2+)$$
$$(S1-, P1+, S2-, P2+) \quad (S1-, P1-, S2-, P2+)$$
$$(S1+, P1+, S2+, P2-) \quad (S1+, P1-, S2+, P2-)$$
$$(S1-, P1+, S2+, P2-) \quad (S1-, P1-, S2+, P2-)$$
$$(S1+, P1+, S2-, P2-) \quad (S1+, P1-, S2-, P2-)$$
$$(S1-, P1+, S2-, P2-) \quad (S1-, P1-, S2-, P2-)$$

When all twelve questions are considered, there are 2^{12} distinct ways the experiment can come out: $2^{12} = 4096$. The reader is urged to discover the method of listing these events as begun above (don't try to complete the list, though!). If you carry the process out a step or beyond the list already developed, you should also be able to see why we have 2 raised to the power 12 for the total number of events in the sample space. In the chapter on sampling and probability we discuss in detail how to determine the number of events in a sample space.

Of the 4096 possible observations only one will be observed in actuality if only one subject is used in the experiment. This is the crucial differentiation between a model and an experiment: The model specifies possible observations, the events of the sample space; the experiment produces one of the events as a result of performing the experiment. So far, however, neither the model nor the experiment (one subject) is very meaningful. We might ask: What difference does it make whether the subject scored $(S1+, P1-, \ldots, P6+)$ and whether there are 4096 events in the sample space?

The answer would have to be "Not much." We need a good deal more before we can answer our psychological question. In particular, we need (1) to know how often we can expect to observe each event in the sample space, (2) to simplify the sample space but retain the essential information about the number of correct answers on each quiz, and (3) to make many more actual (empirical) observations.

We have to postpone consideration of this last point but will now begin to work out the first point by developing methods for simplifying sample spaces.

Imagine that we are not really interested in the specific questions correctly and incorrectly answered. Rather, we want to know if knowledge of sensory physiology and knowledge of personality dynamics "go together." If we measure these types of knowledge by the two quizzes, we need to know only how many correct answers were made on each. It is obvious that these sample spaces for each test are the same:

$$\{0, 1, 2, \ldots, 6\}$$

We can diagram the two sample spaces simultaneously in the following way. Set up a grid of cells with seven columns and seven rows. This grid is illustrated in Figure 4.1. Each column stands for an event in the sensory physiology sample space and each row for an event in the personality dynamics sample space. Each "cell" in this figure represents one of the 49 possible combinations of "scores" (each event being a score). This is somewhat more manageable than the 4096-event nominal sample space; and it is more informative in terms of our basic inquiry. The result of observing a single subject in this experiment corresponds to observing a single cell in the diagram. We might infer an answer to our query by noting the specific event, and hence cell, observed. For example, if the cell observed was near the diagonal running from the lower left to the upper right, we would be led to believe that the two types of knowledge were related—at least for our one subject. If the cell observed was in the upper left or lower right portions of the diagram of the sample space, the opposite conclusion might be suggested.

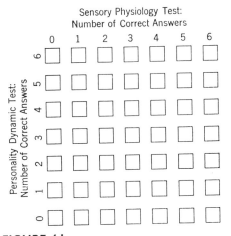

FIGURE 4.1

A grid illustrating the sample space for the number of correct answers in a psychology test.

The reader should realize, however, that any inference at this stage is risky business. It is not sound to let one subject represent all introductory psychology students, even if he is chosen very carefully. We would be grossly over-generalizing if we made any inferences about students on the basis of one such experiment. Methods of dealing with replications of experiments, each having the same sample spaces, must be left until a later chapter.

Another caution is in order. Our sample of test items may not be representative of the sorts of knowledge we want to compare. Unless it is, generalizing from any number of replications of the experiment to knowledge of sensory physiology and personality dynamics would be unwarranted.

Each of the 49 events in the numerical sample space is a combination of events in the nominal sample space of 4096 events. All of the events having two (any two) sensory physiology questions correct are found in the "2" column of the sample space diagram. All events having exactly two (again, any two) sensory questions and five (any five) personality questions correctly answered are represented by the cell in the "2" column and the "5" row. It is clear that we find subsets of the nominal sample space when we redefine the experiment in terms of the "number of correct answers." All of the events in nominal sample space with "X" correct sensory answers and "Y" correct personality answers are included in the (X, Y) event in the numerical sample space. For example, if $X = 2$ and $Y = 5$, then

$$(S1+, P1+, S2-, P2+, S3-, P3+, S4+, P4-, S5-, P5+, S6-, P6+)$$
$$(S1+, P1+, S2+, P2+, S3-, P3-, S4-, P4+, S5-, P5+, S6-, P6+)$$
$$(S1-, P1+, S2-, P2+, S3-, P3+, S4-, P4+, S5+, P5-, S6+, P6+)$$
$$(S1-, P1-, S2+, P2+, S3-, P3+, S4-, P4+, S5+, P5+, S6-, P6+)$$

are included among others in the event corresponding to the cell in column 2 and row 5 of the numerical sample space diagram. The reader should determine how many nominal events have $X = 6$ and $Y = 6$. Comparing this number with what we already know about the number of nominal events with $X = 2$ and $Y = 5$, can we say anything about the probability of the two events in the (X, Y) sample space?

If, as is common in a course, the test of 12 questions was scored by finding $X + Y$, the distinction of the two types of questions would be lost. Under this procedure a new sample space is generated. The nominal sample space is unchanged, but the numerical (total score) sample space is defined by the set

$$\{0, 1, \ldots, 12\}$$

Certain of the events in the (X, Y) numerical sample space are combined in this total-score sample space. All of the events represented in the upper left to lower right diagonals of the diagram have the same total number of correct answers. The total-score sample space can be obtained directly by finding subsets of the nominal sample space or by finding subsets of the sensory-personality, or (X, Y), numerical sample space. The algebra of sets applies in a stepwise fashion (or by successive applications) as well as directly.

We now formalize the process of finding subsets of events in sample spaces. Such subsets are referred to as outcomes. Recalling that sample spaces are sets—sets of the theoretically possible observable events corresponding to an experiment—we invoke the terminology of Chapter 3. A sample space of an experiment is the universal set for the experiment. Each event of a sample space is an element or member of the universal set. Thus, since outcomes are defined as subsets of events in sample spaces, they are subsets of the universal set. There is a direct correspondence between the major concepts of set theory and the theory of sample spaces:

Universal set	Sample space
Element	Event
Subset	Outcome

The importance of this parallel is that the mathematics of sets can be applied to the theory of experiments. We may substitute term for term, sample space for universal set, event for element of a set, outcome for subset. The algebra and formal logic of set theory are applied to the theory of experiments without introducing any new mathematical concepts.

In order to put this in very explicit form, we draw up a list of some observations from the total-score experiment above. For example, take the sample space

$$S = \{0, 1, 2, \ldots, 12\}$$

TABLE 4.1 Set Theory Expressions for Some Results of an Experiment

Result of an Experiment	Set Theory from Sample Space: e Corresponding to r
Some result is observed	$e \in \mathscr{U},\ S = \mathscr{U}$
Result is zero score	$e = E_0,\ e \in S_1,\ e \in S_2$
Result is score greater than six	$e \in \bigcup\limits_{i=6}^{12} E_i,\ e \in S_2,\ e \notin S_1$
Result is score of two	$e \in (S_1 \cap S_2),\ e = E_2$
Result was even score	$e \in (E_0 \cup E_2 \cup E_4 \cup \cdots \cup E_{12}),\ e \in S_2$
Result was score of twenty	$e \in \emptyset$
Result was not even score	$e \in (E_1 \cup E_3 \cup \cdots \cup E_{11}),\ e \in S_2'$

and the thirteen events

$$E_0 = \{0\}, \ E_1 = \{1\}, \ \ldots, \ E_{12} = \{12\}$$

and two arbitrary outcomes

$$S_1 = \{0, 1, 2, \ldots, 6\} \qquad S_2 = \{0, 2, 4, \ldots, 12\}$$

Some results and the corresponding set theory expressions are given in Table 4.1.

Probability in Sample Spaces

In the example developed above it was difficult to interpret an observed result of an experiment. The sample spaces of the experiment were well defined, but the significance of observing a particular event or outcome was not clear. We might ask whether any event was to be expected instead of any other. If a given event occurred, would it be surprising? What could we expect if the student were only guessing? The answers to such questions depend a great deal on how we set out to find an answer. We can develop a method to answer the question based on probability theory and logical analysis. In doing this we take certain risks: (1) other methods, not involving probability theory, may give more adequate answers, or (2) our logic may be erroneous and thus we may be misled. However, if we are correct in our logical analysis of experiments and make no mathematical errors in calculating probabilities, we are able to phrase our scientific-psychological questions in such a way as to provide interesting, informative, and useful answers. Other scientific and quantitative methods can give valuable answers to similar questions when the questions are framed in terms of the respective methodological principles. We cannot take time here to discuss these matters. The reader should refer to books dealing with scientific method and the philosophy of science for a discussion of the choice, merits, inadequacies, and so on, of the various scientific methods.

Probability in Sample Spaces. Recall that probability was defined as a number $P(e_j)$ attached to events e_j such that for a sample space containing n events $e_1, e_2, \ldots, e_n$

$$0 \leq P(e_j) \leq 1$$

$$\sum_{j=1}^{n} P(e_j) = 1$$

The probability of observing some one of the events in a random experiment is the probability that the event observed is in the sample space. Now, it does not matter which of the n events is observed; our convention is that observing any member of a set implies that the set is observed. Thus, by observing one of the n events we observe the sample space. These considerations lead us to make the following definitions.

The probability of the union of events is the sum of probabilities of the events in the union. Let

$$E = \{e_1\} \cup \{e_2\} \cup \cdots \cup \{e_k\}$$

from

$$S = \{e_1, e_2, \ldots, e_n\}$$

Then

$$P(E) = P(e_1) + P(e_2) + \cdots + P(e_k) \qquad (4.1)$$

This implies directly that the probability of any subset of a sample space, that is, an outcome, is the sum of the probabilities of the events making up that subset. This is always true when we are dealing with events.

A special case is the probability of the universal set—sample space. The sample space is composed of all events defined by an experiment, and hence it is the universal set of the experiment. By set theory,

$$S = \mathscr{U} = \{e_1\} \cup \{e_2\} \cup \cdots \cup \{e_n\}$$

and the last definitions give

$$P(S) = P(\mathscr{U}) = P(e_1) + P(e_2) + \cdots + P(e_n)$$
$$= \sum_{i=1}^{n} P(e_i)$$

The completion part of the basic axiom of probability implies

$$P(S) = \sum_{i=1}^{n} P(e_i) = 1$$

Therefore the result is that there is probability of 1.0 of observing one of the events in the sample space in an experiment. Since S is defined as an exhaustive and mutually exclusive set of all possible events of an experiment, one and only one of the events in S must be observed. It is implied that it is necessary to observe the sample space in performing the experiment. From this point on we refer to any event, outcome, sample space, and so on, with probability 1.0 as necessary. Thus the universal set is a necessary set.

At the other extreme we need a second special definition. The probability that none of the events of a sample space of an experiment is observed is zero. One of the events must occur to satisfy the definition of a sample space. By the rule of complementation the failure of any event to occur is $S' = \mathscr{U}'$, and $\mathscr{U}' = \emptyset$.

Therefore, we define the probability of the null set to be zero by virtue of the fact that $P(A) = 1 - P(A')$ and if $P(\mathscr{U}) = 1$ then $P(\mathscr{U}') = 1$ then $P(\mathscr{U}') = 1 - 1 = 0$:

$$P(\emptyset) = 0 \tag{4.2}$$

The null set is the impossible set, and we say that it is impossible to observe an event not in the sample space.

Two warnings are necessary. First, events themselves may have probabilities of 1 and 0. It is consistent to assign $P(e_j) = 1$ and $P(e_i) = 0$, $i = 1, \ldots, n$, for $i \neq j$ where $S = \{e_1, e_2, \ldots, e_n\}$. This assignment does not occur in situations we are likely to encounter in real experiments, but it is consistent with the logic of probability theory. Second, the words "necessary" and "impossible" describe sets with probabilities of 1 and 0, respectively, but do not define the meaning of probability. Probability is an undefined (in the mathematical sense) term governed by the two-part probability axiom. A probability value between 0 and 1 (for example, .4567 or any other number) is not to be interpreted as 45.67% less impossible or 54.33% less necessary than 0 and 1, respectively. The interpretations or meanings of probability, besides the formal meaning, least likely to confuse are those of subjective confidence and relative frequency. Unfortunately, these two meanings both carry overtones of prediction or of "things to come," and these are not inherent in the definition of probability.

As a general example of probability on sample spaces, take $S = \{e_1, e_2, \ldots, e_6\}$. Then

$$S = \{e_1\} \cup \cdots \cup \{e_6\}$$

Suppose that we have the following probabilities for the six events:

$$P(e_1) = .10 \qquad P(e_2) = .20 \qquad P(e_3) = .20$$
$$P(e_4) = .20 \qquad P(e_5) = .15 \qquad P(e_6) = .15$$

These are properly defined probability values (the reader should convince himself of this by checking against the formal definition of probability). Now, by the addition rule for unions of events

$$P(S) = P(e_1) + \cdots + P(e_6) = 1.0$$

and for $S_1 = \{e_1, e_2, e_5\} = \{e_1\} \cup \{e_2\} \cup \{e_5\}$

$$P(S) = P(e_1) + P(e_2) + P(e_5) = .10 + .20 + .15 = .45$$

Take the following contrived case as a concrete example. In an experiment on "perceptual defense" we want to test whether subjects can distinguish between the presentation and nonpresentation of a stimulus. The stimuli (taboo words perhaps) are presented tachistoscopically by a slide projector apparatus. On some of the

Table 4.2 Tabulation of Sample Space and Probabilities for the "Perceptual Defense" Experiment

Event	Symbol for Event	Probability of Event
C, C, C	e_1	$\frac{1}{8}$
C, C, I	e_2	$\frac{1}{8}$
C, I, C	e_3	$\frac{1}{8}$
I, C, C	e_4	$\frac{1}{8}$
C, I, I	e_5	$\frac{1}{8}$
I, C, I	e_6	$\frac{1}{8}$
I, I, C	e_7	$\frac{1}{8}$
I, I, I	e_8	$\frac{1}{8}$

presentations no stimulus is present on the transparency. Suppose that there are three presentations of transparencies and we record only whether or not the subject is correct in determining if a word is projected. For a single experiment we have a sample space made up of all the possible ways combinations of correct and incorrect responses can arise. This is given in Table 4.2, where C stands for correct and I for incorrect and we go from left to right for first, second, and third response. The sample space is composed of the eight ways three responses can be scored:

$$S = \{e_1, e_2, \ldots, e_8\}$$

The probabilities $P(e_j) = \frac{1}{8}$ are assigned to the events in S. This assignment is rationalized under a model to be developed below. For the moment, simply assume these probabilities. The numbers assigned are indeed probabilities; they satisfy all definitional criteria of probabilities:

$$0 < \tfrac{1}{8} < 1$$

and

$$\sum_{i=1}^{8} P(e_i) = 1$$

and they are assigned to events in S. Now take the outcome

$$S_1 = \{e_i \in S \mid e_i \text{ involves two correct responses}\}$$
$$S_1 = \{e_2, e_3, e_4\}$$
$$= \{e_2\} \cup \{e_3\} \cup \{e_4\}$$

and

$$P(S_1) = P(e_2) + P(e_3) + P(e_4) = \tfrac{3}{8}$$

Also, since

$$S = \{e_1, \ldots, e_8\} = \bigcup_{i=1}^{8} \{e_i\}$$

$$P(S) = \sum_{i=1}^{8} P(e_i) = 1$$

and

$$P(\text{two correct } and \text{ two incorrect}) = P(\emptyset) = 0$$

The sample space with equally probable events is a special case of some importance. When a given event in S is as likely to be observed in a random experiment as any other event, we can use the counting function on subsets in S to obtain the probability of those subsets (recall the definition in Chapter 3 of a counting function on a set). The probability assigned to each event under the equal probability rule is $1/n(S) = 1/m$ where there are m events in the sample space. In the detection experiment example, $n(S) = 8$ and $P(e_i) = \frac{1}{8}$ for all events. This assignment satisfies the two requirements on the numbers assigned. Since $n(S) \geq 1$ for any sample space, $0 \leq [1/n(S)] \leq 1$. And since there are $n(S)$ events in S,

$$P(S) = \sum_{i=1}^{n(S)} \frac{n(e_i)}{n(S)} = \underbrace{\frac{1}{n(S)} + \frac{1}{n(S)} + \cdots + \frac{1}{n(S)}}_{n(S) \text{ terms}} = n(S)\frac{1}{n(S)} = 1$$

The probability of any subset or outcome in the sample space is given by the proportion of events in the outcome with respect to the sample space. Let $A = \{e_1, e_2, \ldots, e_k\}$ from $S = \{e_1, \ldots, e_m\}$ where $n(A) = k \leq n(S) = m$. The probability of the outcome A is

$$P(A) = P(e_1, e_2, \ldots, e_k)$$

$$= \underbrace{\frac{1}{n(S)} + \frac{1}{n(S)} + \cdots + \frac{1}{n(S)}}_{n(A) \text{ terms}}$$

$$= \frac{n(A)}{n(S)} = \frac{k}{m}$$

In the detection example, $A = \{e_2, e_3, e_4\}$ and thus $P(A) = \frac{3}{8}$.

This completes the basic theory of probability and sample spaces. We now turn our attention to consequences of these basic principles in terms of probability theorems.

Theorems on Probability

First we prove as a theorem a previous result presented as a definition, namely, that the probability of observing an event in the sample space is 1.0: The probability of observing the sample space is certain.

THEOREM 1. $P(S) = 1$

Proof.   Let

$$S = \{e_1, e_2, \ldots, e_n\}$$
$$= \{e_1\} \cup \{e_2\} \cup \cdots \cup \{e_n\}$$

Hence

$$P(S) = P(e_1) + P(e_2) + \cdots + P(e_n) = \sum_{i=1}^{n} P(e_i)$$

and by definition

$$\sum_{i=1}^{n} P(e_i) = 1$$

One of the special properties of sets and subsets was the property of inclusion. If a subset A is contained in the subset B, both A and B being members of $\mathscr{U}$, that is, $A \subseteq B \subseteq \mathscr{U}$, then the probability of the outcome A is less than or equal to the probability of outcome B. An example of this sort of situation can be taken from the discrimination example. Let A be defined by events in which there are exactly two correct responses and B defined by events in which there are two or more correct responses:

$$A = \{e_2, e_3, e_4\}$$
$$B = \{e_1, e_2, e_3, e_4\}$$

These outcomes have probabilities

$$P(A) = \tfrac{3}{8}$$
$$P(B) = \tfrac{4}{8}$$
$$P(A) \leq P(B)$$

in accordance with the following theorem.

THEOREM 2.   *For outcomes* A *and* B *of sample space* S, *where* A $\subseteq$ B,

$$P(A) \leq P(B)$$

Proof.   By definition of the relation $\subseteq$ all elements $a \in A$ are also elements of B, $a \in B$. However, for some $b \in B$ it possibly is not true that $b \in A$. Therefore, it is possible to find two sets making up B, for example, $E = A$ and $F = B \cap A'$. Then F is composed of all the elements of B not in A. This implies

$$P(B) = P(E) + P(F)$$
$$= P(A) + P(F)$$

Now if F has no elements, $P(F) = 0$. If F is not the null set and if at least one $b \in F$ has nonzero probability, $P(b) > 0$, then $P(F) > 0$. Therefore, $P(B) = P(A)$ in the first case and $P(B) > P(A)$ in the second case: $P(B) \geq P(A)$.

A Venn diagram makes this more intuitively clear. Suppose that we have six events in S arranged as in Figure 4.2, $A = \{e_3, e_4, e_6\}$ and $B = \{e_2, e_3, e_4, e_6\}$. Therefore, $E = \{e_3, e_4, e_6\}$ and $F = \{e_2\}$ and

$$P(B) = P(E) + P(F) = [P(e_3) + P(e_4) + P(e_6)] + [P(e_2)]$$

If $P(e_2) = 0$, $P(B) = P(A) = P(e_3) + P(e_4) + P(e_6)$ and if $P(e_2) > 0$, $P(B) = P(A) + P(e_2) > P(A)$.

Applying this in the special case of equally likely events is illuminating. Arrange the events so that events in A are $e_1, e_2, \ldots, e_k$ and events in B are $e_1, e_2, \ldots, e_k, e_{k+1}, \ldots, e_n$ where $n(A) = k$, $n(B) = n$, $n(S) = m$, and $k \leq n \leq m$. Then

$$P(A) = \frac{n(e_1)}{n(S)} + \cdots + \frac{n(e_k)}{n(S)} = \frac{n(A)}{n(S)} = \frac{k}{m}$$

$$P(B) = \underbrace{\frac{n(e_1)}{n(S)} + \cdots + \frac{n(e_k)}{n(S)}} + \cdots + \frac{n(e_n)}{n(S)} = \frac{n(B)}{n(S)}$$

$$= \underbrace{\frac{k}{m}} + \frac{n(e_{k+1})}{n(S)} + \cdots + \frac{n(e_n)}{n(S)} \geq \frac{k}{m}$$

Implying directly $P(B) \geq P(A)$.

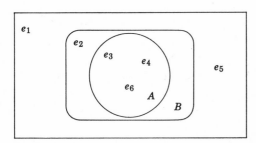

FIGURE 4.2

Venn diagram illustrating an example of Theorem 2.

A third theorem shows that regardless of how outcomes are defined in a sample space, they have a probability between zero and one inclusively.

THEOREM 3.   *If* A *is any outcome,* $0 \leq P(A) \leq 1$.

Proof.   Here $A \subseteq S$ by definition of outcome. By Theorem 2, $P(A) \leq P(S) = 1$. By the definition of the null set we know that $\emptyset \subseteq A$. Theorem 2 gives $P(A) \geq P(\emptyset) = 0$. Putting these two inequalities together gives the desired result $0 \leq P(A) \leq 1$.

We have established the probability of observing one or another (or several or all) of the events in the sample space. Theorem 1 gives this result. Also, by definition, the probability of an outcome is given by the sum of the probabilities of the events defining the outcome. For $A = \{e_1, e_2, \ldots, e_k\}$

$$P(A) = \sum_{i=1}^{k} P(e_i) = \sum_{e \in A} P(e)$$

Now we prove theorems giving the probability of observing events in one or another outcome. The definition of the probability of a union of events does not generalize to the probability of a union of outcomes. The basis of this differentiation lies in the fact that two outcomes may have elements in common whereas, by definition, events are distinct, nonoverlapping elements. We need a preliminary result before we can prove the general theorem.

Let A be composed of two disjoint subsets E and F, $E \cap F = \emptyset$. Then $P(A) = P(E) + P(F)$. Express E and F as unions of events:

$$E = \{e_i\} \cup \cdots \cup \{e_n\} \qquad F = \{f_i\} \cup \cdots \cup \{f_m\}$$

Since by definition $E \cap F = \emptyset$, that is, no $e_i = f_j$ for any i or j, we have

$$E \cup F = \{e_i\} \cup \cdots \cup \{e_n\} \cup \{f_i\} \cup \cdots \cup \{f_m\}$$

which, by the definition of the probability of a union of simple events, is

$$P(E \cup F) = \underbrace{P(e_i + \cdots + P(e_n)}_{P(E)} + \underbrace{P(f_i) + \cdots + P(f_m)}_{P(F)}$$

THEOREM 4.   *Let* A *and* B *be two outcomes. Then*

$$P(A \cup B) = P(A) + P(B) - P(A \cap B)$$

Proof.   (1) Partition A and B separately into nonoverlapping segments. Find two sets of events E and F in A such that $A = E \cup F$ *and* $E = A \cap B'$ and

$F = A \cap B$. Likewise find two sets of events G and H in B such that $B = G \cup H$ and $G = B \cap A'$ and $H = A \cap B$. By the definition of complementation and intersection it follows that $E \cap F = \emptyset$ and $G \cap H = \emptyset$.

(2) By the preliminary result we have

$$P(A) = P(E) + P(F) = P(A \cap B') + P(A \cap B)$$
$$P(B) = P(G) + P(H) = P(A' \cap B) + P(A \cap B)$$

Adding, we have

$$P(A) + P(B) = P(A \cap B') + P(A' \cap B) + 2P(A \cap B)$$

and

$$P(A) + P(B) - P(A \cap B) = P(A \cap B') + P(A' \cap B) + P(A \cap B)$$

(3) The right-hand part of the last equation is the union of A and B. The union, $A \cup B$, is given by (a) events in A not also in B, (b) the events in B not also in A, and (c) the events in both A and B: $(A \cup B) = (A \cap B') \cup (A' \cap B) \cup (A \cap B)$. By definition of intersection and complementation,

$$(A \cap B') \cap (A \cap B) = \emptyset$$
$$(A' \cap B) \cap (A \cap B') = \emptyset$$
$$(A' \cap B) \cap (A \cap B) = \emptyset$$

Therefore, by the preliminary result,

$$P(A \cup B) = P(A \cap B') + P(A' \cap B) + P(A \cap B)$$

(4) Substituting this result into the result of paragraph (2),

$$P(A) + P(B) - P(A \cap B) = P(A \cup B): \quad \text{end of proof.}$$

Take a specific example. Suppose that we have two outcomes A and B in $S = \{e_1, \ldots, e_n\}$:

$$A = \{e_1, e_2, e_3, e_4, e_5\}$$
$$B = \{e_3, e_4, e_5, e_6, e_7\}$$

Note that

$$A \cup B = \{e_1, e_2, e_3, e_4, e_5, e_6, e_7\}$$
$$A \cap B = \{e_3, e_4, e_5\}$$
$$P(A) = P(e_1) + P(e_2) + \cdots + P(e_5)$$
$$P(B) = P(e_3) + \cdots + P(e_7)$$
$$P(A \cup B) = P(e_1) + \cdots + P(e_7)$$
$$P(A \cap B) = P(e_3) + P(e_4) + P(e_5)$$

Now, adding $P(A)$ and $P(B)$,

$$P(A) + P(B) = P(e_1) + \cdots + P(e_5) + P(e_3) + \cdots + P(e_7)$$

Rearranging the right hand side of this,

$$P(A) + P(B) = [P(e_1) + \cdots + P(e_7)] + [P(e_3) + P(e_4) + P(e_5)]$$
$$= P(A \cup B) + P(A \cap B)$$

and thus

$$P(A \cup B) = P(A) + P(B) - P(A \cap B)$$

To soften the harshness of the proof of Theorem 4, we can argue to the same result in a more intuitive way. The sum of the probabilities of the two outcomes includes the probabilities of the events shared by the outcomes, counted twice. This follows from the fact that the events in the intersection of the outcomes are included in the probabilities of both outcomes. Thus, in order to calculate the probability of the union, which includes each event only once, the probabilities included twice must be subtracted from the sum of probabilities of the outcome.

This theorem is specialized for disjoint outcomes.

THEOREM 5. *If two outcomes* A *and* B *are disjoint,* A $\cap$ B $= \emptyset$, *then*

$$P(A \cup B) = P(A) + P(B)$$

Proof. Since $(A \cap B) = \emptyset$, $P(A \cap B) = 0$. Apply Theorem 4.

There remains one more important theorem of probability to be introduced. So far we have been concerned with calculating the probabilities of outcomes, of unions of outcomes, and of intersections of outcomes. We can also determine the probability of the complement of a set. Suppose that we have determined the probability with which a subject in an identification experiment confuses stimulus s_i with one or another of the stimuli s_j used in the experiment. Let this confusion set be the outcome

$$C = \{\text{stimulus identified is } s_j \mid \text{stimulus presented is } s_i \neq s_j\}$$

A certain psychological theory requires $P(C) = .05$. What is the probability of a correct identification according to the theory? If the response a subject can make is either a confusion or a correct identification, then the set of correct identifications is

$$I = \{\text{stimulus identified is } s_j \mid \text{stimulus presented is } s_j\}$$
$$= C'$$

Since the response has to be in I or C, it is easy to see that $P(I) = .95$. This is formalized in the next theorem.

THEOREM 6. *If* A *and* A′ *are complementary outcomes*,

$$P(A') = 1 - P(A)$$

Proof. By the definition of union, intersection, and complementation, $A \cup A' = \mathcal{U}$, $A \cap A' = \emptyset$. By Theorem 5,

$$P(A \cup A') = P(A) + P(A')$$

and $P(\mathcal{U}) = 1$. Therefore

$$P(A) + P(A') = 1$$

The result follows:

$$P(A') = 1 - P(A)$$

SUGGESTED READINGS

Hays, W. (1963), *Statistics for psychologists*, Holt, Rinehart and Winston, New York.

Hodges, J. L., and Lehman, E. L. (1964), *Basic concepts of probability and statistics*, Holden-Day, San Francisco.

Goldberg, S. (1960), *Probability, an introduction*, Prentice-Hall, Englewood Cliffs, N.J.

Mosteller, F., Rourke, R. E., and Thomas, G. B. (1961), *Probability and statistics*, Addison-Wesley, Reading, Massachusetts.

Consult Chapter 2 for additional references.

GLOSSARY

Sample Space. The collection of possible observations of a random experiment is the sample space of the experiment: the set of events of an experiment.

Outcome. An outcome of a random experiment is a subset of events of the sample space of the experiment.

(4.1) $P(E) = P(e_1) + \cdots + P(e_k)$ when $E = \{e_1\} \cup \{e_2\} \cup \cdots \cup \{e_k\}$

(4.2) $P(\emptyset) = 0$

Theorem 1. $P(S) = 1$

Theorem 2. For outcomes A and B, respectively, of S where $A \subseteq B$

$$P(A) \leq P(B)$$

Theorem 3. If A is any outcome $0 \leq P(A) \leq 1$.

Theorem 4. If A and B are outcomes,

$$P(A \cup B) = P(A) + P(B) - P(A \cap B)$$

Theorem 5. If outcomes A and B are disjoint, $A \cap B = \emptyset$, then

$$P(A \cup B) = P(A) + P(B)$$

Theorem 6. Let outcomes A and B be complementary, $A = B'$, then

$$P(A) = 1 - P(B)$$

PROBLEMS

1. Specify *several* appropriate sample spaces for the following experiments and indicate the sorts of questions one can ask of each of the sample spaces.
   (a) A student is chosen at random from a class.
   (b) The children of a family of three children are observed.
   (c) A student takes a four-item true-false test.

2. For 1(c) above,
   (a) Determine all the elements in S if "number of correct answers" is the sample space.
   (b) Show the correspondence of this sample space to one defined by the pattern of correct and incorrect responses on the respective questions.
   (c) Could a sample space be defined by the responses regardless of whether or not they were correct? If it is possible, list the elements.
   (d) How can we go from the sample space in (c) to that in (b)?

3. A probability experiment involves tossing three coins (penny, nickel, and quarter) simultaneously.

   (a) List the events in this experiment.
   (b) Is this similar to the sample space for "toss a coin three times"? Explain.
   (c) What other ways can a sample space be defined for the experiment?
   (d) Find the outcomes in this space corresponding to events in a sample space defined by "number of heads showing on the three coins."

4. Assuming equal probability for each event in the sample space of 2(c) above, find the probabilities in the sample space for 2(a) above.

5. Prove that $P(e) \leq 1$ when probability of event e is defined as a number $P(e)$ associated with events such that

$$0 \leq P(e)$$

and

$$\sum_{e \in s} P(e) = 1$$

Comment on the difference in this definition and the one given in the text.

6. Theorem 4 requires $(A \cup B) = (A \cap B) \cup (A' \cap B) \cup (A \cap B)$. This was not proved in the proof of the theorem. Construct a proof.

7. John Watson applies for graduate fellowships from agencies A and B. The probability of a fellowship from A is .7, and the probability that his application will be rejected by B is .5. The probability of at least one of his applications being rejected is .6. What is the probability that he will receive at least one of the fellowships?

8. Prove that if A and B are any outcomes,

$$P(A \cap B) \leq P(A) \leq P(A \cup B)$$

9. Theorem 4 can be generalized to three outcomes A, B, and C:

$$P(A \cup B \cup C) = P(A) + P(B) + P(C) - P(A \cap B) - P(A \cap C)$$
$$= P(B \cap C) + P(A \cap B \cap C)$$

Prove this generalization—if not rigorously, at least by an intuitive argument and Venn diagrams.

10. A sample space $S = \{e_1, e_2, e_3, e_4\}$ has probabilities

$$P(e_1) = P(e_2) \quad \text{and} \quad P(e_3) = P(e_4) = 2P(e_1)$$

Find $P(e_1 \cup e_3)$. What are the values of $P(e_2)$ and $P(e_4)$?

CHAPTER 5

Conditional Probability
and Independence

The theorems and principles just discussed are concerned with the unions of events. We were able to calculate the probabilities of outcomes in the sample space of events, of unions of outcomes, and of complements of outcomes. We need to be able, also, to calculate the probabilities of intersections of outcomes. The question often arises: What is the probability of the occurrence of *both* outcome A and outcome B? Also, an important part of probability theory deals with questions regarding the probability of one outcome if it is known that another outcome has been observed: What is the probability of A conditional to the fact that the observed event is a member of B? For example, what is the probability of a person developing lung cancer if the person smokes forty cigarettes a day? The answer to this question is obtained by calculating conditional probabilities, which in turn depend on the probabilities of the intersection of sets defined by the population, smokers, cancer victims, and so on. This chapter develops the mathematical and conceptual tools for this kind of analysis.

Probability of Intersections

Two special cases are involved. The first case is where A is a subset of B; $A \subseteq B$. Thus

$$\text{if } e \in A \text{ then } e \in B$$

For example, the set of all sophomores at a given university, A, is included in the set of students in the university, B.

Theorem 2 of Chapter 4 covers one aspect of this case of intersection of outcomes:

$$P(A) \leq P(B) \text{ for } A \subseteq B$$

This is not a very strong statement relating the two outcomes. We would like to know more about the probabilities than their relative sizes. In particular, we might inquire about the probability of observing the outcome B if we knew that the event observed was a member of the set comprising outcome A. Or we may reverse the roles of A and B in the question and ask, "What is the probability also of observing A *given* that we observe B?"

The same sort of questions can be asked about outcomes only partially overlapping. Imagine outcomes E and F such that some elements of E are included in F and some elements of F are included in E. It is not true necessarily that if $e \in E$ then $e \in F$. There may be some $e \in E$ such that $e \in F$ and vice versa. Only those elements in the set $(E \cap F)$ are in both.

The probabilities of intersections of events in the sample space can be found by using the mutual-exclusion property of events. Thus, since $(A \cap B)$ is a set,

$$P(A \cap B) = \sum_e P(e), \quad \text{for all } e \in (A \cap B)$$

However, this requires listing the elements in $(A \cap B)$ with their probabilities. If we have the sample space and the probabilities for each event in the sample space, we can obtain all the basic relationships among the outcomes in the space. Imagine two outcomes E and F, then, for example

$$P(E) = \sum_e P(e) \quad \text{for all } e \in E \tag{5.1}$$

$$P(F) = \sum_e P(e) \quad \text{for all } e \in F$$

$$P(F \cap E) = \sum_e P(e) \quad \text{for all } e \in (F \cap E) \tag{5.2}$$

$$P(E \cup F) = P(F) + P(E) - P(F \cap E)$$

$$P(E') = 1 - P(E)$$

$$P(F') = 1 - P(F)$$

$$P(E' \cup F') = P[(E \cap F)'] = 1 - P(E \cap F)$$

Other relationships also may be stated. The reader is encouraged to demonstrate the validity of the equations shown.

Conditional Probability

Another question was asked: What is the probability of an outcome A given that outcome B also is observed? Imagine the sets V and W where

$$V = \{\alpha \in A \mid A \text{ is the alphabet, } \alpha \text{ is a vowel}\}$$
$$W = \{\text{first six letters of the alphabet: } a, b, c, d, e, f\}$$

We write
$$P(W \mid V)$$

"The probability of observing W given that V is observed," or equivalently, probability of W given V. This is called the conditional probability of W, given V, or the probability of W given V.

Example. Take the sample space resulting from three tosses of a coin.

$$S_1 = \{HHH, HHT, HTH, THH, HTT, THT, TTH, TTT\}$$

Suppose that we know that a particular sequence of tosses resulted in exactly two heads and we want to know the probability that the two heads were observed on successive tosses. Note that we have declared part of the sample space S_1 impossible, since we already know that exactly two heads occurred; we know that

$$S_2 = \{HHH, HTT, THT, TTH, TTT\}$$

is a subset of events that could not possibly have occurred. Thus we are, in effect, dealing with a sample space

$$S_3 = \{HHT, HTH, THH\}$$

Since these events are equally likely and one of them *has* to be observed, we have the result that

$$P(HHT) = \tfrac{1}{3}, \quad P(THH) = \tfrac{1}{3}, \quad P(HTH) = \tfrac{1}{3}$$

The fact that one of the three has to be observed implies

$$\sum_e P(e) = 1, \quad e \in S_3$$

and the equally likely requirement can be satisfied only by breaking 1.0 into three equal parts: $\tfrac{1}{3}$.

This application of the logic of equally likely chances may seem like sleight of hand. Another application may help clarify the meaning of conditional probability. Let us agree that the probability of the occurrence of an outcome E is given by the ratio of the number of events in E to the number of events in the sample space S:

$$P(E) = \frac{n(E)}{n(S)}$$

This is tantamount to the equally likely case but with a more "reasonable" sounding basis. Thus, every event in S has a probability $1/n(S)$. Now let us condition our sample space on some characteristic (such as, all events are conditional in the sense that they are known to have two heads). This gives us a new, or conditional sample space (for example, $S_3 = \{HHT, HTH, THH\}$). Now we may rewrite the conditional probability E of the heads occurring in consecutive tosses:

$$P(E \mid S_3) = \frac{n(E)}{n(S_3)} = \frac{2}{3}$$

If we had not assumed, at earlier points in our discussion, that the sample space (that is, the $\mathscr{U}$ of a particular experiment) was understood, we would have had to write for $P(E)$, where $\mathscr{U} = S_1$,

$$P(E) = P(E \mid \mathscr{U}) = \frac{n(E)}{n(\mathscr{U})} = P(E \mid S_1) = \frac{n(E)}{n(S_1)}$$

Another approach is to note that when we are given a sample space S, with a probability distribution satisfying the definition of probabilities, we compute $P(E)$ by finding $\sum_e P(e)$, $e \in E$. Since $P(S) = 1$ and $(E \cap S) = E$, we can write

$$P(E) = \frac{P(E \cap S)}{P(S)}$$

indicating that $P(E)$ is the ratio of that part of E included in S, in this case E itself, to $P(S) = 1$.

In terms of Venn diagrams, when we condition outcomes A, B, and C on the outcome C, we change from the $\mathscr{U}$ and the A, B, C of Figure 5.1a to the space indicated by the heavily drawn line of Figure 5.1b, which now corresponds to our universe. Notice that C' is now outside the universe C and that parts of A and B are also excluded. Nevertheless, parts of A and B are still "possible" (all C' is in the set of impossible events). These "possible" outcomes are specified by $(A \cap C)$ and $(B \cap C)$.

If we have "redefined" our sample space to coincide with the set of elements in C, we must reassign probabilities to those elements so that the newly assigned

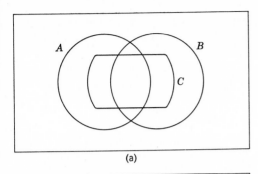

(a)

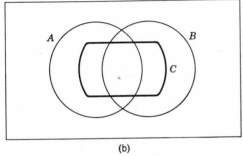

(b)

FIGURE 5.1

Venn diagrams illustrating the conditioning of an outcome space.

probabilities will total 1. To do this we divide the probabilities of elements in C by $P(C)$ and set the probabilities in C' to zero. This gives us

$$\frac{P(e)}{P(C)} = P(e \mid C) \qquad \text{for all } e \in C \tag{5.3}$$

and

$$P(e \mid C) = 0 \qquad \text{for all } e \notin C \tag{5.4}$$

Now, since

$$P(C) = \sum_e P(e), \qquad e \in C$$

then

$$\sum_e \frac{P(e)}{P(C)} = \frac{\sum_e P(e)}{P(C)} = 1, \qquad e \in C$$

By Theorem 2 of Chapter 4, if an e is contained in C,

$$P(e) \leq P(C)$$

and thus

$$0 \leq \frac{P(e)}{P(C)} \leq 1$$

Thus, $P(e \mid C) = P(e)/P(C)$ is a permissible probability assignment. The probability of an outcome (or set of events) is the sum of the probabilities of the events that make up the outcome. Thus we can write

$$P(A \mid C) = \sum_e \frac{P(e \mid C)}{P(C)} = \frac{P(A \cap C)}{P(C)}, \qquad e \in A$$

Where it is noted that $P(e \mid C) = 0$ for all $e \notin C$, we can rewrite

$$P(A \mid C) = \frac{P(A \cap C)}{P(C)} \qquad (5.5)$$

This is the definition we set out to rationalize. Thus the definition of conditional probability can be "derived" from the definitions of sample spaces, events, and probability distributions.

A general restriction must be placed on the definition of conditional probability. It is always necessary for the conditioning outcome C to be possible, that is, $P(C) \neq 0$. If this condition is not met, a division by zero is implied in the definition of $P(A \mid C)$ and, consequently, the conditional probability is undefined mathematically.

If we draw a card and have a friend tell us its suit—clubs, for example—what is the probability that it is an ace? In the unconditional situation we know that $P(\text{ace}) = 4/52 = 1/13$. We also know that only one ace of clubs is in the deck, and hence $P(\text{ace} \cap \text{club}) = 1/52$.

$$P(\text{club}) = \frac{13}{52}$$

$$P(\text{ace} \mid \text{club}) = \frac{1/52}{13/52} = \frac{1}{13}$$

In this example $P(\text{ace} \mid \text{club}) = P(\text{ace})$.

Toss two fair coins. This gives $S = \{HH, HT, TH, TT\}$ with equally probable events, $P(e) = 1/4$. Take two outcomes:

E, not more than one head

F, at least one of each face

This gives
$$P(E) = \tfrac{3}{4}; \qquad P(F) = \tfrac{2}{4}; \qquad P(E \cap F) = \tfrac{2}{4}$$

The student is to make sure that these values are correct and that

$$P(E \mid F) = \frac{2/4}{2/4} = 1$$

$$P(F \mid E) = \frac{2/4}{3/4} = \frac{2}{3}$$

If we had three fair coins instead of two and the same E and F but in $S = \{HHH, HHT, \ldots, TTT\}$ with equally likely events, $P(e) = \frac{1}{8}$, then

$$P(E) = \tfrac{4}{8} = \tfrac{1}{2}, \qquad P(F) = \tfrac{6}{8} = \tfrac{3}{4}, \qquad P(E \cap F) = \tfrac{3}{8}$$

$$P(E \mid F) = \frac{3/8}{6/8} = \frac{1}{2}$$

$$P(F \mid E) = \frac{3/8}{4/8} = \frac{3}{4}$$

This example points out that the specific sample space we are dealing with will determine different conditional probabilities of outcomes, even when the outcomes have the same general meaning.

This example leads us to consider another application. Notice, in the last example, that

$$P(E) = P(E \mid F)$$

$$P(F) = P(F \mid E)$$

When this special condition obtains, we say that E and F are independent outcomes.

Independence and Causality

Much scientific effort is directed at discovering how things are dependent on one another. In many of the physical sciences we encounter so-called deterministic relationships:

> If we have a given value of variable X in this process, we also observe a particular specific value of variable Y where $Y = f(X)$.

An example is the motion of a falling body. If an object is falling for time t (free fall in gravitational field of earth), then the distance traveled is $D = at^2$ where a is a constant related to the force of gravity. Thus, for any value of $t \geq 0$, D is a specific value, that is, at^2.

In the behavioral sciences we have few such simple structures. We have to deal with different kinds of relationships between events—dependencies in a probabilistic sense. Even in the physical sciences it is recognized that the behavior of physical systems is to some degree probabilistic. Depending on the level of analysis, physical laws appear deterministic or appear probabilistic. A probabilistic law has the following general form:

If we have observed a given outcome E, from a sample space, the probability of observing another outcome F at the same time is

$$P(F \mid E) = \frac{P(E \cap F)}{P(E)}$$

The two examples above are special cases. In the example of two tosses of a coin, $P(E \mid F) = 1$. In the other example, with three tosses of a coin

$$P(E \mid F) = \frac{3/8}{6/8} = \frac{P(E \cap F)}{P(F)} = \frac{1}{2}$$

For two tosses

$$P(E) \neq P(E \mid F)$$

For three tosses

$$P(E) = P(E \mid F)$$

In tossing two coins, it makes a difference if we observe F or not in terms of whether we observe E or not. However, in tossing three coins, $P(E)$ is the same as $P(E \mid F)$—thus, observing E given that we had observed F is the same as where we know nothing. In the situation where $P(E) \neq P(E \mid F)$ the probability of observing E is dependent on having or not having the knowledge of F. In the situation where $P(E) = P(E \mid F)$, E is said to be independent of F. Two outcomes E and F are probabilistically independent if $P(E) = P(E \mid F)$ and probabilistically dependent if $P(E) \neq P(E \mid F)$. In accordance with this definition we would say that (1) if two tosses are made, observing "not more than one head" is probabilistically dependent on observing "at least one of each face"; (2) if three tosses are made, observing "not more than one head" is probabilistically independent of observing "at least one of each face."

A special relationship enables us to calculate $P(E \cap F)$ for any sets E and F when we have $P(E)$, $P(F)$, and $P(E \mid F)$ or $P(F \mid E)$:

$$P(E \mid F) = \frac{P(E \cap F)}{P(F)}$$

Multiplying by $P(F)$, we get

$$P(F)P(E \mid F) = P(E \cap F) \tag{5.6}$$

This is referred to as the multiplication rule for the probability of intersection of outcomes.

The multiplication rule can be extended to several outcomes. We can find $P(E \cap F \cap G)$. First substitute D for $(F \cap G)$, that is, $D = (F \cap G)$. Now the multiplication rule gives

$$P(D \cap E) = P(E)P(D \mid E)$$

Note that

$$P(D \mid E) = P(F \cap G \mid E)$$

where E takes the place of the previously implicit universal set $\mathcal{U}$. In more detailed notation we would write $P(A)$ as $P(A \mid \mathcal{U})$, and thus $P(D \mid E)$ is written $P(D \mid E \cap \mathcal{U})$. This last expression implies that E has replaced $\mathcal{U}$; that is, since $E \cap \mathcal{U} = E$, $P(D \mid E \cap \mathcal{U}) = P(D \mid E)$. Following the same logic, we write

$$P(F \cap G \mid E) = P(G \mid E \cap F)P(F \mid E)$$

Substituting, we have

$$P(E \cap F \cap G) = P(E)P(F \mid E)P(G \mid E \cap F) \tag{5.7}$$

The student should define $P(E \cap F \cap G \cap H)$ in a similar fashion.

When we have two independent outcomes E and F, we can apply the multiplication rule and get a special result. The independence of E and F implies

$$P(E) = P(E \mid F)$$

and by the multiplication rule

$$P(E \cap F) = P(F)P(E \mid F)$$

Substituting, we have

$$P(E \cap F) = P(F)P(E) \tag{5.8}$$

This is called the multiplication rule for independent events.

Implied in this last derivation is a special consequence which may not be obvious but is of some importance. If the outcome E is independent of the outcome F, we know that

$$P(E) = P(E \mid F)$$

We also know that if F is independent of E, then

$$P(F) = P(F \mid E)$$

We can prove this relationship quite simply.

THEOREM 1. *If* E *is independent of* F, *then* F *is independent of* E.

Proof. By the condition in the theorem and the definition of the intersection of outcomes,

$$P(E \cap F) = P(F)P(E \mid F) = P(F)P(E)$$

But we have also

$$P(E \cap F) = P(E)P(F \mid E)$$

and by the nature of the relation "=" these two equations imply

$$P(F)P(E) = P(E)P(F \mid E)$$

This, in turn, implies

$$P(F \mid E) = P(F)$$

the condition of F being independent of E.

The results we have already obtained, along with our definitions, lead to three other useful theorems. We shall prove the last of these theorems and leave the proofs of the remaining two to the student.

> THEOREM 2. *If* E *and* F *are independent outcomes, then* E *and* F′ *are independent.*

> THEOREM 3. *If* E *and* F *are independent outcomes, then* E′ *and* F *are independent.*

> THEOREM 4. *If* E *and* F *are independent outcomes, then* E′ *and* F′ *are independent.*

Proof.   By the definition of independence, complementation, and the condition of the theorem we must show that if

$$P(E \cap F) = P(E)P(F)$$

then

$$P(E' \cap F') = P(E')P(F')$$

By an earlier derivation in set theory and De Morgan's law,

$$P(E' \cap F') = P[(E \cup F)'] = 1 - P(E \cup F)$$

But

$$P(E \cup F) = P(E) + P(F) - P(E \cap F)$$

where by independence of E and F given in the theorem,

$$P(E \cup F) = P(E) + P(F) - P(E)P(F)$$

Thus

$$P(E' \cap F') = 1 - P(E) - P(F) + P(E)P(F)$$
$$= [1 - P(E)][1 - P(F)]$$
$$= P(E')P(F')$$

The language we have been using—"independence," "dependence," "function," "probabilistic relation," and so on—may lead the student to believe that we are discussing causal connections in the real world. We are not!

If two events E and F are dependent [that is, $P(E \mid F) \neq P(E)$], there *may or may not* be a direct causal relationship involved. For example, imagine that schizophrenics are shown to have a "higher" level of a chemical X in their blood systems than persons in the general population, that is, $P(\text{high } X \mid \text{schizophrenic}) \neq P(\text{high } X)$. It is tempting to say that X in high concentrations tends to produce schizophrenia, but it is not necessarily so. All we know is that these two observations are associated or related. There could be another factor producing both high levels of X and schizophrenia, or they could be related by some distant genetic factor that is no longer directly active.

A complete discussion of the nature of causality, statistical or probabilistic evidence, and functional relations is not appropriate here. The interested student would do well to read M. R. Cohen and Ernst Nagel's book, *An introduction to logic and scientific method*, New York: Harcourt Brace, 1934.

Independence of Several Outcomes

A complication arises when we have more than two outcomes to consider with respect to independence. We must define two kinds of independence: pairwise and mutual.

The outcomes A_1, A_2, and A_3 are pairwise independent if all three pairs (A_1, A_2), (A_1, A_3), and (A_2, A_3) are independent as pairs. This condition holds when

$$P(A_1 \mid A_2) = P(A_1) \quad \text{or} \quad P(A_2 \mid A_1) = P(A_2)$$
$$P(A_1 \mid A_3) = P(A_1) \quad \text{or} \quad P(A_3 \mid A_1) = P(A_3)$$
$$P(A_2 \mid A_3) = P(A_2) \quad \text{or} \quad P(A_3 \mid A_2) = P(A_3)$$

or, equivalently, when

$$P(A_1 \cap A_2) = P(A_1)P(A_2)$$

$$P(A_1 \cap A_3) = P(A_1)P(A_3)$$

$$P(A_2 \cap A_3) = P(A_2)P(A_3)$$

This is generalized for k outcomes $A_1, \ldots, A_k$ when, if

$$P(A_i \cap A_j) = P(A_i)P(A_j)$$

for all $i, j = 1, \ldots, k$ $i \neq j$, we say the set of outcomes is composed of pairwise independent events.

The fact of pairwise independence tells us nothing about higher orders of independence. Assume pairwise independence in A_1, A_2, and A_3. Now, do we know whether A_1 and $(A_2 \cap A_3)$ are independent (remember, $(A_2 \cap A_3)$ is an outcome under our definitions)? We would have to answer: "Not necessarily." An example below points this out specifically. Pairwise independence does not cover the range of possible relationships. However, we define mutual independence to cover a wide range of combinations of events.

Outcomes A_1, A_2, and A_3 are said to be mutually independent if and only if the multiplication rule holds for all combinations of the outcomes—that is, the conditions of pairwise independence plus

$$P(A_1 \cap A_2 \cap A_3) = P(A_1)P(A_2)P(A_3)$$

In general, we say that the outcomes $A_1, A_2, \ldots, A_n$ are stochastically independent *if and only if* the multiplication rule holds for all combinations of two or more events:

$$P(A_i \cap A_j) = P(A_i)P(A_j) \qquad 1 \leq i \leq j \leq n$$
$$P(A_i \cap A_j \cap A_k) = P(A_i)P(A_j)P(A_k) \qquad 1 \leq i \leq j \leq k \leq n$$
$$\vdots \qquad\qquad \vdots$$
$$P(A_1 \cap A_2 \cap \cdots \cap A_n) = P(A_1) \cdots P(A_n)$$

We state without proof that there are $(2^n - n - 1)$ such combinations. In the next part we demonstrate that there are 2^n combinations of n things (outcomes here). However, in this instance we must exempt from consideration the combinations of *no* things and of *one* thing. Hence there are $(n + 1)$ combinations of outcomes not involved in the definition of mutual or stochastic independence.

Consider an example utilizing these considerations. Five astronauts are all on a standby status for a space flight. In order to be put in this status they had to pass (among other requirements) three human-factors tests: (1) intelligence, (2) physical fitness, (3) mental stability. However, even among astronauts there are individual differences, and above the passing level on all three tests there is, say, a superior level of performance. We call our five men A, B, C, D, and E. Imagine that A and E were superior on test 1 only, B was superior on test 2 only, C was superior on test 3 only, and D was a "superastronaut," being superior on all three tests. According to their agreement, the astronaut making the flight is chosen by lot where each man has the same chance as any other.

We can define several sample spaces from the random experiment the astronauts execute when they draw the lots: (1) the five events $e_1 =$ man A is chosen, $e_2 =$ man B is chosen, $\ldots$, $e_5 =$ man E is chosen; (2) the number of tests passed at the superior level by the man chosen, $e_1 = 1$, $e_2 = 2$, $e_3 = 3$; and (3) the events $e_1 =$ man chosen is superior in test number 1 (intelligence), $e_2 =$ man chosen is superior in test number 2 (physical fitness), $e_3 =$ man chosen is superior in test number 3 (mental stability). This last sample space can be broken down in more detail by noting all the possible combinations of superior–not-superior on the tests. Letting S_1 stand for superior performance on test 1, S_2 for superior performance on test 2, and S_3 for superior performance on test 3, and similarly N_1, N_2, and N_3 for not superior on the three tests, produces a sample space of eight events:

$$N_1 N_2 N_3 = e_1$$
$$S_1 N_2 N_3 = e_2$$
$$N_1 S_2 N_3 = e_3$$
$$N_1 N_2 S_3 = e_4$$
$$S_1 S_2 N_3 = e_5$$
$$S_1 N_2 S_3 = e_6$$
$$N_1 S_2 S_3 = e_7$$
$$S_1 S_2 S_3 = e_8$$

Notice that since selection of a given astronaut is equivalent to selection of one of the events in this space and since we have only five men, two of whom are alike with respect to this sample space, only four of the events are possible. The correspondence is

$$\text{men } A \text{ and } E = S_1 N_2 N_3 = e_2$$
$$\text{man } B \quad\;\; = N_1 S_2 N_3 = e_3$$
$$\text{man } C \quad\;\; = N_1 N_2 S_3 = e_4$$
$$\text{man } D \quad\;\; = S_1 S_2 S_3 = e_8$$

Now we can assign probabilities to the events in the sample space. The experiment "pick a *man* by lot" means that

$$P(e_1) = P(e_5) = P(e_6) = P(e_7) = 0$$

because we cannot pick a man with those patterns of S and N. Also, we know that the "lots" method of picking the man implies that the probability of picking any given man is the same as that of picking any other man. This implies that the probability of observing one of the possible events (e_2, e_3, e_4, or e_8) is proportional to the number of men in the group with the respective pattern of N's and S's. This leads to

$$P(e_3) = P(e_4) = P(e_8) = P$$

where P is the probability of any given man being chosen, and

$$P(e_2) = 2P$$

We also know that all values are positive and that

$$P(e_3) + P(e_4) + P(e_8) + P(e_2) = 1$$
$$P + P + P + 2P = 1$$
$$5P = 1$$
$$P = \tfrac{1}{5}$$

and, finally, in summary,

$$P(e_3) = P(e_4) = P(e_8) = \tfrac{1}{5}$$
$$P(e_2) = \tfrac{2}{5}$$
$$P(e_1) = P(e_5) = P(e_6) = P(e_7) = 0$$

The reader should check this assignment of probabilities against our definition of probabilities on events in sample spaces. Is this assignment proper?

Notice that we have defined our sample space by events and not outcomes. Yet our topic in this chapter, conditional probability, has been presented exclusively in terms of outcomes. This was done for two reasons: (1) single element outcomes are equivalent to the events making up the outcomes, and we cover the theory of conditional probability of events as a special case of outcomes; (2) the conditional probability of *any* event, given *any* other event, is zero. The proof of this last statement is simple. By definition

$$P(e_i \mid e_j) = \frac{P(e_i \cap e_j)}{P(e_j)}$$

However, by definition of events,

$$e_i \cap e_j = \emptyset$$

and $P(\emptyset) = 0$, which gives

$$P(e_i \mid e_j) = \frac{0}{P(e_j)} = 0$$

Now we apply the theory of conditional probability to the astronaut example. Define some outcomes:

$$F_1 = \{e \mid \text{exactly two superior tests}\}$$
$$F_2 = \{e \mid \text{exactly one superior test}\}$$
$$F_3 = \{e \mid \text{two or more superior tests}\}$$
$$F_4 = \{e \mid \text{test 1 superior}\}$$
$$F_5 = \{e \mid \text{test 2 superior}\}$$
$$F_6 = \{e \mid \text{test 3 superior}\}$$

The reader should at this point specify all the outcomes like $F_2 = \{e_2, e_3, e_4\}$. Also construct a Venn diagram of the sample space and the outcomes.

The probabilities of the outcomes are

$$P(F_1) = P(\{e_5, e_6, e_7\}) = P(e_5) + P(e_6) + P(e_7) = 0$$
$$P(F_2) = P(\{e_2, e_3, e_4\}) = \tfrac{2}{5} + \tfrac{1}{5} + \tfrac{1}{5} = \tfrac{4}{5}$$

and similarly

$$P(F_3) = \tfrac{1}{5} \qquad P(F_5) = \tfrac{2}{5}$$
$$P(F_4) = \tfrac{3}{5} \qquad P(F_6) = \tfrac{2}{5}$$

A special case is illustrated by the probability of choosing the astronaut so that he was superior on test 3 given that he was superior on exactly two tests. This is

$$P(F_6 \mid F_1) = \frac{P(F_6 \cap F_1)}{P(F_1)}$$

However, $P(F_1) = 0$ and it is not permissible to divide by zero; and this special case violates the requirement that the conditioning outcome must be possible.

Let us ask some special questions about the relationships of the three tests. First, is mental stability of our five men independent of intelligence? This is tantamount to asking if

$$P(F_4 \cap F_6) = P(F_4)P(F_6)$$

or if

$$P(F_6 \mid F_4) = P(F_6)$$

First, recall

$$P(F_4) = \tfrac{3}{5} \qquad P(F_6) = \tfrac{2}{5}$$

and note that

$$F_4 \cap F_6 = \{e_2, e_5, e_6, e_8\} \cap \{e_4, e_6, e_7, e_8\}$$
$$= \{e_6, e_8\}$$

and hence

$$P(F_4 \cap F_6) = P(e_6) + P(e_8) = \tfrac{1}{5}$$

Thus

$$P(F_6 \mid F_4) = \frac{1/5}{3/5} = \frac{1}{3}$$

or one out of three that if the astronaut chosen had a superior score on mental stability he also had a superior score on intelligence. Likewise we see that

$$P(F_6 \mid F_4) = \tfrac{1}{3} \neq P(F_6) = \tfrac{2}{5}$$

and that

$$P(F_4 \cap F_6) = \tfrac{1}{5} \neq P(F_4)P(F_6) = \tfrac{6}{25}$$

Thus, F_6 and F_4 are dependent events; they are not independent. We can add a cautious note about the extent of the lack of independence or the degree of dependency. To the extent that

$$P(F_6 \mid F_4) > P(F_6)$$

it follows that we have a better chance of choosing a man with superior intelligence (given $\mathscr{U}$, that we are talking only about the five astronauts). In our example, however,

$$P(F_6 \mid F_4) = \tfrac{5}{15} < \tfrac{6}{15} = P(F_6)$$

and the opposite conclusion follows.

If we had only the first four men, A, B, C, and D, then the assignment of probabilities would be different:

$$P(e_1) = P(e_5) = P(e_6) = P(e_7) = 0$$
$$P(e_2) = P(e_3) = P(e_4) = P(e_8) = \tfrac{1}{4}$$

This implies that

$$P(F_4) = P(F_5) = P(F_6) = \tfrac{1}{2}$$

and

$$P(F_4 \cap F_5) = P(F_4 \cap F_6) = P(F_5 \cap F_6) = \tfrac{1}{4}$$

and

$$P(F_4 \cap F_5 \cap F_6) = \tfrac{1}{4}$$

(the reader should verify these facts). From these values we have

$$P(F_4 \cap F_5) = P(F_4)P(F_5) = \tfrac{1}{4}$$
$$P(F_4 \cap F_6) = P(F_4)P(F_6) = \tfrac{1}{4}$$
$$P(F_5 \cap F_6) = P(F_5)P(F_6) = \tfrac{1}{4}$$

and thus, F_4, F_5 and F_6 are pairwise independent. However,

$$P(F_4 \cap F_5 \cap F_6) = \tfrac{1}{4} \neq \tfrac{1}{8} = P(F_4)P(F_5)P(F_6)$$

implying that pairwise independence does not assure stochastic independence of the outcomes involved.

This last set of computations illustrates the general principle that the specific definition of an experiment determines the character of the probability structures of the experiment. Change any aspect of the experiment (drop one astronaut out of five, for example, but otherwise leave the experiment unchanged), and the entire probability results may be changed also.

Algebra of Conditional Probability

Since conditional probabilities satisfy all the requirements of the definition of probabilities in general, we can apply the theorems of probability to conditional probabilities. These theorems are not proved here, and the reader is urged to work the proofs out for himself. Assume that $P(A)$, $P(B)$ are greater than zero.

1. $P(A \mid A) = 1$ (5.9)
2. $P(\emptyset \mid A) = 0$ (5.10)
3. If $A \subseteq B$, then $P(A \mid C) \leq P(B \mid C)$ (5.11)
4. $P(A' \mid B) = 1 - P(A \mid B)$ (5.12)
5. $P(A \cup B \mid C) = P(A \mid C) + P(B \mid C) - P(A \cap B \mid C)$ (5.13)
6. If $P(C) = 1$, then $P(A \mid C) = P(A)$ (5.14)
7. If $A \cap B = \emptyset$, then $P(A \mid B) = 0$ (5.15)

Independence, Dependence, and Intersection

This final section of this chapter is a warning section: Beware of confusing the concepts of intersection and dependence. Frequently, students confusedly conclude from their not-too-careful thinking through of probability that for two outcomes to be independent they must not share any events. Not so! The converse is true. *In order for two possible outcomes to be independent they must share some events.* In order for an outcome A to be independent of an outcome B, where $P(A) > 0$, the proportion of A's probability in $\mathscr{U}$ as a whole must be exactly the same as the proportion of A's probability in B as a subset of $\mathscr{U}$. That is the basic meaning of the definition of independence:

$$A \text{ is independent of } B \text{ if}$$

$$P(A) = \frac{P(A \cap B)}{P(B)} = P(A \mid B)$$

This cannot be true if $(A \cap B) = \emptyset$ because then

$$P(A \mid B) = \frac{P(\emptyset)}{P(B)} = 0$$

and $P(A) \neq 0$.

If $P(A) \neq 0$, then $(A \cap B) = \emptyset$ implies that A is dependent on B.

SUGGESTED READING

The texts listed at the end of the previous chapter are particularly appropriate for the materials in this chapter.

GLOSSARY

(5.1)
$$P(E) = \sum_e P(e) \text{ for all } e \in E$$

(5.2)
$$P(F \cap E) = \sum_e P(e) \text{ for all } e \in (E \cap F)$$

Conditional Probability. The conditional probability $P(E \mid F) = P(E \cap F)/P(F)$ is the probability of observing the outcome E given that outcome F is observed.

(5.3)
$$P(e \mid C) = P(e)/P(C) \text{ for all } e \in C$$

(5.4)
$$P(e \mid C) = 0 \text{ for all } e \notin C$$

(5.5)
$$P(A \mid C) = P(A \cap C)/P(C)$$

Independence. Two outcomes E and F are probabilistically independent if $P(E) = P(E \mid F)$.

(5.6)
$$P(F)P(E \mid F) = P(E \cap F)$$

(5.7)
$$P(E \cap F \cap G) = P(E)P(F \mid E)P(G \mid E \cap F)$$

(5.8)
$$P(E \cap F) = P(F)P(E) \text{ for } E \text{ and } F \text{ independent}$$

Theorem 1. If E is independent of F, then F is independent of E.

Theorem 2. If E and F are independent, E and F' are independent.

Theorem 3. If E and F are independent, E' and F are independent.

Theorem 4. If E and F are independent, E' and F' are independent.

Mutual Independence. Outcomes A_1, A_2, $\ldots$, A_k are mutually independent if and only if all $(2^k - k - 1)$ combinations of outcomes are independent.

(5.9) $\qquad P(A \mid A) = 1$

(5.10) $\qquad P(\emptyset \mid A) = 0$

(5.11) $\qquad$ If $A \subseteq B$, then $P(A \mid C) \leq P(B \mid C)$

(5.12) $\qquad P(A' \mid B) = 1 - P(A \mid B)$

(5.13) $\qquad P(A \cup B \mid C) = P(A \mid C) + P(B \mid C) - P(A \cap B \mid C)$

(5.14) $\qquad$ If $P(C) = 1$, then $P(A \mid C) = P(A)$

(5.15) $\qquad$ If $A \cap B = \emptyset$, then $P(A \mid B) = 0$

PROBLEMS

1. Prove the following statements:

$$P(X \cup Y) = P(X) + P(Y) - P(X \cap Y)$$
$$P(E') = 1 - P(E)$$
$$P(E' \cup F') = 1 - P(E \cap F)$$

2. Extend equation (5.7) to the intersection of four outcomes.

3. Prove Theorems 2 and 3 of this chapter.

4. Imagine selecting stimuli at random for an experiment. Two types of stimuli are used, emotionally charged words and neutral words. Two lists are available, each with both types of words. List I contains three emotionally charged words and five neutral words. List II contains five emotionally charged words and three neutral words. Select a list with equal probability for the two lists and a word from that list at random. What is the probability that an emotionally charged word is selected?

5. For problem 4 imagine that list I contained one emotionally charged word and two neutral words and that list II contained two emotionally charged words and one neutral word. Select the list with equal probability for the two lists. Select a word from the list with equal probability for the words. Select a word a second time (from all of the words of the list). Are the two selections independent in the process of selection of a list and then two words? Are the selections independent within the list from which they are made?

6. Construct proofs of the following statements:

$$P(A \mid A) = 1$$
$$P(\emptyset \mid A) = 0$$
If $A \subseteq B$, then $P(A \mid C) \leq P(B \mid C)$
$$P(A' \mid B) = 1 - P(A \mid B)$$
$$P(A \cup B \mid C) = P(A \mid C) + P(B \mid C) - P(A \cap B \mid C)$$
If $P(C) = 1$, then $P(A \mid C) = P(A)$
If $A \cap B = \emptyset$, then $P(A \mid B) = 0$

7. Seven events are defined in a probability model of an experiment. What is the probability of observing the results corresponding to E_1, E_2, or E_3 given that neither of the results corresponding to the events E_6 or E_7 is observed. Assume that initially the events were equally likely.

8. In a problem-solving experiment three subjects are given information leading to the problem, giving each subject a different probability of solving the problem. Subject A has a probability of .4 of solving the problem, subject B a probability of .3, and subject C a probability of .2. The winner is the subject to solve the problem first. A subject can withdraw from the game at any time before the solution and avoid points against himself in a total score. If subject C withdraws, what is the probability that B will solve the problem before A ?

CHAPTER 6

Sampling and Probability

In this chapter we examine in detail several methods of determining the probability of events and outcomes in sample spaces. All of these methods are based on a logical analysis of the sample space and the method of performing the experiment defining the sample space. In other words, we study the probability implications of the methods by which we observe or sample events. From the analysis of how we sample from the sample space we can determine the probability of observing each event in the sample space.

Up to this point we have been somewhat casual about the methods of assignment of probabilities to events in a sample space. Now we shall formalize this procedure.

Bear in mind that we are discussing the probability models of experiments. Thus, when we use the term "observe an event" we refer to conceptual or "abstract" observations, not to real or empirical experiments giving tangible results. If this distinction is maintained clearly in the mind, fewer difficulties will develop.

Populations and Samples

In the previous chapters we always assumed a list of observations or events that could be observed when an experiment was performed. This list of possible observations is sometimes called the population of the experiment. It is distinguished from the sample space in that the sample space is defined by possible samples taken from the population where more than one element in the population may be involved. A collection of five astronauts A, B, C, D, and E is a population. The collection of men selected for three flights is a sample and is not a member of the population. Only when the sample consists of one element from the population do the sample space and the population coincide (the student should develop an argument showing that this is true).

In the work to come we restrict our definition of populations to collections of the objects and events from which we sample. Sample spaces are restricted to the collection of possible samples from the population and depend on the sampling models used (see below). Populations are not dependent on sampling methods.

The word "population" is used in ordinary language to refer to the inhabitants of a country or geopolitical region. This is included in our usage of the word. However, we use "population" in the widest sense: Any collection of objects or events is a population. Examples of populations are easy to think up. All the examples of sets we thought of are examples of populations. Some new ones might be: all the subscribers to the local newspaper at this time; all the subscribers to the local newspaper ever; the members of the Democratic Party; the inmates of a given prison who committed murder; the members of the genus *Muscus norwegicus albinus*; the scores on an intelligence test (yes, the scores, not the *persons* getting the scores); and so on.

All of these examples are of discrete populations, and only one is quantitative. However, it is possible to define populations that are continuous and infinite; indeed in all ways we found we could define sample spaces.

In a very real sense, populations are more fundamental than sample spaces: The definition of a sample space depends both on the definition of a population and on a sampling model. However, we generally find it to be more convenient, and without loss of generality, to deal with sample spaces directly. Always there is a population involved, but unless it is necessary for the sake of clarity, we do not explicitly attend to the population.

We now turn to the analysis of random experiments from the standpoint of sampling models. In this analysis we restrict the letter e to stand for events in sample spaces and the letter a to stand for elements of populations.

Repeated Experiments

The first task in the logical analysis of a random experiment is to reduce the experiment to its simplest form consistent with the information desired from the experiment (that is, the questions asked of the experiment or that can be answered by the experiment). Many of the experiments we have discussed have been more complicated than the simplest form of the experiment. For example, we at one time "tossed a coin" three times and observed "heads" or "tails." The heads and tails were recorded as they were observed, with the side observed on the first toss written down first, and so on. The list of possible events in the sample space of this experiment is

$$HHH, \ HHT, \ HTH, \ THH, \ HTT, \ THT, \ TTH, \ TTT$$

The probability $\frac{1}{8}$ was assigned to each of these events on the basis of the argument that any one combination of heads-tails would be observed as often as any other combination. This argument has a certain intuitive appeal, but the assignment of equal probabilities must be more firmly grounded. The first step is to note that we really have three repetitions of a basic experiment: Toss a coin.

Now, in this basic experiment the argument that a head is as probable as a tail and hence $P(H) = P(T) = .5$ is convincing because of the physical symmetry and basic identity of the two sides of the coin. The contention that $P(HHH) = P(HTT)$ is harder to support in this way. However, by arguing that $P(H) = P(T) = .5$ we are able to prove that $P(HHH) = P(HTT) = \frac{1}{8}$.

The first step is to conceive of the larger experiment as three repetitions of the simple experiment. It is customary to call these three repetitions trials rather than experiments to avoid the confusion of terms (all three experiments are the same, but to say "three experiments" may mislead us to think there were three different experiments involved). The population is the two elements heads and tails.

The second step is to perform the experiment in such a way that the elements of the population observed on one trial do not change the probability of observing the same elements on another trial. This is formalized in our coin tossing experiment by

$$P(H) = P(H \mid H \text{ on previous trials})$$
$$= P(H \mid T \text{ on previous trials})$$

and

$$P(T) = P(T \mid H \text{ on previous trials})$$
$$= P(T \mid T \text{ on previous trials})$$

That is, the trials are independent with respect to the probability of elements in the population observed on the trials.

This leads to the following definition. If two trials are independent, then the probability of observing element a_i on one trial and a_j on another is

$$P(a_i a_j) = P(a_i)P(a_j)$$

In general, the probability of observing the succession of elements $a_i, a_j, \ldots, a_k$ on n mutually independent trials is

$$\underbrace{P(a_i, a_j, \ldots, a_k)}_{n \text{ elements}} = \underbrace{P(a_i)P(a_j) \cdots P(a_k)}_{n \text{ terms}}$$

This holds true regardless of the specific nature of the elements $a_i, a_j, \ldots, a_k$ or the probabilities associated with them.

In the case of the three trials of the coin-tossing,

$$P(HHH) = P(H)P(H)P(H) = (\tfrac{1}{2})^3 = \tfrac{1}{8}$$
$$P(HHT) = P(H)P(H)P(T) = (\tfrac{1}{2})^3 = \tfrac{1}{8}$$

$$\cdot \qquad \cdot \qquad \cdot$$
$$\cdot \qquad \cdot \qquad \cdot$$
$$\cdot \qquad \cdot \qquad \cdot$$

$$P(TTT) = P(T)P(T)P(T) = (\tfrac{1}{2})^3 = \tfrac{1}{8}$$

This of course assumes that $P(H) = \tfrac{1}{2} = P(T)$ and that the three trials are independent. It is clear that the probabilities assigned to the combined trials are consistent with our axioms of probability. Having established a permissible assignment of probabilities to the set of population elements selected by repeating independent trials, we can assign probabilities to the events in the sample space of the more complicated experiment. This of course assumes that the more complicated experiment is strictly a repetition of the simple experiment.

It is easy to prove in the general case that the probabilities assigned to the sample spaces by this method are appropriate, that is, that they satisfy the probability axiom. We do this now for two independent repetitions of the same experiment with n population elements and unspecified probabilities. We write (a_i, a_j) when a_i is selected on the first repetition and a_j on the second. By definition of the experiment,

$$P(a_i, a_j) = P(a_i)P(a_j)$$

for $i = 1, \ldots, n$ and $j = 1, \ldots, n$. The entire collection of (a_i, a_j) makes up a sample space of n^2 events defined by two successive selections from a population of n elements. The probabilities of this sample space can be written down as a two-way array:

$$
\begin{array}{cccc}
P(a_1, a_1) & P(a_1, a_2) & \cdots & P(a_1, a_n) \\
P(a_2, a_1) & P(a_2, a_2) & \cdots & P(a_2, a_n) \\
\cdot & \cdot & \cdots & \cdot \\
\cdot & \cdot & \cdots & \cdot \\
\cdot & \cdot & \cdots & \cdot \\
P(a_n, a_1) & P(a_n, a_2) & \cdots & P(a_n, a_n)
\end{array}
$$

Summing the rows gives, for the ith row,

$$\sum_{j=1}^{n} P(a_i, a_j) = \sum_{j=1}^{n} P(a_i)P(a_j)$$

$$= P(a_i)\sum_{j=1}^{n} P(a_j) = P(a_i)$$

because

$$\sum_{j=1}^{n} P(a_j) = 1$$

by the probability axiom. This is true for all a_i, $i = 1, \ldots, n$. Hence, summing all n^2 values is given by

$$\sum_{i=1}^{n}\sum_{j=1}^{n} P(a_i, a_j) = \sum_{i=1}^{n} P(a_i) = 1$$

and since

$$\sum_{i=1}^{n}\sum_{j=1}^{n} P(a_i, a_j) = P(S)$$

we have shown that this product-rule assignment of probabilities is permissible.

Four Types of Experiments

In the repeated experiments model we assumed that each repetition was independent of the others. We also used the fact that in this model each element of the population can reoccur on succeeding trials (the student should see that this property and independence are related). We also assumed that we knew which population element occurred on what repetition—that is, we knew the order of the observation of the events.

In this section we consider experiments having complementary characteristics. We ask: What are the probabilities of events in the sample spaces of "repeated experiments" where (1) the probabilities of observations of population elements in successive repetitions depend on the observations in preceding repetitions and (2) the order in which specific population elements (in the simple experiments) are observed is not heeded? This suggests that there are four main alternatives. These are indicated as cells in Figure 6.1.

To simplify the discussion we limit our attention to equal probability sampling. Each element of the population that can be selected on a given trial is assumed to have the same probability of being selected as any other element. If we do sample

	Dependent Trials	Independent Trials
Ordered Observations	Ordered without replacement	Ordered with replacement
Unordered Observations	Unordered without replacement	Unordered with replacement

FIGURE 6.1

Scheme for classification of random experiments.

from the population so that each member has the same chance of being sampled as any other member, we say the sampling is equal probability sampling. If there are N elements in the population, then an equal probability sample of one element means that each of the N elements will be selected with probability $1/N$.

We induce dependency of trials by assuming that once a given population element is selected it can never be selected again. That is, selecting an element removes it from the population in the random experiment. This is called sampling without replacement. The equal probability assumption and this nonreplacement assumption assure that regardless of $P(a_i)$ on a trial in which it is selected (sampled), the probability of sampling that element is zero, $P(a_i) = 0$, on every repetition after that. Once an element a_j has been selected it is no longer available, because it is removed from the population of elements. Other mechanisms will also produce dependent samples, but we restrict our attention here to the nonreplacement mechanism.

The converse is the case where a specific element may be selected more than once in a number of repetitions of the experiment. This case is called sampling with replacement. Notice that $P(a_j)$ is constant from repetition to repetition under this method of sampling and the equal probability sampling rule.

Ordered Sampling with Replacement. This is the case discussed under the heading "Repeated Experiments" above. In the astronaut example the man sent on a flight was chosen by lot. Imagine that there are to be two flights and that the four men $A, B, C,$ and D are involved. What is the sample space of possible one-man crews for the two flights? The basic experiment is: Pick a man. This is repeated (two experiments or trials). The single experiment has a sample space

$$S_1 = \{A, B, C, D\}$$

with probabilities $P(A) = \cdots = P(D) = .25$. The two-part experiment has the sample space

$$S_2 = \{AA, AB, AC, AD, BA, BB, BC, BD, CA, CB, CC, CD, DA, DB, DC, DD\}$$

with probabilities $P(AA) = \cdots = P(DD) = \frac{1}{16}$. If there were to be three flights, then

$$S_3 = \{AAA, AAB, AAC, \ldots, DDD\}$$

with probabilities $P(AAA) = \cdots = P(DDD) = \frac{1}{64}$. The equal probability rule in our model implies $P(e) = 1/n(S)$, and by writing out the sample spaces S_1, S_2, and S_3, we see $n(S_1) = 4$, $n(S_2) = 16$, and $n(S_3) = 64$. However, even $n(S_3)$ is difficult to find this way. It is not very happy work to enumerate the entire list of events in S_3. Fortunately we do not need to write the list out to find $n(S_3)$, or for that matter $n(S_m)$ where there are m repetitions of the basic experiment.

In ordered sampling with replacement, with m *trials and* N *elements in the population, there are* N^m *different events in the sample space* S_m:

$$n(S_m) = N^m \tag{6.1}$$

This result may be derived by noting that on the first repetition, N different selections can be made. For each of these N distinct selections there are N possible selections on the second repetition. Thus, there are $(N \times N = N^2)$ different ways the experiment can be repeated twice. If for each of these ways we repeat the experiment again, there are N ways the repetition can come out, and there are $(N \times N^2 = N^3)$ total events in the three-part experiment. In general, if the experiment has been repeated k times there are N^k events in the sample space of the k-part experiment. For each of these there are N ways of adding the $(k + 1)$th experiment for a total of

$$N \times N^k = N^{k+1}$$

events.

The assignment of $P(e) = 1/N^m$ to each event in the sample space generated by repeating an experiment m times with N population elements is consistent with the multiplication rule for repeated experiments. For $m = 1$ the sample space is the sample space of the basic experiment, that is, the population. In this special case there are N events (the same as the elements of the population) and by the equal probability rule,

$$P(e) = \frac{1}{N}$$

Repeat the experiment and the events are all possible pairs of population elements: where a_1 is the element sampled on the first experiment, a_2 on the second,

$$P(a_1, a_2) = P(a_1)P(a_2) = \frac{1}{N} \times \frac{1}{N} = \frac{1}{N^2}$$

But $m = 2$, which gives $N^m = N^2$ and hence where e is the corresponding event from the sample space of the twice-repeated basic experiment, that is, $e = (a_1, a_2)$,

$$P(e) = \frac{1}{n(\mathcal{U})} = \frac{1}{N^2} = P(a_1, a_2)$$

Ordered Sampling Without Replacement. In the preceding section we replaced events in the population when the experiment was repeated. How many ways may we sample if the population is decreased on every sample?

If there are N elements to start with and we sample one of them on the first trial, there are N ways our first trial can come out. This leaves $(N - 1)$ elements left over, because by definition in this model it is not possible to select an element on any trial after it has once been selected. For each of the N elements we could select on trial 1 there are $(N - 1)$ elements that can be selected on trial 2. Thus, for a sample of two, $m = 2$, where S_2 is the resulting sample space,

$$n(S_2) = \underbrace{N(N - 1)}_{m = 2 \text{ terms}}$$

For $m = 3$ we need only note that for each of the $[N(N - 1)]$ ways the first two trials can come out there are $(N - 2)$ elements that could be selected on trial 3. Thus

$$n(S_3) = \underbrace{N(N - 1)(N - 2)}_{m = 3 \text{ terms}}$$

In general, ordered sampling without replacement, with a population of N elements and m trials, $N \geq m$, produces a sample space S_m where

$$n(S_m) = \underbrace{N(N - 1)(N - 2) \cdots (N - m + 1)}_{m \text{ terms}}$$

The student should prove to himself that for the case where $m > N$, $n(S_m) = 0$.

In the astronaut example, if a man was limited to one flight, the number of ways the crews could be selected for three flights under this model would be

$$n(S_3) = 4 \times 3 \times 2 = 24$$

This is considerably less than the $n(S_3) = 64$ we calculated under the model for ordered sampling with replacement. There are 40 (of the 64) events in S_3 having an astronaut flying two or three times. In the "without replacement" model, events such as AAA, ABA, ACC, CBC, and so on are not possible.

Under the equal probability rule, where N events are possible on the first trial, we assign to each event in the sample space the probability

$$P(e) = \frac{1}{N(N-1)\cdots(N-m+1)}$$

This corresponds to the probability we would assign under the multiplication rule for dependent outcomes. On the first trial each of the elements in the population has the same probability: $1/N$. On the second trial each of the remaining $(N-1)$ elements has the probability $1/(N-1)\ldots$. On the mth trial each of the $(N-m+1)$ remaining elements has probability $1/(N-m+1)$ of being sampled. But these are conditional probabilities (conditional on the trial number). Let a_j be an element that can be sampled on the jth trial; then

$$P(a_j \mid \text{1st trial, 2nd trial} \cdots (j-1)\text{th trial})$$
$$= P(a_1)P(a_2 \mid a_1) \cdots P(a_{j-1} \mid a_1, a_2, \ldots, a_{j-2})$$
$$= \frac{1}{N} \times \frac{1}{(N-1)} \times \frac{1}{(N-2)} \times \cdots \times \frac{1}{(N-j+1)}$$
$$= \frac{1}{N(N-1)\cdots(N-j+1)}$$

A great deal of simplification can be achieved by introducing the factorial notation here. Remember from college algebra that

$$N(N-1)(N-2)\cdots(N-N+2)(1) = N!$$

or "N factorial." (If this is unfamiliar to the student, he should consult a text on college algebra. Also, he would not be remiss to take this equality as a simple definition of $N!$) Notice that if we were to cancel the last $(N-m)$ terms in $N!$ we would obtain the number of events in S_m. That is,

$$n(S_m) = N(N-1)(N-2)\cdots(N-m+1)$$
$$= \frac{N(N-1)(N-2)\cdots(N-m+1)(N-m)\cdots(1)}{(N-m)\cdots(1)}$$

where the last terms in the numerator are canceled by the denominator. By the definition of the factorial we have then

$$n(S_m) = \frac{N!}{(N-m)!}$$

as the number of events in S_m.

Ordered sampling without replacement from a population of N *elements and* m *repetitions leads to* S_m *with*

$$n(S_m) = \frac{N!}{(N-m)!} \tag{6.2}$$

The probability of any one of the samples in S_m is given by

$$\frac{1}{N!/(N-m)!} = \frac{(N-m)!}{N!}$$

In the astronaut example with ($N = 6$) astronauts and ($m = 3$) flights there would be

$$\frac{6!}{(6-3)!} = \frac{6 \times 5 \times 4 \times 3 \times 2 \times 1}{3 \times 2 \times 1} = 6 \times 5 \times 4 = 120$$

different ordered events from ordered sampling without replacement.

Unordered Sampling Without Replacement. Each of the samples in the last section was defined in part by the order in which the elements of the population were selected. That is, the three events ABC, CBA, and BCA were treated as distinct. We now ask: How many unordered samples without replacement are there in m repetitions of the experiment with an N element population? We find the answer in a slightly roundabout way.

First take any single element, such as A, and note that it can occur in only one order. Two elements A and B can occur in two orders AB and BA. Three elements A, B, and C can occur in six orders ABC, ACB, CAB, BAC, BCA, and CBA. Four elements can occur in 24 orders (list them for A, B, C, and D). And in general, N things can occur in $N!$ orders. The logic is simple. There are N things that can occur first (in the first position of the ordering) and ($N - 1$) things in the second for $[N(N - 1)]$ ways of defining order 1 and order 2 in combination. There are ($N - 2$) ways of adding the third object in the third place, and so on. Thus there are

$$N! = N(N-1)(N-2) \cdots \quad (1)$$

ways of placing N objects in distinct orders. (The student should show that this is the number of ordered samples without replacement when $N = m$.)

This fact and the number of samples possible in ordered sampling without replacement lead to the answer of our question. If each of the distinct collections of m objects in the sample can be ordered in $m!$ different ways, then

$$n\binom{\text{Samples: ordered}}{\text{without replacement}} = (m!)n\binom{\text{Samples: unordered}}{\text{without replacement}}$$

$$\frac{N!}{(N-m)!} = (m!)n\binom{\text{Samples: unordered}}{\text{without replacement}}$$

$$\frac{N!}{(N-m)!\, m!} = n\binom{\text{Samples: unordered}}{\text{without replacement}}$$

This number is denoted $\binom{N}{m}$, the *binomial coefficient*: The number of unordered samples without replacement is given by the binomial coefficient

$$\binom{N}{m} = \frac{N!}{(N-m)!\,m!} \tag{6.3}$$

Suppose that we wish to select four men out of seven to participate in a certain task. It makes no difference in which order they are selected; we are interested only in the group of four selected. The number of ways in which we can select the four is given by

$$\binom{N}{m} = \binom{7}{4} = \frac{7!}{(7-4)!\,4!} = \frac{7 \times 6 \times 5 \times 4 \times 3 \times 2 \times 1}{3 \times 2 \times 1 \times 4 \times 3 \times 2 \times 1} = 7 \times 5 = 35$$

Under the equal probability rule we assign each of these different collections of objects or samples the probability

$$\frac{1}{\binom{N}{m}}$$

It is not meaningful to apply the repeated-experiments reasoning here directly to find probabilities. The sampling procedure is most simply seen as a simultaneous selection of the m elements from the N. However, if each sample under this model is seen as the union of all of the ordered samples having the same collection (unordered) of objects, then we can apply the repeated-experiments logic to check our probability:

$$\frac{1}{\binom{N}{m}}$$

For example, if $N = 5$, $m = 2$, we repeat the experiment twice starting with five possible selections on the first experiment. The elements a_1 on the first experiment have probabilities of $\frac{1}{5}$ and the ones remaining, a_2, on the second experiment have probabilities of $\frac{1}{4}$. Therefore, the probability of *any* pair of elements a_1 and a_2, that is, any event in S_2, has

$$P(a_1, a_2) = P(a_1)P(a_2) = \tfrac{1}{5} \times \tfrac{1}{4} = \tfrac{1}{20}$$

There are two events in S_2 that have the same objects. The union of these events has a probability equal to the sum of the probabilities of the events (by definition of events and probability of unions of events). Since they are equal, the probability is $2(\frac{1}{20}) = \frac{1}{10}$. This is the probability, then, of an unordered collection of two distinct objects. This value checks out correctly.

$$\frac{1}{\binom{5}{2}} = \frac{1}{(5 \times 4 \times 3 \times 2 \times 1)/(3 \times 2 \times 1 \times 2 \times 1)} = \frac{1}{10}$$

A more formal proof is not given here.

A great deal of importance is attached to the binomial coefficients in the next few chapters. This model of sampling is found quite often in scientific work and leads, in the following discussions, to the most widely used model in statistical theory and practice.

Unordered Sampling with Replacement. We shall not elaborate on this model (interested readers should refer to D. W. Stilson, *Fundamentals of probability and statistics for psychologists*, San Francisco: Holden-Day, 1965, or to W. Feller, *An introduction to probability theory and its applications*, New York: Wiley, 1960). If we sample with replacement but are unconcerned with order, then we have simple groups of objects in which the same object may be represented more than once. The number of ways of selecting m objects out of N without ordering and with replacement is

$$\frac{(N + m - 1)!}{m!\,(N - 1)!}$$

Summary. We can summarize these equations by filling the table of sampling with the equations for number of events in the sample spaces S_m. We begin with a population of N elements for an experiment and sample m of them (or repeat the experiment m times, as the case may be). The number of events in S_m, $n(S_m)$, as given under the four models is indicated in Figure 6.2.

The general scheme of things is to define (1) the number N of elements in the population, (2) the number of objects selected from the N as m, (3) the sample

	Replacement	Nonreplacement
Ordered Samples	N^m	$\dfrac{N!}{(N - m)!}$
Unordered Samples	$\dfrac{(N + m - 1)!}{m!\,(N - 1)!}$	$\dfrac{N!}{m!\,(N - m)!} = \binom{N}{m}$

FIGURE 6.2

Number of events in sample spaces for four experimental models.

space S defined by the sampling method, N and m, is composed of the $n(S)$ different ways the sampling can come out, that is, the number of events in S, (4) the probability of each event in S, $P(e)$. The given experiment and sampling method specifies the values involved. Under the general condition of random sampling, every selection possible at a given stage of the experiment has the same probability as any other possible selection and we have four basic models. The three that we discussed in detail are the following.

1. Ordered sampling without replacement. Select m objects from N one at a time, not allowing the same object to be selected a second time:

$$n(S) = \frac{N!}{(N-m)!} \qquad P(e) = \frac{(N-m)!}{N!}$$

2. Ordered sampling with replacement. Select m objects from N one at a time, allowing an object to be selected as often as it comes up:

$$n(S) = N^m \qquad P(e) = N^{-m}$$

3. Unordered sampling without replacement. Select m objects from N simultaneously (without respect to order):

$$n(S) = \frac{N!}{(N-m)!\, m!} = \binom{N}{m} \qquad P(e) = \frac{(N-m)!\, m!}{N!}$$

SUGGESTED READING

David, F. N., and Barton, D. E. (1962), *Combinatorial chance*, Hafner, New York.

Feller, W. (1950), *An introduction to probability theory and its applications*, Wiley, New York.

Goldberg, S. (1960), *Probability theory, an introduction*, Prentice-Hall, Englewood Cliffs, N.J.

Parzen, E. (1960), *Modern probability theory and its applications*, Wiley, New York.

Stilson, D. (1965), *Fundamentals of probability and statistics for psychologists*, Holden-Day, San Francisco.

GLOSSARY

Population. The collection of things that can be observed on a repetition of an experiment; the sample space of a single repetition of an experiment; the collection of objects and events from which a sample is made.

Sampling with Replacement. In repeated sampling with replacement an element selected from the population on a previous trial is in the population and may be selected on successive trials.

Sampling Without Replacement. In repeated sampling without replacement an element selected from the population on a previous trial is no longer in the population and may not be selected on successive trials.

Ordered Samples. In sampling, the elements selected from the population are recorded in the order in which they are selected.

Unordered Samples. In sampling, the elements selected are a simple collection, the order of selection not being considered.

Ordered Sampling with Replacement. In ordered sampling with replacement, with m trials and N elements in the population, there are N^m different events in the sample space, S_m.

Ordered Sampling Without Replacement. In ordered sampling without replacement from a population of N elements and m repetitions there are $[N!/(N-m)!]$ elements in the sample space S_m.

Unordered Sampling Without Replacement. In unordered sampling without replacement from a population of N elements with m element samples, the number of samples is the binomial coefficient $\binom{N}{m} = \dfrac{N!}{(N-m)!\,m!}$.

Unordered Sampling with Replacement. In unordered sampling with replacement the number of samples in the sample space is $(N+m-1)!/[(m!\,(N-1)!]$.

(6.1) $n(S_m) = N^m$, ordered sampling with replacement.

(6.2) $n(S_m) = N!/(N-m)!$, ordered sampling without replacement.

(6.3) $\binom{N}{m} = \dfrac{N!}{(N-m)!\,m!}$, binomial coefficient, unordered sampling without replacement.

PROBLEMS

1. The power set of a set is the collection of all subsets of the set, including the set itself and the null set. How many subsets are there in a power set of a set with n elements? Argue from the following statement: There are two conditions for each element in the set; it is in the subset or it is not in the subset. Hence, there are two possibilities for a given subset for all n elements.

2. Evaluate the following:

(a) $8!$

(b) $\dfrac{6!}{2!\,4!}$

(c) $\dbinom{6}{2}$

(d) $\dbinom{6}{4}$

(e) $\dbinom{8}{8}$

(f) $\dbinom{8}{0}$

(g) $\dbinom{6}{3}\dbinom{8}{6}$

(h) $\dbinom{7}{3}\dbinom{7}{4}$

3. The number $\dbinom{N}{m}$ was called the binomial coefficient. Illuminate this by evaluating the number for $N = 5$ and $m = 0, 1, 2, \ldots, 5$. Compare these numbers with the coefficients in $(p + q)^5$.

4. A favorite trick of probability theorists is to calculate the opposite of what is required and then proceed to figure the required result. Often it is possible to calculate the number of ways a certain combination will not be observed and the number of combinations in total. The difference gives the number of ways the certain combination will occur. Use this information to solve the following problem. Four dormitory wings each house 25 students. Four of these in each wing are graduate student counselors. From each wing, one student is selected to participate in a survey concerning attitudes on dormitory life. What is the probability that the four selected students are all graduate counselors?

5. In the expansion of the binomial expression $(p + q)^n$ let $p = 1$ and $q = x$. This gives

$$(1 + x)^n = \sum_{r=0}^{n} \binom{n}{r} x^r = \binom{n}{0} + \binom{n}{1} x^1 + \binom{n}{2} x^2 + \cdots + \binom{n}{n} x^n$$

Show that if $x = 1$ the resultant sum is equal to 2^n. This is interpretable as showing that the number of subsets that can be formed out of a set of n elements is equal to 2^n.

6. Prove the following equalities:

(a) $\dbinom{n}{r} = \dbinom{n-1}{r-1} + \dbinom{n-1}{r}$

(b) $\dbinom{n}{r} = \dbinom{n}{n-r}$

(c) $r\dbinom{n}{r} = n\dbinom{n-1}{r-1}$

(Hint: Rewrite the binomial coefficients in terms of factorials and reduce the right-hand side to the left-hand side.)

7. A jar contains 3 white and 7 red balls. Two balls are drawn with replacement (that is, the first ball drawn is replaced before the second ball is drawn). Assuming that all the 100 possible choices for the first and second ball are equally likely, find the probability that the second ball is red.

8. From a group of 5 women and 15 men, W women and M men are selected at random (equally probable chances for each of the 20 people). Does every person in the group have the same probability of being included in the sample if (a) $W = 2$, $M = 2$, (b) $W = 1$, $M = 3$?

9. A psychologist is studying problem-solving ability differences in children in the fourth, sixth, and ninth grades. He plans to submit the following task to each subject in each group. The child is given a deck of seven cards, on each of which there is a different picture of some farm animal. The cards will be shuffled by the psychologist and placed in a row (picture up) in front of the child. Two of the cards will have black spots on the back sides. The child will be told that he must choose the cards with the black spots on the back. The child then turns over cards until both spots show up. The procedure is repeated for N trials. At first the child will turn over cards "randomly," and the psychologist needs to know the probabilities associated with the different events. Noting that a card is left turned over after a child chooses it, what is the probability that (a) the first card turned over has a spot on it, (b) the first three cards turned over do not have spots on them, and (c) only one of the first three cards has a spot on the back side?

10. A graduate student in social psychology wishes to conduct an experiment. However, because of the dearth of subjects available he must use subjects who participated in an experiment conducted by his adviser. There are 9 subjects available. However, two of these subjects are undergraduate "stooges" used in the adviser's research. The adviser will not tell the graduate student who the stooges are. The experimenter needs M subjects and selects them at random from the subject pool available by one of three procedures: (1) without replacement, (2) with replacement, (3) after each selection the subject's name is replaced twice in the subject population. For each of these three procedures answer the following two questions. If $m = 2$, what are the probabilities that the sample contains two stooges, one stooge, and no stooges (3 problems)? Calculate the probabilities when $m = 3$.

11. There are four clinical psychologists, Freud, Watson, Hull, and Ebbinghous, at a state hospital. Because of the expense of processing incoming patients, only one of the psychologists is able to classify each patient. The superintendent has ordered a follow-up study of the classification procedures. The follow-up is performed on 16 patients, 4 for each diagnostician. Out of a total of 4 wrong diagnoses, 3 were attributed to Dr. Freud. Is Dr. Freud justified in stating that his high frequency of misclassification is due to chance? Why?

12. Of a sequence of ten responses in a two-choice learning experiment the four correct responses all occurred on adjacent trials. Is there any reason to believe that this is not simply a chance occurrence? Frame your argument in terms of the probability of observing such an event if the correct responses occurred at random.

CHAPTER 7

Random Variables and Probability Distributions

In the preceding chapters we have specified events by some quantitative charac-teristics, for example, the number of heads in three tosses of a coin. If we dealt directly with these numbers, it would be unnecessary to deal with the nominal sample space events one at a time in order to solve the problem. The fact that $\{HHT\}$, $\{HTH\}$, and $\{THH\}$ individually satisfy the condition "two heads" is largely irrelevant. It makes no difference whether we observe $\{HHT\}$ instead of $\{THH\}$. What we are concerned with is that two heads are observed. To be sure, we could use the fact that there are three separate events (out of eight) having two heads in determining the probability of observing "2" (that is, two heads). How-ever, there are simpler and more elegant ways of calculating that

$$P(\text{number of heads} = 2) = \tfrac{3}{8}$$

This chapter is concerned with developing the theory and methods of dealing with the numerical characteristics of random experiments and the probabilities of observing the respective numbers. That is, we develop the theory of random variables, probability distributions of random variables, and methods of calculating probability distributions of random variables.

Random Variables

A random variable X is a quantity such that its value is determined by a random experiment. For each event in the sample space (that is, each different outcome of the random experiment) we determine a specific value of the quantity X. The same value may be associated with more than one of the events.

The concept of function can be used to define a random variable in a more satisfactory way. *A random variable is a single-valued function of the events in a sample space:* X *is a rule attaching to each event* e *in the sample space* S *some value* x, $X(e) = x$. The function is single valued in the sense that one and only one value is attached to a given event.

The collection of values that the function X associates with the events in S is the value set of the random variable. This set is designated

$$\{\text{values assigned by } X\} = \{x \mid X(e) = x, e \in S\}$$

or, where it is understood that $e \in S$,

$$\{x \mid X(e) = x\}$$

Our discussion of random variables here is limited to *discrete* random variables. All of the comments about discrete numerical sample spaces in Chapter 4 apply here. All quantitative, that is, numerical, sample spaces are random variables. The limitation to discrete variables is carried one step further, for simplicity, in the examples. Only integers are used. All of the sample spaces, random variables, of examples will be composed of sets of the class of numbers $(0, 1, 2, \ldots)$. In Chapter 9 *continuous* random variables are introduced.

Examples

In testing intelligence with a ten-item test, the sample space may be specified by listing all the possible combinations of true and false answers. However, the specific questions answered correctly are of little concern. What is of concern is the *number* of correct answers. Thus, we do not have to deal with the $2^{10} = 1024$ different possible events. The 11 different "scores," the number of correct answers, serve to specify everything we need to know. A random variable X is defined in this way: Each event in the sample space is assigned the value equal to the number of correct answers in the event; $X(e) = x = $ number of correct answers in e and the value set of X is given by $\{x \mid X(e) = x\} = \{0, 1, 2, \ldots, 10\}$.

The number of different random variables that can be defined for a given sample space is limited virtually only by the imagination of the theorist. For example, we can construct ten different random variables on the sample space of the example just given by the following device. If the ith question is answered correctly in the event e, then $X_i(e) = 1$; otherwise $X_i(e) = 0$. Thus, for X_1 only the answer to question 1 is considered. If e is composed of correct and incorrect answers such that question 1 is answered correctly, then $X_1(e) = 1$. If for the same e the answer to question 1 had been false, then $X_1(e)$ would have been equal to zero. This defines a random variable $X_1(e)$ with value set $\{x \mid X_1(e) = x\} = \{1, 0\}$.

Now the same thing is done over again substituting question 2 where question 1 was involved in $X_1(e)$. This gives a second random variable $X_2(e)$ where $X_2(e) = 1$ when question 2 is answered correctly and $X_2(e) = 0$ when question 2 is answered incorrectly. The value set of X_2 is again $\{0, 1\}$. The random variables X_1 and X_2 are not the same even though their value sets are identical. The values in the value set are assigned to the events according to different rules, and thus the random variables are different. The events to which specific values are assigned make up different sets for X_1 and X_2.

Following this scheme for each of the ten questions defines ten random variables $X_1, X_2, \ldots, X_{10}$. Many more random variables can be defined in S. The student is urged to think of several more.

Indicator Random Variables

The random variables $X_1, X_2, \ldots, X_{10}$ in the example above each had the value set $\{1, 0\}$. Any random variable with that value set is called an indicator random variable. In an intuitive sense an indicator random variable "indicates" whether or not (1 or 0, yes or no) the event belongs to a class. Any time a random variable assigns values to events by a counting process, for example, number of heads in five tosses of a coin, or the number of questions answered correctly, then the random variable can be analyzed as a sum of indicator random variables. This feature of "counting random variables" is used in examples worked out in later chapters.

Probability and Random Variables

The entire structure of probability theory is directly applicable to the theory of random variables. Since random variable values are associated with events, the probabilities associated with these values are easily determined. Recall the basic theorem: For any outcome

$$E = \{e_1, e_2, \ldots, e_k\}, \; P(E) = P(e_1) + P(e_2) + \cdots + P(e_k)$$

If E is defined such that it is composed of all the events to which a specific value x of the random variable X is attached, then the probability of observing that value when the experiment is performed is equal to the probability of E. That is, E is defined as $\{e \mid X(e) = x\}$, which includes all events such that x is the value of the random variable attached to the event. Then $P(X = x) = P(E)$. For example, if $S = \{e_1, e_2, e_3, e_4, e_5, e_6\}$ and

$$X(e_1) = 4 \qquad X(e_2) = 4 \qquad X(e_3) = 6$$
$$X(e_4) = 5 \qquad X(e_5) = 4 \qquad X(e_6) = 2$$

and we set $x = 4$, then $E = \{e \mid X(e) = 4\} = \{e_1, e_2, e_5\}$ and

$$P(E) = P(e_1) + P(e_2) + P(e_5)$$

The probability of observing the random variable value $x = 4$ is the probability of observing any one or another of the probabilities of the events to which the random variable attaches the value 4. Imagine that $P(e_1) = .10$, $P(e_2) = .30$, and $P(e_5) = .15$; then $P(E) = .10 + .30 + .15 = .55$. The probability that the observed value of the random variable is 4 is $P(X = 4) = P(E) = .55$.

The definition of random variables specifies that a given event be assigned one and only one value and that each event is assigned a value from the value set. These features of random variables, together with the method of calculating the probability of the values of the random variables, lead to two theorems.

THEOREM 1. *For the random variable* X, $P(X = x) \geq 0$ *for each* $x \in \{x \mid X(e) = x\}$.

Proof. Since $P(X = x) = \sum_e P(e)$ for $e \in \{e \mid X(e) = x\}$ and since $0 \leq P(e)$, the smallest value $P(X = x)$ is achieved when $P(e) = 0$ for all $e \in \{e \mid X(e) = x\}$, which gives $P(X = x) = 0$ since $\sum_e P(e) = (0 + 0 + \cdots + 0) = 0$.

THEOREM 2. *For the random variable* X, $\sum_x P(X = x) = 1$ *for* $x \in \{x \mid X(e) = x\}$.

Proof. Since each $e \in S$ is assigned one and only one value $x \in \{x \mid X(e) = x\}$, then for each value of the value set $\{x_1, x_2, \ldots, x_m\}$ we have outcomes $E_1 = \{e \mid X(e) = x_1\}, \ldots, E_m = \{e \mid X(e) = x_m\}$ where all E's are mutually exclusive, and $E_1 \cup E_2 \cup \cdots \cup E_m = S$. It follows that $P(X = x) = \sum_{i=1}^m P(E_i) = \sum_{e \in S} P(e)$ by definition of probability for events in S.

TABLE 7.1 Organization of a Model for the Experiment "Toss a Coin Three Times"

S Sample Space	Probability Distribution of S	$X(e)$ = Number of Heads	Value Set of X	Probability Distribution of X
$e_1: HHH$	$P(e_1) = \frac{1}{8}$	$X(e_1) = 3$	$x = 3$	$P(X = 3) = P(e_1) = \frac{1}{8}$
$e_2: HHT$	$P(e_2) = \frac{1}{8}$	$X(e_2) = 2$		
$e_3: HTH$	$P(e_3) = \frac{1}{8}$	$X(e_3) = 2$	$x = 2$	$P(X = 2)$
				$= P(e_2) + P(e_3) + P(e_4) = \frac{3}{8}$
$e_4: THH$	$P(e_4) = \frac{1}{8}$	$X(e_4) = 2$		
$e_5: HTT$	$P(e_5) = \frac{1}{8}$	$X(e_5) = 1$		
$e_6: THT$	$P(e_6) = \frac{1}{8}$	$X(e_6) = 1$	$x = 1$	$P(X = 1)$
				$= P(e_5) + P(e_6) + P(e_7) = \frac{3}{8}$
$e_7: TTH$	$P(e_7) = \frac{1}{8}$	$X(e_7) = 1$		
$e_8: TTT$	$P(e_8) = \frac{1}{8}$	$X(e_8) = 0$	$x = 0$	$P(X = 0) = P(e_8) = \frac{1}{8}$

Thus the theory of the probability of events in sample spaces carries over directly to the theory of random variables. In particular, the probability assignment in the sample space S is directly involved in the probability assignments on the random variable. Where probability in a sample space is the set of probabilities of events, the probability assignments for the random variable is the set of probabilities associated with the values of the random variable. Table 7.1 shows a model for the experiment of three tosses of a coin, illustrating these relationships.

The collection of probabilities assigned to the values in the value set of a random variable is referred to as the probability distribution of the random variable. The usage has a natural but nontechnical interpretation: The probability is distributed among the values in the value set. In the example of Table 7.1 the value set of the random variable is $\{0, 1, 2, 3\}$. The probabilities associated with these values, that is, $P(X = 0) = \frac{1}{8}, \ldots, P(X = 3) = \frac{1}{8}$, are terms in the probability distribution of the random variable.

A Special Case: The Binomial Random Variable

If an experiment is defined so that (1) it has only two possible events, (2) it is repeated a number of times, (3) each repetition is independent of the others, and (4) the probability of each observation does not change from one repetition to the next, the experiment is called a binomial experiment. Each repetition of the experiment is called a trial. For N repetitions, the experiment is said to consist of N

binomial trials. One of the two possible outcomes of a binomial trial is customarily called "success," and the other "failure." This usage has the value of not depending on specific wording of what a "success" is in an experiment.

The usual convention is to denote the probability of success as p: $P(\text{success}) = p$. Since there are only two observations possible, it follows that $P(\text{failure}) = 1 - p = q$. Applying this to a coin-tossing experiment, we see that for "heads" taken as "success" $P(\text{heads}) = p$, $P(\text{success}) = p = .5$ and $P(\text{tails}) = P(\text{failure}) = 1 - p = q = .5$. If we toss the coin five times in succession, the possible observations satisfy all the requirements of binomial trials (the student should verify this). Imagine one of the specific events of the experiment. Denoting a success on the ith trial as S_i and a failure on the kth trial as F_k, we might observe the event

$$\{S_1, F_2, S_3, S_4, F_5\}$$

Since the trials are independent, the multiplication law for independent events implies

$$P(S_1, F_2, S_3, S_4, F_5) = P(S_1) \times P(F_2) \times P(S_3) \times P(S_4) \times P(F_5)$$

Furthermore, because of the constancy of p and q over trials it follows that

$$P(S_1, F_2, S_3, S_4, F_5) = p \times q \times p \times p \times q = p^3 q^2$$

This probability checks with the result obtained by earlier methods of calculating probabilities. The event $\{HTHHT\}$ is one of $2^5 = 32$ different events in the sample space arising from the experiment. From earlier work we get $P(HTHHT) = \frac{1}{32}$. This, however, is equal to $p^3 q^2 = (\frac{1}{2})^3 (\frac{1}{2})^2 = \frac{1}{32}$.

Notice that in the derivation of the probability of $\{S_1, F_2, S_3, S_4, F_5\}$ there lies a general method for calculating the probability of any combination of successes and failures out of the number of trials involved. Thus, for the event $\{S_1, S_2, S_3, F_4, S_5\}$, $P(S_1, S_2, S_3, F_4, S_5) = p^4 q^1$. In general, for n trials of which any s are successes (and consequently the remaining $(n - s)$ are failures), the probability of any single specified event e is $P(e) = p^s q^{n-s}$.

The coin-tossing experiment involves a special case with $p = .5$. In this case $P(e) = p^s q^{n-s} = p^n = (\frac{1}{2})^n$, the same result achieved by assuming equally likely cases in the sample space. However, if $p \neq .5$, this relationship no longer holds. Imagine that the intelligence test of the example above is a multiple-choice test with four choices for each of the ten questions, only one choice being correct. If we assume that the questions are written so that they are independent (that is, the way one question is answered does not influence the way another is answered), we see that when a subject is guessing on all questions the test as a whole is a set of ten binomial trials, with $p = \frac{1}{4}, q = \frac{3}{4}$. The probability that a combination of responses will be correct on only the third, fourth, sixth, and tenth questions is

$$P(F_1, F_2, S_3, S_4, F_5, S_6, F_7, F_8, F_9, S_{10})$$
$$= P(F_1) \times P(F_2) \times P(S_3) \times P(S_4) \times P(F_5)$$
$$\times P(S_6) \times P(F_7) \times P(F_8) \times P(F_9) \times P(S_{10})$$
$$= (\tfrac{3}{4}) \times (\tfrac{3}{4}) \times (\tfrac{1}{4}) \times (\tfrac{1}{4}) \times (\tfrac{3}{4}) \times (\tfrac{1}{4}) \times (\tfrac{3}{4}) \times (\tfrac{3}{4}) \times (\tfrac{3}{4}) \times (\tfrac{1}{4})$$
$$= (\tfrac{1}{4})^4 \times (\tfrac{3}{4})^6$$

The number of successes, S, in binomial trials is the binomial random variable. Thus the binomial random variable is a counting random variable that ignores the order (or trial number) of successes and failures.

For ten binomial trials there is only one way for all of them to be successes. The probability of this event is $P(10 \text{ successes}) = p^{10}$. However, there are $\binom{10}{2}$ different events for which there are two successes (unordered sampling without replacement—sample from the ten trials to get two for the success trials). Thus, to compute the probability of observing two successes we follow the addition rule for events and sum the probabilities of each of the $\binom{10}{2}$ different events having two successes. The probability for *any* event composed of a combination of two successes and eight failures is p^2q^8. Thus the probability of observing one or another of these $\binom{10}{2}$ events is $\binom{10}{2}p^2q^8$. Doing the same thing for each different possible number of successes, 0, 1, 2, . . . , 10, gives

$$P(\text{no successes}) = \binom{10}{0} p^0 q^{10}$$

$$P(\text{one success}) = \binom{10}{1} p^1 q^9$$

$$P(\text{two successes}) = \binom{10}{2} p^2 q^8$$

$$P(\text{three successes}) = \binom{10}{3} p^3 q^7$$

$$\vdots \qquad \qquad \vdots$$

$$P(k \text{ successes}) = \binom{10}{k} p^k q^{10-k}$$

$$\vdots \qquad \qquad \vdots$$

$$P(\text{ten successes}) = \binom{10}{10} p^{10} q^0$$

In general, $P(S = s \mid n$ binomial trials with probability p of success) is given by the expression

$$P(S = s) = \binom{n}{s} p^s q^{n-s} \tag{7.1}$$

This set of probabilities, for $\{0, 1, 2, \ldots, n\}$, is the binomial probability distribution of the binomial random variable. Each term of this distribution corresponds to a term in the binomial expansion of $(p + q)^n$. The student should review the algebra of the binomial expansion at this point. Table B of the appendix at the back of the book gives the probabilities for $n = 2, \ldots, 20$ and p of .05 through .50 in steps of .05.

The binomial probability distribution depends on two quantities p and n in the sense that each term of the distribution $P(S = s)$ for $\{s \mid S(e) = s\} = \{0, 1, \ldots, n\}$ can be calculated if and only if p and n are known. No other information is necessary or used to calculate $P(S = s)$. To simplify the problem of notation we refer to the binomial probability distribution by writing $B(n, p)$, implying $B(n, p) = \{P(S = 0), P(S = 1), P(S = 2), \ldots, P(S = n)\}$.

Probability Distributions of Random Variables: Probability Functions

Thus far the discussion of probability for random variables has been limited to statements of the probability of given values in the value set of the random variable. The concept of probability distributions on random variables can be formalized and generalized. We start with the same basic concept used to define probabilities on sample spaces: To each element in the value set $\{x \mid X(e) = x\}$ of the random variable X is attached a value given by the function $f(x)$: This function is called the probability function and corresponds to the following rule. For each value x of $\{x \mid X(e) = x\} = \{x_1, x_2, \ldots, x_n\}$, for example x_i, find the outcome in the sample space on which X is defined, with E such that

$$E = \{e \mid X(e) = x_i\} = \{e_{i1}, e_{i2}, \ldots, e_{im}\}$$

The set E is thus composed of all (and only) events to which the value x_i of the random variable X is attached. And we know from earlier results that

$$P(E) = P(e_{i1}) + P(e_{i2}) + \cdots + P(e_{im}) = \sum_{j=1}^{m} P(e_{ij})$$

This can be written

$$P(e \mid X(e) = x_i)$$

and for simplicity of notation we substitute the function symbol to get

$$f(x_i) = P(e \mid X(e) = x_i)$$

To clarify this definition and to produce an example, apply it to the experiment of rolling a six-sided die twice, recording the number of points on the upper face after each roll. This experiment gives the sample space

$$S = \{(1, 1), (1, 2), \ldots, (2, 1), (2, 2), \ldots, (6, 1), \ldots, (6, 6)\}$$

Now, define a random variable X on S by the following function of the outcomes in S:

$$X(e) = \text{number of points total}$$

This produces a value set

$$\{x \mid X(e) = x\} = \{2, 3, 4, \ldots, 12\}$$

By counting the number of ways each of these values, x, can occur (that is, $n(E)$, $E = \{e \mid X(e) = x\}$ for $x = 2, 3, \ldots, 12$), we get the probability function of Table 7.2. Such a table is called a probability table.

The function $f(x)$ can be plotted against x in a two-coordinate graph. In this case we speak of a probability graph and plot only the points themselves (lines are drawn from x to the point $[x, f(x)]$ to accent the graph). The graph of the die-tossing probability function is given in Figure 7.1. Some values $\{x \mid X(e) \neq x\}$ are shown in Figure 7.1 to emphasize the fact that if a value x' is not in the value set of X, then $P(X = x') = 0$.

A more widely used method of plotting a probability function is the histogram. If we represent the probability, $f(x) = P(X = x)$, associated with x as a bar above x, our representation is a histogram. For example, take $f(4)$ in the probability graph of Figure 7.1. If we draw in a bar, the sides of which are vertical lines touching at $x = 3.5$ and $x = 4.5$ and the height of which is equal to $f(4)$, then the area of the box is equal to $f(4)$. Thus the probability of observing $x = 4$ is represented as an area under a steplike curve representing $f(x)$. The adjacent bars have common sides, and the total area under the curve is equal to 1. The student should demonstrate this to his own satisfaction. A histogram corresponding to the probability graph of Figure 7.1 is given in Figure 7.2.

TABLE 7.2 Probability Table for the Experiment in Which the Sum of Points on Two Rolls of a Die Defines the Variable X

x	2	3	4	5	6	7	8	9	10	11	12
$f(x)$	$\frac{1}{36}$	$\frac{2}{36}$	$\frac{3}{36}$	$\frac{4}{36}$	$\frac{5}{36}$	$\frac{6}{36}$	$\frac{5}{36}$	$\frac{4}{36}$	$\frac{3}{36}$	$\frac{2}{36}$	$\frac{1}{36}$

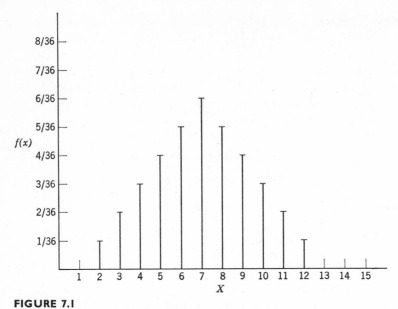

FIGURE 7.1

Probability graph of the probability function of Table 7.2.

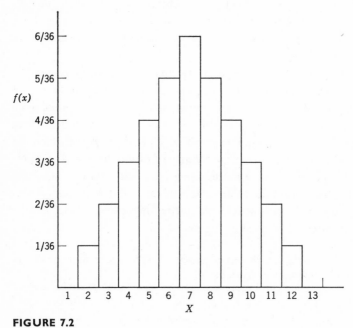

FIGURE 7.2

Histogram corresponding to the probability function in Table 7.2 and Figure 7.1.

The definition of the probability function of a random variable leads to two basic results. These theorems are the direct analogs to the defining characteristics of probabilities.

THEOREM 3.
$$f(x) \geq 0 \quad for \quad x \in \{x \mid X(e) = x\}$$
$$f(x) = 0 \quad for \quad x \in \{x \mid X(e) = x\}$$

THEOREM 4.
$$\sum_x f(x) = 1 \quad where \quad x \in \{x \mid X(e) = x\}$$

Probability Distributions of Random Variables: Distribution Functions

Frequently (and generally in the theory of statistics) we are concerned with collections of values of a random variable. For example, we might want to know the probability of observing six or more points in two rolls of a die, or the probability of correct answers on five or fewer test items. Questions of this sort can be answered easily and quickly when the distribution function of the random variable is known. The distribution function, denoted $F(x)$, is sometimes called the cumulative probability distribution, because as x increases $F(x)$ is the cumulation of the probabilities $f(x)$ of all values of X equal to or less than x.

For a given probability function $f(x)$ the distribution function is defined by

$$F(x) = P(x \mid X(e) \leq x) = \sum_{x' \leq x} f(x')$$
$$F(x) = P(X \leq x)$$

The distribution function value of x is the sum of probabilities of all values x' of X equal to or less than x. An equivalent definition in terms of the elements in the sample space is as follows. First, the set $E = \{e \mid X(e) \leq x\}$ is defined; it is all elements in S that have random variable values equal to or less than x. Then $P(E) = F(x) = \sum_{e \in E} P(e)$. Of course, where we have the probability function of the random variable the definition of $F(x)$ in terms of simple events would not be used in calculating $F(x)$.

Table 7.3 gives $F(x)$ in the die-tossing experiment. Note that each succeeding term $F(x)$ is larger by a value $f(x)$ than the preceding value. That is,

$$F(2) = f(2)$$
$$F(3) = F(2) + f(3) = f(2) + f(3)$$
$$F(4) = F(3) + f(4) = f(2) + f(3) + f(4)$$
$$F(5) = F(4) + f(5) = f(2) + f(3) + f(4) + f(5)$$

and so on.

TABLE 7.3 Distribution Function for the Die-Tossing Experiment—Corresponds to the Density Function Indicated in Table 7.2

x	2	3	4	5	6	7	8	9	10	11	12
$F(x)$	$\frac{1}{36}$	$\frac{3}{36}$	$\frac{6}{36}$	$\frac{10}{36}$	$\frac{15}{36}$	$\frac{21}{36}$	$\frac{26}{36}$	$\frac{30}{36}$	$\frac{33}{36}$	$\frac{35}{36}$	$\frac{36}{36}$

A graph of the distribution function is made by plotting the value $F(x)$ against x in two coordinates, as in Figure 7.3. The probability $F(x)$ remains constant for the values of x' where $x < x' < x + 1$, so that we draw a horizontal line at a height $F(x)$ between the points x and $(x + 1)$ to represent the fact that no probability is "added" anywhere between x and $(x + 1)$. In other words, the probability comes in "chunks" and thus $F(x)$ is a step function.

Several features of the distribution function are worthy of special mention. Since $\sum_x f(x) = 1$ where $x \in \{x \mid X(e) = x\}$, then for the largest x, say x', $F(x') = 1$. In the example, Figure 7.3, $F(12) = 1$. More generally, if

$$y \geq (\text{largest } x \mid X(e) = x)$$

then $F(y) = 1$. In our example any value of y equal to or greater than $x = 12$,

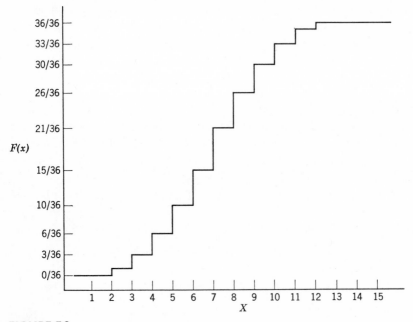

FIGURE 7.3

Graph of the distribution function of Table 7.3.

such as $x' = 20$, gives $F(x') = 1$. This follows from the fact that $f(13) = f(14) = \cdots = f(20) = 0$ or in terms of the simple events the set $E = \{e \mid X(e) > 12\} = \emptyset$ and $P(E) = P(\emptyset) = 0$.

At the other extreme, and essentially by the same reasoning, if $y <$ (smallest $x \mid X(e) = x$), then $F(y) = 0$.

The student should prove the following theorems concerning distribution functions, where X is a random variable and x, x', y', and y are values.

THEOREM 5. *If* y $\geq$ (*largest* x $\mid$ X(e) = x), F(y) = 1.0.

THEOREM 6. *If* y $<$ (*smallest* x $\mid$ X(e) = x), F(y) = 0.

THEOREM 7. f(x) = F(x) $-$ F(y) *for* y = x $-$ 1 (*where* x *and* y *are integers*).

THEOREM 8. *If* y $\geq$ x, *then* F(y) $\geq$ F(x).

THEOREM 9. P(y $<$ X $\leq$ y') = F(y') $-$ F(y)

THEOREM 10. P(X $>$ x) = 1 $-$ F(x)

The student should plot a distribution function and the implied probability function, and attempt to discover the graphic intuitive meaning of the above theorems in terms of these functions.

An Application—Concept Learning

An interesting contemporary development in the field of learning is the emergence of formal models. One of the most successful and engaging such models is the Bower and Trabasso model of concept learning. In this model a concept is a collection of attributes defining a stimulus set. For example, the stimuli might be composed of colored figures of various sizes. The colors might be red and green, the figures squares and circles, and the sizes large and small. A given stimulus would be, for example, a large red circle. A concept is equivalent to a class of stimuli defined in terms of one or another of the stimulus dimensions—all red figures, for example. In this example the large red circle, the small red circle, the large red square, and the small red square are exemplars of the concept. The other stimuli are nonexemplars of the concept: the large green circle, the small green circle, the large green square, and the small green square.

A concept-learning task consists of a series of presentations of individual stimuli from the collection, a guess on the part of the learner regarding the classification of the stimulus (is it an exemplar or a nonexemplar of the concept?), and an indication from the experimenter regarding the correct classification of the stimulus. The experiment consists in repeated presentation of the collection of stimuli until a number of errorless trials is observed. The data of interest are the sequences of correct and incorrect responses.

In the theory the attributes are called cues; that is, color is the relevant cue for distinguishing exemplars and nonexemplars in the example above. In the simplest version of the theory it is proposed that a subject, in the learning situation described, will select a cue and try it out until it leads to an error. If the cue color were selected, the subject would not make any errors and he would maintain this cue as the basis of his responses. However, if he were using size (for example, small) as the cue, he might be correct for a few trials because of the coincidence of red and small in a series of exemplars and green and large in a series of nonexemplars. However, when the small green figure (one of them) was encountered, the guess would be wrong and the subject would give up that cue. It is presumed, in addition, that the process of learning corresponds to selecting a cue from the collection of attributes (in the example, size, color, and shape) at random when an error is made. The collection of cues remains the same on each selection. Thus an abandoned cue might be selected for the trial following an error. If the subject has not selected the relevant cue, the probability of a correct response (guess) with the stimulus set described above is $q = .5$. The subject will continue to make correct responses with probability $q = .5$ for a series of trials. Until an error is made, with probability $(1 - q = .5)$ on any given trial, the subject will not resample from the set of cues. On an error trial (or rather before the next trial) a new cue is sampled. Imagine that the probability of selecting the relevant cue is c, which may be related to attributes such as the salience or attractiveness of the cue. The probability c is presumed to be constant over all of the trials.

Applying some probability considerations, it can be shown that the probability of an error on a given trial and the sampling of the correct cue for the following trials are independent, and consequently for the presolution trials

$$P(\text{solve} \mid \text{error}) = c$$
$$P(\text{solve} \mid \text{correct before solution}) = .0$$
$$P(\text{error}) = q$$

After the solution trial

$$P(\text{error}) = 0$$

$$P(\text{correct response}) = 1.0$$

Because of the complementary character of correct and error trials before solution,

$$P(\text{solve}) = P(\text{solve} \cap \text{error}) + P(\text{solve} \cap \text{correct})$$

The multiplication rule for intersections gives

$$P(\text{solve}) = P(\text{solve} \mid \text{error})P(\text{error}) + P(\text{solve} \mid \text{correct})P(\text{correct})$$

$$= qc + .0 = qc$$

Prior to solution the learning process is said to be in a presolution state, for example, S'. On sampling the relevant cue, the learning process is said to be in a solution state, such as S. The learning process starts in the presolution state. The probability of making an error in the presolution state is q. The probability of making an error *and* then sampling the relevant cue is qc. The probability of making an error *and* sampling some cue other than the relevant cue *or* of not making an error is $(1 - qc)$. The probability of making an error when in the solution state is 0, and the probability of not making an error when in the solution state is 1.0.

This development is sufficient to permit an evaluation of several random variables defined within the framework of the experiment. Define a random variable X_n such that $X_n = 0$ indicates a correct response on trial n and $X_n = 1$ indicates an error in trial n where trial n is a presolution trial. In order for the process to remain in the presolution state, the trial has to result in an error and a sampling of some cue other than the relevant cue or not making an error even though an irrelevant cue is sampled. This has a probability of

$$P(S_2') = 1 - qc$$

The probability of remaining in the presolution state for two trials is given by the probability of being there after the first trial and either not making an error on the basis of an irrelevant cue or making an error and sampling an irrelevant cue times the probability of beginning the second trial in the presolution state:

$$P(S_3') = (1 - qc)(1 - qc) = (1 - qc)^2$$

Arguing by induction, the probability of being in the presolution state on the nth trial is given by

$$P(S_n') = (1 - qc)^{n-1}$$

Only if the process is in the presolution state will an error occur and then only with probability q. Hence

$$P(X_n = 1) = qP(S_n')$$
$$P(X_n = 0) = (1 - q)P(S_n') + P(S_n)$$

and by substitution for $P(S_n')$ from above,

$$P(X_n = 1) = q(1 - qc)^{n-1}$$
$$P(X_n = 0) = (1 - q)(1 - qc)^{n-1} + 1 - (1 - qc)^{n-1}$$

Evaluating the distribution of X for a given n involves knowledge of q and c. A value for q can be determined by the method of presenting the stimuli and the attributes of the stimuli. For the example given above, $q = .5$ is a reasonable assignment. On the basis of some data from similar experiments, the value of c might be approximately .4, that is, $c = .4$. With these two probabilities the distribution of X_n for any value of n may be determined; a selection of these distributions is given in Table 7.4. On each trial the indicator variable X for error indication has a different probability distribution—dependent on the initial parameters specified by the model and the trial number.

One of the more useful and common statistics dealt with in learning experiments is the total number of errors made during the learning. Let the number of total errors be denoted T. If the process begins in state S, there must be one error. However, there is no necessary upper limit on the number of total errors. Consequently the value set for T is 1, 2, 3, 4, ... Deriving the probability for each value in the value set consists in applying the principles of the model. The basic principle in the model is that learning will occur only on error trials and that errorless trials have no effect on learning. Also, the probability that a given error is the last depends on sampling the relevant cue on that trial—this has a probability

TABLE 7.4 Probability: Distributions for "Correct," "Incorrect" on the First Five Trials in the Bower-Trabasso Model

Trial	Correct $X_i = 0$	Incorrect $X_i = 1$
$i = 1$	.500	.500
$i = 2$	.600	.400
$i = 3$	.680	.320
$i = 4$	.744	.256
$i = 5$	.795	.205

denoted c. Thus the probability that $T = 1$ is the probability that the subject learns on the first error, and $P(T = 1) = c$. The probability that $T = 2$ is given by the probability that the subject did not learn on the first error $(1 - c)$, times the probability that the subject learns on the second error, c. Thus, $P(T = 2) = c(1 - c)$. Similarly, to have $T = 3$ the first two errors must not result in learning and the third error must have resulted in learning: $P(T = 3) = c(1 - c)^2$. For $P(T = m)$ there must have been $(m - 1)$ error trials on which the subject did not learn before the mth error trial on which the subject learned. Consequently, the distribution may be written as

$$P(T = m) = c(1 - c)^{m-1} \qquad \text{for} \quad m = 1, 2, 3, \ldots$$

This equation defines the probability distribution for the variable T, the total number of errors. That this is a probability distribution can be shown by showing that each term in the distribution is nonnegative and that the terms sum to 1.0. Since these demonstrations would lead us afield of our primary purpose, the reader is referred to the original literature.

The probability distribution of T is tabled in Table 7.5 and graphed in Figure 7.4. The value of c is presumed to be .4. The figure does not indicate the entire

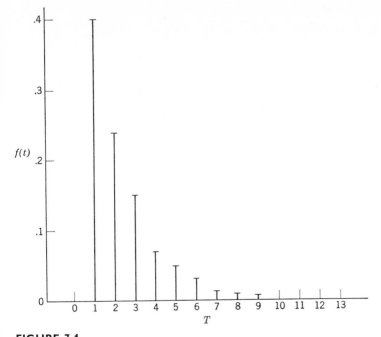

FIGURE 7.4

Probability function $f(t)$ for distribution in Table 7.5.

TABLE 7.5 Probability Function and Distribution Function of Total Number of Errors in the Bower-Trabasso Concept-Learning Model

t	1	2	3	4	5	6	7	8	9
$f(t)$	.400	.240	.149	.086	.052	.031	.019	.011	.007
$F(t)$	.400	.640	.784	.870	.922	.953	.972	.983	.990

value set, but only the values of T that have some appreciable probability. The distribution function of T is shown in Figure 7.5. Again, the distribution function for values of T having little probability are not indicated in the figure.

The probabilities in Table 7.5 are instructive regarding the meaning of probability of values in a random variable. The probability that $T = 9$ is very small, but larger than zero. Consequently, we would expect to observe $T = 9$ with a small

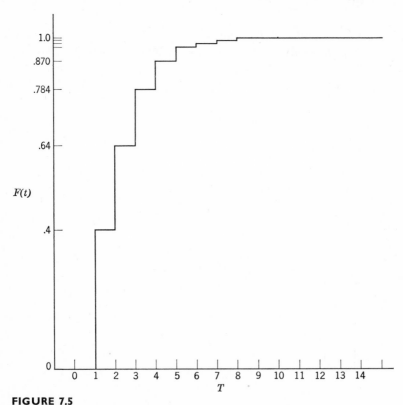

FIGURE 7.5

Distribution function $F(t)$ for distribution in Table 7.5.

frequency in even a large number of repetitions of the experiment of running a subject in the concept-learning experiment. However, it should be clear that observing the value $T = 9$ does not indicate that the model or the parameters q and c are not correct. It is quite possible to observe unusual things on any given repetition of the experiment.

The reader will note that $F(9) = .990$ indicates that $P(T > 9) = .010$ as a result of the definition of the distribution function. Similarly, it is evident from the distribution function that the preponderant number of errors made to solution are four or less, $F(4) = .870$.

The results of derivations—that is, Table 7.5—indicate the probability model of the observations that would be made if the experiment were done a large number of times. The model is dependent on several considerations. First, of course, are the psychological or process assumptions made about concept learning. These assumptions are stated in a rather straightforward way in terms of the equations fundamental in the derivations. Second are the assumptions made when the values of q and c were selected. If the assumptions of the model and the parameter estimation are correct for human concept learning, then experimental data should correspond to the figures implied by the probability distributions derived above. If the model or the parameters are not adequate, then the data from an experiment will not correspond to the distributions derived. Many complications and more subtle considerations must be taken into account before these bald statements are completely acceptable. However, the essence of the usefulness of the derivations is expressed in the statements.

More is said about the methods of deciding the support value or nonsupport value of data for a model when the topic of statistical decision theory is introduced.

SUGGESTED READING

Atkinson, R. C., Bower, G. H., and Crothers, E. J. (1965), *An introduction to mathematical learning theory*, Wiley, New York.

GLOSSARY

Random Variable. A random variable is a single-valued function of the events in a sample space: X is a rule attaching to each event e in the sample space S some value x, $X(e) = x$.

Value Set of a Random Variable. The value set of a random variable is the collection of values that the function X associates with the events in S.

Indicator Random Variable. An indicator random variable is a random variable with two values, one value indicating the presence of some attribute in an event, the other indicating an absence of that attribute.

Theorem 1. For the random variable X, $P(X = x) \geq 0$ for each $x \in \{x \mid X(e) = x\}$.

Theorem 2. For the random variable X, $\Sigma_x P(X = x) = 1$ for $x \in \{x \mid X(e) = x\}$.

Binomial Experiment. A binomial experiment is an experiment that has only two possible events for each experiment and is repeated a number of times; each repetition is independent of the others, and the probability of each observation does not change from one repetition to the next.

Binomial Trial. A binomial trial is one repetition of a binomial experiment.

Binomial Random Variable. The number of successes out of N repetitions of a binomial experiment.

$$(7.1) \qquad P(S = s) = \binom{n}{s} p^s q^{n-s}$$

Probability Function or Probability Distribution. A probability function or probability distribution of a random variable is the set of probability values associated with the values in the value set of the random variable, $f(x) = P(X = x)$.

Theorem 3.

$$f(x) \geq 0 \text{ for } x \in \{x \mid X(e) = x\}$$
$$f(x) = 0 \text{ for } x \notin \{x \mid X(e) = x\}$$

Theorem 4.

$$\sum_x f(x) = 1 \text{ for } x \in \{x \mid X(e) = x\}$$

Distribution Function. The distribution function of a random variable (cumulative probability function) is the cumulative probability function of the random variable:

$$F(x) = P(x \mid X(e) \leq x) = \sum_{x'} (x') \text{ for } x' \leq x$$
$$= P(X \leq x)$$

Theorem 5. If $y \geq$ (largest $x \mid X(e) = x$), $F(y) = 1.0$.

Theorem 6. If $y <$ (smallest $x \mid X(e) = x$), $F(y) = 0$.

Theorem 7. $f(x) = F(x) - F(y)$ for $y = x - 1$ where x and y are integers.

Theorem 8. If $y \geq x$, then $F(y) \geq F(x)$.

Theorem 9. $P(y < x \leq y') = F(y') - F(y)$.

Theorem 10. $P(X > x) = 1 - F(x)$.

PROBLEMS

1. Construct a probability table like that of Table 7.1 for four repetitions of the basic experiment.

2. Evaluate each of the terms in the binomial probability distributions $B(6, .5)$, $B(6, .8)$, $B(6, .2)$, and $B(6, .1)$. Prepare probability tables for each.

3. Draw graphs of the probability functions and distribution functions of the binomial distributions in problem 2.

4. What is the special relationship between the distributions $B(6, .8)$ and $B(6, .2)$? (Hint: Compare the upper values with the lower values across the two distributions.)

5. Prepare histograms for the distributions specified in problem 2.

6. Work out proofs, or at least explicit rationalizations, of Theorems 4 through 10 of this chapter.

7. Evaluate the following values for the concept-learning model:

$$F(12)$$
$$F(1)$$
$$P(T < 10)$$
$$P(T \leq 10)$$
$$P(3 < T < 6)$$
$$P(3 \leq T < 6)$$
$$P(3 < T \leq 6)$$
$$P(3 > T, \text{ or } 6 < T)$$
$$P(3 \leq T, \text{ but } 6 > T)$$

8. What is the value set of the random variable number of points on (a) a toss of three dice, (b) a toss of k dice?

9. For an experiment defined by three tosses of a fair coin, let X be the number of heads and Y be the number of tails. Show that these two random variables have the same distribution. Show that these two random variables are not the same random variables.

10. In an experiment on ESP the "clairvoyant" is to pick out marked cards from a set of eight cards three of which are marked. He picks three cards at a time. Let N denote the number of marked cards picked out by the subject. Assuming that the subject cannot really do better than a nonclairvoyant, determine the probability function of the random variable N. Construct a graph of the probability function and the corresponding distribution function.

CHAPTER 8

Expectations and Variances of Random Variables

It is often desirable to specify some feature of the distribution of a random variable without specifying all of the values in the value set and the associated probabilities. For example, what is the most likely value of a binomial random variable with distribution $B(n, p)$? How big a difference between two values of X might we expect to observe if two values were sampled? The answers of these questions involve the methods of description of random variables developed here. Each of the concepts introduced in this chapter is concerned with some distinctive aspect of random variables. All of these concepts have their parallels in empirical observations. The usefulness of a concept in statistics depends on being able to find its analog in the observations made in an empirical setting. However, as in the foregoing chapters, the distinction between a model of an experiment and the experiment must be maintained clearly or confusion will result. This chapter is devoted to the development of methods of dealing with some of the salient features of random variables and their distributions. Two properties of random variables, expectation and variance, are major theoretical elements in statistical methods.

"Location" of a Random Variable

How large are the values of a random variable?

Without some sort of reference point a reasonable answer to this question is not possible. In fact, two reference points (or frames of reference) are desirable. The first is suggested by the fact that random variables are real quantities. It is convenient to use the distinctive value zero (in an intuitive and measurement sense, the anchor point of all values of a quantity) as the reference point for the value set of a random variable. Each value of the random variable is, in this way, as "large" as it is different from zero in the real number system. This, however, is identical with the value of the quantity. If any other value had been used as the point of reference, this would not have been true. If, for example, we had used 100 as the point of reference, then the *magnitude* (how large?) of the quantity (random variable) would have been the value of the quantity minus 100. That is, if X is a quantity and x is a value of X, then (the magnitude of X) = $(x - 100)$. This would be perfectly legitimate, but somewhat complicated. Instead, we agree that

$$\text{(the magnitude of } X) = (x - 0) = x$$

The values of a random variable are used as a measure of its magnitude. In addition, this convention leads to a terminology that has an intuitive appeal. A quantity is "located" on the real line at a "distance" from zero equal to the value of the quantity. Thus the magnitude of a quantity (random variable) is associated directly with its location or position in the real number system.

The reader should have been sensitive in the preceding paragraphs to a serious difficulty of this method in describing the magnitude of a random variable. By definition there are a number of values of a random variable, and hence we may argue that the random variable has a number of locations or magnitudes. Unless we can devise a means of avoiding this problem, we have gained little. We develop here four ways to measure or represent the location of a random variable by a single value or at most a small number (such as two) of values.

Extreme Values. The two most extreme values of a value set of a random variable indicate the general location of all the other values. If the extremes are denoted as X_{min} and X_{max}, for the minimum or smallest value of X and the maximum or largest value of X, then for any value x_i of X, it follows that

$$X_{min} \leq x_i \leq X_{max}$$

This "locates" the random variable in the sense that all of the values of the quantity lie between two points, the locations of which are known precisely: X_{min} and X_{max}.

The magnitude of the quantity is thus bounded within an interval. For example, the magnitude of the random variable "number of points in two tosses of a die" is limited by $X_{\min} = 2$ and $X_{\max} = 12$. Any given value of X cannot be greater than 12 or smaller than 2.

At times this measure of location of a random variable may be satisfactory, or perhaps even all that we may be able to say with precision. However, we cannot be satisfied in general with the extreme-values definition of the location of a random variable. Three characteristics of this measure are particularly objectionable. First, only two values are involved, all the others ignored. This gives a misleading representation when one value is a black sheep, a value very unlike any of the others. Imagine a value set of a random variable $X = \{1, 2, 3, 4, 5, 6, 7, 8, 9, 10, 1000\}$. The values $(X_{\min} = 1)$ and $(X_{\max} = 1000)$ certainly do not represent the value set very well. Second, the probability of the values in X are ignored. The situation of the value set just written down is an example. Suppose that in the subset $\{1, 2, 3, 4, \ldots, 10\}$ of the value set, each value had a probability .09999 and that $P(X = 1000) = .0001$. To claim that the value set was located between $(X_{\max} = 1000)$ and $(X_{\min} = 1)$ obscures the fact that if X were observed, only 1 in 10,000 times would a value greater than 10 be observed (on the average). Third, it is very ungainly to manipulate the location of the random variable algebraically when it is defined in this way. It is not a quantity or a set or anything we know how to manipulate in an algebraic fashion. This limits its usefulness severely.

Perhaps the most telling of all the inadequacies of the extreme-values measure of location is the lack of desirable statistical properties of empirical estimates of this measure. In the random variable there is no uncertainty about the values $X_{\max}$ and $X_{\min}$. However, in a small number of observations on X in an empirical experiment the values of $X_{\max}$ and $X_{\min}$ may not be observed. Consequently, they have to be estimated from the sample of observations actually made. These estimates do not compare favorably with other estimates of other measures of location. In brief, the extreme-values measure of location is inadequate with respect to certain criteria involving statistical considerations. The statistical considerations are introduced in Chapter 11.

The Most Likely Values. Another measure of the location of the random variable is the most likely value of the value set of a random variable. This value is called the mode (the most "popular" value). In the case of the binomial random variable with distribution $B(4, .5)$ the mode is equal to 2 because $P(S = 2) = .3750$ is larger than the probability of any other value.

The mode is a measure of location in the sense that it is the value of the quantity most often observed. Probability is *the* determiner of the specific value "locating" the distribution—but only one probability. All values of the random variable that do not have the highest probability of all values are ignored. Thus, this measure of location, generally, is based on a small part of the value set. On the other hand, there may be many values of location of the random variable under this definition. For example, in $B(3, .5)$ the largest probability is associated with two values, $S = 1$ and $S = 2$, because $P(S = 1) = P(S = 2) = .375$. An even more extreme example showing the inadequacy of this measure is any random variable with equally probable values (uniformly distributed). All values have the same probability, and hence there are as many modes as values.

The mode is not easily manipulated algebraically. The value of the mode is not determined by a succinct algebraic expression that lends itself to manipulation. Also, the mode has failings similar to those of the extreme-values measure with regard to statistical properties.

The Frequency Center. A third measure of location, the median, is defined as the value of the quantity Mdn that falls at a point where $P(X < \text{Mdn}) = P(X > \text{Mdn}) = .5$. That is, the median is the value such that observations above the median would occur with the same frequency as observations below the median. Finding the median is complicated by several rules for interpolating when there is no single value obviously in the frequency center of the distribution. We do not reproduce here the "recipes" necessary to cook up medians. If you remember that they are all different ways of interpolating to satisfy certain peculiar value-set and probability distribution conditions *and* if you remember the following basic definition, you can derive the formulas easily. The median is the *smallest* value, say Mdn, for which $P(X \leq \text{Mdn}) \geq .5$.

The median suffers from the inadequacies of the mode as well as those just indicated.

Mathematical Expectation. Imagine that the value set of a random variable is a population. The probability distribution of the random variable indicates for each value the probability that, if a sample of one observation were drawn, the particular value would be drawn.

Now, repeat the experiment (with replacement) of drawing a sample of a single element from the population. Do this a very large number of times. The average of the values observed is informative as to how large the values of the random variable are. In fact, this average has some very nice properties. A discussion of these properties must be postponed, however. In the meantime we explore the implications of this suggestion for the location of the random variable.

Say that four scores on a psychological test are possible: 1, 2, 3, and 4. Also, assume that $P(X = 1) = .2$, $P(X = 2) = .3$, $P(X = 3) = .25$, and $P(X = 4) = .25$. If we observed these scores in a very large sample, we would observe approximately 20% ones, 30% twos, 25% threes and fours. For the sake of simplicity here (we do not make this assumption in the general case), assume that relative frequency and probability agree. Imagine that for 1000 observations the data of Table 8.1 were obtained.

The average of the 1000 observations, tabulated in Table 8.1, X_i for $i = 1, \ldots, 1000$, is given by

$$\bar{X} = \frac{1}{1000} \sum_{i=1}^{1000} x_i$$

$$= \frac{1}{1000} [200(1) + 300(2) + 250(3) + 250(4)]$$

This last expression indicates simply that the summation of all the observations includes 200 ones, 300 twos, 250 threes, and 250 fours and no more. Dividing through by the 1000 gives

$$\bar{X} = .2(1) + .3(2) + .25(3) + .25(4)$$

The values in this expression are the relative frequencies, not the probabilities. However, if the probabilities were substituted for the relative frequencies, the calculation and the value would be the same. This is true *only* when probability and relative frequency are equal numerically. Therefore, we want to distinguish between the equation, above, with relative frequencies and the average $\bar{X}$, and the corresponding equation with probabilities. When we put probabilities in the right-hand side of the last equations, we say that the resulting summation is the "expectation" of the random variable, $E(X)$:

$$E(X) = P(X = 1)(1) + P(X = 2)(2) + P(X = 3)(3) + P(X = 4)(4)$$

TABLE 8.1 Some Illustrative Data and a Model

Score	Frequency	Relative Frequency	Probability
1	200	.20	.20
2	300	.30	.30
3	250	.25	.25
4	250	.25	.25

This provides a description of the random variable parallel to the average of a sample of observations. It is idealized in the sense that it involves (1) the entire value set of the random variable and (2) the probability of each value. Here $\bar{X}$ involves relative frequency, and in general there is no assurance that the sample represents *all* the values in the value set of the random variable.

The mathematical expectation, $E(X)$ of a random variable X with value set $\{x_1, x_2, \ldots, x_k\}$ is formally defined by

$$E(X) = P(X = x_1)x_1 + \cdots + P(X = x_k)x_k$$

$$= \sum_{i=1}^{k} P(X = x_i)x_i \tag{8.1}$$

Recalling the definition of $P(X = x)$ and X, expectation can also be defined in terms of the events in a sample space, $S = \{e_1, \ldots, e_n\}$:

$$E(X) = X(e_1)P(e_1) + \cdots + X(e_n)P(e_n)$$

$$= \sum_{i=1}^{n} X(e_i)P(e_i) \tag{8.2}$$

Consider a binomial random variable. "What is the expectation of X with $B(10, .4)$?" "How many successes in a binomial experiment of ten trials with probability of .4 of success may be expected?" In the ordinary use of the word "expect," we can expect to observe one of the 11 possible numbers of successes. However, a description of the entire random variable with respect to the numbers of successes is given by the mathematical expectation: $E(S)$ is a measure of the location or magnitude of the random variable S. In the binomial example illustrated in Table 8.2 four successes out of the ten trials might be expected. The .005 discrepancy is due to rounding error.

This may be interpreted as meaning that if S is distributed as $B(10, .4)$, then in a long run of repetitions of the experiment (ten binomial trials is the experiment) the average number of successes will be approximately 4. Chapter 11 presents an extensive analysis of the relationship of the average and expectation.

The mathematical expectation of a random variable as a measure of location has many desirable properties. In particular, the serious defects of the three other measures we have discussed are avoided. All the values of the random variable and the associated probabilities are utilized. It is always single valued and it is algebraically manipulable. This last characteristic leads to the possibility of an exploration of the mathematical properties of expectation. Of particular interest are some statistical properties of expectation.

TABLE 8.2 Illustration of the Calculation of the Expectation of a Binomial Random Variable

Number of Successes s	$P(S = s)$ with $B(10, .4)$	$sP(S = s)$
0	.0060	.0000
1	.0403	.0403
2	.1209	.2418
3	.2150	.6450
4	.2508	1.0032
5	.2007	1.0035
6	.1115	.6690
7	.0425	.2975
8	.0106	.0848
9	.0016	.0144
10	.0001	.0010
$E(S) = \sum_{s=0}^{10} sP(S = s)$		4.0005

Properties of Expectation

Expectation as a Center of Gravity. One of the popular ways of attempting to give intuitive meaning to expectation is to demonstrate that it is the "balance" point of the distribution of the random variable. This demonstrates that the expectation is the "center" of the random variable (and hence expectation is sometimes referred to as a measure of central tendency).

Imagine a weightless plank. If you place a collection of objects on the plank, then some particular point on the plank is a balance point. If the weight of an object A is twice that of an object B and if they are placed on the plank at two different places, then the balance point is at a place one-third of the distance from A to B, as illustrated in Figure 8.1. The situation is more complex with several

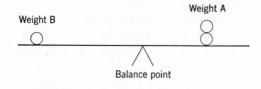

FIGURE 8.1

Illustration of a balance point.

objects. In order to keep track of the position of the objects and their weights, the plank is marked with a set of values (as, for example, the value set of a random variable). The magnitudes of these values are assumed to be defined relative to a value zero. The balance point of the plank now corresponds to some magnitude. That magnitude is determined both by the value (location) of the objects and by the weight of the objects. The balance point magnitude BP is given by

$$BP = \sum_{o} m(o)w(o)$$

where $m(o)$ is the magnitude (value) of the object and $w(o)$ is its weight. Now if we take as the objects the values of the value set of a random variable and assign them magnitudes equal to their values and weights equal to the probabilities associated with the values, the magnitude of the balance point is given by

$$BP = \sum_{x \in X} xP(X = x)$$

But for this special case $BP = E(X)$ and the expectation of a random variable with respect to its probability distribution is its balance point, or more precisely, the magnitude of its balance point.

Expectation as a Deviation Center. One of the consequences of the balance point characteristics of $E(X)$ is that the amount of weighted value on one side of the expectation is the same as on the other side of it. More is said about this property of expectation in the context of the discussion of measures of dispersion of random variables.

Laws of Expectation

The algebraic form of mathematical expectation leads to some general properties of expectation that are of value in later sections. These laws are introduced as theorems and proved.

THEOREM 1. *For any random variables* X *and* Y, E(X + Y) = E(X) + E(Y).

Before proving this theorem, we need to clarify what is meant by the sum of random variables. The most straightforward meaning can be found by investigating how the sum associates values with events in the sample space. Each random variable taken singly is defined by the association of values from its value set with

TABLE 8.3 Illustration of Two Random Variables and Their Sum

e	$X(e)$	$Y(e)$	$Z = X + Y$	$P(e)$
e_1	2	6	8	.20
e_2	2	9	11	.40
e_3	3	6	9	.30
e_4	3	9	12	.10

the events. The sum of random variables is taken to mean simply the association of the sum of the values with the events. Imagine the sample space $S = \{e_1, e_2, e_3, e_4\}$ and two random variables X and Y, as shown in Table 8.3. The sum $(X + Y)$ is such that where $Z = X + Y$,

$$Z(e_1) = X(e_1) + Y(e_1) = 8$$
$$Z(e_2) = X(e_2) + Y(e_2) = 11$$
$$Z(e_3) = X(e_3) + Y(e_3) = 9$$
$$Z(e_4) = X(e_4) + Y(e_4) = 12$$

This operation does not change the probabilities associated with the events. The probability of a given value $(X + Y)$ is the probability of the outcome (collection of events) associated with that value.

Proof of Theorem 1. Let $S = \{e_1, e_2, \ldots, e_n\}$ and $(X + Y) = X(e_i) + Y(e_i)$ for each event in S, $i = 1, \ldots, n$. Then

$$E(X + Y) = [X(e_1) + Y(e_1)]P(e_1) + \cdots + [X(e_n) + Y(e_n)]P(e_n)$$
$$= P(e_1)X(e_1) + P(e_1)Y(e_1) + \cdots + P(e_n)X(e_n) + P(e_n)Y(e_n)$$
$$= [P(e_1)X(e_1) + P(e_2)X(e_2) + \cdots + P(e_n)X(e_n)]$$
$$+ [P(e_1)Y(e_1) + P(e_2)Y(e_2) + \cdots + P(e_n)Y(e_n)]$$
$$= E(X) + E(Y)$$

In the example

$$E(X) = 2(.2) + 2(.4) + 3(.3) + 3(.1)$$
$$= 2.4$$

$$E(Y) = 6(.2) + 9(.4) + 6(.3) + 9(.1)$$
$$= 7.5$$

$$E(X + Y) = 8(.2) + 11(.4) + 9(.3) + 12(.1)$$
$$= 9.9$$

$$E(X) + E(Y) = 2.4 + 7.5 = 9.9$$

THEOREM 2. *If a random variable* X *has only one member* c *in its value set, that is,* X = {c}, *then it is a constant random variable and*

$$E(X) = E(c) = c$$

Proof. Since the outcome of the sample space to which c is assigned by X is the sample space $S = \{e \mid X(e) = c\}$, then $P(X = c) = 1$. By definition of expectation,

$$E(X) = P(X = c)c = c$$

This is a degenerate case in the sense that constants cannot really be random variables. However, we act as if they were and find $E(c) = c$. This result is useful in the next proofs and in later sections.

THEOREM 3. *For any random variable* X *and any constant* c,

$$E(cX) = cE(X)$$

Proof. The value of cX for any event e is simply $cX(e) = cx$. Thus, for $X = \{x_1, x_2, \ldots, x_k\}$, $cX = \{cx_1, cx_2, \ldots, cx_k\}$ and $P(X = x_i) = P(cX = cx_i)$ and

$$\begin{aligned}
E(cX) &= cx_1 P(X = x_1) + \cdots + cx_k P(X = x_k) \\
&= c[x_1 P(X = x_1) + \cdots + x_k P(X = x_k)] \\
&= cE(X)
\end{aligned}$$

THEOREM 4. *For any random variable* X *and any constant* c,

$$E(X + c) = E(X) + c$$

Proof. By treating c as a constant random variable, this follows directly from Theorems 1 and 2.

Expectation of the Binomial Random Variable

An important application, and also an elegant one, of this theory is made to the binomial random variable. First we need the expectation of an indicator random variable.

Let I be an indicator random variable; $I = \{0, 1\}$. Recalling the earlier convention,

$$I = 1 \text{ if we observe success}$$
$$= 0 \text{ if we observe failure}$$

on an experiment defining I. Then let

$$P(I = 1) = p$$
$$P(I = 0) = 1 - p = q$$

and it follows that

$$E(I) = P(I = 1) \times 1 + P(I = 0) \times 0$$
$$= P(I = 1) = p$$

Recall that a binomial random variable is expressible as (or decomposable into) a sum of indicator random variables, one for each binomial trial. For the ith trial let

$$I_i = 1 \text{ if success on trial } i$$
$$= 0 \text{ if failure on trial } i$$

This gives

$$S = I_1 + I_2 + \cdots + I_n$$

for n binomial trials. Here S is $B(n, p)$ and

$$P(I_i = 1) = p$$

for all $i = 1, \ldots, n$. By substituting the sum of indicators for S,

$$E(S) = E(I_1 + I_2 + \cdots + I_n)$$

By Theorem 1,

$$E(I_1 + I_2 + \cdots + I_n) = E(I_1) + \cdots + E(I_n)$$
$$= \underbrace{p + \cdots + p}_{n \text{ terms}}$$
$$= np$$

THEOREM 5. *The expectation of a binomial random variable* S *with probability distribution* B(n, p) *is*

$$E(S) = np$$

Measures of Dispersion or Spread of a Random Variable

How variant are the values of the random variable?

We approach this question in the manner of the discussion of the magnitude of a random variable. We often need to make statements about the spread of a random variable as well as its location. And, following the same sorts of arguments about the magnitude of a random variable, a measure of the magnitude of the dispersion or spread of a random variable is developed.

Imagine two quite different random variables X and Y that are contrived as having the same expectations. Let X be the binomial random variable with $B(10, .5)$ and $Y = Z - 10$ where Z is the binomial random variable with $B(30, .5)$. The graphs of the density functions of these two random variables are shown in Figure 8.2. Note that

$$E(X) = np = 5$$
$$E(Y) = E(Z - 10) = E(Z) - 10$$
$$= 15 - 10 = 5$$

and $E(X) = E(Y)$.

This example makes it clear that even in the special case of binomial random variables the distributions may be quite different even when their expectations are the same. Thus, in describing random variables, we need to have at least measures of location and dispersion.

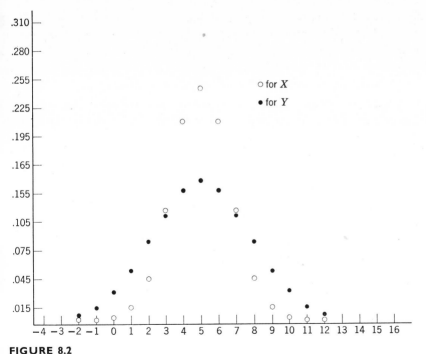

FIGURE 8.2

Graphs of two random variables, X and Y where X is $B(10, .5)$ and $Y = Z - 10$ where Z is $B(30, .5)$.

Range and a Random Variable. The extreme values measure of location might also be taken as a measure of the dispersion of the random variable. In the sense that X_{max} and X_{min} are the values of X most distant from each other, they measure the spread of X. Furthermore, we can get a single value to represent the magnitude of the dispersion. This measure is called the range, R, and is defined by

$$R = X_{max} - X_{min}$$

The value R depends only on two values of the random variable and hence may be a poor measure of dispersion. For example, take X and Y:

$$X = \{1, 2, 3, 4\}$$
$$P(X = 1) = \cdots = P(X = 4) = .25$$
$$Y = \{1, 2, 3, 4, 100\}$$
$$P(Y = 1) = \cdots = P(Y = 4) = .24 \qquad P(Y = 100) = .04$$

The range of X, $R_X = 3$ while the range of Y, $R_Y = 99$. However, this is somewhat misleading because of the heavy weight given to the value $Y = 100$, a small probability value. It would be better to use all the values in X and to attend to their probabilities. All the derogatory things said about the extreme-values measure of location apply to the range as a measure of dispersion.

Expected Deviation. One way of conceptualizing the dispersion of a random variable is in terms of all the differences among the values of the value set. That is, to determine how the values differ among themselves. It can be shown that all of these differences (there are $[(n-1)n/2]$ distinct differences) need not be involved. In fact, comparing each value with some single value conveys just as much information as all the $[(n-1)n/2]$ differences. It is particularly instructive to use the expectation of the random variable as the reference value. Thus, each value in the value set is compared with the expectation of the random variable. For each value of X, for example x_i, the appropriate comparison is given by $[x_i - E(X)]$. However, since there are many values of X, x_i, $i = 1, \ldots, n$, we need to find some way of combining $[x_1 - E(X)]$, $[x_2 - E(X)]$, $\ldots$, $[x_n - E(X)]$ to give a single index as a measure of the spread of the distribution.

We apply the same logic as in the case of expectation. Let each value $[x_i - E(X)]$ enter into the measure proportionately to the probability that $X = x_i$. Since $E(X)$ is a constant, the probability of observing a deviation from $E(X)$, $[x_i - E(X)]$, is the probability of observing x_i. Thus the desired expression is

$$P(X = x_1)[x_1 - E(X)] + \cdots + P(X = x_n)[x_n - E(X)]$$

This corresponds to the definition of the expectation of the random variable X minus a constant $E(X)$. Thus, this expression can be written

$$E[X - E(X)]$$

By Theorems 4 and 2 on expectation,

$$E[X - E(X)] = E(X) - E[E(X)]$$
$$= E(X) - E(X) = 0 \qquad (8.3)$$

The expected deviation of the random variable from its expectation is zero. Since no assumptions are made about the random variable or its distribution, this result is true for any random variable. This result makes the apparently elegant solution to the dispersion problem worthless. All random variables have the same (zero) expected deviation from their expectations.

The fact that $E[X - E(X)]$ is zero for all random variables X is informative about $E(X)$. Recall that the magnitude $E(X)$ corresponds to the balance point of the random variable with respect to the probability distribution. There is as much of the weighted (by probability) value of the random variable on one side of $E(X)$ as on the other. Thus, X can be divided into two parts, $X' = \{x \mid x - E(X) < 0\}$ and $X'' = \{x \mid x - E(X) \geq 0\}$, and

$$\sum_{x \in X'} P(X = x)[x - E(X)] = \sum_{x \in X''} P(X = x)[x - E(x)]$$

Expected Absolute Deviation. The dilemma can be solved two ways. The first is to use the absolute values of $x - E(X)$, that is, $|x - E(X)|$, in the expectation. Thus, we would calculate as a measure of dispersion

$$P(X = x_1)(|x_1 - E(X)|) + \cdots + P(X = x_n)(|x_n - E(X)|)$$

the expected absolute deviation.

This is not a very good solution. The absolute values are difficult to work with mathematically. Also, this expectation is dependent on the difference between the median and expectation of the random variable. This latter defect is the more serious of the two, but we cannot take time here to discuss it in detail.

Variance of the Random Variable. The generally accepted solution to the problem is powerful in the sense that it has many desirable properties. In order to simplify our notation a new symbol μ is defined. This is simply the convention used for the expectation of a random variable:

$$E(X) = \mu$$

Instead of taking the absolute value of $(x - \mu)$ in the measure of dispersion, the square $(x - \mu)^2$ is used. This has the effect of avoiding the sign of the deviation and is more tractable algebraically than the absolute value. Since the deviation of the value x_i from μ, that is, $x_i - \mu$, occurs with probability $P(X = x_i)$, it follows that

$$E[(X - \mu)^2] = \sum_{x \in X} P(X = x)(x - \mu)^2$$

or

$$E[(X - \mu)^2] = \sum_{i=1}^{n} P(X = x_i)(x_i - \mu)^2$$

This is a very important concept in statistical theory. It is called the variance of the random variable X:

$$\text{var}(X) = E[(X - \mu)^2] \tag{8.4}$$

For example, take two random variables with the same expectations and different degrees of dispersion:

$$X = \{9, 10, 11\}$$
$$P(X = 9) = .25, \qquad P(X = 10) = .50, \qquad P(X = 11) = .25$$
$$Y = \{8, 10, 12\}$$
$$P(Y = 8) = .25, \qquad P(Y = 10) = .50, \qquad P(Y = 12) = .25$$
$$E(X) = .25(9) + .5(10) + .25(11) = 10 = \mu_X$$
$$E(Y) = .25(8) + .5(10) + .25(12) = 10 = \mu_Y$$

Each step in calculating $\text{var}(X)$ and $\text{var}(Y)$ is indicated in Table 8.4. The variance of Y is four times the variance of X in these two examples.

TABLE 8.4 Illustrations of the Calculation of Variances

X	9	10	11
$P(X = x)$	.25	.50	.25
$(x - \mu_x)^2$	1	0	1
$P(X = x)(x - \mu_x)^2$	.25	0	.25

$$\text{var}(X) = .25 + 0 + .25 = .5$$

Y	8	10	12
$P(Y = y)$	.25	.50	.25
$(y - \mu_y)^2$	4	0	4
$P(Y = y)(y - \mu_y)^2$	1	0	1

$$\text{var}(Y) = 1 + 0 + 1 = 2$$

The formula for the variance of a random variable can be simplified considerably. The first step is to square the $(X - \mu_X)$ term and then apply the laws of expectation:

$$
\begin{aligned}
\mathrm{var}(X) &= E[(X - \mu_X)^2] \\
&= E[X^2 - 2\mu_X X + \mu_X{}^2] \\
&= E(X^2) - E(2\mu_X X) + E(\mu_X{}^2) \\
&= E(X^2) - 2\mu_X E(X) + \mu_X{}^2 \\
&= E(X^2) - 2\mu_X{}^2 + \mu_X{}^2 \\
&= E(X^2) - \mu_X{}^2
\end{aligned}
\tag{8.5}
$$

In words, $\mathrm{var}(X)$ is the expectation of the square of the random variable minus the squared expectation.

Properties of the Variance

The interpretation of the variance of a random variable is not as simple as in the case of expectation. Here $E(X)$ is the balance point (or center of gravity). Now, for the student with a knowledge of elementary physics we can point out that the variance is the moment of inertia of a distribution of mass. Two other interpretations may help give some insight into the formal nature of the variance.

This section was begun by suggesting that all possible pairs of values of the random variable could be compared to find a measure of dispersion. The variance is in fact proportional to the weighted squared difference between all pairs of values in the value set of the random variable.

One of the fundamental and important properties of the variance is related to the probability of observing values of the random variable at extremes from the expectation. If the dispersion of the random variable is great, then the probability of observing a value quite different from the expectation is large. The very definition of variance, the probability-weighted sum of values, implies this. Thus the larger the variance, the larger is the probability of observing extreme values.

Before formalizing this, we need to introduce another concept: the standard deviation of a random variable. The variance is in terms of the squares of the values of the random variable and as such cannot be compared directly with the values of the random variable. For example, when we want to compare the dispersion with the location of a random variable, the expectation is in first-power terms and the variance in second-power terms. A comparison of expectation and variance would be like comparing length with area. In the sense that there is no meaningful comparison of four square inches (area) with four inches (length), the comparison of values of the value set of a random variable and the variance does not always make sense. In the sections to follow, there are other first-power features of random variables we want to compare with the dispersion, and thus we need a first-power measure of dispersion. Such a measure is the standard deviation,

$$SD(X) = \sqrt{\operatorname{var}(X)} \tag{8.6}$$

or simply the square root of the variance.

To simplify our notation we introduce a symbol. Let the standard deviation of the random variable X be denoted as σ_X:

$$SD(X) = \sigma_X$$

In turn, then,

$$\operatorname{var}(X) = \sigma_X{}^2$$

This conversion from variance to standard deviation has another desirable result. Changes of the scale of measurement are reflected more directly in σ_X than in $\sigma_X{}^2$. Suppose that we have two measurements of reaction time, one in seconds and one in minutes. Let the first be X (seconds) and the second be Y (minutes); then

$$X = 60\,Y$$

and $\sigma_X{}^2 = 60^2\sigma_Y{}^2$ by the rule that $\operatorname{var}(cX) = c^2\operatorname{var}(X)$ (see below). Thus the scale of Y is expanded 3600 times in terms of variance but only 60 times in terms of the magnitude of X. We need a measure of dispersion reflecting in an absolute way the change in scale of the random variable. The standard deviation does this (when the scaling change has a positive value):

$$\sigma_{cX} = c\sigma_Y$$

In terms of the reaction time situation,

$$\sigma_X = 60\sigma_Y$$

The scaling constant is represented directly in the standard deviations.

Chebyshev's Theorem. The probability that an observation of X will be a value x where $-k\sigma_X \le (x - \mu_X) \ge k\sigma_X$ is at least as large as $1 - 1/k^2$, where $k \ge 1.0$.

Another way of stating this theorem is as follows. Find a subset of the value set of X such that the values are less extreme than $\mu + d$ and $\mu - d$ for $d = k\sigma_X$. Call this set of values T:

$$T = \{x \mid (\mu - d) \le x \le (\mu + d)\}$$

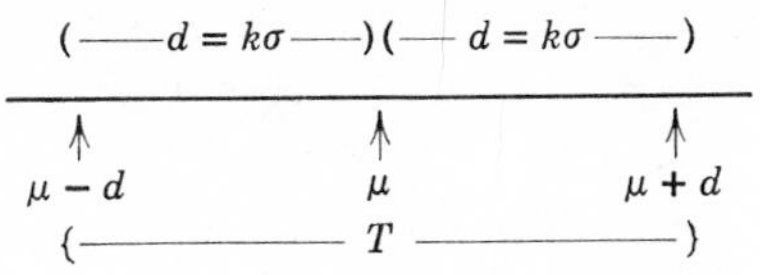

$$(\text{------} d = k\sigma \text{------})(\text{-----} d = k\sigma \text{------})$$

$$\mu - d \qquad \mu \qquad \mu + d$$
$$(\text{---------------} T \text{---------------})$$

FIGURE 8.3

Illustrations of the set T in Chebyshev's theorem.

The set T can be graphed as in Figure 8.3. The probability of observing a value $x \in T$ is given by

$$P(T) \geq 1 - \frac{1}{k^2}$$

For example, we may have a random variable with $\mu = 10$ and $\sigma = 5$. Suppose that we want to know the probability that x is 3 or less or 17 or greater, that is,

$$P(x \leq 3 \text{ or } 17 \leq x)$$

This means that $d = 7$ and hence $d = k\sigma$ implies $7 = k5$ or $k = \frac{7}{5}$. By the Chebyshev rule, then,

$$P(3 \leq x \leq 17) \geq 1 - \frac{1}{(7/5)^2} = 1 - \frac{25}{49}$$

$$P(T) = 1 - \frac{25}{49}$$

$$P(T') < \frac{25}{49}$$

or the probability that we would observe a value more extreme than 3 or 17 is about .5, or less.

As d becomes larger, the probability of observing values as extreme as $(\mu + d)$ and $(\mu - d)$ becomes smaller. Take two other cases, $d = 10$ and $d = 20$, where $\mu = 10$ and $\sigma = 5$. For $d = 10$, $k = \frac{10}{5} = 2$ and $P(0 \leq x \leq 20) \geq 1 - \frac{1}{4} = .75$; for $d = 20$, $k = \frac{20}{5} = 4$ and $P(-10 \leq x \leq 30) \geq 1 - \frac{1}{16} = .938$. These probabilities have a clear interpretation. The probability of observing values as extreme or more extreme than two standard deviations ($k = 2$) from the expectation is at most .25. The probability of observing values as extreme or more extreme than four standard deviations from the expectation is at most .062. Conversely, the probability of a value between plus and minus two standard deviations from μ is .75 or greater. The probability of a value between plus and minus four standard deviations from μ is .938 or more.

Most of a probability distribution is massed within the $\pm 4\sigma$ distance of μ. Hence, σ gives strong limits to the range or dispersion of X with respect to the probability distribution of X.

Notice that no assumptions are made in Chebyshev's theorem about the distribution of the random variable. *Any* distribution is characterized by this relationship. Therefore our measures of dispersion, the variance and the standard deviation, are very powerful concepts in studying the characteristics of random variables.

We do not prove Chebyshev's theorem here. We use it again in Chapter 9, and a proof of a special case is given there.

When special assumptions are made about the distribution, more exact statements of probability of observing extreme values can be made. However, these are never in conflict with the Chebyshev probabilities; that is, they can only be bounded by the Chebyshev values.

Laws of Variance

Since the variance is an expectation, it is no surprise that it follows laws similar to those of the expectation $E(X)$. The fact that the $E(X^2)$ is involved in var(X) makes the laws take slightly different forms.

THEOREM 6. *For a constant random variable* c,

$$var(c) = 0$$

THEOREM 7. *For any random variable* X *and any constant* c,

$$var(X + c) = var(X)$$

THEOREM 8. *For any random variable* X *and any constant* c,

$$var(cX) = c^2\,var(X)$$

The proofs of these theorems are simple; they are left to the student as exercises. The next theorem is no more difficult to prove, but a new concept can be introduced in the proof and so the proof is worked out here.

THEOREM 9. *For any two random variables* X *and* Y *the variance of their sum is*

$$var(X + Y) = var(X) + var(Y) + 2\,cov(X, Y)$$

where cov(X,Y) *is the covariance of the random variables.*

We only comment briefly on covariance here and develop the full theory of the covariance (and correlation) of variables in Chapter 20. First we prove Theorem 9.

Proof of Theorem 9. Let $Z = X + Y$. Then $E(Z) = E(X + Y) = \mu_X + \mu_Y$ and

$$\begin{aligned}
\text{var}(Z) &= E\{[Z - E(Z)]^2\} \\
&= E\{[(X + Y) - (\mu_X + \mu_Y)]^2\} \\
&= E\{[(X - \mu_X) + (Y - \mu_Y)]^2\} \\
&= E[(X - \mu_X)^2 + 2(X - \mu_X)(Y - \mu_Y) + (Y - \mu_Y)^2] \\
&= E[(X - \mu_X)^2] + 2E[(X - \mu_X)(Y - \mu_Y)] + E[(Y - \mu_Y)^2] \\
&= \text{var}(X) + 2\text{cov}(X, Y) + \text{var}(Y)
\end{aligned}$$

The middle term in this last expression is called the covariance of X and Y. The covariance is a measure of the degree of dependence of X and Y and the variances of X and Y. That is, if X and Y are independent, $\text{cov}(X, Y)$ will be zero. And if $\text{cov}(X, Y) \neq 0$, X and Y are dependent. However, the value or magnitude of $\text{cov}(X, Y)$ depends on the magnitude of σ_X^2 and σ_Y^2, and thus we must exercise caution in interpreting $\text{cov}(X, Y)$. The entire concept of dependence and independence in random variables is discussed at length in Chapter 20, but a brief discussion is in order here.

The random variables X and Y are independent if and only if for every pair (x_i, y_j)

$$P(X = x_i \mid Y = y_j) = P(X = x_i)$$

or, in terms of the corresponding multiplication rule,

$$P(X = x_i \quad and \quad Y = y_j) = P(X = x_i)P(Y = y_j)$$

In order for this to be defined, of course, X and Y must be defined for the same sample space. Now, when X and Y are independent, then $\text{cov}(X, Y) = 0$ and

$$\text{var}(X + Y) = \text{var}(X) + \text{var}(Y) \tag{8.7}$$

This relationship provides a means of calculating $\text{cov}(X, Y)$. Note that from above,

$$\text{var}(X + Y) - \text{var}(X) - \text{var}(Y) = 2\text{cov}(X, Y)$$

Thus, if we assign to each event in S the value $X(e) + Y(e) = Z(e)$ and calculate the variance of Z, σ_Z^2, and also calculate σ_X^2 and σ_Y^2, it follows that

$$2\text{cov}(X, Y) = \sigma_Z^2 - \sigma_X^2 - \sigma_Y^2$$

If X and Y are independent, then $\text{cov}(X, Y) = 0$. However, a note of caution is necessary. If $\text{cov}(X, Y) = 0$, it does not necessarily follow that X and Y are independent. The source of this asymmetry is discussed in Chapter 20.

Variance of a Binomial Random Variable

The variance of the binomial random variable with distribution $B(n, p)$ is developed in much the same way as was its expectation. First we find the variance of an indicator random variable:

$$I = \{0, 1\}$$
$$P(I = 1) = p$$
$$P(I = 0) = 1 - p = q$$
$$E(I) = p$$
$$\text{var}(I) = E(I^2) - [E(I)]^2$$

Notice that $I^2 = \{1, 0\} = I$ and hence $E(I^2) = E(I)$. Therefore

$$\text{var}(I) = E(I) - [E(I)]^2 = p - p^2$$
$$= E(I)[1 - E(I)] = p(1 - p)$$
$$= p[1 - p]$$
$$= pq$$

Extending this to n independent indicator random variables, that is, to the binomial trials with $B(n, p)$,

$$S = I_1 + I_2 + \cdots + I_n$$
$$\text{var}(S) = \text{var}(I_1 + \cdots + I_n)$$
$$= \text{var}(I_1) + \cdots + \text{var}(I_n)$$
$$= \underbrace{pq + \cdots + pq}_{n \text{ terms}}$$
$$= npq$$

The standard deviation of S is simply $\sigma_S = \sqrt{npq}$.

SUGGESTED READING

In addition to sources already cited, refer to:

Edwards, A. L. (1964), *Expected values of discrete variables and elementary statistics*, Wiley, New York.

GLOSSARY

Extreme-Values Measure of Location. The extreme-values measure of location is the two most extreme values in the value set of X. If for all of x in X, $X_{min} \leq x \leq X_{max}$, the value pair (X_{min}, X_{max}) is the measure.

Mode. The mode is a measure of location of a random variable given by the value(s) with the greatest probability.

Median. The median, Mdn, is a measure of location of a random variable. The median is defined by $P(x \leq \text{Mdn}) \geq .5$, where Mdn is the smallest possible value satisfying both inequalities.

Mathematical Expectation. The mathematical expectation of a random variable is a measure of location of the random variable defined as a weighted sum of the values of the value set of the random variable, the probabilities of the respective values being the weights.

(8.1) $$E(X) = \sum_{x} P(X = x)x$$

(8.2) $$E(X) = \sum_{e} P(e)X(e)$$

Theorem 1. For any random variable X and Y, $E(X + Y) = E(X) + E(Y)$.

Theorem 2. If a random variable X has only one value c, that is, $X = \{c\}$, in its value set, then it is a constant random variable and $E(X) = E(c) = c$.

Theorem 3. For any random variable X and any constant c, $E(cX) = cE(X)$.

Theorem 4. For any random variable X and constant c, $E(X + c) = E(X) + c$.

Theorem 5. The expectation of a binomial random variable S with probability distribution $B(n, p)$ is $E(S) = np$.

Range of a Random Variable. The range of a random variable, $R = X_{max} - X_{min}$ is a measure of the dispersion of the random variable.

Expected Deviation. The expectation of the deviation of the values of a random variable from a common reference point.

(8.3) $$E(X - E(X)) = .0$$

Variance. The variance of a random variable is a widely used measure of the dispersion of a random variable. The variance of X is the expected squared deviation of X and $E(X)$.

(8.4) $$\text{var}(X) = E[(X - \mu)^2]$$

(8.5) $$\text{var}(X) = E(X^2) - \mu_X^2$$

Chebyshev's Theorem. The probability that an observation of X will be a value x where $-k\sigma_X \leq (x - \mu) \leq k\sigma_X$ is at least as large as $1 - 1/k^2$, where $k \geq 1.0$.

Theorem 6. For a constant random variable c, $\text{var}(c) = .0$.

Theorem 7. For any random variable X and any constant c, $\text{var}(X + c) = \text{var}(X)$.

Theorem 8. For any random variable X and any constant c, $\text{var}(cX) = c^2 \, \text{var}(X)$.

Theorem 9. For any random variables X and Y, $\text{var}(X + Y) = \text{var}(X) + \text{var}(Y) + 2\text{cov}(X, Y)$.

$$(8.6) \qquad SD(X) = \sqrt{\text{var}(X)}$$

$$(8.7) \qquad \text{var}(X + Y) = \text{var}(X) + \text{var}(Y) \text{ for independent } X \text{ and } Y$$

Standard Deviation. The standard deviation is a measure of dispersion of a random variable in the units of measurement of the random variable. The standard deviation is the square root of the variance of a random variable.

PROBLEMS

1. Show that the expected deviation of X about c, a constant, is as worthless as a measure of dispersion as the expectation of X about $E(X)$.

2. Prove Theorems 6, 7, and 8 by methods suggested in proof of Theorem 9.

3. What is the expected number of points on a toss of a fair die? What is the expected total number of points in two tosses of a die?

4. A random variable X has a value set consisting of all of the integer values $1, 2, \ldots,$ 100. The probability distribution on this random variable is the uniform (equal) probability distribution. What is the expectation of this random variable? Do not calculate your answer in the usual sense, but make a logical argument leading to the value.

5. The sum of the first N integers can be calculated by the equation

$$S_1 = 1 + 2 + \cdots + N = \frac{N(N + 1)}{2}$$

The sum of the first N squared integers can be calculated by the equation

$$S_2 = 1^2 + 2^2 + \cdots + N^2 = \frac{N(N + 1)(2N + 1)}{6}$$

These relationships can be used to calculate the expectation and variance of any random variable consisting of m consecutive integers with uniform probability distribution. Calculate the expectation and variance of the random variable $Y = \{5, 6, \ldots, 9\}$ with the uniform probability distribution. Use only algebra and evaluations of S_1 and S_2 above. Repeat this process for other random variables defined in similar ways.

6. You are asked to play a game with dice. You are asked to pay $1.00 for the privilege of playing. You will receive $2.00 if one six appears in a roll of two fair dice, $4.00 if two sixes appear, and nothing if no six appears. Let your net gain be X. Find $E(X)$. What would be the charge to you for the privilege of playing if you wanted to insure that $E(X) = 0$?

7. Two numbers are selected at random with replacement from the value set 1, 2, . . . , 10. If X is the larger of the two numbers selected, what is $E(X)$?

8. Out of 100 students of which 70 are known to favor some proposal and the remaining 30 are opposed, one is selected by lottery. Letting $X = 1$ indicate "favor" and $X = 0$ indicate "not favor," calculate the expectation and variance of X.

9. The expectation and variance of a random variable Y are 100 and 16, respectively. What is the expectation of Y^2? What is the variance of $4Y - 10$? What is the standard deviation of $4Y - 10$? What are the expectation and variance of $-Y$?

10. Recall the definition of S_1 and S_2 of problem 5. Let X denote the number selected when one number is selected at random (equal probability) from the numbers 1, 2, 3, . . . , N. Show in general, that is, for any N, that

$$E(X) = \frac{N + 1}{2} \quad \text{var}(X) = \frac{N^2 - 1}{12}$$

11. Prove that $E(Z - W) = E(Z) - E(W)$. This is not given directly by Theorem 1.

12. Prove that $E(cZ + dW) = cE(Z) + dE(W)$.

13. If $X_1, X_2, \ldots, X_N$ are random variables having common expectation μ, prove that the average of single observations on each of the variables is equal to μ.

CHAPTER 9

Continuous Random Variables

$\mathbf{W}$e have been concerned with two kinds of sample spaces: those defined by nominal events and those defined by numerals. All of the numerals we have dealt with in detail have been a special sort, the natural numbers and zero—0, 1, 2, ...

It should be clear to the student that a numeral sample space is a special case of nominal sample spaces. To each event in S we assign some numeral from a value set N. The sample space thus obtained is called a random variable. The random variable is said to have a value set defined by the values of N. If N is a finite set (that is, containing a limited number of elements), it is referred to as a finite discrete random variable.

We extend our discussion here to another, more general, special case in which any real number may be a member of the value set. Such a random variable is called a continuous random variable. Examples of this type of random variable are easy to think up: the amount of pressure exerted on a hand dynamometer; the time between a signal and a reaction of a subject to the signal; the strength of association between stimulus and response; the concentration of a given drug administered; the amount of weight gained on a weight-reducing regime (negative, the subject hopes); the proportion of light of a given wavelength in a light source.

The primary distinction is in the value set of the variable: Finite discrete random variables have limited value sets; continuous random variables have value sets with an unlimited number of values. Although the natural numbers are parts of the real number system, there are "gaps" between the members of each pair of contiguous numbers. The numbers are in that sense discrete. On the other hand, continuous random variables have value sets without "gaps" between the values.

This distinction is easily understood in the framework of measurement. In the random variables so far discussed—all of them discrete—we have been concerned primarily with counting. We were concerned with such things as the number of objects of a certain class observed at a given time, or the number of successes in so many trials. This leads to value sets of natural numbers $0, 1, 2, \ldots, n$. We never had to deal with $1\frac{1}{2}$ successes,* 2.324 children in families of middle income, and so forth. For continuous random variables, on the other hand, we can in theory observe all values on the real line. In measuring the pressure exerted on a hand dynamometer, we could possibly observe $1, 2, \ldots, n$ ounces. We might also observe 1.46324 ounces, or any other such value. Conceivably we could distinguish between values in this variable to any degree of difference—for example, 1.463214326 ounces as contrasted with 1.463214325 ounces.

This mathematically weak introduction to continuous variables (real variables) must suffice in this presentation. The rigorous (and elegant) theory of real variables is not difficult, but limitations on space and time dictate that the student discover this from other sources.

We must pass quickly to the implications of this development for the theory of probability.

Probability Density Functions

For the sake of simplicity we work for a time with a small part of the real line—the interval of zero through 1.0—the unit interval. Thus the random variable X has a value set

$$\{X = x \mid 0 \leq x \leq 1\}$$

We might picture this as a segment (interval) on the line as shown in Figure 9.1. If we imagine an experiment in which each of the values in this interval is as likely

-.5   .0   1.0   2.0

FIGURE 9.1

The unit interval as a segment of the real line.

* The expectation and variance of a discrete random variable could of course have such a value, and we are used to having $E(X)$ and $\mathrm{var}(X)$ be values not in the value set of the random variable.

(to be observed) as any other given value, we are led to a uniform probability distribution on this interval: The probability of each value is the same as any other. This leads to an enigma: There are an unlimited number of values in the sample space, and hence each has a probability approaching zero. To see this more clearly we divide the interval [0, 1] into two equal parts A_1 and A_2:

$$A_1 = \{x \mid 0 \leq x < .5\}$$
$$A_2 = \{x \mid .5 \leq x \leq 1\}$$

This gives $P(A_1) = \frac{1}{2} = P(A_2)$. Now divide A_1 into two parts, B_1 and B_2, and A_2 into two parts, B_3 and B_4, such that

$$B_1 = \{x \mid 0 \leq x < .25\}$$
$$B_2 = \{x \mid .25 \leq x < .5\}$$
$$B_3 = \{x \mid .5 \leq x < .75\}$$
$$B_4 = \{x \mid .75 \leq x \leq 1\}$$

This gives $P(B_1) = P(B_2) = P(B_3) = P(B_4) = \frac{1}{4}$. Imagine doing this subdivision ten times over, so that there are $2^{10} = 1024$ intervals, each of equal length and consequently with equal probabilities of $\frac{1}{1024} = .00098$. Divide each of these intervals in half ten times again and 1,048,576 intervals are obtained, each with probability of approximately .000001. Since this subdivision can be extended indefinitely, the "final" stage probabilities are essentially equal to zero (they are zero in the limit). Thus, no point in the value set of X has a probability greater than zero.

The enigma is that $P(S) = 1$ but no single element in S has $P(e) > 0$. For an experiment in which the random variable is defined, some value must be observed, that is, $P(S) = 1$, yet no single value has a probability greater than zero. The model we have been using for integer-valued random variables is inadequate to deal with continuous-valued random variables. Clearly some method for dealing with continuous random variables must be found.

We take our cue from the graphic representation of the probability distribution of an integer-valued random variable. A histogram is constructed so that the probability of each value is represented as an area in a rectangle (bar). The histogram is constructed with each rectangle having a base of length one and a height equal to $f(x) = P(X = x)$. If we think of the tops of the rectangles as forming a graph of a function $f(x)$ of the real numbers, the probability function of X, the probability of X falling in any given interval on the real line is given by the area under the graph of the function.

Applying this to the continuous-variable case, we set out to construct a function $f(x)$ that permits us to define the probability of observing a value in an interval as an area under the graph of the function. In the example above, where $P(A_1) = P(A_2) = \frac{1}{2}$ we need a function $f(x)$ that is constant over all A_1 and constant over all A_2, and so that the areas are equal to $\frac{1}{2}$. An area is defined here as the length of the interval times the value of the function, $f(x)$. Let the length of the interval be b and the area be $a = P(X = x)$. Thus the area is given by

$$a = bf(x)$$

implying

$$f(x) = \frac{a}{b}$$

From the definition of the problem with the value set of the random variable being the unit interval, broken into two equal halves of length $\frac{1}{2}$, and areas of $\frac{1}{2}$ for each subinterval,

$$f(x) = \frac{1/2}{1/2} = 1$$

Taking four equal subintervals produces

$$a = \tfrac{1}{4} \qquad b = \tfrac{1}{4}$$

and hence for each subinterval

$$f(x) = \frac{1/4}{1/4} = 1$$

With 1024 intervals

$$a = \frac{1}{1024} \qquad b = \frac{1}{1024}$$

giving

$$f(x) = \frac{1/1024}{1/1024} = 1$$

It can be shown in general that this reasoning leads to a function

$$f(x) = 0 \qquad \text{for all } x < 0$$
$$f(x) = 1 \qquad \text{for all } 0 \leq x \leq 1$$
$$f(x) = 0 \qquad \text{for all } x > 1$$

A graph of this function is shown in Figure 9.2.

Whereas $f(x)$ was a probability with integer-valued variables, $f(x)$ is not a probability when defined for a continuous variable. The fact that for integer-valued variables each interval has length $b = 1$ implies $f(x) = a = P(X = x)$. Where $b \neq 1$ this relationship does not hold.

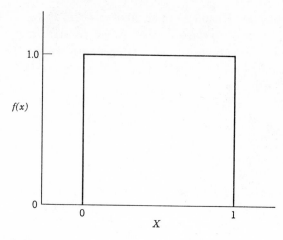

FIGURE 9.2

Graph of a uniform probability density function on the unit interval.

The function $f(x)$ for continuous variables is called a probability density function. Any nonnegative function of a continuous random variable with total area equal to 1.0 under the curve of the function is a probability density function of the random variable. The example above certainly qualifies.

The values of $f(x)$ do not conform to the restraints on probabilities except in the case of an integer-valued random variable. More specifically, the values $f(x)$ are not probabilities in the case of noninteger-valued random variables. Where X is integer valued, $f(x)$ values are probabilities. In other instances of uniformly distributed random variables the value $f(x)$ is defined by the equation

$$f(x) = \frac{a}{b}$$

where a is defined as the probability of the interval and b is defined as the width of the interval.

A second example is instructive. Imagine a function

$$f(x) = 0, \qquad x < 0$$
$$f(x) = 4x, \qquad 0 \leq x \leq .5$$
$$f(x) = 4(1 - x), \qquad .5 < x \leq 1$$
$$f(x) = 0 \qquad x > 1$$

This is a completely positive function with total area of 1.0 under its curve. A graph of this function is shown in Figure 9.3. At $x = .5$, $f(x) = 2$. At $x = .25$, $f(x) = 1.0$. The probability in a given interval may be found by calculating the area under

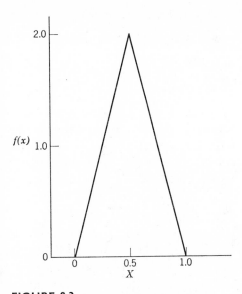

FIGURE 9.3

Graph of a probability density function.

the curve of the function above that interval. A bit of simple ingenuity with areas of triangles permits the calculation of the probability of intervals such as $.7 \leq x \leq .9$ and $.3 \leq x \leq .7$.

The more interesting and useful examples of continuous random variables do not have density functions as simple as the two examples above. In these examples, knowledge of the simple plane geometry of rectangles and triangles is sufficient to calculate the probability of intervals from the density functions. However, these methods are not satisfactory with density functions that are not composed of simple straight-line segments.

An Introduction to the Integral Calculus

This problem leads us to introduce a new set of considerations of a mathematical nature. These ideas come from the theory of integral calculus. We now turn our attention to conceptualizing probability as an area in the general case of an arbitrary density function $f(x)$.

Calculation of areas is a familiar operation to anyone who has had high school mathematics. That is, we know how to calculate areas for certain types of simple geometric forms. For example, the area involved in the density function

$$f(x) = \begin{cases} 0 & \text{if } x < 0 \\ 1 & \text{if } 0 \leq x \leq 1 \\ 0 & \text{if } x > 1 \end{cases}$$

is simply the area of the region bounded by $f(x)$, the X axis, $x = 0$ and $x = 1$. This region is a square of area 1.0. This area corresponds simply to the base length times the height. In more general terms, the distance between $(x = 1)$ and $(x = 0)$ times the distance between $f(x) = 0$ and $f(x) = 1$ is the area with which we are concerned. Other types of two-dimensional figures are familiar in terms of ways of calculating areas. The circle has an area equal to πR^2 where R is the radius of the circle. Likewise a right triangle has an area equal to $[\frac{1}{2} \times$ (length of leg 1) $\times$ (length of leg 2)], and so on. The area of the region between the curve and the X axis between $x = 0$ and $x = 1$ of the function shown in Figure 9.4a is not so familiar a concept. This area cannot be found exactly by the ordinary methods of algebra and plane geometry. However, we can approximate the solution by "cutting up" the figure into countless rectangles standing along the X axis. Each of these rectangles has a height equal to the height of the figure, $f(x)$, for some value of X included in the base of the rectangle. Taking a piece of the figure, we can see the detail of such a construction in Figure 9.4b. For the sake of convenience we set up the rectangles with the base extending from $x_1 = a$ to $x_2 = b$ and having a height equal to the smaller of $f(x_1)$ and $f(x_2)$. Since $f(b) \leq f(a)$, the area of this rectangle is defined by

$$A = (b - a)f(b)$$

The distance along the base is simply $(b - a)$ and the height $f(b)$. However, the area below the curve of $f(x)$ between a and b is larger than A by an amount equal to the area of the shaded region above the rectangle (in Figure 9.4b). We can improve our estimate of the area of this segment by introducing several rectangles between a and b, as shown in Figure 9.4c. Having subdivided the segment of X into six intervals, $a = x_1$ to $x_2, \ldots, x_6$ to $x_7 = b$, we have six areas. The total area bounded by $f(x) = 0$ (the X axis), a, b, and the curve $f(x)$ extending from a to b has a value approximately equal to the sum of the separate areas of the rectangles we have constructed. Where A_i is the area of the ith rectangle, $A_i = (x_i - x_{i-1})f(x_i)$, we have:

$$A \cong \sum_{i=2}^{7} A_i = (x_2 - x_1)f(x_2) + (x_3 - x_2)f(x_3)$$
$$+ (x_4 - x_3)f(x_4) + (x_5 - x_4)f(x_5)$$
$$+ (x_6 - x_5)f(x_6) + (x_7 - x_6)f(x_7)$$

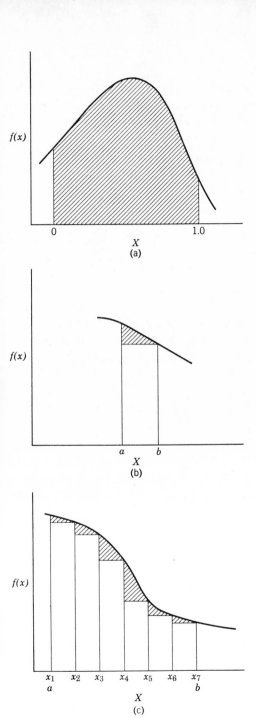

FIGURE 9.4

Graphs of an illustrative function.

The summation of the A_i terms is just smaller than the area in which we are concerned. Let the area under the curve between a and b, that is, for $a \leq x \leq b$, be A. We might call the error ε (the Greek letter epsilon) and write

$$A = \sum_{i=2}^{7} A_i + \varepsilon$$

Extending this argument to 12 intervals defined by equally spaced points $a = x_1, x_2, x_3, x_4, \ldots, x_{13} = b$, we have

$$A = \sum_{i=2}^{13} A_i + \varepsilon$$

where ε is now smaller. (Draw a graph to demonstrate that part of the shaded area of the graph with six intervals is now included in the region defined by the 12 rectangles: Argue that ε for 12 intervals is thus smaller than ε for six intervals.) Continuing this process, it can be shown that if the interval ($a \leq x \leq b$) is divided into n equal subintervals and x_i' is the mid-value of X in the ith subinterval defined by x_i and x_{i+1}, then

$$\lim_{n \to \infty} \sum_{i=1}^{n} f(x_i')(x_{i+1} - x_i) = A$$

Note that we must use the notation $\lim_{n \to \infty}$ to insure the equality. The meaning of this mathematical designation is that if we let n grow without bound (∞ is not a number, it is a symbol implying the concept of limitlessness; "$n \to \infty$" means n increases without limit). The consequence of letting $n \to \infty$ is that in the limit $(x_{i+1} - x_i) \to 0$ (note $(x_{i+1} - x_i) \neq 0$) and hence $\varepsilon \to 0$.

This limit is known as the definite integral of $f(x)$ from a to b, and we write

$$\int_a^b f(x)\, dx = \lim_{n \to \infty} \sum_{i=1}^{n} f(x_i')(x_{i+1} - x_i)$$

where $x_1 = a$ and $x_{n+1} = b$. The dx replaces the difference $(x_{i+1} - x_i)$ in the limiting form.

This highly conceptualized introduction to the definite integral carries with it some specific points that can be generalized. The generalization is interesting and necessary for a rigorous presentation of the definite integral. However, this special case suffices for the conceptual basis necessary for the purposes of this discussion.

The definite integral, being equal to the limit on a summation, follows theorems corresponding to those of summation.

THEOREM 1.

$$\int_a^b f(x)\,dx + \int_b^c f(x)\,dx = \int_a^c f(x)\,dx$$

THEOREM 2.

$$\int_a^b [f(x) + g(x)]\,dx = \int_a^b f(x)\,dx + \int_a^b g(x)\,dx$$

THEOREM 3.

$$\int_a^b cf(x)\,dx = c\int_a^b f(x)\,dx$$

Some additional notation conventions must be introduced to deal with some special intervals on X. In the arguments above, a and b are lower and upper bounds of the interval for which we found the area between $f(x)$ and the X axis. This can be extended to the interval indefinitely in both (or either) direction. If a is an indefinitely large negative value of X, then we write

$$\int_{-\infty}^b f(x)\,dx$$

as the area bounded between $f(x)$ and the X axis below b. At the other extreme, if b is an indefinitely large positive number, $b \to \infty$ and we write

$$\int_a^\infty f(x)\,dx$$

as the area bounded by $f(x)$, the X axis above a. If no limits are placed on a or b, then we write

$$\int_{-\infty}^\infty f(x)\,dx$$

as the area between $f(x)$ and the X axis. For ease of verbal communication we read this last expression as "the integral of $f(x)$ with respect to X from minus infinity to plus infinity."

Application to Probability Density Functions

The integral $\int_a^b f(x)\,dx$ for a given density function $f(x)$ is a variable and a function of a and b, not of x. If $a = b$, then $\int_a^b f(x)\,dx = 0$, and $\int_{-\infty}^\infty f(x)\,dx = 1$. In general, if we deal with $\int_{-\infty}^{b} f(x)\,dx$, then we have a function of b in the following sense. Let b_1 and b_2 be two numbers such that $b_1 < b_2$. Then by definition of the integral,

$$\int_{-\infty}^{b_1} f(x)\,dx \le \int_{-\infty}^{b_2} f(x)\,dx \qquad (9.1)$$

In general, $\int_{-\infty}^{b} f(x)\,dx$ is a nondecreasing function of b.

Going back to the uniform density function $f(x) = 1$ on the interval $0 \le x \le 1$, we find

$$\int_{-\infty}^{b} f(x)\,dx = 0 \qquad \text{for} \quad b < 0$$

$$\int_{-\infty}^{b} f(x)\,dx = 1 \qquad \text{for} \quad b \ge 1$$

and hence

$$\int_{-\infty}^{\infty} f(x)\,dx = 1$$

These two relationships are easy to rationalize by considering the cumulative nature of integration. Clearly this is a function of b. The common designation is

$$F(b) = \int_{-\infty}^{b} f(x)\,dx$$

This is as far as we take this development here. None of the following materials require the technical apparatus of finding the function $F(x)$ from $f(x)$. The development of $F(x)$ from the calculus is quite simple and revealing about the logic and methods of mathematics. Interested students can find this developed clearly and meaningfully in references given at the end of the chapter. The basic concepts of the integral developed here give us the necessary intuitive knowledge to work with continuous random variables and their distribution functions and density functions. To be sure, we must skim over many derivations and proofs that we could develop in detail if the calculus were an available tool. However, the lack of calculus does not impede understanding of the basic results in the theory of random variables and statistics for continuous variables.

The Distribution Function

Where X is a continuous random variable and $f(x)$ is its probability density function, then

$$F(b) = \int_{-\infty}^{b} f(x)\,dx$$

is its distribution function. This is directly analogous to the distribution function for discrete random variables discussed in the preceding chapter. There, the distribution function was defined as the cumulative probability distribution of a random variable. In the case of the continuous random variable the "cumulation" is not accomplished by summing but by integration; that is,

$$\int_{-\infty}^{x'} f(x)\,dx \qquad \text{replaces} \qquad \sum_{x \leq x'} f(x)$$

The sense of $F(x)$ in both cases is the same:

$$F(x) = P(X \leq x)$$

The distribution function of continuous random variables obeys the same rules that were true for discrete random variable distribution functions. They are as follows.

THEOREM 4. *If* $x' \geq$ *(largest* x *in* X*),* $F(x') = 1.0,$ $\int_{-\infty}^{x'} f(x)\,dx = 1.0.$

THEOREM 5. *If* $x' <$ *(smallest* x *in* X*),* $F(x') = .0,$ $\int_{-\infty}^{x'} f(x)\,dx = .0.$

THEOREM 6. *If* $a \geq b$ *then* $F(a) \geq F(b),$ *and if* $a = b + h$ *then*

$$\int_{-\infty}^{a} f(x)\,dx = \int_{-\infty}^{b} f(x)\,dx + \int_{b}^{b+h} f(x)\,dx$$

THEOREM 7. *If* $a \leq x \leq b,$ $P(a \leq x \leq b) = F(b) - F(a),$ *then*

$$P(a \leq x \leq b) = \int_{-\infty}^{b} f(x)\,dx - \int_{-\infty}^{a} f(x)\,dx$$

THEOREM 8. *Since* $\int_{-\infty}^{\infty} f(x)\,dx = 1,$ *then* $\int_{a}^{\infty} f(x)\,dx = 1 - \int_{-\infty}^{a} f(x)\,dx.$

All of this is abstract, "theoretical," and poorly connected with the activities of the empirical researchers. It would have no place in a book of this nature were it not for the fact that the most widely used, and useful, models of statistical analysis are concerned with continuous random variables. In order to deal effectively with these models an understanding of the character of $f(x)$ and $F(x)$ is necessary. The three points most important to be remembered, where X is a continuous random variable, are the following.

(1) $f(x)$ is not a probability: $f(x) \geq 0$, but $f(x) > 1$ is possible.

(2) $F(x)$ is a probability: $P(X \leq a) = \int_{-\infty}^{a} f(x)\, dx = F(a)$.

(3) The probability of every single point in X is zero, but the probability of intervals $(a < x < b)$ may be greater than zero,

$$P(a < x < b) = F(b) - F(a) = \int_{a}^{b} f(x)\, dx \geq 0.$$

Expected values of continuous random variables are calculated by equations parallel to the summation equations for discrete random variables:

$$\mu = \int_{-\infty}^{\infty} x f(x)\, dx \tag{9.3}$$

$$\sigma^2 = \int_{-\infty}^{\infty} (x - \mu)^2 f(x)\, dx \tag{9.4}$$

All rules of μ and σ^2 derived for discrete random variables hold for continuous random variables.

Observations on Continuous Random Variables

Switching the frame of reference from the abstract to the concrete, it should be clear from the definition of "continuous" that it is impossible ever to observe all the values in the value set of a continuous random variable. In general, only a limited number of values with a limited degree of accuracy can be observed (for example, two decimal points on a scale of force applied in a string pulling experiment). Thus, making experimental (that is, empirical) observations on a continuous random variable defines a result set that is discrete. If measured force varies from .000 to 10.000 by units of .001, there are 10,001 separate and distinct values that might be observed even if there is an unlimited (continuous) set of "real" forces. Clearly we could not observe all of the measurable force value with fewer than 10,001 observations, and many more might be required before all are observed (the same value, for example 4.361, might be observed many times before 4.362 is observed).

The device that avoids having to observe extremely large samples and still obtains a good representation of $f(x)$ and $F(x)$ from a sample is to group the values of X and to treat all the various values observed from each group as though they were the same. Thus, groups of values might be defined by successive equal intervals on X. For example,

$$A_1 = \{x \mid .0 \leq x \leq 1.0\}$$
$$A_2 = \{x \mid 1.0 < x \leq 2.0\}$$
$$A_2 = \{x \mid 2.0 < x \leq 3.0\}$$

$$. \qquad . \qquad . \qquad .$$
$$. \qquad . \qquad . \qquad .$$
$$. \qquad . \qquad . \qquad .$$

$$A_{10} = \{x \mid 9.0 < x \leq 10.0\}$$

These sets of potential observations are treated as discrete outcomes. Each set is assigned a value corresponding to one of the values in the set. The most usual convention is to use the midpoint of an interval as the constant value representing the entire set. Suppose that these values are $Y_1, \ldots, Y_{10}$:

$$Y(A_1) = .5$$

$$Y(A_2) = 1.5$$

$$. \qquad .$$
$$. \qquad .$$
$$. \qquad .$$

$$Y(A_{10}) = 9.5$$

This, in effect, is a rescoring of the variables as a discrete random variable. The grouped values can be treated as if they formed a discrete random variable. Some error is involved, but where the continuous values in a given interval have approximately the same probability the error is negligible (about as much error in one direction as in the other in each interval).

The similarity of the approximation in this example to the measurement approximation in such areas as questionnaires, IQ, rating techniques, should be apparent to the reader. What attitude, ability, psychophysical judgment process, and so on, is integer valued? Few if any are, although the measurements made by standard tests and procedures are almost always integer valued and always discrete. The measurements are taken to be quantitative approximations to the objects and processes we wish to observe.

There may be too many values in the value set to preserve simplicity and clarity of the frequency distribution of a set of observations. In such an instance wider class intervals are defined and treated exactly as the unit length intervals were treated in the example above. For example, if X is a continuous random variable with upper and lower limits of 500 and 100, respectively, and we wish ten class intervals, the interval length is given by

$$\frac{\text{upper limit} - \text{lower limit}}{\text{number of intervals}} = \frac{500 - 100}{10} = 40$$

Thus, ten sets of values are defined:

$$A_1 = \{x \mid 100 \leq x \leq 100 + 40\}$$
$$A_2 = \{x \mid 140 < x \leq 140 + 40\}$$

$$\cdot \qquad \cdot \qquad \cdot$$
$$\cdot \qquad \cdot \qquad \cdot$$
$$\cdot \qquad \cdot \qquad \cdot$$

$$A_i = \{x \mid 100 + (i - 1)40 < x \leq 100 + (i)40\}$$

$$\cdot \qquad \cdot \qquad \cdot$$
$$\cdot \qquad \cdot \qquad \cdot$$
$$\cdot \qquad \cdot \qquad \cdot$$

$$A_{10} = \{x \mid 460 < x \leq 500\}$$

Observations having values belonging to these sets are assigned the midpoints of the sets as the "grouped" value. That is,

$$Y(A_1) = 120$$
$$Y(A_2) = 160$$

$$\cdot \qquad \cdot$$
$$\cdot \qquad \cdot$$
$$\cdot \qquad \cdot$$

$$Y(A_{10}) = 480$$

where the values of Y are to the nearest integer.

Graphs of Observations on Continuous Random Variables

In Chapter 7 the probability density function of a random variable was represented graphically by points plotted in the $[x, f(x)]$ plane. Lines were drawn from $[x, f(x)]$ to X at each value to accentuate the visual properties of the graph. The notion of histogram or bar graph was also introduced. These concepts generalize to samples of observations on continuous random variables. The observations

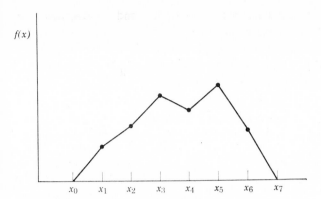

FIGURE 9.5

Illustration frequency polygon.

themselves are discrete and are dealt with as such. In cases where class intervals of observed values are plotted, it is customary to use the histogram with the bar limits defining the upper and lower limits on the class intervals, and bar height given by $f(x) = a/b$ where a is the relative frequency and b the width of the bar.

Another graphic device is the frequency polygon. This figure, illustrated in Figure 9.5, is constructed by connecting the points in $[x, f(x)]$ of the frequency function. The points for adjacent values of X are connected with each other by a

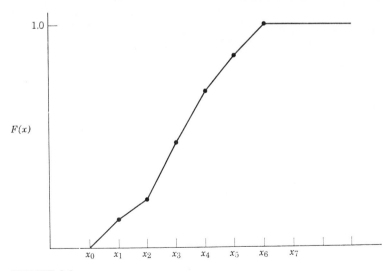

FIGURE 9.6

Illustration of cumulative frequency polygon.

straight line rather than drawing a line from the point to the base of the graph. Note that the polygon is "closed" by connecting it to the X axis at the two extreme midpoints or the first zero-frequency midpoint beyond the most extreme nonzero-frequency midpoint.

Cumulative frequency distributions are dealt with in a fashion similar to that used in discrete random variables. In the ogive, lines are drawn from point to point in $[x, F(x)]$ rather than the stepwise way used in discrete random variables. This type of graph is illustrated in Figure 9.6. Note that the ogive is closed at the largest midpoint where $F(x) = .0$.

Computation

With the availability of automatic calculating machinery, of both the desk calculator and the computer varieties, it is recommended that no calculations be performed on the classified values (that is, the midpoints). If these values are used and the grouping is coarse, there is a substantial degree of error. If the grouping is fine, little is gained by the grouping.

For those with a need to use the procedures, certain corrections for calculational formulas should be familiar. These corrections are discussed in many standard introductory statistics texts (for example, Guilford and McNemar).

SUGGESTED READINGS

Kleppner, D. and Ramsey, N. (1965), *Quick calculus*, Wiley, New York.

Lewis, Donald (1960), *Quantitative methods in psychology*, McGraw-Hill, New York.

Luchins, A. S., and Luchins, E. H. (1965), *Logical foundations of mathematics for behavioral scientists*, Holt, Rinehart and Winston, New York.

Nahikian, H. M. (1964), *A modern algebra for biologists*, University of Chicago Press.

Richmond, D. E. (1950), *Fundamentals of the calculus*, McGraw-Hill, New York.

Riggs, D. S. (1963), *The mathematical approach to physiological problems*, Williams and Wilkins, Baltimore.

Sagan, Hana (1962), *Integral and differential calculus: an intuitive approach*, Wiley, New York.

Thomas, G. B. (1953), *Calculus and analytic geometry*, Addison-Wesley, Reading.

Vance, E. P. (1963), *An introduction to modern mathematics*, Addison-Wesley, Reading.

GLOSSARY

Finite Discrete Random Variable. A finite discrete random variable with finite value set; values with adjacent magnitudes are separated by gaps.

Continuous Random Variable. A continuous random variable is a random variable with an infinite, real value, value set; every real point in the interval defined by the random variable is a member of the value set of the random variable.

Definite Integral. The definite integral of $f(x)$ from a to b is

$$\int_a^b f(x)\,dx = \lim_{n \to \infty} \sum_{i=1}^n f(x')(x_{i+1} - x_i)$$

where x' is the midpoint between x_i and x_{i+1}, $x_1 = a$ and $x_{n+1} = b$.

Density Function. Any positive function $f(x)$ of a random variable X with

$$\int_\infty^{-\infty} f(x)\,dx = 1.0$$

is a density function of the random variable.

Distribution Function. The distribution function of a random variable is the cumulative probability density function of the random variable defined at each point a in the value set in the random variable:

$$F(a) = \int_{-\infty}^a f(x)\,dx$$

Theorem 1.

$$\int_a^b f(x)\,dx + \int_b^c f(x)\,dx = \int_a^c f(x)\,dx$$

Theorem 2.

$$\int_a^b [f(x) + g(x)]\,dx = \int_a^b f(x)\,dx + \int_a^b g(x)\,dx$$

Theorem 3.

$$\int_a^b cf(x)\,dx = c \int_a^b f(x)\,dx$$

(9.1)

$$\int_{-\infty}^b f(x)\,dx \le \int_{-\infty}^a f(x)\,dx, \quad b < a$$

(9.2)

$$F(b) = \int_{-\infty}^b f(x)\,dx$$

Theorem 4. If $x' \geq$ (largest x in X), $F(x') = 1.0$, $\displaystyle\int_{-\infty}^{x'} f(x)\, dx = 1.0$.

Theorem 5. If $x' <$ (smallest x in X), $F(x') = .0$, $\displaystyle\int_{-\infty}^{x'} f(x)\, dx = .0$.

Theorem 6. If $a \geq b$ then $F(a) \geq F(b)$, and if $a = b + h$ then

$$\int_{-\infty}^{a} f(x)\, dx = \int_{-\infty}^{b} f(x)\, dx + \int_{b}^{b+h} f(x)\, dx$$

Theorem 7. If $a \leq x \leq b$, $P(a \leq x \leq b) = F(b) - F(a)$,

$$P(a \leq x \leq b) = \int_{-\infty}^{b} f(x)\, dx - \int_{-\infty}^{a} f(x)\, dx$$

Theorem 8.

$$\int_{a}^{\infty} f(x)\, dx = 1 - \int_{-\infty}^{a} f(x)\, dx$$

(9.3)
$$\mu = \int_{-\infty}^{\infty} x f(x)\, dx$$

(9.4)
$$\sigma^2 = \int_{-\infty}^{\infty} (x - \mu)^2 f(x)\, dx$$

PROBLEMS

1. For the uniform probability density function on the interval $(3 \leq x \leq 14)$, find the probabilities for the following subintervals:

   (a) $3 \leq x \leq 7$
   (b) $2.7 \leq x \leq 3.0$
   (c) $4 \leq x \leq 10$ or $11 \leq x \leq 14$
   (d) $4 \leq x \leq 10$ or $3 \leq x \leq 7$
   (e) $4 \leq x \leq 10$ and $7 \leq x \leq 14$
   (f) $4 \leq x \leq 10$ and $11 \leq x \leq 14$
   (g) $1 \leq x \leq 20$

2. For the probability distribution

$$f(x) = \begin{cases} 4x, & 0 \leq x \leq .5 \\ 4(1 - x), & .5 \leq x \leq 1 \\ 0, & \text{elsewhere} \end{cases}$$

find the probabilities for the following subintervals:

(a) $x \leq .5$

(b) $x \geq .5$

(c) $.25 \leq x \leq .5$

3. For the density function of problem 2 the areas under the graph of the function for intervals from 0 to x for $x \leq .5$ are given by

$$\tfrac{1}{2}(x \times 4x) = 2x^2$$

Use this result to evaluate

(a) $\displaystyle\int_{-\infty}^{0.1} f(x)\, dx$

(b) $\displaystyle\int_{-\infty}^{0.25} f(x)\, dx$

(c) $P(.2 \leq x \leq .25)$

4. For the density function specified in problem 2 the distribution function is

$$F(x) = \begin{cases} 2x^2 & \text{for } 0 \leq x \leq .5 \\ 1 - 2(1 - x)^2 & \text{for } .5 \leq x \leq 1.0 \end{cases}$$

Try to reconstruct the argument leading to these equations. Evaluate the following for the density function of problem 2:

(a) $P(x \leq .75)$

(b) $P(.25 \leq x \leq .75)$

(c) $P(x \leq .75)$

(d) $P(x \leq .25 \text{ or } x \geq .75)$

(e) $P(x \leq .1 \text{ or } x \geq .9)$

(f) $P(x \leq .2 \text{ and } x \geq .8)$

5. By evaluating a few points in the distribution function of the uniform probability density function on the unit interval, make a conjecture as to its equation. Now construct a mathematical argument regarding the distribution function: Let the base of an arbitrary interval with a lower limit of 0 and an upper limit, x less than 1.0, be x; the area is height times base; the base is simply x; and the height is 1 for all points.

6. Which of the functions graphed in the figure below could be probability density functions? Construct graphs of the distribution functions of the functions that are density functions.

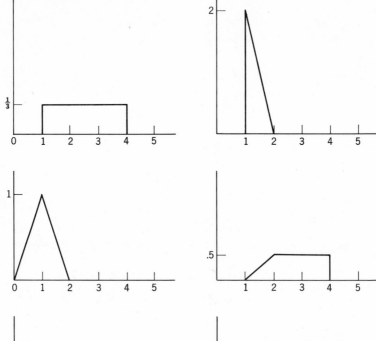

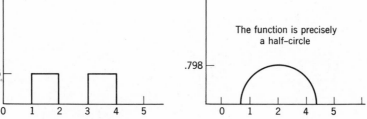

The function is precisely a half-circle

CHAPTER 10

A Continuous Random Variable: The Normal Distribution

Texts on probability and statistics frequently allude to the "remarkable" fact that so many empirical frequency functions seem to have the same general form. And it is true that many empirical frequency distributions are characterized by (1) heavy concentration of frequency in a narrow range of observed values; (2) symmetry about this center of concentration; (3) gradual decrease in frequency, followed by rapid decrease, followed in turn by gradual decrease as observed values more and more removed from the center of concentration are considered; (4) rarity of values extremely removed from the center of concentration. Other forms of frequency functions are not rare, but in many areas of behavioral science this so-called normal or Gaussian or bell-shaped form is familiar. For example, biological, aptitude, and performance measures are largely distributed in this form; psychophysical functions, learning measures, physiological responses, and so on, frequently have this form.

In addition, it has been shown by rigorous mathematical deduction that, under certain very general conditions, averages of observations on random variables have the "normal" form of distribution. That is, if we had many means from samples of observations on a given random variable, they *must* have a frequency function that is approximately normal in form. This result holds regardless of the form of the probability distribution or nature of the random variable observed, within certain very broad restraints.

The reader may have anticipated the point that the form of frequency function described in the first paragraph of this chapter is roughly the same as the probability distribution of the binomial random variable. It can be shown rigorously that this similarity is more than skin deep. When the number of binomial trials is large, the binomial distribution is approximated by the normal distribution. Thus, when n is large enough to insure that the approximation is a good one, the normal distribution may be substituted with nominal error (as determined by the degree of approximation). Other probability distributions have the normal probability function as an approximating "limit" as well as the binomial. Examples of these develop out of our work in statistical tests in the chapters to follow.

The seeming diversity of conditions and experimental situations in which the normal probability distribution and "normal" frequency functions are found is illusory. The examples of situations where the normal distribution is found all have a unifying characteristic: The variables involved are all composites of many other variables or can be traced to the mutual operation of many factors or components. This unifying principle is spelled out in detail later in this chapter.

The plan of this chapter is first to discuss the case of the normal approximation to the binomial distribution. This serves to illustrate the characteristics of the distribution in terms of an already familiar distribution. The formal properties of the distribution develop out of this discussion in a natural way. The concepts of standardization and standard normal distributions come out of this presentation in a natural way also.

Following this discussion, substantive psychological considerations are presented, rationalizing the assumption of the normality of distributions of observed phenomena. That is, we argue from psychology that normally distributed random variables can be taken as models of behavior under certain circumstances.

In many applications of the normal theory, arguments from psychology are not needed. The justification of statistical procedures can be made from the central limit theorem, that is, on mathematical grounds alone. This theorem states that means of large samples of observations of random variables are normally distributed. A nonmathematical presentation of this theorem is made.

The Normal Approximation to the Binomial

Recall that the expectation and variance of the binomial random variable S with distribution $B(n, p)$ are $\mu = np$, and $\sigma^2 = np(1 - p)$. These values become large as a function of the size of n. For S_1 with $B(100, .5)$, S_2 with $B(1000, .5)$, and S_3 with $B(10000, .5)$ the respective expectations and variances are

$$\mu_1 = 50 \qquad \sigma_1^2 = 25$$

$$\mu_2 = 500 \qquad \sigma_2^2 = 250$$

$$\mu_3 = 5000 \qquad \sigma_3^2 = 2500$$

As $n \to \infty$, μ and σ^2 also increase without limit.

The increase of μ and σ^2 as a function of increase in n poses a problem in comparing distributions for different values of n. The locations and dispersions of binomial random variables vary as functions of n. Figure 10.1 illustrates the progressive changes in $B(n, p)$ with $p = .5$ and $n = 2$, $n = 8$, and $n = 16$.

A change in the units of the value set of the random variable can be made to bring the distributions into a common "focus." The change to be made modifies both the location and the dispersion of the distribution. The first step is to change S into a random variable S' so that $E(S') = 0$. That is, we shift the location of S to zero and call the relocated random variable S'. This is accomplished by subtracting $E(S)$ from S: $S' = S - E(S)$. This shift is represented in the upper portions of Figure 10.2. The differences of adjacent values in the value sets of S' and S are identical, but the value magnitudes are modified. The reader should note that we have given up a real origin for an arbitrary origin and, as a consequence, the magnitudes represented in S' are "arbitrary."

The second step is to change the unit of the value set. The value sets of S and S' have as units the value associated with one success. In place of the natural units we wish to use the units of dispersion. We simply find the number of successes in terms of the natural units relative to our measure of dispersion. The measure of dispersion in the same units as the natural set is the standard deviation, σ_S. We replace the value of σ_S by the value 1.0 and adjust all other values in the original value set accordingly. This is accomplished by finding S^* for every value in S': $S^* = S'/\sigma_S$. This step is represented in the lower portion of Figure 10.2. Replacing S' with $S - E(S)$ as per the definition of S' gives

$$S^* = \frac{S - E(S)}{\sigma_S} \tag{10.1}$$

which is called the standardized binomial random variable.

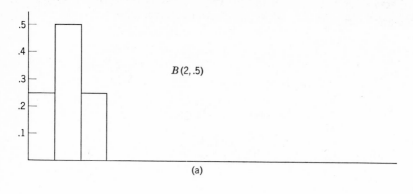

(a)

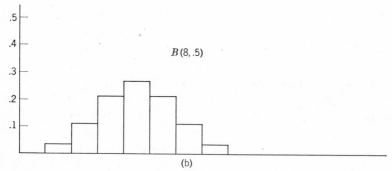

(b)

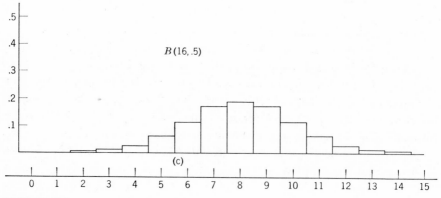

(c)

FIGURE 10.1

Three binomial random variables.

The expectation and variance of S^* can be calculated by the theorems on expectation and variance. Since $1/\sigma_S$ and $E(S)$ are constants,

$$E(S^*) = E\left[\frac{S - E(S)}{\sigma}\right]$$

$$= \frac{1}{\sigma} E[S - E(S)]$$

$$= \frac{1}{\sigma} [E(S) - E(S)]$$

$$= 0$$

$$\text{var}(S^*) = \text{var}\left[\frac{S - E(S)}{\sigma}\right]$$

$$= \frac{1}{\sigma^2} \text{var}[S - E(S)]$$

$$= \frac{1}{\sigma^2} \text{var}(S)$$

$$= \frac{\sigma^2}{\sigma^2} = 1$$

Thus S^*, the standardized binomial, has expectation of zero and variance of 1. The values n and p play no role in the expectation and variance of S^*. However, n and p are vital in finding S^* from S.

Nothing has been said throughout this discussion of standardization regarding the probability distribution of S, S', or S^*. The transformations are executed on the value sets of the random variables, not on the probability distributions. Since there is a one-to-one identity of specific values in S and S^*, the nominal sample spaces are not different and hence for all $s \in S$ and the corresponding $s^* \in S^*$,

$$P(S = s) = P\left[s^* = \frac{s - E(S)}{\sigma}\right]$$

However, a striking thing happens to the histogram of the distribution when we go from S to S^*. The bar on a histogram for S covers the interval from $(s - .5)$ to $(s + .5)$ and thus has a base of unit width. However, the limits of the corresponding interval on S^* are

$$\frac{s - .5 - E(S)}{\sigma} \quad \text{to} \quad \frac{s + .5 - E(S)}{\sigma}$$

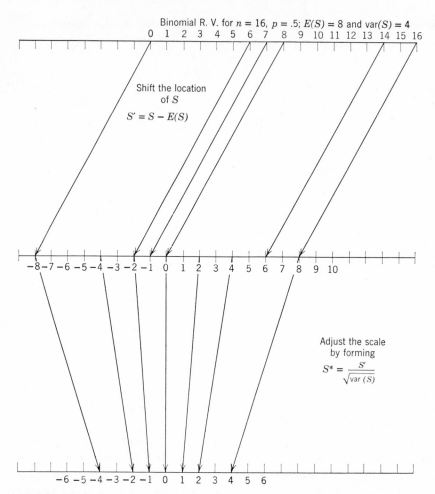

FIGURE 10.2

Illustration of the standardization of a random variable.

The intervals have lengths given by

$$\frac{s + .5 - E(S)}{\sigma} - \frac{s - .5 - E(S)}{\sigma} = \frac{s - s + .5 + .5 - E(S) + E(S)}{\sigma}$$

$$= \frac{1}{\sigma}$$

which is in general not equal to 1. Thus, if the area enclosed by the bars of the histograms is to represent the probabilities involved, the height must be different from the probabilities. The height of the bar over s^* must be equal to $f(s^*)$, which must satisfy the height-times-base equation

$$\text{(height)} \times \text{(base)} = f(s^*) \times \left(\frac{1}{\sigma}\right) = P(S^* = s^*)$$

for probability represented as an area, and consequently

$$f(s^*) = \sqrt{npq}\, P(S^* = s^*) \tag{10.2}$$

A direct evaluation of this when n is large is not practical. Furthermore, when n is large the difficulty in calculating $P(s_a{}^* \leq S^* \leq s_b{}^*)$ with several terms s_a, $s_{a+1}, \ldots, s_b$ is compounded. Fortunately, when n is large a well-tabled approximation is available.

It can be shown that where $n \to \infty$

$$f(s^*) = \frac{1}{\sqrt{2\pi}} e^{-.5s^{*2}}$$

Any real-valued variable X that has a density function

$$f(x) = \frac{1}{\sqrt{2\pi}} e^{-.5x^2}$$

is called a normally distributed random variable with $\mu = 0$ and $\sigma^2 = 1$, or briefly a unit normal random variable. A special symbol is set aside to denote this important distribution; the density function of a normally distributed random variable X at a value x is denoted $\phi(x)$ where

$$\phi(x) = \frac{1}{\sqrt{2\pi}} e^{-.5x^2} \tag{10.3}$$

Applying the concept of area under a curve in this case to get $P(s_a{}^* \leq S^* \leq s_b{}^*)$ we have, for $I = \{s^* \mid s_a{}^* \leq s^* \leq s_b{}^*\}$ and where $s^* = (s - np)/\sqrt{npq}$,

$$\frac{1}{\sqrt{npq}} \sum_{s^* \in I} f(s^*) = \sum_{s^* \in I} \binom{n}{s} p^s q^{n-s}$$

which can be shown to have the limit

$$\frac{1}{\sqrt{2\pi}} \int_{x_a}^{x_b} e^{-.5x^2}\, dx = \int_{x_a}^{x_b} \phi(x)\, dx$$

where

$$x_a = \frac{s_a - .5 - np}{\sqrt{npq}}$$

$$x_b = \frac{s_b + .5 - np}{\sqrt{npq}}$$

The integral of the normal density function is represented by the special symbol $\Phi(x)$ where

$$\Phi(x') = \int_{-\infty}^{x'} \phi(x)\, dx = \frac{1}{\sqrt{2\pi}} \int_{-\infty}^{x'} e^{-.5x^2}\, dx \tag{10.4}$$

and is called the normal distribution function. Thus, for x_a and x_b

$$P(s_a{}^* \le S^* \le s_b{}^*) = P(x_a \le x \le x_b)$$
$$= \frac{1}{\sqrt{2\pi}} \int_{x_a}^{x_b} e^{-.5x^2}\, dx$$
$$= \frac{1}{\sqrt{2\pi}} \int_{-\infty}^{x_b} e^{-.5x^2}\, dx - \frac{1}{\sqrt{2\pi}} \int_{-\infty}^{x_a} e^{-.5x^2}\, dx$$
$$= \Phi(x_b) - \Phi(x_a)$$

We turn now to a brief discussion of the function $\Phi(x)$ and its properties. Then we demonstrate the approximation of the binomial by the normal.

Normal Distribution and Density Functions

The graph of $\phi(x)$ is given in Figure 10.3 and the graph of $\Phi(x)$ is given in Figure 10.4. A table of the areas under the normal density function, Table C, is given in

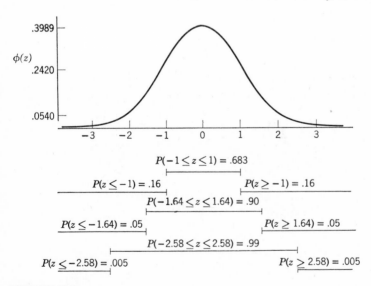

FIGURE 10.3

Graph of the normal density function, showing probability of certain regions.

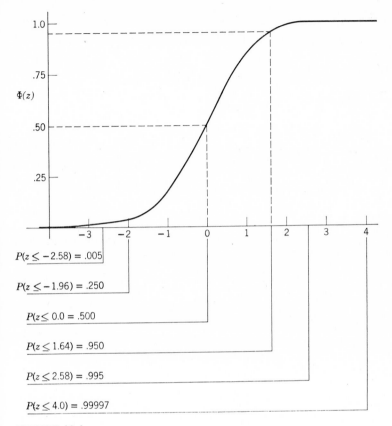

$P(z \leq -2.58) = .005$

$P(z \leq -1.96) = .250$

$P(z \leq 0.0 = .500$

$P(z \leq 1.64) = .950$

$P(z \leq 2.58) = .995$

$P(z \leq 4.0) = .99997$

FIGURE 10.4

Graph of the normal distribution function, showing the probability of certain regions.

the back of the book in the Appendix. The table gives $\Phi(x)$ for the values of $x \geq 0$ only. Figure 10.5 illustrates the area $\Phi(x)$ tabled for the value x.

Several specific characteristics of the normal distribution and density function particularly important are the following.

(1) The density function $\phi(x)$ is symmetric about $x = 0$, that is,

$$\phi(x) = \phi(-x) \tag{10.5}$$

(2) The symmetry of $\phi(x)$ implies that

$$\int_{-\infty}^{-a} \phi(x)\, dx = \int_{a}^{\infty} \phi(x)\, dx = 1 - \int_{-\infty}^{a} \phi(x)\, dx$$

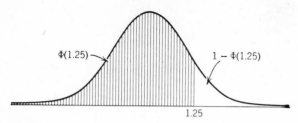

FIGURE 10.5

Graph of the normal density function, illustrating the relationship between area and the distribution function.

or

$$\Phi(-a) = 1 - \Phi(a) \tag{10.6}$$

(3) The probability for extreme values on a normally distributed random variable are very rare, indeed:

$$\Phi(4) \geq .9999$$

(4) The probability of regions or intervals on the random variables are calculated by the familiar use of the distribution function values

$$P(a \leq x \leq b) = \Phi(b) - \Phi(a) \tag{10.7}$$

If $a < 0$, substitute $1 - \Phi(-a)$, giving

$$\Phi(b) - [1 - \Phi(-a)] = \Phi(b) + \Phi(-a) - 1 \tag{10.8}$$

An example of this type of interval is given in Figure 10.6. For $P(x \geq b$ or $x \leq a)$, $a \leq b$ the formula is

$$1 - \Phi(b) + \Phi(a)$$

An example of this type of region (two parts are involved) is given in Figure 10.7.

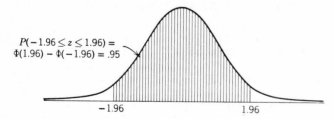

FIGURE 10.6

Probability of an interval on the normal variable, expressed as an area and as a function of Φ.

$\Phi(-1.96) = .025$ $1 - \Phi(1.96) = .025$

FIGURE 10.7

Probabilities of intervals in the tails of a normal random variable, expressed as areas and as function of Φ.

Standardization and Nonstandard Normal

In the discussion of the unit normal random variable we developed the logic of the standardized binomial random variable

$$S^* = \frac{s - np}{\sqrt{npq}}$$

which has zero expectation and unit variance. The density and distribution functions $\phi(x)$ and $\Phi(x)$ illustrated in Figures 10.3 through 10.7 were shown centered at zero. Since the distribution of S^* is approximately normal for $n \to \infty$ and since

$$E(S^*) = 0 \qquad \text{var}(S^*) = 1$$

we might expect a normally distributed random variable X to have

$$E(X) = 0 \qquad \text{var}(X) = 1$$

This is borne out by evaluating first

$$\mu = \int_{-\infty}^{\infty} x\phi(x)\, dx$$

and then

$$\sigma^2 = \int_{-\infty}^{\infty} (x - \mu)^2 \phi(x)\, dx$$

These evaluations require calculus not available to us here, but give

$$\mu = 0 \qquad \sigma^2 = 1$$

To convey all of this information about X in a simple statement, we say the X is a normally distributed random variable with zero expectation and unit variance, or more succinctly that X is a unit normal distributed random variable. This is symbolized by

$$X: N(0, 1)$$

There is a large family of random variables that have distributions similar to $N(0, 1)$ except for values of μ and σ^2. In fact, there is an unlimited number of variables Y such that, when we calculate

$$Z = \frac{Y - \mu}{\sigma}$$

we obtain a normally distributed random variable Z, where $Z: N(0, 1)$.

We say that Y is normally distributed with μ and σ^2 when

$$Z = \frac{Y - \mu}{\sigma}$$

is normally distributed, $Z: N(0, 1)$. The symmetry of the definitions of Z and Y lead to the notation

$$Y: N(\mu, \sigma^2)$$

where μ and σ^2 may be any values ($\sigma^2 > 0$). Thus, if $\mu_Y = 10$ and $\sigma_Y{}^2 = 42$, then

$$Y: N(10, 42)$$

if

$$Z = \frac{Y - 10}{\sqrt{42}}$$

is such that $Z: N(0, 1)$. Then Y is said to be normally distributed but not standard normal, and Z is a standard normal random variable. Figure 10.8 shows graphs

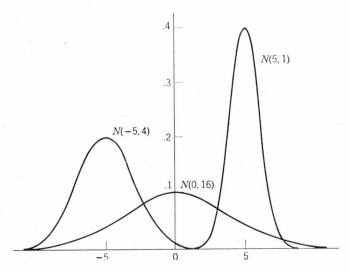

FIGURE 10.8

Graph of three nonstandard normal density functions.

of three nonstandard normally distributed random variables, X_1: $N(0, 16)$, X_2: $N(5, 1)$, and X_3: $N(-5, 4)$.

The reader is cautioned that

$$Z = \frac{X - E(X)}{\sigma_X}$$

gives Z: $N(0, 1)$ only when X is $N[E(X), \sigma_X{}^2]$. If X is not normally distributed, then Z is not normally distributed even though $E(Z) = 0$ and $\sigma_Z{}^2 = 1$. In the case where Z is not normal it is referred to as a standardized variable, but standardization does not imply normality of distribution.

Goodness of Approximation of Normal to Binomial

The degree of accuracy in the normal approximation to the binomial is a function primarily of the parameter n of $B(n, p)$. When p is extreme, however, larger values of n are required to achieve the level of accuracy achieved with less extreme values of p.

Figure 10.9 shows the density functions of the binomial random variable and the normal approximations for $n = 6$ and $n = 30$ for each of $p = .1$ and $p = .5$.

The Distribution of Sums of Random Variables

Much of the work in developing tests of statistical hypotheses is based on an extremely general result in the theory of random variables. This result states that if a random variable is expressible as the sum of a large number of independent random variables, then it is normally distributed.

Imagine $S_n = X_1 + X_2 + \cdots + X_n$ where for all $i = 1, \ldots, n$; $\mu_i = E(X_i)$ and $\sigma_i{}^2 = \text{var}(X_i)$. The random variable S_n has

$$E(S_n) = \mu_1 + \cdots + \mu_n \qquad \text{var}(S_n) = \sigma_1{}^2 + \cdots + \sigma_n{}^2$$

by the theorems on expectation and variance of sums of random variables. Note that the independence condition is used in obtaining σ^2.

We do not attempt a proof here, but it can be shown that S_n is approximately normally distributed as $n \to \infty$. This is stated as the central limit theorem (CLT).

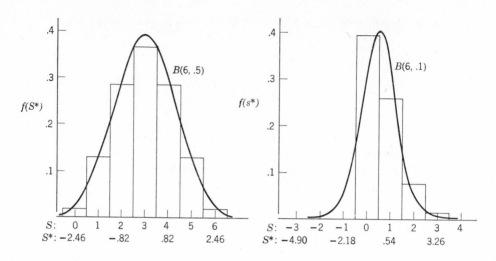

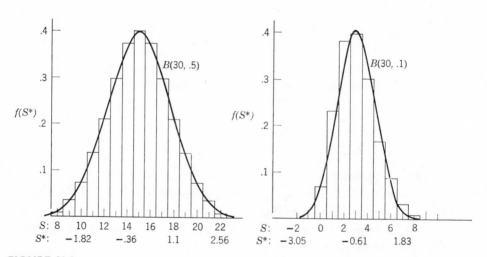

FIGURE 10.9

Graphs illustrating the goodness of approximation of the normal distribution to the binomial distribution.

Central Limit Theorem (CLT). For any arbitrary but independent random variables X_1, X_2, ..., X_n the sum

$$S_n = X_1 + X_2 + \cdots + X_n$$

has a limiting normal distribution with $n \to \infty$ with expectation

$$E(S_n) = \mu_1 + \cdots + \mu_n$$

and variance
$$\text{var}(S_n) = \sigma_1^2 + \cdots + \sigma_n^2$$

This theorem implies that for every real number $a < b$

$$\lim_{n \to \infty} P\left[a < \frac{S_n - E(S_n)}{\sqrt{\text{var}(S_n)}} < b \right] = \Phi(b) - \Phi(a)$$

That is, $[S_n - E(S_n)]/\sqrt{\text{var}(S_n)}$ is $N(0, 1)$.

The special case where X_i is the binomial random variable for a single repetition already has been demonstrated. The binomial random variable with n trials is nothing more or less than the sum of n binomial random variables with one trial. Hence we can conclude from the CLT that where S_n is binomial with $B(n, p)$, then $S^* = (S_n - np)/\sqrt{npq}$ is $N(0, 1)$.

In the following chapter we use the CLT to show that averages of observations on random variables are normally distributed in the limit regardless of the distributions of the random variables observed.

Sources of Normal Distributions in Psychology

In this section we consider two ways of conceptualizing the source of behavioral observations that lead to the development of normal distribution models. These two points of view are set in opposition in much of the literature on psychological measurement. We treat them separately, but this treatment should not be taken as an indication of agreement with the either-or views. The possibility of other sources of observations and of mixtures of sources must be reserved. However, for our purposes we view measurements as (1) the outcome of multiple (or partial) but specific causal factors in nature or (2) the combination of some "true" state of nature and errors of observation.

Polygenic Determination of Behavior. Imagine a behavioral observation where the source of the behavior can be linked with the activity or operation of basic biological units such as muscle fibers, neurons, genes, synaptic connections, biochemical reactions in the bloodstream, and so on. It is not unreasonable to assume that each activity or operation at some primitive level is all or none in the sense that it either happens or does not. A given genetic composition occurs or it does not; a nerve cell transmits or it does not; a muscle cell contracts or it does not; a molecule of ACh is hydrolized in the synaptic transport mechanism or it is not. Many such discrete events combine to produce overall behavior.

If this seems too atomistic or reductionistic, the reader might prefer to base the following development on Hebbian cell assemblies, reaction thresholds, behavior tendencies, and so on, that are active or not, are surpassed or not, are acted on or not, and so on. The only things we must have are a large number of elements, each contributing to the observed behavior or property being studied. We take as our example a genetic basis for the behavior or property being studied. The genetic terms might well be replaced with neurological terms, physiochemical terms, and so on. With minor modifications owing to specifics in each subject, the argument we cast would be different. However, the general form of the argument holds in many topics.

It is generally accepted genetical principal that a wide variety of phenotypic characteristics are determined by multiple pairs of genes. Each of the gene pairs exerts a small but additive effect on the phenotype. The simplest gene action is to contribute a discrete two-valued condition, present or absent usually. However, the two-state gene action is the simplest sort, and multiple-state gene action is also possible. The contribution of each gene action-state is some specific value. For example, two individuals (parents) contribute genes to an offspring at a given chromosomal locus that can have four states, x_1, x_2, x_3, x_4. For an individual selected at random from a population where the alleles are randomly assorted from parents contributing one each, the individuals observable are defined by genotypes:

$$(x_1, x_1), (x_1, x_2), (x_1, x_3), (x_1, x_4), (x_2, x_2),$$

$$(x_2, x_3), (x_2, x_4), (x_3, x_3), (x_3, x_4), (x_4, x_4)$$

Assume that the phenotype is determined in an additive fashion from the values associated with the states x_1, x_2, x_3, and x_4. Illustratively we let

$$x_1 = 0, \qquad x_2 = .1, \qquad x_3 = .2, \qquad x_4 = .3$$

Then we have the following value set for the phenotype characteristic, V:

$$V(x_1, x_1) = x_1 + x_1 = 0 \quad \text{can occur one way}$$
$$V(x_1, x_2) = x_1 + x_2 = .1 \quad \text{can occur two ways}$$
$$V(x_1, x_3) = x_1 + x_3 = .2 \quad \text{can occur two ways}$$
$$V(x_1, x_4) = x_1 + x_4 = .3 \quad \text{can occur two ways}$$
$$V(x_2, x_2) = x_2 + x_2 = .2 \quad \text{can occur one way}$$
$$V(x_2, x_3) = x_2 + x_3 = .3 \quad \text{can occur two ways}$$
$$V(x_2, x_4) = x_2 + x_4 = .4 \quad \text{can occur two ways}$$
$$V(x_3, x_3) = x_3 + x_3 = .4 \quad \text{can occur one way}$$
$$V(x_3, x_4) = x_3 + x_4 = .5 \quad \text{can occur two ways}$$
$$V(x_4, x_4) = x_4 + x_4 = .6 \quad \text{can occur one way}$$

Gathering the specific possible individual values and noting the total number of ways each value may occur, we have the phenotype and frequencies of Table 10.1. Thus, for this phenotype determined by a single gene locus, a random individual from the population of offspring from parents in which x_1, x_2, x_3, and x_4 are contributed at random (as in random mating) would have $v = .2$ with probability $\frac{3}{16}$.

It can be shown by arguments similar to those developed above in showing that $B(n, p)$ was approximately $N(np, npq)$ that the distribution of the phenotype values in these examples are approximated by a normal distribution. Our example is a special case of the multinomial (many-state as contrasted with bistate binomial random variable). As n increases in the multinomial V with $m(n, p_1, p_2, \ldots, p_r)$, the form of the distribution of V approaches the form of a normally distributed random variable. Consequently, if we standardize V by

$$V^* = \frac{V - E(V)}{\sigma_V}$$

V^* has a distribution approaching $N(0, 1)$.

The reader will have noticed that the binomial is a special case of the multinomial. Instead of multiple states the binomial has only two states on each trial—in this case a combination of genes through mating.

These examples are the simplest sort. More complex examples are possible with the same general result. However, care must be taken to avoid concluding that every normally distributed behavior is genetically determined, or that all genetically determined behavior is normally distributed. The most general conclusion warranted is that a normal model *may* be an appropriate model for behavior when a polygenetic basis for the behavior can be established.

It may be objected that our development of the normal limit for the distribution of genetically determined behaviors assumed (tacitly) independence of the genes. This point is well taken, but we need not make any such assumption. The CLT in its most general form holds for large classes of dependent random variables. In addition, if a majority of the random variables are independent, then the CLT as presented above is adequate. Thus, if a quantitative phenotype is expressible as the mean of a large number of variables, whatever their distribution, the phenotype is

TABLE 10.1 Phenotype Values and Probabilities

V	0	.1	.2	.3	.4	.5	.6
$n(V = v)$	$\frac{1}{16}$	$\frac{2}{16}$	$\frac{3}{16}$	$\frac{4}{16}$	$\frac{3}{16}$	$\frac{2}{16}$	$\frac{1}{16}$

normally distributed. Each pair of genes influencing the phenotype corresponds to an observation on one of the variables. In the two-state case the variables have two values, as in the binomial, although the values need not be zero and one.

The Normal Distribution as an Error Distribution. The traditional derivation of the normal distribution is in terms of the distribution of errors. The quantities observed are said to be a combination of some true value and the chance occurrence of potentially many discrete and separate errors in the process of observation. For example, a finger maze is said to have a fixed value X of difficulty. Measures Y of this value are viewed as a function g of the fixed value, and an error term

$$Y = g(X) = X + \mathscr{E}$$

where $\mathscr{E}$ is the error term.

The error term $\mathscr{E}$ results from the operation of many separate "partial errors." These partial errors are attributed to such factors as momentary fluctuations of attention, motivation, noise in the testing environment, chance motions into correct alleyways. Each of these factors contributes a specific value, such as e, to the overall value observed, Y. If the values of e obey certain simple rules, Y is distributed approximately normally.

This model of experimental observations is widely used in experimental psychology where the major interest lies in stimulus, instruction, situational, and other conditions under which the observations are made. The biological differences in individual subjects (persons, animals, and so on) are dealt with as contributors to the error terms. This handling of individual differences is not universal, and often different strains of animals, different populations of people, and so on are treated as basic factors leading to differences in "true" values of observable properties. In this case, strain differences, for instance, are taken as the fundamental factor and all intrastrain differences are assumed to be random and lumped with the error of observation.

Imagine a population of subjects highly homogeneous with respect to a large number of biological factors that would influence accuracy of judgments of the length of a line presented visually under some standard condition. A subject is presented the line and he estimates (verbally, for example) its length X in centimeters. The subject's estimate is a random variable Y:

$$Y = g(X) = X + \mathscr{E}$$

In the process of inspecting, judging, and reporting the estimated X, there are many possibilities of error, some of which lead to a response value Y larger than X and some of which lead to response value Y smaller than X.

Many physical and physiological sources of error can be postulated: fluctuations in light falling on the stimulus, diffraction and scattering of light reflected and entering the eye, physiological nystagmus, fluctuations in retinal thresholds, neural mishaps in transmission, response production vagaries.

First, assume that the magnitude of each partial error is small in comparison with the overall magnitude of the total error, regardless of the direction of the partial error. This permits us to consider all partial errors to be of the same magnitude (with respect to the total error). For example, if the total error is accumulated over 100,000 units and the separate errors range (in absolute value) from .001 and .006 units it makes little difference that some errors are six times as large as others. All are about the same order of magnitude with respect to 100,000 units.

Second, assume that the sign of an error magnitude is just as likely to be positive as it is to be negative. Let m be the average magnitude of a single error and the probability that $m > 0$ be p and $P(m \leq 0) = 1 - p$.

Third, we assume that there are n errors contributing to the judgment. If all n errors are positive, the observed value is

$$Y = X + nm$$

If $(n - 1)$ errors are positive and 1 error negative, then

$$Y = X + (n - 1)m - m$$
$$= X + (n - 2)m$$

If this is carried out completely, finally producing n negative errors and no positive errors, we obtain the $(n + 1)$ values of Y given as rows of Table 10.2. The probability of observing each of these values is easily calculated by a familiar formula. The probability of all n errors being positive is simply p^n: Each of the n errors must be positive and if each error occurs independently of the others, the multiplication law for independent events gives p^n. By the same reasoning, one particular set of $(n - 1)$ positive errors and 1 negative error is $p^{n-1}q$. However, there are $\binom{n}{1}$ ways to have one of n errors positive and $(n - 1)$ negative. Thus,

$$P[Y = X + (n - 2)m] = \binom{n}{1} p^{n-1}q$$

These probabilities are familiar as the probabilities of the binomial random variable. The values of Y are not equal to the values of the binomial. However, the probability of $(n - k)$ successes out of n binomial trials is the same as $(n - k)$ positive

TABLE 10.2 Tabulation of an Error Distribution

Number of Negatives and Positives	Observed Value of Y	Probability of Observing the Value
n are $+$ 0 are $-$	$Y_n = X + nm$	p^n
$(n-1)$ are $+$ 1 is $-$	$Y_{n-1} + X + (n-2)m$	$\binom{n}{1} p^{n-1} q$
$(n-2)$ are $+$ 2 are $-$	$Y_{n-2} + X + (n-4)m$	$\binom{n}{2} p^{n-2} q^2$
$\cdot$ $\cdot$ $\cdot$	$\cdot$ $\cdot$ $\cdot$	$\cdot$ $\cdot$ $\cdot$
$(n-r)$ are $+$ r are $-$	$Y_{n-r} + X + (n-2r)m$	$\binom{n}{n-r} p^{n-r} q^r$
$\cdot$ $\cdot$ $\cdot$	$\cdot$ $\cdot$ $\cdot$	$\cdot$ $\cdot$ $\cdot$
2 are $+$ $(n-2)$ are $-$	$Y_2 = X + 2m - [(n-2)m]$ $= X - (n-2)m$	$\binom{n}{2} p^2 q^{n-2}$
1 is $+$ $(n-1)$ are $-$	$Y_1 = X + m - [(n-1)m]$ $= X - (n-2)m$	$\binom{n}{1} p q^{n-1}$
0 are $+$ n are $-$	$Y_0 = X - nm$	q^n

errors out of n total errors. Thus

$$P[S_n = (n-k) \mid S_n : B(n,p)] = P[Y = X + (n-2k)m \mid p]$$

$$= \binom{n}{n-k} p^{n-k} q^k$$

The tabulation of these probabilities is given in Table 10.2.

The consecutive values of Y have equal differences:

$$Y_{n-k+1} - Y_{n-k} = X + [(n-2)(k-1)]m - [X + (n-2k)m]$$
$$= X - X + nm - nm + 2km - 2m(k-1)$$
$$= 2mk - [2mk - 2m]$$
$$= 2m$$

The area enclosed in this interval is set equal to the probability of observing Y by letting

$$P(Y_{n-k}) = (\text{height})(\text{base}) = f(Y_{n-k})2m$$

Since

$$P(Y_{n-k}) = \binom{n}{n-k} p^{n-k} q^k$$

$$f(Y_{n-k}) = \frac{1}{2m} \binom{n}{n-k} p^{n-k} q^k$$

Now we find ourselves at a point in the derivation of the normal approximation to the binomial distribution, except that there the constant $\frac{1}{2}m$ was the constant $\sqrt{npq}$. Since the value of the constant term does not materially affect the form of the function, the error distribution is seen to have the same limiting form as the binomial. Following the same logic, we find that

$$Y^* = \frac{Y - E(Y)}{\sigma_Y}$$

is $N(0, 1)$. The values of $E(Y)$ and σ_Y depend on X, m, p, and q, and it serves no useful purpose to develop them here. Note, however, that if $p = .5$, the distribution of Y has an expectation equal to the true value X. Where $p > q$ the expected value is greater than X, and where $p < q$ the expected value is less than X.

Thus, where there are $(n \to \infty)$ minute and independent sources of errors contributing to an observed variable, the distribution of observations is normally distributed. When n is large (in practice n need not be much more than 20 or 30), the empirical distribution is approximated by the normal distribution. Hence the normally distributed random variable may serve as an appropriate model for the experimental observations.

SUGGESTED READING

Of the previously listed sources, see particularly the books by Cramér, Hays, Lewis, and Stilson.

GLOSSARY

Standardization. A random variable is a standard random variable if its expectation is 0 and its variance is 1. Standardization is achieved by dividing the difference of the random variable and its expectation by the standard deviation of the variable.

(10.1)
$$S^* = \frac{S - E(S)}{\sigma_S}$$

(10.2)
$$f(s^*) = \sqrt{npq}\, P(S^* = s^*)$$

(10.3)
$$\phi(x) = \frac{1}{\sqrt{2\pi}}\, e^{-.5x^2}$$

(10.4)
$$\Phi(x') = \int_{-\infty}^{x'} \phi(x)\, dx = \frac{1}{\sqrt{2\pi}} \int_{-\infty}^{x'} e^{-.5x^2}\, dx$$

(10.5)
$$\phi(x) = \phi(-x)$$

(10.6)
$$\Phi(-a) = 1 - \Phi(a)$$

(10.7)
$$P(a \leq x \leq b) = \Phi(b) - \Phi(a)$$

(10.8)
$$\Phi(b) - [1 - \Phi(-a)] = \Phi(b) + \Phi(-a) - 1$$

Unit Normal Random Variable. A unit normal random variable has a normal density function with $\mu = 0$ and $\sigma^2 = 1$: $N(\mu, \sigma^2) = N(0, 1)$.

Central Limit Theorem (CLT). For any arbitrary but independent random variables, $X_1, X_2, \ldots, X_n$, the sum

$$S_n = X_1 + \cdots + X_n$$

has a limiting normal distribution with $n \to \infty$ with expectation

$$E(S_n) = \mu_1 + \cdots + \mu_n$$

and variance

$$\mathrm{var}(S_n) = \sigma_1^2 + \cdots + \sigma_n^2$$

that is,

$$[S_n - E(S_n)]/\sqrt{\mathrm{var}(S_n)} \text{ is } N(0, 1).$$

PROBLEMS

1. From the table of the binomial at the back of the book, construct graphs of $B(10, .3)$ and $B(10, .7)$. For each of these two random variables construct a graph of a normal distribution having the expectation and standard deviation of the respective binomial distribution. Plot the normal graphs on the same paper as the binomial graphs. Write an analysis on the differences among the binomial distributions, among the normal distributions, and the relationships between the binomial and normal distributions.

What is the primary way in which the normal distributions fail as approximations to the binomial distributions?

2. Using the table of the unit normal distributed random variable, construct (a) a graph of the unit normal density function, and (b) a graph of the unit normal distribution function. Transform the unit normal distribution to give (1) $N(10, 2)$, (2) $N(-10, 2)$, (3) $N(10, .5)$, and (4) $N(20, 10)$. Plot the density functions and distribution functions on the respective graphs of the unit normal distribution. Comment on the effect of the differences of expectation and variance in the graphs.

3. A random variable X has expectation of 50 and standard deviation of 10. Find the standardized (expectation 0, standard deviation 1) values corresponding to $x_1 = 50$, $x_2 = 33$, $x_3 = 70$, $x_4 = 55$, and $x_5 = 0$. Find the values of $x_6, x_7, \ldots, x_{10}$ corresponding to the standardized values $x_6' = -3.3$, $x_7' = -1.7$, $x_8' = 0$, $x_9' = 5$, and $x_{10}' = 1.65$.

4. Let X be a random variable with distribution $N(0, 1)$. Calculate the probabilities of the following intervals:

   (a) $A = \{x \mid -1.0 \le x \le 1.5\}$
   (b) $B = \{x \mid 0 \le x \le 1.65\}$
   (c) $C = \{x \mid -1.65 \le x \le 0\}$
   (d) $D = \{x \mid 1.65 \le x \text{ and } x \le -1.65\}$
   (e) $E = B' \cup C'$
   (f) $F = D'$
   (g) $G = B \cup C$
   (h) $H = \{x \mid -.25 \le x\}$
   (i) $I = \{x \mid x < -.25\}$
   (j) $J = H \cup I$
   (k) $K = \{x \mid 1.0 < x\}$

5. Find the probabilities of the following, assuming that X is distributed as indicated, $X: N(\mu_X, \sigma_X^2)$:

   (a) $P(0 \le x \le 1)$ in $N(0, 1)$
   (b) $P(0 \le x \le 1)$ in $N(0, 4)$
   (c) $P(-20 \le x \le 10)$ in $N(10, 400)$
   (d) $P(16.5 \le x)$ in $N(0, 100)$
   (e) $P(|x - 10| < 4)$ in $N(10, 4)$
   (f) $P(36 \le x \le 44)$ in $N(40, 4)$
   (g) $P(36 \le x \le 44)$ in $N(40, 9)$

6. If X is $N(\mu_X, \sigma_X^2)$ in each of the following, find c such that

   (a) $P(x \ge \mu_X + c\sigma_X) = .05$
   (b) $P(\mu_X - c\sigma_X \le x \le \mu_X + c\sigma_X) = .95$
   (c) $P(x \le \mu_X + c\sigma_X) = .01$
   (d) $P(\mu_X - c\sigma_X \le x \le \mu_X + c\sigma_X) = .99$

(e) $P(x \leq \mu_X - c\sigma_X) = .05$

(f) $P(x \geq \mu_X + c\sigma_X) = .01$

7. In 10,000 tosses of a fair coin exactly 5,081 heads were observed. What is the approximate probability of observing this event or an event with more heads out of the 10,000 tosses? What would the approximate probability be if the coin were biased so that $P(\text{head}) = .52$?

8. Find the approximate probability that in 3,600 binomial trials, with the probability of success p, the number of successes is between $(3600p - 20)$ and $(3600p + 20)$ inclusive if (a) $p = .5$ and (b) $p = .3$.

9. Reexpress each of the probability statements in problems 4, 5, and 6 in terms of the unit normal distribution Φ: for example, in 4(a),

$$P(A) = \Phi(1.5) - \Phi(-1.0) = \Phi(1.5) + \Phi(1.0) - 1.0$$

CHAPTER 11

Means and Variances of Samples

$\mathbf{A}$t this point we "shift gears" slightly and turn our attention from the purely mathematical nature of probability models. We have developed many results dealing with sample spaces, probabilities of events and outcomes, random variables and their distributions. Some of these results were obtained by considering special types of experimental definitions, subject matter, and so on. However, all of our work was in terms of probabilities and events. The possibility of actually performing the experiments modeled seemed remote at times, or so it may have seemed to the reader.

Our attention from now on is directed more to the application of probability theory in (1) describing the results of experiments and (2) comparing the actual results with models based on theory.

The level of discussion from here on is not significantly different from that preceding. The work is primarily logical and mathematical, just as before. The major difference is that we are more directly concerned with what to do with actual observations in an experiment. In order to accomplish this task smoothly and with honesty we must introduce some few notions and mathematical concepts.

The plan of this chapter is first to introduce some general notions and then to apply those notions in two types of experimental procedures. First, we introduce the distinction between a parameter and an estimate of the parameter and then develop the formal properties of unbiasedness and efficiency of estimates of parameters. These concepts are applied to binomial experiments as an illustration. Two psychologically relevant conceptualizations for experiments—sampling from populations, and measurement—are next analyzed to give general results regarding the estimation of expectations and variances of random variables.

Parameters and Estimates

A basic distinction between the values descriptive of a probability model and the experiment modeled is inherent in the use of the terms "model" and "modeled." We need to introduce other terms that reflect this distinction. The descriptive values with which we have worked most closely in probability models are (1) n, the number of trials in a model; (2) the probability of events and values; (3) the expectation of a random variable, $E(X)$ or μ_X; and (4) the variance of a random variable, $\text{var}(X)$ or $\sigma_X{}^2$. Other characteristics of probability models might be of interest, but we have not been greatly concerned with them.

Some properties of a random variable, for example, n, p, $P(X = x)$, μ, and σ^2, are said to be parameters of the distribution of the random variable. The properties of a distribution, which, when known, specify the distribution, are called the parameters of the distribution. Thus, if n and p are known in a binomial model, all the details of the distribution of the specific binomial random variable may be determined. In this example $B(n, p)$ is the set of values of $P(S = s)$ each depending only on n and p. When each value of $P(S = s)$, for the entire value set of S, is known, then everything relevant to the model is determined, including μ and σ^2. Thus, specifying that S is $B(n, p)$ completely specifies the probability model—that is, it is a binomial model—and n and p are said to be the parameters of the model.

A model involving a normally distributed random variable can be specified completely by the two parameters μ and σ^2, since if X is normal and has μ and σ^2 as expectation and variance, we know the probability of each value interval in the probability distribution of X,

$$P\left(x_i \leq \frac{X - \mu}{\sigma} \leq x_j\right) = \Phi(x_j) - \Phi(x_i)$$

where $\Phi(a)$ is the integral

$$\int_{-\infty}^{a} \frac{1}{\sqrt{2\pi}} e^{-.5x^2} \, dx$$

Knowing μ and σ^2 permits calculation of the probabilities for any set of values of the random variable, and hence μ and σ^2 are the parameters of a normal model.

There may be several properties of a distribution that are parameters of a given probability model, as in the case of binomial models where p and n are the parameters in $B(n, p)$ or in normal models where μ and σ^2 are parameters in $N(\mu, \sigma^2)$. In later chapters we have occasion to deal with models depending only on the number of experimental trials (or observations).

For our purposes at this point we deal with single parameters regardless of how many parameters are involved in a distribution. First, we develop the general theory of estimation of parameters and then special instances of parameters, for example, estimates for μ and σ^2 in $N(\mu, \sigma^2)$. As a convention we let Greek letters stand for parameters and English letters for estimates of parameters. For example, imagine a model with a parameter θ and an estimate T of that parameter.

When a parameter is known there is no reason to estimate it, of course. However, much of the work we do in statistics and most of the empirical applications of probability and statistics involve situations where one or more of the parameters of the relevant distributions are unknown. In order to specify the complete model these parameters must be obtained. If logical argument and psychological theory do not suffice, we must resort to empirical evaluation of parameter values.

Empirical methods are limited by their very nature. The values obtained are always associated with accidental circumstances; chance factors may make the observations a poor representation of the population; only a limited number of observations are possible; and so forth. Thus, empirical evaluation leads to a value that is at best an approximation to the value of the parameter. The value of the approximation is called an estimate of the parameter. For an estimate of the parameter θ several observations, $X_1, X_2, \ldots, X_N$, of the random variable X are made. The estimate is some function g of these observed values:

$$T = g(X_1, X_2, \ldots, X_N) \tag{11.1}$$

A common estimate of the parameter μ_X of a random variable X is the average or algebraic mean of N observations on X:

$$g(X_1, X_2, \ldots, X_N) = \frac{X_1 + X_2 + \cdots + X_N}{N}$$

Other estimates, such as median and mode, are also used, and thus

$$g(X_1, X_2, \ldots, X_N) = \begin{cases} t \text{ such that } t \text{ falls at the point separating the } 50\% \\ \text{largest values from the } 50\% \text{ smallest values of } X_1, \\ X_2, \ldots, X_N \end{cases}$$

$$g(X_1, X_2, \ldots, X_N) = \begin{cases} t \text{ where the relative frequency of } t \text{ is the greatest} \\ \text{relative frequency of all values represented in } X_1, \\ X_2, \ldots, X_N \end{cases}$$

Two types of estimates are relevant to this discussion: point estimates and interval estimates. If the estimate T is a single value from the set of all possible values that the parameter theoretically may have, T is called a point estimate (after the point on the number line corresponding to the value). However, if the estimate is stated as a collection of values of which the parameter is thought to be a member, then the estimate is called an interval estimate. It is natural, but not necessary, for the collection of values in an interval estimate to be defined by adjacent points on the number line. Where a point estimate is some specific value, for example $t = 40.327$, an interval estimate corresponds to a set of values, such as $E = \{t \mid x_1 \leq t \leq x_2, t \in T\}$ or $E = \{t \mid 38.1 \leq t \leq 42.37, t \in T\}$. In this chapter we discuss only point estimates. The concept of interval estimate is developed in the next chapter.

Point estimates of parameters calculated from samples of empirical observations lead us to consider an estimate as a random variable itself. For example, if we draw one sample from a binomial random variable with $B(10, p)$ where p is unknown, we observe some value X, the number of successes out of the ten binomial trials of the observation. Suppose that this is 3, that is, $X = 3$. This value, 3, corresponds to the frequency of successes in 10 trials. Converting to a relative frequency we have $\frac{3}{10} = .3$. Taking this as an estimate of p seems natural, so we say that

$$t = .3$$

However, we might easily have observed $X = 4$ or $X = 2$, and so on, resulting in a value set of possible values of T:

$$T = \{.0, .1, .2, \ldots, 1.0\}$$

The value actually observed depends on the parameter p: The probabilities of observing the various values of X depend on p. Thus, T is a random variable where

$$P(T = t) = \binom{10}{X} p^X (1 - p)^{10-X}$$

In general terms, an estimate T of a parameter θ is a random variable whose distribution depends on the value of θ. The implication is that $\theta \neq T$ in general.

Since we deal with estimates of parameters as random variables, it is natural to be interested in the expectation and variance of the estimates. This chapter is devoted to the study of $E(T)$ and $\text{var}(T)$ for a limited number of parameters. This interest in estimates is widened in the next chapter, where we develop the theory of the distribution of estimates. The order of presentation may seem backward, but we wish first to show that sample estimates have certain properties and then to deal with the estimates in a more detailed way.

Since estimating θ from empirical observations is a chancy affair, we want to impose two restrictions on the estimate T. First, we want T to be equal to θ on the average, and second, we want T to have as small a dispersion as possible.

Other restrictions are also desirable, but we do not include them in our discussion because we make no use of the properties involved. Briefly, the two restrictions are that (1) the estimate should be consistent and (2) the estimate should be sufficient. The consistency requirement means that the value of the estimate should have a higher probability of being closer to the value of the parameter as sample size is increased. The sufficiency requirement means that no data in the sample are neglected if those data are relevant to the estimation of the parameter.

Unbiased Estimators. For any parameter there may be many estimators (for example, mean, median, mode for μ), some of which are equal, on the average, to the parameter value. Whenever an estimator T of the parameter θ is distributed so that

$$E(T) = \theta \tag{11.2}$$

T is said to be unbiased. If $E(T) \neq \theta$ then T is said to be biased $[E(T) < \theta$ or $E(T) > \theta]$. In the binomial example above it can be shown that $T = X/10$ is an unbiased estimator of p in $B(10, p)$. In order to show this we need to know the value of θ, which of course we would not know in practice. Imagine that $p = .4$. Then we would have, where $T = X/10$,

$$P(T = t) = \binom{10}{X}(.4)^X(.6)^{10-X}$$

Recall that X is the number of successes observed in 10 trials. Thus, X is a binomial random variable with $B(10, p)$, which, by hypothesis, is $B(10, .4)$. The expectation of T is found by the following calculations:

$$E(T) = \sum_{t \in T} tP(T = t)$$

$$= \sum_{x=0}^{10} \frac{X}{10} \binom{10}{X}(.4)^X(.6)^{10-X}$$

$$= \frac{1}{10} \sum_{x=0}^{10} X\binom{10}{X}(.4)^X(.6)^{10-X}$$

$$= \frac{1}{10} E(X)$$

but since X is $B(10, .4)$, $E(X) = 10(.4) = 4$ and hence

$$E(T) = \tfrac{1}{10}E(X) = \tfrac{1}{10}(4) = .4$$

Thus, for the binomial,

$$T = \frac{\text{number of successes observed}}{\text{number of trials }(n)}$$

is an unbiased estimate of p in $B(n, p)$.

An example of a biased estimator of p is given by the numerator of the estimator above. If we take T as the number of successes observed, we discover that

$$E(T) = np \neq p$$

and hence T is biased unless $n = 1$.

Efficiency of Estimators. For the moment we restrict our attention to the class U of estimators T of a parameter θ which contains only unbiased estimators. Assume the class is given by the set

$$U = \{T_1, T_2, \ldots, T_k\}$$

Then we say that for $T_i \in U$ and $T_j \in U$, T_i is more efficient than T_j if

$$E[(T_i - \theta)^2] < E[(T_j - \theta)^2] \qquad (11.3)$$

Since U contains only unbiased estimators, $E(T_i) = E(T_j) = \theta$, the inequality can be rewritten

$$E([T_i - E(T_i)]^2) < E([T_j - E(T_j)]^2)$$

or

$$\sigma_{T_i}{}^2 < \sigma_{T_j}{}^2$$

The estimator having the smaller variance is the more efficient. If $\sigma_{T_i}{}^2 = \sigma_{T_j}{}^2$, T_i and T_j are equally efficient, and we would choose between them on other grounds.

Recall the bionomial example of the paragraphs above: Observe X, where X is $B(10, p)$. Let T_1 be the relative frequency of successes on the first trial, T_2 be the relative frequency of successes on the first two trials, $\ldots$, T_{10} be the relative frequency of successes on the ten trials. The value sets of these random variables are

$$T_1 = \{0, 1\}$$
$$T_2 = \{0, \tfrac{1}{2}, 1\}$$
$$T_3 = \{0, \tfrac{1}{3}, \tfrac{2}{3}, 1\}$$
$$T_4 = \{0, \tfrac{1}{4}, \tfrac{2}{4}, \tfrac{3}{4}, 1\}$$

$$\vdots \qquad \vdots$$

$$T_{10} = \{0, \tfrac{1}{10}\, \tfrac{2}{10}, \ldots, 1\}$$

Calculating the expectations of these is equivalent to calculating the expectations of binomial random variables multiplied by the constants $\frac{1}{1}, \frac{1}{2}, \frac{1}{3}, \ldots, \frac{1}{10}$, respectively, for $B(1, p), B(2, p), \ldots, B(10, p)$. Thus

$$E(T_i) = \frac{1}{i} E[S \mid S: B(i, p)] = \frac{1}{i} ip$$

$$= p$$

for $i = 1, 2, \ldots, 10$. This implies that all these estimators are unbiased.

However, the variances for the estimators are quite different:

$$\sigma_{T_i}{}^2 = E[(T_i - p)^2] = \frac{1}{i^2} ip(1 - p)$$

$$= \frac{p(1 - p)}{i}$$

Since p is a constant, $p(1 - p)$ is a constant, then

$$\sigma_{T_1}{}^2 = p(1 - p)$$

$$\sigma_{T_2}{}^2 = \frac{p(1 - p)}{2}$$

$$\sigma_{T_3}{}^2 = \frac{p(1 - p)}{3}$$

$$\cdot \qquad \cdot$$
$$\cdot \qquad \cdot$$
$$\cdot \qquad \cdot$$

$$\sigma_{T_{10}}{}^2 = \frac{p(1 - p)}{10}$$

and

$$\sigma_{T_1}{}^2 > \sigma_{T_2}{}^2 > \cdots > \sigma_{T_{10}}{}^2$$

Thus, T_{10} is the most efficient, and T_1 is the least efficient of this class of unbiased estimators of p.

Efficiency Versus Unbiasedness. The implications of the example used in the preceding section are clear. If a family of unbiased estimators is known, the one with maximum efficiency should be used. It gives a narrower range of values in the calculation of the estimated value. The accuracy of the estimate is directly a function of the efficiency of the estimate. Figure 11.1 shows such a set of estimators.

Consider the following situation. Two estimators T_1 and T_2 for a parameter θ are known; T_1 is unbiased and T_2 is biased. However, the efficiency of T_2 is better than that of T_1. Which estimate should be used? Figure 11.2 shows the sort of situation that calls for a carefully thought-out answer.

Unbiasedness is not enough to guarantee good estimation, and accuracy may be better achieved by using a highly efficient but biased estimator. The relative merits of these two properties are difficult to assess in a vacuum. The costs involved in errors and the costs of obtaining the estimates are the primary bases on which

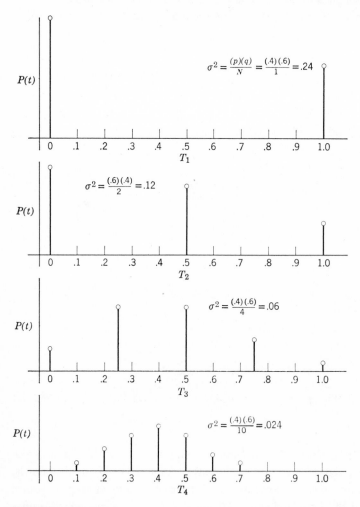

FIGURE 11.1

Illustration of several unbiased estimators with different degrees of efficiency.

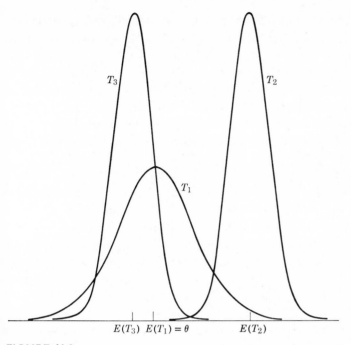

$$E(T_3) \quad E(T_1) = \theta \qquad\qquad E(T_2)$$

FIGURE 11.2

Three distributions of estimators of the parameter θ. The estimator T_1 is unbiased but is inefficient. The estimator T_2 is efficient but very biased. The estimator T_3 is efficient but slightly biased.

an estimator must be chosen. In most psychological research these costs are difficult if not impossible to assess.

We do not make any more of this problem, largely because it does not offer much in the way of different practical alternatives for the psychological researcher.

From the formal, conceptual properties of estimates we turn now to developing a rationale for estimating expectations and variances. Two special models for estimating μ are presented: (1) a sampling model, and (2) a measurement model. The law of large numbers is used to develop an estimate of σ^2.

Estimation of μ in Sampling

In the chapter on the normal random variable we developed two types of content considerations leading to random variables with normal distributions. One of these assumed each behavior or property of bio-social units was determined by multiple factors. The diversity of factors and the possibility that certain factors might not be present in a given bio-social unit introduced variety or variability in the population of units involved. We can cite several examples. One is the intelligence of an individual; for each individual in a given population there is a connivance of heredity and environment to produce a certain "quantity" of intelligence. A similar argument can be made for any of a very large number of individual difference variables—aptitude, personality, temperament. The affluence of a family unit is affected by many factors that play on all family units in similar ways: The set of families in an economic system makes up a "homogeneous" population with respect to that system.

Consider a population composed of N objects (such as people, schools, families, hospitals, trials in choice experiments, conditioned responses, stimulus-response associations). For each object we assume a value of x from the set of values possible. The sampling model begins with the assumption that when an object is sampled from the population the value x for that object is sampled without any uncertainty as to what the value is. No measurement error or other source of error or lack of knowledge about x is involved. Another way of looking at this is that x is a stable, constant, quantitative attribute of the object that is observed without error. Suppose that we wish to estimate the expectation and variance of the distribution of the values in the population. If we had the entire set of values for the population, their average value would be equivalent to the expected value of the variable in the following sense. Let one of the objects of the population be sampled at random; that is, each object is selected with probability $1/N$. If the N objects have values $x_1, x_2, x_3, \ldots, x_N$, then

$$E(X) = \frac{1}{N} x_1 + \frac{1}{N} x_2 + \cdots + \frac{1}{N} x_N$$

$$= \frac{x_1 + x_2 + \cdots + x_N}{N}$$

$$= \mu_X$$

The population variance is given in a similar argument by

$$\text{var}(X) = \frac{1}{N} (x_1 - \mu_X)^2 + \frac{1}{N} (x_2 - \mu_X)^2 + \cdots + \frac{1}{N} (x_N - \mu_X)^2$$

$$= \frac{1}{N} \sum_{i=1}^{N} (x_i - \mu_X)^2 = \frac{1}{N} \sum_{i=1}^{N} x_i^2 - \mu_X^2$$

$$= \sigma_X^2$$

Now we draw an ordered random sample of size k from the population: X_1 the first value, X_2 the second, $\ldots$, X_k the kth. If an ordered random sample of size k is drawn from a population of N objects, the probability that any particular individual object is selected is $1/N$. Thus the draws on the k samples correspond to k random variables having the same distributions, in particular

$$E(X_1) = E(X_2) = \cdots = E(X_k) = \mu_X$$

and

$$\operatorname{var}(X_1) = \operatorname{var}(X_2) = \cdots = \operatorname{var}(X_k) = \sigma_X{}^2$$

The form of the expectation of X in the population is suggestive of a way of estimating that expectation from the sample. Let the value

$$\bar{X} = \frac{1}{k}(X_1 + X_2 + \cdots + X_k)$$

be called the mean of the values sampled. Since this is the mean of observed values of random variables, then

$$E(\bar{X}) = \frac{1}{k} E(X_1 + X_2 + \cdots + X_k)$$

$$= \frac{1}{k} [E(X_1) + \cdots + E(X_k)]$$

$$= \frac{1}{k} [\mu_X + \mu_X + \cdots + \mu_X]$$

$$= \frac{1}{k} [k\mu_X]$$

$$= \mu_X \tag{11.4}$$

Hence, $\bar{X}$ is an unbiased estimate of the population expectation μ. It can be shown that this is true whether sampling is with or without replacement. However, estimating the variance of $\bar{X}$ from the sample depends on the sampling method. For sampling with replacement, the random variable corresponding to each sample observation has a variance $\sigma_X{}^2$. Thus

$$\operatorname{var}(\bar{X}) = \operatorname{var}\left[\frac{1}{k}(X_1 + X_2 + \cdots + X_k)\right]$$

$$= \frac{1}{k^2} \operatorname{var}(X_1 + X_2 + \cdots + X_k)$$

$$= \frac{1}{k^2} [k \operatorname{var}(X_i)]$$

$$= \frac{k}{k^2} \sigma_X{}^2$$

$$= \frac{\sigma_X{}^2}{k} \tag{11.5}$$

Suppose that we randomly sample k objects without replacement from a population of N objects where each object has some value x. Let the values of objects so sampled be denoted $X_1, X_2, \ldots, X_k$. If we consider $X_1, X_2, \ldots, X_k$ to be k random variables, we have, as before, in estimating μ,

$$\text{var}(X_1 + X_2 + \cdots + X_k)$$
$$= k\sigma^2 + 2[\text{cov}(X_1, X_2) + \text{cov}(X_1, X_3) + \cdots + \text{cov}(X_{k-1}, X_k)]$$

Each of the $\binom{k}{2}$ pairs of variables has the same covariance, since all of the variables have the same distribution and dependencies brought about by sampling without replacement. The value of the covariance is, for example, ξ. Thus

$$\text{var}(X_1 + X_2 + \cdots + X_k) = k\sigma^2 + 2\binom{k}{2}\xi$$

A trick is used to evaluate ξ. Let $k = N$ and $X_1 + X_2 + \cdots + X_N$ is some fixed value (all the population is included). Hence

$$\text{var}(X_1 + X_2 + \cdots + X_N) = 0$$

and

$$N\sigma^2 + 2\binom{N}{2}\xi = 0$$

Noting that

$$2\binom{N}{2} = N(N - 1)$$

we have

$$N\sigma^2 + N(N - 1)\xi = 0$$
$$\xi = \frac{-\sigma^2}{(N - 1)}$$

Substituting k for N and $-\sigma^2/(N - 1)$ for ξ in the equation above,

$$\text{var}(X_1 + X_2 + \cdots + X_k) = k\sigma^2 - k(k - 1)\frac{\sigma^2}{(N - 1)}$$
$$= k\sigma^2\left(1 - \frac{k - 1}{N - 1}\right)$$
$$= k\sigma^2\left(\frac{(N - 1) - (k - 1)}{N - 1}\right)$$
$$= \frac{N - k}{N - 1}k\sigma^2$$

If k is small with respect to N, the correction factor is $(N - k)/(N - 1)$, practically

1.0, and it may be ignored. This is tantamount to sampling with replacement, where $\xi = 0$.

Substituting in var($\overline{X}$), we have

$$\text{var}(\overline{X}) = \frac{1}{k^2} \text{var}(X_1 + \cdots + X_k)$$

$$= \frac{1}{k^2} \left(\frac{N-k}{N-1}\right) k\sigma^2$$

$$= \left(\frac{N-k}{N-1}\right) \frac{\sigma^2}{k} \tag{11.6}$$

Notice that when k is small in comparison with N, these two sampling models give the same variance, that is, $(N-k)/(N-1) \cong 1$. Also notice that if $k = N$, $\overline{X}$ *must* be exactly equal to μ_X and hence var($\overline{X}$) $= 0$. This is clear from the fact that when $k = N$, $(N-k) = 0$, and hence

$$\text{var}(\overline{X}) = \frac{0}{N-1} \frac{\sigma_X^2}{k} = 0$$

It can be shown that any other estimator of μ_X has at least as large a variance as $\overline{X}$. Hence, $\overline{X}$ is the most efficient unbiased estimator of μ_X.

Estimation of μ with Measurements

The sampling model in the preceding section parallels the polygenic model of behavior determination leading to the normal distribution. Thus, we might expect that the experimental error model of Chapter 10 would lead to an estimator of μ as did the sampling model.

In the experimental (or measurement) error model it is assumed that each observation of a quantity is accompanied by certain amounts of error. We view this as an observation of the sum of the constant quantity and a value determined by a strictly random process, the process being the same on each observation. We call this composite value X, which has a value set $x_1, x_2, \ldots$ with probabilities $P(X = x_1) = p_1, P(X = x_2) = p_2, \ldots$. By the definition of expectation and variance,

$$\mu = E(X) = x_1 p_1 + x_2 p_2 + \cdots$$

and

$$\sigma^2 = \text{var}(X) = (x_1 - \mu)^2 p_1 + (x_2 - \mu)^2 p_2 + \cdots$$

Thus, a single observation has an expected value μ and variance σ^2. This is our model for a measurement. An example would be determination of the responsiveness of a strain of mice to a drug that is said to increase level of activity. The measurement of the activity of a specific animal under the influence of the drug is determined by the strain responsiveness, chance factors of metabolism, temperature in the environment, slight inaccuracies in dosage, and so on. Regardless of how closely all of these factors are controlled in the observation, a certain amount of error of measurement is involved. However, for the value observed we have

$$E(X) = \mu$$
$$\text{var}(X) = \sigma^2$$

as defined above. Now, if we repeat the experiment, with another individual from the strain, the same true value of strain responsiveness is involved and, more importantly, the same sources of error (not necessarily the same values) are present as random processes. To distinguish the first and second observations, we denote them X_1 and X_2, respectively. Since X_1 and X_2 both sample the same random process, they have the same value set and the probabilities of the values are the same. Hence

$$E(X_1) = E(X_2) = \mu$$
$$\text{var}(X_1) = \text{var}(X_2) = \sigma^2$$

For N such measurements it follows that

$$E(X_1) = E(X_2) = \cdots = E(X_N) = \mu$$
$$\text{var}(X_1) = \text{var}(X_2) = \cdots = \text{var}(X_N) = \sigma^2$$

The values of $p_1, p_2, \ldots$ are generally unknown, and hence μ and σ^2 are unknown.

It is particularly of interest to estimate μ. Under the conditions of the derivation of the normal distribution of measurements (all partial errors relatively equal in absolute magnitude, equal probability of negative and positive partial errors) μ is the value of the "true" responsiveness, or the "true" population value. The quantity

$$\bar{X} = \frac{X_1 + X_2 + \cdots + X_N}{N}$$

is commonly used as the estimate of μ. This has the property that

$$E(\bar{X}) = E\left(\frac{X_1 + X_2 + \cdots + X_N}{N}\right)$$

$$= \frac{1}{N} E(X_1 + X_2 + \cdots + X_N)$$

$$= \frac{1}{N} [E(X_1) + E(X_2) + \cdots + E(X_N)]$$

$$= \frac{1}{N} [\mu + \mu + \cdots + \mu]$$

$$= \frac{1}{N} [N\mu]$$

$$= \mu$$

and

$$\text{var}(\bar{X}) = \text{var}\left(\frac{X_1 + X_2 + \cdots + X_N}{N}\right)$$

$$= \frac{1}{N^2}\text{var}(X_1 + X_2 + \cdots + X_N)$$

$$= \frac{1}{N^2}[\text{var}(X_1) + \text{var}(X_2) + \cdots + \text{var}(X_N)]$$

$$= \frac{1}{N^2}(N\sigma^2)$$

$$= \frac{\sigma^2}{N}$$

Thus, in the measurement model, $\bar{X}$ is an unbiased estimate of μ with variance σ^2/N. These two properties of $\bar{X}$ are the same as we found in the sampling model, providing we were sampling with replacement or if $(N - k)/(N - 1) \cong 1$. Thus, in general, we estimate μ in the same way both for sampling models and for measurement models. The mean value $\bar{X}$ of the sample of values or the collection of measurements is an unbiased estimate of μ in both models.

We can show easily that the mean of the squared deviations of the observations about an arbitrary value is greater than the mean squared deviation about the mean. Stating this symbolically,

$$\frac{1}{N}\sum_{i=1}^{N}(X_i - \bar{X})^2 \leq \frac{1}{N}\sum_{i=1}^{N}(X_i - C)^2 \tag{11.7}$$

for any C whatsoever. When and only when $C = \bar{X}$ does the equality hold. Since C and $\bar{X}$ are both real numbers, we can always find a value a where

$$C = \bar{X} + a$$

When $a = .0$, $C = \bar{X}$. Rewriting the inequality using $C = \bar{X} + a$, we obtain

$$\frac{1}{N}\sum_{i=1}^{N}(X_i - \bar{X})^2 \leq \frac{1}{N}\sum_{i=1}^{N}[X_i - (\bar{X} + a)]^2$$

$$\frac{1}{N}\sum_{i=1}^{N}X_i^2 - \bar{X}^2 \leq \frac{1}{N}\sum_{i=1}^{N}X_i^2 - \frac{2(\bar{X} + a)}{N}\sum_{i=1}^{N}X_i + \frac{1}{N}\sum_{i=1}^{N}(\bar{X} + a)^2$$

$$\frac{1}{N}\sum_{i=1}^{N}X_i^2 - \bar{X}^2 \leq \frac{1}{N}\sum_{i=1}^{N}X_i^2 - 2\bar{X}^2 - 2a\bar{X} + \bar{X}^2 + 2a\bar{X} + a^2$$

$$\frac{1}{N}\sum_{i=1}^{N}X_i^2 - \bar{X}^2 \leq \frac{1}{N}\sum_{i=1}^{N}X_i^2 - \bar{X}^2 + a^2$$

Subtracting $(1/N) \sum_{i=1}^{N} X_i^2 - \bar{X}^2$ from both sides, we have

$$0 \leq a^2$$

which is true because a^2 must always be nonnegative. The actual magnitude of the excess in the mean squared deviation with C is a^2.

The Algebra of Means

The estimator $\bar{X}$, the average or algebraic mean, which is an unbiased maximally efficient estimator of μ, has some useful algebraic properties. These properties parallel the properties of expectations as defined in Chapter 8. We repeat the definition: The mean of a sample of N observations on a random variable X is

$$\bar{X} = \frac{1}{N} \sum_{i=1}^{N} X_i \tag{11.8}$$

In general, there are several different observed values of X. However, we have situations in which all values observed are the same. That is, X is a constant value, say C. In this case

$$\bar{X} = \frac{1}{N} \sum_{i=1}^{N} X_i = \frac{1}{N} \sum_{i=1}^{N} C = \frac{1}{N} NC = C \tag{11.9}$$

A number of operations we perform on observations involve multiplying each observation by some constant value c (or dividing by k, hence $c = 1/k$). The mean in this case is given by $\bar{Y}$ where $Y = cX$:

$$\bar{Y} = \frac{1}{N} \sum_{i=1}^{N} cX_i = c \frac{1}{N} \sum_{i=1}^{N} X_i = c\bar{X} \tag{11.10}$$

Likewise, we often deal with $Y = X + c$, in which case

$$\bar{Y} = \frac{1}{N} \sum_{i=1}^{N} (X_i + c) = \frac{1}{N} \left(\sum_{i=1}^{N} X_i + \sum_{i=1}^{N} c \right)$$

$$= \frac{1}{N} \sum_{i=1}^{N} X_i + \frac{1}{N} (Nc)$$

$$= \bar{X} + c \tag{11.11}$$

Putting these two operations together, we have for $Y = a + bX$, where a and b are constants, the general result

$$\bar{Y} = a + b\bar{X}$$

If each object observed (persons, subjects, mazes, drugs, and so on) is assigned two values—that is, two random variables are observed for each object—and the value of interest is the total value, say $Z = X + Y$, then

$$\bar{Z} = \frac{1}{N}\sum_{i=1}^{N}(X_i + Y_i)$$

$$= \frac{1}{N}\sum_{i=1}^{N}X_i + \frac{1}{N}\sum_{i=1}^{N}Y_i$$

$$= \bar{X} + \bar{Y} \tag{11.12}$$

In general, if k values are observed for the ith object, $X_{i1}, X_{i2}, \ldots, X_{ik}$, then when $Y_i = X_{i1} + X_{i2} + \cdots + X_{ik}$,

$$\bar{Y} = \frac{1}{N}\sum_{i=1}^{N}Y_i = \frac{1}{N}\sum_{i=1}^{N}(X_{i1} + X_{i2} + \cdots + X_{ik})$$

$$= \frac{1}{N}\sum_{i=1}^{N}X_{i1} + \frac{1}{N}\sum_{i=1}^{N}X_{i2} + \cdots + \frac{1}{N}\sum_{i=1}^{N}X_{ik}$$

$$= \bar{X}_1 + \bar{X}_2 + \cdots + \bar{X}_k$$

Putting all of these rules together, we have the result that if $X_1, X_2, \ldots, X_k$ are combined in a linear fashion with coefficients $a, b_1, b_2, \ldots, b_k$ to form Y,

$$Y = a + b_1X_1 + b_2X_2 + \cdots + b_kX_k$$

then

$$\bar{Y} = a + b_1\bar{X}_1 + b_2\bar{X}_2 + \cdots + b_k\bar{X}_k \tag{11.13}$$

Estimation of σ^2

We take a different tack in developing an estimate of the variance of a random variable, with two conditions: (1) where the parameter μ is known, and (2) where μ is not known but the estimate $\bar{X}$ of μ is known. This section depends first on the result often called the law of large numbers. We do not prove this result, but simply state it.

Let $n(x)$ be the number of times a value x of a random variable X is observed out of m observations. The relative frequency of the observation of x is thus

$$f_m(x) = \frac{n(x)}{m}$$

where the subscript m on f_m indicates that f is possibly different with a different number of trials. That is, $f_m(x)$ is a function of m as well as X. For any positive number ε, regardless of how small, it can be shown that

$$\lim_{m \to \infty} P(|f_m(x) - P(X = x)| > \varepsilon) = 0 \tag{11.14}$$

That is, there is some m at which there is essentially zero probability of observing an absolute difference of probability and relative frequency greater than ε.

If we let $f_m(x)$ replace $P(X = x)$ in the definition of variance

$$\text{var}(X) = \sum_{x \in X} P(X = x)(x - \mu)^2$$

we have

$$V_X = \sum_{x \in X'} f_m(x)(x - \mu)^2 \tag{11.15}$$

where X' is the set of values actually observed. Because of the law of large numbers,

$$\lim_{m \to \infty} V_X = \text{var}(X) \tag{11.16}$$

Thus, in the limit the estimate of the variance is a good one. However, this result is weak. How large m must be to have this limit hold is not specified (although Chebyshev's inequality can be applied here to give an m for some tolerable error). What we want is to show that $E(V_X) = \text{var}(X)$. This can be proved by the following argument. Rewriting V_X, we have

$$V_X = \sum_{x \in X'} f_m(x)(x^2 - 2x\mu + \mu^2)$$

$$= \sum_{x \in X'} f_m(x)x^2 - 2\mu \sum_{x \in X'} f_m(x)x + \mu^2$$

We have already shown that

$$\sum_{x \in X'} f_m(x)x = \frac{1}{N} \sum_{i=1}^{N} X_i = \bar{X}$$

and it should be clear that

$$\sum_{x \in X'} f_m(x)x^2 = \frac{1}{N} \sum_{i=1}^{N} X_i^{\,2}$$

Taking the expectation, we have

$$E(V_X) = E\left(\frac{1}{N} \sum_{i=1}^{N} X_i^{\,2}\right) - 2\mu E(\bar{X}) + \mu^2$$

$$= E\left[\frac{1}{N} \sum_{i=1}^{N} E(X_i^{\,2})\right] - \mu^2$$

$$= \frac{1}{N} \sum_{i=1}^{N} E(X_i^{\,2}) - \mu^2$$

If each sample value X_i is an observation on the same random variable X, then

$$E(X_1^2) = E(X_2^2) = \cdots = E(X_N^2) = E(X^2)$$

and

$$\begin{aligned} E(V_X) &= \frac{1}{N}[NE(X^2)] - \mu^2 \\ &= E(X^2) - \mu^2 \\ &= \sigma^2 \end{aligned}$$

The last step is given by the definition of σ^2.

Thus, where μ is known the estimator

$$\begin{aligned} V_X &= \sum_{x \in X'} f_m(x)(x - \mu)^2 \\ &= \frac{1}{N}\sum_{i=1}^{N}(X_i - \mu)^2 \end{aligned} \tag{11.17}$$

is unbiased.

When the unbiased estimator of μ, $\overline{X}$, is substituted into the equations above, we have

$$S^2 = \sum_{x \in X'} f_m(x)(x - \overline{X})^2$$

Expressing this in terms of the observations, we have

$$S^2 = \frac{1}{N}\sum_{i=1}^{N}(X_i - \overline{X})^2 \tag{11.18}$$

The expectation of S^2 is given by the following reasoning:

$$\begin{aligned} E(S^2) &= E\left(\frac{1}{N}\sum_{i=1}^{N}X_i^2\right) - E(\overline{X}^2) \\ &= \frac{1}{N}\sum_{i=1}^{N}E(X_i^2) - E(\overline{X}^2) \end{aligned}$$

Recall that

$$\sigma^2 = E(X_i^2) - \mu^2$$

and thus

$$E(X_i^2) = \sigma^2 + \mu^2$$

so that

$$\begin{aligned} E(S^2) &= \frac{1}{N}\sum_{i=1}^{N}(\sigma^2 + \mu^2) - E(\overline{X}^2) \\ &= \sigma^2 + \mu^2 - E(\overline{X}^2) \end{aligned}$$

Note that

$$\sigma_{\bar{X}}^2 = E(\bar{X}^2) - \mu^2$$

and hence

$$E(\bar{X}^2) = \sigma_{\bar{X}}^2 + \mu^2$$

so that

$$E(S^2) = \sigma^2 + \mu^2 - \sigma_{\bar{X}}^2 - \mu^2 = \sigma^2 - \sigma_{\bar{X}}^2 \tag{11.19}$$

and S^2 is biased! The degree of the bias is $-\sigma_{\bar{X}}^2$.

The upshot of this development is that if μ is not known, substituting $\bar{X}$ in place of μ gives a biased estimate of σ^2 even though $\bar{X}$ is an unbiased estimate of μ. This result should serve as warning not always to trust innocent-looking substitutions. The result may be disastrous.

Since we have been able to find an expression for the exact degree of bias in S^2, it is natural to try to find a way of correcting for it. Recall that we defined

$$\sigma_{\bar{X}}^2 = \frac{\sigma^2}{N}$$

Thus

$$E(S^2) = \sigma^2 - \frac{\sigma^2}{N} = \frac{N-1}{N} \sigma^2 \tag{11.20}$$

And S^2 is too small by a factor of $(N-1)/N$. Multiplying S^2 by the inverse of this factor corrects its bias. Since N is known, we have

$$\hat{S}^2 = \frac{N}{N-1} S^2 \tag{11.21}$$

as the corrected, or unbiased, estimate of σ^2 when $\bar{X}$ is used as an estimate of μ. Checking this result, we have

$$E(\hat{S}^2) = E\left(\frac{N}{N-1} S^2\right) = \frac{N}{N-1} E(S^2)$$

$$= \frac{N}{N-1} \frac{N-1}{N} \sigma^2 = \sigma^2$$

The two values S^2 and $\hat{S}^2$ estimating σ^2 are known by different names. The biased estimator S^2 is the sample variance. No attempt is made in S^2 to achieve an unbiased estimate of σ^2. Rather, the theoretical definition of var(X) is specialized to a sample of observation using relative frequency in place of probability and observed values in place of the value set of the random variable. Thus, S^2 is descriptive of the sample.

The estimate $\hat{S}^2$ is a corrected description. As such it is not an accurate description of the sample. It is unbiased as an estimate of σ^2 and hence is called the unbiased estimate of variance.

The Algebra of Sample Variances

The algebraic properties of the variance of a random variable were discussed in Chapter 8. The parallel form of the definitions of var(X) and S^2 suggests that the rules for var(X) hold for S^2. It can be shown rigorously that the rules in fact are the same in form. The proof of these rules or theorems is left to the student. The proofs follow the same form we have used in proving the corresponding rules for var(X) and similar rules in the algebra of $\bar{X}$.

If X and Y are observations on random variables and $Z = X + Y$, then

$$S_Z{}^2 = S_X{}^2 + S_Y{}^2 + 2S_{XY}{}^2 \qquad (11.22)$$

where $S_{XY}{}^2$ is the sample covariance of X and Y. If X and Y have zero covariance, then

$$S_Z{}^2 = S_X{}^2 + S_Y{}^2$$

This result is generalized to observations on many random variables. Take the specific case of three variables X, Y, and W where $Z = X + Y + W$. Then

$$S_Z{}^2 = S_X{}^2 + S_W{}^2 + S_Y{}^2 + 2S_{XY}{}^2 + 2S_{XW}{}^2 + 2S_{YW}{}^2$$

If X is an observation on a random variable and c is some constant, then for $Y = cX$

$$S_Y{}^2 = c^2 S_X{}^2 \qquad (11.23)$$

If $Y = X + c$, then

$$S_Y{}^2 = S_X{}^2 \qquad (11.24)$$

That is, the variance of a constant times observations on a random variable is the variance of the observations times the squared constant. The variance of a sample of observations plus a constant is just the sample variance.

If X is a constant random variable or if all values observed have the same value c, then

$$S_X{}^2 = S_c{}^2 = 0 \qquad (11.25)$$

That is, the variance of constant-valued observations is always .0.

The general case is the variance of a linear function of observations on several random variables. For

$$Z = a + bX + cY + dW$$

$$S_Z{}^2 = b^2 S_X{}^2 + c^2 S_Y{}^2 + d^2 S_W{}^2 + 2bc S_{XY}{}^2 + 2b\,dS_{XW}{}^2 + 2c\,dS_{YW}{}^2 \qquad (11.26)$$

In the preceding section we defined S^2 as

$$S^2 = \frac{1}{N}\sum_{i=1}^{N} X_i{}^2 - \bar{X}^2$$

the mean square minus the squared mean. A simpler formula can be found for calculational purposes. Writing out the mean in summation gives

$$S^2 = \frac{1}{N}\sum_{i=1}^{N} X_i^2 - \frac{(\sum_{i=1}^{N} X_i)^2}{N^2}$$

Multiplying and dividing by N^2,

$$S^2 = \frac{N\sum_{i=1}^{N} X_i^2 - (\sum_{i=1}^{N} X_i)^2}{N^2} \qquad (11.27)$$

requiring only the values

$$\sum_{i=1}^{N} X_i^2 \quad \text{and} \quad \sum_{i=1}^{N} X_i$$

as calculated values. The round-off error and the additional labor of two separate divisions are avoided. This gives a simple formula for $\hat{S}^2$ also, by multiplying by $N/(N-1)$:

$$\hat{S}^2 = \frac{N}{N-1} S^2 = \frac{N}{N-1} \frac{N\sum_{i=1}^{N} X_i^2 - (\sum_{i=1}^{N} X_i)^2}{N^2} = \frac{N\sum_{i=1}^{N} X_i^2 - (\sum_{i=1}^{N} X_i)^2}{N(N-1)}$$

$$(11.28)$$

Sample Standard Deviation

The standard deviation of a sample is defined as the square root

$$S_X = \sqrt{S_X{}^2} \qquad (11.29)$$

This does *not* generalize to the unbiased estimate of σ^2, $\hat{S}^2$. It can be shown that

$$E(\hat{S}) \neq \sigma$$

This point is an unresolved problem, but it is of little significance because in situations in statistical inference in which $\hat{S}$ is used, the bias is so small as to be inconsequential. In statistical situations where $\hat{S}$ is biased to a greater degree, special techniques not depending on $\hat{S}$ are available.

In an early section of this chapter we showed that the variance of the distribution of means was given by

$$\sigma_{\bar{x}}{}^2 = \frac{\sigma^2}{N}$$

where N observations are involved. In order to estimate this from a single sample of observations, the value σ^2 must be estimated from the sample. We follow the convention here rather than develop the estimate in detail. Let $\hat{S}_{\bar{x}}{}^2$ be the estimated variance of the random variable $\bar{X}$; then we define

$$\hat{S}_{\bar{X}}^{\,2} = \frac{\hat{S}^2}{N} \tag{11.30}$$

In terms of the sample variance, the estimate of $\sigma_{\bar{X}}^2$ is

$$\hat{S}_{\bar{X}}^{\,2} = \frac{N}{N-1}\frac{S^2}{N} = \frac{S^2}{N-1} \tag{11.31}$$

The standard deviation $\sigma_{\bar{X}}$ is estimated by the square root of the estimate of $\sigma_{\bar{X}}^2$. This gives

$$\hat{S}_{\bar{X}} = \frac{S}{\sqrt{N-1}} \tag{11.32}$$

which is not unbiased, but the degree of bias is slight in situations where $\hat{S}_{\bar{X}}$ is used.

The two estimates $\bar{X}$ and $\hat{S}_{\bar{X}}$ are important in the theory of statistics. This importance is reflected in the fact that they have been assigned special names, the mean and the standard error (of the mean), respectively. The value $\bar{X}$, in effect, is the best we can do in estimating μ from a sample. The value $\hat{S}_{\bar{X}}$, in effect, tells us how good (or bad) the estimate of μ is. The larger the standard error, the greater is the probability of observing an $\bar{X}$ greatly removed from μ.

Chebyshev's Inequality for $\bar{X}$ and $\hat{S}_{\bar{X}}$

We see that for

$$\bar{X} = \frac{1}{N}(X_1 + X_2 + \cdots + X_N)$$

where $X_1, X_2, \ldots, X_N$ are independent observations on a random variable,

$$E(\bar{X}) = \mu_{\bar{X}} = \mu \quad \text{and} \quad \operatorname{var}(\bar{X}) = \sigma_{\bar{X}}^2 = \frac{\sigma^2}{N}$$

An estimate for $\sigma_{\bar{X}}^2$ in the sample is given by $\hat{S}_{\bar{X}}^{\,2}$.

This information allows us to make some powerful statements about the accuracy of estimate of the expectation with the sample mean in terms of the difference between μ and $\bar{X}$. Let c be any value greater than 1.0. Applying Chebyshev's inequality, we have

$$P(\mu - c\hat{S}_{\bar{X}} \leq \bar{X} \leq \mu + c\hat{S}_{\bar{X}}) \geq 1 - \frac{1}{c^2} \tag{11.33}$$

Thus the probability that $\bar{X}$ is within two standard errors of μ is .75. The probability that $\bar{X}$ is within three standard errors of μ is .8889 approximately.

This result has a special form that is a special case of the law of large numbers. By simple algebra we can reexpress the Chebyshev inequality (where $k = cS_{\bar{x}}$) as

$$P(\mu - k \leq \bar{X} \leq \mu + k) \geq 1 - \frac{\hat{S}_{\bar{x}}^2}{k^2}$$

which, when we note that $\hat{S}_{\bar{x}}^2 = \hat{S}^2/N$, is

$$P(\mu - k \leq \bar{X} \leq \mu + k) \geq 1 - \frac{\hat{S}^2}{Nk^2}$$

For a fixed value of k and a fixed value of $\hat{S}^2$ (which has an expectation of σ^2 when N is large) this probability can be as close to 1.0 as we desire. Let $\hat{S}^2 = 5$ and $k = 1$. This gives

$$P(\mu - 1.0 \leq \bar{X} \leq \mu + 1.0) \geq 1 - \frac{5}{N}$$

This inequality makes sense for probability only when $N > 5$ (calculate it for some $N \leq 5$ to check this). If we wanted this probability to be .95 or greater, then $5/N$ must be at most .05, that is, $N \geq 100$. If $N = 100$, $1 - 5/N = .95$. If $N = 1000$, then the inequality gives

$$P(\mu - 1.0 \leq \bar{X} \leq \mu + 1.0) \geq 1 - \frac{5}{1000} = .995$$

This result has a very direct interpretation. The standard error of estimate of the mean decreases with increasing sample size. The larger the sample size, the smaller

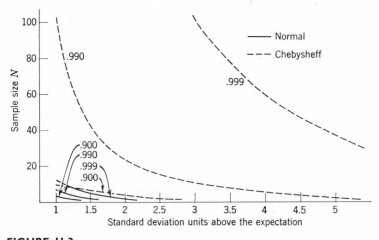

FIGURE 11.3

Sample size as a function of standard deviate units above the expectation.

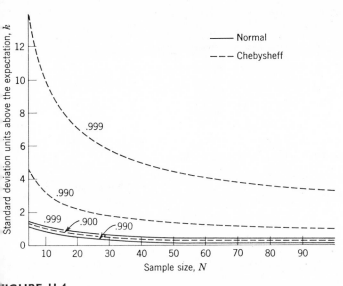

FIGURE 11.4

Standard deviate scores as a function of sample size.

are the possible errors in estimating the expectation of the random variable (population) being observed. It follows that for any arbitrary small value of k,

$$\lim_{N \to \infty} P(\mu - k \le \bar{X} \le \mu + k) = 1.0 \qquad (11.34)$$

This result is the law of large numbers for sample means.

We can view the application of Chebyshev's inequality to $\bar{X}$ in two ways. (1) How large would N have to be to have $P(\mu - k \le \bar{X} \le \mu + k) \ge .99$ for any value k, $0 \le k \le 6$? (2) How small can k be to have $P(\mu - k \le \bar{X} \le \mu + k) \ge .99$ for some fixed N? Graphs of the answers to these questions are given in Figures 11.3 and 11.4 for a selection of values.

In the use of Chebyshev's inequality no assumptions are made regarding distributions. However, this is wasteful of information about the distribution of $\bar{X}$. We know from the CLT that $\bar{X}$ is approximately normally distributed with μ and σ^2/N as means and variances. This gives us a special advantage, because we know that

$$P\left(\frac{-k}{\sqrt{\sigma^2/N}} \le \frac{\bar{X} - \mu}{\sqrt{\sigma^2/N}} \le \frac{+k}{\sqrt{\sigma^2/N}}\right) \simeq \Phi\left(\frac{+k}{\sqrt{\sigma^2/N}}\right) - \Phi\left(\frac{-k}{\sqrt{\sigma^2/N}}\right) \qquad (11.35)$$

For samples so large that the CLT holds, we can calculate the same graphs of Figures 11.3 and 11.4 using the normal probability distribution. The same assumption regarding the value of $\hat{S}^2$ is made for all of these curves.

SUGGESTED READINGS

In particular, refer to Edwards, Hays, Goldberg, and Hodges and Lehman.

GLOSSARY

Parameter. A parameter is a property of a distribution, which, when known, specifies (along with other parameters perhaps) the entire distribution.

Estimate. An estimate of a parameter is a function of sample observations that is intended to give an approximation to the value of the parameter.

$$(11.1) \qquad T = g(X_1, X_2, \ldots, X_N)$$

Point Estimate. An estimate that is a single value is a point estimate.

Interval Estimate. An estimate that is a set of values, generally defining an interval, is an interval estimate.

Unbiased Estimate. Any estimate that has an expectation equal to the parameter estimated is an unbiased estimate.

$$(11.2) \qquad E(T) = \theta$$

Estimator Efficiency. Between two unbiased estimators, that estimator that has the smallest variance is said to be the more efficient.

$$(11.3) \quad E[(T_i - \theta)^2] < E[(T_j - \theta)^2], \text{ for } T_i \text{ more efficient than } T_j$$

$$(11.4) \quad E(\bar{X}) = \mu_X$$

$$(11.5) \quad \text{var}(\bar{X}) = \frac{\sigma_X{}^2}{k} \text{ sampling with replacement}$$

$$(11.6) \quad \text{var}(\bar{X}) = \left(\frac{N-k}{N-1}\right)\frac{\sigma^2}{k} \text{ sampling without replacement}$$

$$(11.7) \quad \frac{1}{N}\sum_{i=1}^{N}(X_i - \bar{X})^2 \leq \frac{1}{N}\sum_{i=1}^{N}(X_i - C)^2 \text{ for any constant } C$$

$$(11.8) \quad \bar{X} = \frac{1}{N}\sum_{i=1}^{N}X_i$$

$$(11.9) \quad \bar{X} = \frac{1}{N}\sum_{i=1}^{N}X_i = \frac{1}{N}\sum_{i=1}^{N}C = \frac{1}{N}NC = C \text{ for } X_i \text{ a constant, } C$$

$$(11.10) \quad \bar{Y} = \frac{1}{N}\sum_{i=1}^{N}cX_i = c\frac{1}{N}\sum_{i=1}^{N}X_i = c\bar{X} \text{ for } c \text{ a constant}$$

(11.11) $\bar{Y} = \bar{X} + c$ for c a constant and $Y_i = X_i + c$

(11.12) $\bar{Z} = \bar{X} + \bar{Y}$ for $Z_i = X_i + Y_i$

(11.13) $\bar{Y} = a + b_1\bar{X}_1 + \cdots + b_k\bar{X}_k$ for constants $a, b_1, \ldots, b_k$

$$\text{in } Y_i = a + b_1X_1 + \cdots + b_kX_k$$

(11.14) $\lim_{m \to \infty} P(|f_m(x) - P(X = x)| > \varepsilon) = 0$

(11.15) $V_X = \sum_{x \in X'} f_m(x)(x - \mu)^2$

(11.16) $\lim_{m \to \infty} V_X = \text{var}(X)$

(11.17) $V_X = \sum_{x \in X'} f_m(x)(x - \mu)^2 = \dfrac{1}{N}\sum_{i=1}^{N}(X_i - \mu)^2$

(11.18) $S^2 = \dfrac{1}{N}\sum_{i=1}^{N}(X_i - \bar{X})^2$

(11.19) $E(S^2) = \sigma^2 + \mu^2 - \sigma_{\bar{X}}^2 - \mu^2 = \sigma^2 - \sigma_{\bar{X}}^2$

(11.20) $E(S^2) = \sigma^2 - \dfrac{\sigma^2}{N} = \dfrac{N-1}{N}\sigma^2$

(11.21) $\hat{S}^2 = \dfrac{N}{N-1}S^2$

(11.22) $S_Z^2 = S_X^2 + S_Y^2 + 2S_{XY}^2$ where $Z = X + Y$

(11.23) $S_Y^2 = c^2 S_X^2$ for $Y = cX$ where c is a constant

(11.24) $S_Y^2 = S_X^2$ where $Y = X + c$, c a constant

(11.25) $S_X^2 = S_c^2 = 0$ where $X = c$, c a constant

(11.26) $S_Z^2 = b^2 S_X^2 + c^2 S_Y^2 + d^2 S_W^2 + 2bc S_{XY}^2 + 2bd S_{XW}^2 + 2cd S_{YW}^2$

$$\text{where } a, b, c, \text{ and } d \text{ are constants in } Z = a + bX + cY + dW$$

(11.27) $S^2 = \dfrac{N\sum_{i=1}^{N}X_i^2 - \left(\sum_{i=1}^{N}X_i\right)^2}{N^2}$

(11.28) $\hat{S}^2 = \dfrac{N\sum_{i=1}^{N}X_i^2 - \left(\sum_{i=1}^{N}X_i\right)^2}{N(N-1)}$

(11.29) $S_X = \sqrt{S_X^2}$

$$(11.30) \quad \hat{S}_{\bar{X}}^2 = \frac{\hat{S}^2}{N}$$

$$(11.31) \quad \hat{S}_{\bar{X}}^2 = \frac{S^2}{N-1}$$

$$(11.32) \quad \hat{S}_{\bar{X}} = \frac{S}{\sqrt{N-1}}$$

$$(11.33) \quad P(\mu - c\hat{S}_{\bar{X}} \le \bar{X} \le \mu + c\hat{S}_{\bar{X}}) \ge 1 - \frac{1}{c^2}$$

$$(11.34) \quad \lim_{N \to \infty} P(\mu - k \le \bar{X} \le \mu + k) = 1.0$$

$$(11.35) \quad P\left(\frac{-k}{\sqrt{\sigma^2/N}} \le \frac{\bar{X} - \mu}{\sqrt{\sigma^2/N}} \le \frac{+k}{\sqrt{\sigma^2/N}}\right) \cong \Phi\left(\frac{+k}{\sqrt{\sigma^2/N}}\right) - \Phi\left(\frac{-k}{\sqrt{\sigma^2/N}}\right)$$

PROBLEMS

1. Imagine that the table of random numbers in the back of the book is a table of th
results of observing some behavior. Each value observed is a single digit (ignore th
spacing of the digits in the table). Performing a smaller experiment would correspon
to observing a smaller number of values than the number listed in the table. Becaus
of the way the numbers in the table were selected, we know that the "expectation" c
the numbers is 4.5 and the "standard deviation" of the numbers is 2.89. Demonstrat
that the sample mean is a random variable by repeating the following experiment
number of times. Sample from the table of random numbers ten times, each tim
taking 15 random numbers. Use the first few random numbers in the table to deter
mine where each sample will be taken from the table. Calculate the mean an
standard deviation for each sample. Plot the relative-frequency distribution of th
sample values, the mean of the sample values, and the standard deviation of th
sample values. Compare the mean of the sample means and sample standard devia
tions with the parameter values given above. How do the results of this exercise relat
to the statement that $\bar{X}$ is a random variable?

2. It was shown in the text that the relative frequency of successes in binomial trials i
an unbiased estimate of the probability of success, p. The demonstration was in term
of a specific number of trials and an example value of p, 10 and .4 respectively
Construct a general proof using n and p instead of specific values.

3. Prove that the number of successes in binomial trials $n > 1$ is a biased estimator of p, the probability of success.

4. Pick ten pairs of digits from the random number table at the end of the book. Call one member of each pair X_i, and the other member Y_i, for $i = 1, 2, \ldots, 10$. Calculate $\bar{X}, \bar{Y}, \bar{Z}$ for $Z_i = X_i + Y_i$, and $\bar{W}$ for $W_i = 3.0 + 2.0\,X_i - 4.0\,Y_i$. Check your answers against the values required by equations (11.12) and (11.13).

5. The fact that the mean of a sample of observations tends to be less variable about the expectation of the variable being observed, as sample size is increased, can be demonstrated empirically. Using the table of random numbers, sample one number (a single digit) at a time, keeping a running mean for the entire sample with each successive number sampled. That is, take the mean of the first number sampled, then the mean of the first two numbers sampled, then the mean of the first three numbers sampled, and so on. Plot the running mean as a function of the sample size. In order to judge the discrepancy between the sample mean and the expected value (in the case of the random numbers, 4.5), draw a horizontal line across the graph at the value of the expectation.

6. For the data of problem 1 calculate the unbiased estimates of variances for each time you calculated a sample variance. Compare the differences in magnitude of the respective values.

7. Using the data of problem 4, demonstrate equations (11.22) and (11.26). How much would your answers be off if you ignored the covariance terms in the equations? The covariance of two variables X and Y is calculated most readily by

$$C_{XY} = \frac{1}{N}\sum_{i=1}^{N} X_i Y_i - \bar{X}\bar{Y}$$

8. Construct graphs like those in Figures 11.3 and 11.4 for probability .95.

9. The incomes of twenty people are given in the list below.

Income	Number of People
$6000	2
7000	5
7500	5
8000	3
8500	2
9000	2
9500	1

Calculate the mean and variance of the sample. In order to simplify the calculation, subtract a constant from the income and divide the difference by a constant and use the resulting values in the calculation, transforming the answer back into the original scale of income as the last steps (select the two constants carefully).

10. Construct an argument that the mean of a sample of observations on a variable, $\bar{X}$, is greater or equal to the smallest value observed and smaller or equal to the largest value observed.

11. If the sample of N observed values, $X_1, X_2, \ldots, X_N$ is composed of k values equal to c (a constant), and $(N - k = m)$ values equal to d (a constant), then

$$\bar{X} = \frac{kc + md}{k + m}$$

Verify this by algebraic means and demonstrate it with a numerical example.

12. Construct a problem similar to that of problem 11 for the sample variance.

CHAPTER 12

Sampling Distributions of Means and Variances

The preceding chapter introduced the expectation and variance of the sample mean by methods not involving a specification of the value sets or probability distribution of the random variables. This chapter develops the theory of the distribution of the sample mean and the distribution of the sample variance.

The distribution of sample means is a special case of the distribution of sample statistics. A sample statistic is any function of the sample of observations of a random variable. The mean is a function of a sample of observations and hence is a sample statistic. In addition to the sampling distribution of $\overline{X}$, two other special cases are developed: a chi square distributed statistic, and a statistic called Student's t. These statistics are not clearly estimates of any parameter in the same way that $\overline{X}$ is an estimate of μ.

Since S^2 and $\hat{S}^2$ are functions of observations of random variables, it follows that they are statistics. The sampling distributions of S^2 and $\hat{S}^2$ are of interest because of their use in the theory of statistical hypothesis testing.

Once the distribution of a sample statistic is known, it is possible to make strong statements about the range of probable values of parameters estimated by the statistic. Such statements are known as interval estimates. In this chapter we develop the theory of interval estimates and the special cases of interval estimates based on $\bar{X}$ and $\hat{S}^2$ for μ and σ^2.

Distribution of the Sample Mean with σ^2 Known

In the previous chapter we developed a relatively extensive picture of the behavior of the sample mean. In Chapter 10 the central limit theorem was presented as a mechanism leading to the development of normally distributed random variables. In this section we combine the two developments to find the sampling distribution of sample means.

The central limit theorem is specialized here to the case where all N random variables sampled are the same variable, and hence each has the same distribution, expectation, and variance. Instead of taking the sum of the random variables, we take the mean. Each observation in the sample is an observation on the same random variable. Consequently the N observations correspond to N random variables with the same distribution, expectation, variance, and so on. Let the random variables corresponding to the N independent observations be $X_1, X_2, \ldots, X_N$, each with expectation μ_X and variance σ_X^2. By the rules of expectation and variance,

$$\bar{X} = \frac{1}{N}(X_1 + X_2 + \cdots + X_N)$$

has an expectation

$$\mu_{\bar{X}} = \frac{1}{N} N\mu_X = \mu_X$$

and variance

$$\sigma_{\bar{X}}^2 = \frac{1}{N^2} N\sigma_X^2 = \frac{\sigma_X^2}{N}$$

This result obtains regardless of the magnitude of N. However, as soon as N is large, $N \to \infty$ in the limit, the random variable $\bar{X}$ is normally distributed:

$$\bar{X}: N\left(\mu_X, \frac{\sigma_X^2}{N}\right)$$

This result is known as the central limit theorems for means. It should be stressed that no restrictions on the value sets, distributions, parameters, and so on, are made to obtain this result. The only restrictive condition is that the number of observations be large and that the observations be independent. The independence is assured in random sampling with replacement from the population, or by the condition that the population be large compared with the sample.

Figures 12.1, 12.2, and 12.3 show theoretical and empirical distributions relevant to illustrate the operation of the central limit theorem in three special cases. In the first figure the random variable being sampled has a uniform distribution on the unit interval (.0 to 1.0). The expectation and variance of this distribution are given by

$$\mu = \int_0^1 xf(x)\, dx = .5$$

$$\sigma^2 = \int_0^1 (x - .5)^2 f(x)\, dx = \tfrac{1}{12}$$

Applying the central limit theorem to samples of size $N = 4$, $N = 10$, and $N = 25$ results in distributions of means approximated by $N(.5, 1/48)$, $N(.5, 1/120)$ and $N(.5, 1/300)$, respectively. These are graphed in Figure 12.1.

In addition, Figure 12.1 displays smoothed frequency functions of the results of experiments corresponding to the probability models graphed. These experiments were conducted on an electronic computer in the following way. For $N = 1$, 3000

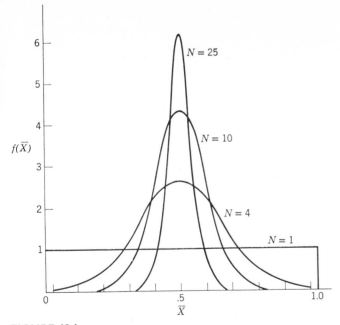

FIGURE 12.1

Graphs of frequency distributions of means of samples of $N = 1$, $N = 4$, $N = 10$, and $N = 25$, where the random variable being sampled is uniformly distributed on the unit interval as shown for $N = 1$.

samples of size $N = 1$ were selected by generating a pseudo-random number x with a uniform distribution in the range $(0 \leq x \leq 1)$. These numbers were arranged into 100 class intervals of length .01. The relative frequency of random numbers in each of these class intervals was calculated and plotted. For $N = 4$ the random numbers were generated in groups of four, the average of the four numbers was calculated, and the process repeated 3000 times. Thus, 3000 means for $N = 4$ were obtained. The same graphing procedure was followed for the frequency distribution of these samples. The frequency distribution graphs for samples of $N = 10$ and $N = 25$ were obtained in the same general manner. The degree of approximation of the normal distribution to the frequency distributions obtained is quite good for $N = 10$ and $N = 25$ but poor for $N = 4$.

The same general procedures were used to obtain the frequency distributions of Figures 12.2 and 12.3. Figure 12.2 is based on a population in which the density function of a value in the interval $(0 \leq x \leq 1)$ had a J-shaped form as indicated

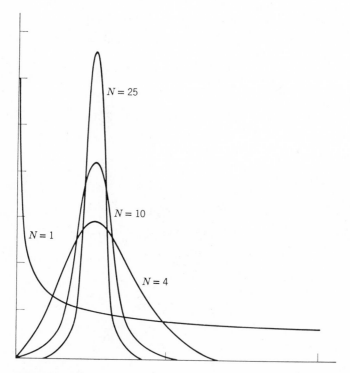

FIGURE 12.2

Graphs of frequency distributions of means of samples with $N = 1$, $N = 4$, $N = 10$, and $N = 25$, where the random variable is distributed as shown for $N = 1$.

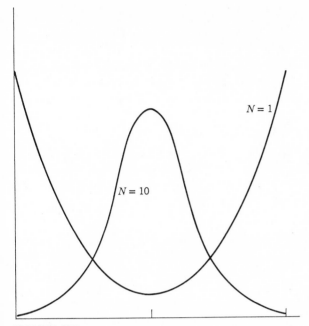

FIGURE 12.3

Graphs of frequency distributions of means of samples with $N = 1$ and $N = 10$, where the random variable is distributed as shown in the graph for $N = 1$.

in the figure for $N = 1$. Figure 12.3 is based on a population in which the density function of a value in the interval $(0 \leq x \leq 1)$ had a U-shaped form as indicated in the figure.

From the degree of goodness of the normal approximation to the frequency distribution of the sample mean indicated by the three figures above, it is clear that the number of observations required in a sample in order for the CLT for means to apply is relatively small. The usual convention is that if $N \geq 25$, or perhaps more conservatively if $N \geq 30$, then the sampling distribution of $\bar{X}$ is so nearly normal that the normal distribution may be taken as its model.

The smaller the number of observations in the calculation of $\bar{X}$, the poorer the normal approximation is, except under one condition. If the random variable X is normally distributed in the population, then the sample mean is normally distributed regardless of the number of observations involved in $\bar{X}$. If the population distribution of a random variable is even roughly normal, the sample size N need not be large to have the sampling distribution of $\bar{X}$ approximately normal. Thus the sample size necessary to insure the approximate normality of the distribution of $\bar{X}$ depends on the distribution of the random variable sampled.

There is an area of uncertainty that causes a good deal of trouble in statistical theory. When the distribution of the random variable being observed is not known and N is small, the sampling distribution of $\bar{X}$ cannot be specified. Special techniques for dealing with these types of sampling distributions are developed in the context of statistical problem discussions in the chapters to come.

In most empirical applications the variance of the observed random variable is not known. The results just presented are not appropriate in those applications, since it is presumed that σ^2 is known. The problem thus posed is solved, in part, by using an estimate, $\hat{S}^2$, of σ^2. However, as we see below, simply substituting $\hat{S}^2$ for σ^2 is insufficient and we must develop new sampling distributions to solve the problem. The sampling distribution of $\bar{X}$ with unknown σ^2 is best developed after having dealt with the sampling distribution of sample variances.

Sample Variances as Random Variables

Since S^2 is a function of a sample of observations of a random variable in exactly the same sense as $\bar{X}$, the interpretation of S^2 as a random variable follows the arguments developed for $\bar{X}$ as a random variable.

In order to derive the distribution of S^2, we need to develop a new theoretical distribution. The development of this abstract model clarifies its relevance to the distribution of S^2.

Chi Square Distributed Random Variables

Let X be $N(\mu, \sigma^2)$. We state as a definition that a single observation on X when standardized and squared, that is,

$$Z^2 = \left(\frac{X - \mu}{\sigma}\right)^2$$

has a density function called the chi square, χ^2, distribution. Although we are unable to prove it here, the χ^2 distribution is not dependent on μ or σ. We do not attempt a derivation of the χ^2 distribution here, nor shall we display the equation for its density function. The derivation of this equation and its interpretation require mathematical tools and concepts we do not have time or space to develop. However, several properties of this density function can be made clear.

First, the standardization of X in the above definition is familiar. In particular, we know that if X is $N(\mu, \sigma^2)$, then $Z = (X - \mu)/\sigma$ is $N(0, 1)$. Hence Z^2 is distributed as χ^2.

Second, $Z^2 \geq 0$, that is, the sample space of the random variable is composed of all the nonnegative real numbers. Also, since X is continuous, Z^2 is continuous. The value set of Z^2 is thus a real-valued set:

$$Z^2 = \{u \mid 0 \leq u < \infty\}$$

Third, the probability for an interval is not changed by transforming the interval values. Hence, since we know $P(-1 \leq Z \leq 0\} \cong .34$ and $P(0 < Z \leq 1) \cong$ .34, then $P(Z^2 \leq 1) = P(-1 \leq Z \leq 1) = .68$ by the addition rule for probabilities of disjoint sets. If we followed the same procedures for more finely defined intervals, we would get the entries in the table of the χ^2 distribution at the end of the book, Table D, and the graph of Figure 12.4. In Figure 12.4 the transformation from Z to Z^2 is illustrated for two symmetrical intervals on Z: $(-.3 \leq Z \leq -.2)$ and

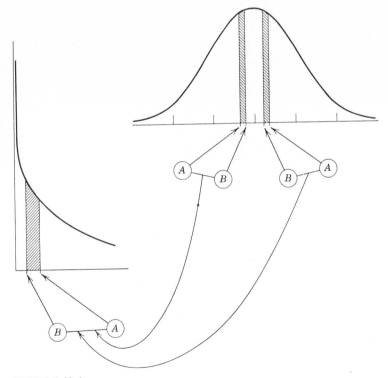

FIGURE 12.4

Graph of unit normal density function and a χ^2 density function, showing the mapping of probability from the normal function to the χ^2 function.

$(.2 \leq Z \leq .3)$. Squaring Z maps these intervals into a single interval, $(.04 \leq Z^2 \leq .09)$. Because of the rule for finding the probability of a union of disjoint sets, we have

$$P(.04 \leq Z^2 \leq .09) = P(-.3 \leq Z \leq -.2) + P(.2 \leq Z \leq .3)$$

Note that, as in dealing with any continuous random variable, $f(z^2)$ is not a probability.

This result can be generalized to several random variables. If X_1 is $N(\mu_1, \sigma_1^2)$, X_2 is $N(\mu_2, \sigma_2^2), \ldots, X_k$ is $N(\mu_k, \sigma_k^2)$, we can form the following family of random variables:

$$Z_1^2 = \frac{(X_1 - \mu_1)^2}{\sigma_1^2}$$

$$Z_2^2 = \frac{(X_1 - \mu_1)^2}{\sigma_1^2} + \frac{(X_2 - \mu_2)^2}{\sigma_2^2}$$

$$\vdots \qquad \qquad \vdots$$

$$Z_k^2 = \frac{(X_1 - \mu_1)^2}{\sigma_1^2} + \cdots + \frac{(X_k - \mu_k)^2}{\sigma_k^2} \qquad (12.1)$$

Since each of the random variables X_i is normally distributed, the components of the sums are $N(0, 1)$ and hence the sums are sums of χ^2 distributed random variables. We claim without proof that the sum of k independent random variables with χ^2 distributions has a distribution χ_k^2, chi square with k degrees of freedom (see below). Consequently, we call the basic χ^2 distribution χ_1^2 or chi square with one degree of freedom. Before engaging in a discussion of this new concept "degrees of freedom," we inspect the variable Z_k^2 and its distribution.

First we derive the expectation and variance of Z_k^2. If we let

$$V_1 = \frac{X_1 - \mu_1}{\sigma_1}$$

$$V_2 = \frac{X_2 - \mu_2}{\sigma_2}$$

$$\vdots \qquad \qquad \vdots$$

$$V_k = \frac{X_k - \mu_k}{\sigma_k}$$

then

$$Z_k^2 = V_1^2 + V_2^2 + \cdots + V_k^2$$

Since the V's are independent and have expectations of zero and variances of 1.0,

$$
\begin{aligned}
E(Z_k^2) &= E(V_1^2) + \cdots + E(V_k^2) \\
&= \text{var}(V_1) + \cdots + \text{var}(V_k) \\
&= 1 + \cdots + 1 \\
&= k
\end{aligned}
\tag{12.2}
$$

Thus, as the number of terms are added in a chi square distributed variable (increasing the degrees of freedom, k), the expected value grows as a direct one-to-one function of the number of terms. Thus, Z_{142}^2 would have an expected value of 142. This can be made clear by noting that we are adding together, not averaging, the k variables. There are more large values in such a sum with higher probability than in the individual terms in the sum.

The variability or dispersion of Z_k^2 is given by the usual formula for the variance of sums of independent random variables

$$
\text{var}(Z_k^2) = \text{var}(V_1^2) + \text{var}(V_2^2) + \cdots + \text{var}(V_k^2)
$$

Since $V_1, V_2, \ldots, V_k$ are independent. The variance of any individual term V_j, is given by

$$
\begin{aligned}
\text{var}(V_j^2) &= E(V_j^2 - 1)^2 \\
&= E(V_j^4 - 2V_j^2 + 1) \\
&= E(V_j^4) - 2E(V_j^2) + 1 \\
&= 3 - 2 + 1 = 2
\end{aligned}
$$

and hence

$$
\text{var}(Z_k^2) = 2k
\tag{12.3}
$$

This result depends on the fact that the expected fourth power of a $N(0, 1)$ random variable is equal to 3.0.

Thus, Z_k^2 is distributed as χ_k^2, which increases in location as k increases, and increases in dispersion at twice the rate. When $k = 8$, $E(Z_8^2) = 8$ and var $(Z_8^2) = 16$. Several χ^2 density functions are graphed in Figure 12.5.

Because the χ^2 distributed variable Z^2 is defined as a simple sum, it follows that χ^2 variables are additive in a special way. Recall the definition of the family of variables $Z_1^2, Z_2^2, \ldots, Z_k^2$. Here Z_k^2 was the Z_{k-1}^2 variable plus the square of the standardized kth random variable. By the law of summation this generalizes to

$$
\begin{aligned}
Z_k^2 &= \sum_{i=1}^{k} \left(\frac{X_i - \mu_i}{\sigma_i} \right)^2 = \sum_{i=1}^{m} \left(\frac{X_i - \mu_i}{\sigma_i} \right)^2 + \sum_{i=m+1}^{k} \left(\frac{X_i - \mu_i}{\sigma_i} \right)^2 \\
&= Z_m^2 + Z_{k-m}^2
\end{aligned}
\tag{12.4}
$$

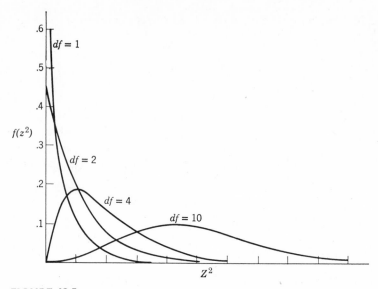

FIGURE 12.5

Graphs of several χ^2 distributions.

If $Z_a{}^2$, $Z_b{}^2$, $Z_c{}^2$, ..., $Z_f{}^2$ are distributed respectively as $\chi_a{}^2$, $\chi_b{}^2$, ..., $\chi_f{}^2$, then $Z_n{}^2 = Z_a{}^2 + Z_b{}^2 + \cdots + Z_f{}^2$ is distributed as $\chi_n{}^2$ where $n = a + b + \cdots + f$.

The family of χ^2 distributions is determined completely by the number k of independent terms entering each of the specific (member of the family) $\chi_k{}^2$ distributions. For each value of k the distribution is distinct. No two χ^2 distributions are the same, and for small values of k even the standardized $\chi_k{}^2$ distributed random variables do not have similar graphs. Thus, unlike normal distributions, where we could deal with the standard normal, the χ^2 distributions must be dealt with separately. This leads to a χ^2 table of rather ponderous size. For each value of k there is a unique set of values $P(Z_k{}^2 \leq z_k{}^2)$ to be tabled. In addition, the values of $z_k{}^2$ that have salient values of $P(Z_k{}^2 \leq z_k{}^2)$ are different for different values of k. As k becomes large, so does $E(Z_k{}^2)$, and hence the probabilities $P(Z_k{}^2 \leq \mathbf{z_k}^2)$ for large values of $z_k{}^2$ are large (relatively).

In tables of the standard normal random variable the values of $P(X \leq x)$ are given for $(0 \leq x \leq 4)$, generally in steps of .01. This gives a table of about 400 entries. Should we construct such a table for each $\chi_k{}^2$ distributed random variable for just $(1 \leq k \leq 30)$, we would have to have 30 pages of table with 400 entries to a page. Tables of this magnitude and more are available for detailed use of the χ^2 distribution. However, for the purposes of most statistical work such extensive tables are of little use. The extremes of the distributions are the most important regions in statistical applications, and hence, in general, only the extremes are tabled in much detail. This opens the questions regarding the definition of extremes. A moment's reflection reveals that extreme values in Z_1^2 are not the same as the extreme values in Z_{30}^2: $z^2 = 1$ is extremely small in Z_{30}^2 but not in Z_1^2; $z^2 = 30$ is extremely large in Z_1^2 but not in Z_{30}^2. If "extreme" is defined as a probabilistic concept, then a value z_k^2 that has $P(Z_k^2 \leq z_k^2) = .99$ would be an "extremely" large value (Z_k^2 has values in excess of z_k^2 with probability .01), regardless of k.

Construction of the χ^2 table begins with a selection of some important or salient values for $P(Z_k^2 \leq z_k^2)$. These values have to be very small values as well as very large values because of the asymmetry of the χ^2 distribution. The values selected in the table for this text are .005, .01, .025, .05, .1, .25, .5, .75, .9, .975, .99, .995. For these probability values p the equation $P(Z_k^2 \leq z_k^2) = p$ was solved for z_k^2 for values of $(1 \leq k \leq 30)$. The values of z_k^2 are the values in Table D.

The graphs of χ^2 distributed variables of Figure 12.5 suggest that as k becomes large χ_k^2 approaches a normal distribution. Reflection on the CLT leads to the same conclusion. In fact, for large values of k, Z_k^2 is approximately normally distributed with expectation k and variance $2k$. However, this result requires rather large values of k before the approximation of χ_k^2 by $N(k, 2k)$ is a good one.

Often a transformation of a random variable has a distribution more nearly normal. The simplest such transformation for χ^2 distributed random variables is that developed by R. A. Fisher. For $k \geq 30$

$$U = \sqrt{2\chi_k^2} \text{ is approximately } N(\sqrt{2k - 1}, 1)$$

Hence

$$U^* = \frac{\sqrt{2\chi_k^2} - \sqrt{2k - 1}}{1} \text{ is approximately } N(0, 1)$$

Thus, for some value z_k^2 from Z_k^2

$$P(Z_k^2 \leq z_k^2) \simeq \Phi(\sqrt{2z_k^2} - \sqrt{2k - 1}) \qquad (12.5)$$

Sampling Distribution of Sample Variances

If we have N independent observations $X_1, X_2, \ldots, X_N$ on a normally distributed random variable with expectation μ and variance σ^2, the statistic

$$V = \frac{1}{N} \sum_{i=1}^{N} (X_i - \mu)^2 \qquad (12.6)$$

has a distribution that is a simple function of a $\chi_N{}^2$ distributed random variable. Since

$$Z_N{}^2 = \frac{\sum_{i=1}^{N}(X_i - \mu)^2}{\sigma^2}$$

is $\chi_N{}^2$, then

$$\frac{N}{\sigma^2}V = \frac{N\dfrac{1}{N}\sum_{i=1}^{N}(X_i - \mu)^2}{\sigma^2} = \sum_{i=1}^{N}\frac{(X_i - \mu)^2}{\sigma^2}$$

$$= Z_N{}^2$$

is $\chi_N{}^2$. This gives us a direct method for determining the sampling distribution of V. Let

$$W = \frac{N}{\sigma^2}V$$

We know that W is $\chi_N{}^2$. Hence

$$P(W \leq w) = P(Z_N{}^2 \leq z_N{}^2)$$

and thus

$$P\left(\frac{N}{\sigma^2}V \leq \frac{N}{\sigma^2}v\right) = P(Z_N{}^2 \leq z_N{}^2)$$

But since N and σ^2 are constants,

$$P\left(\frac{N}{\sigma^2}V \leq \frac{N}{\sigma^2}v\right) = P(V \leq v)$$

Hence

$$P(V \leq v) = P(Z_N{}^2 \leq z_N{}^2) \tag{12.7}$$

where the correspondence between values of v and z^2 is given by

$$z_N{}^2 = \frac{N}{\sigma^2}v \tag{12.8}$$

Thus, knowledge of $P(V \leq v)$ depends on knowledge of σ^2, which may not be available in empirical samples. However, in the statistical applications of χ^2 distributions, this does not cause any embarrassment.

The case of the sample variance S^2 when μ is not known is more realistic:

$$S^2 = \frac{1}{N}\sum_{i=1}^{N}(X_i - \bar{X})^2$$

We may express $(X_i - \mu)$ as a sum of additive components: (1) the difference of X_i and $\bar{X}$, and (2) the difference in $\bar{X}$ and μ. That is,

$$(X_i - \mu) = (X_i - \bar{X}) + (\bar{X} - \mu) \tag{12.9}$$

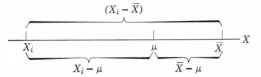

FIGURE 12.6

Illustration of additive components in $(X_i - \mu)$.

as illustrated in Figure 12.6. Thus

$$(X_i - \mu)^2 = [(X_i - \bar{X}) + (\bar{X} - \mu)]^2$$
$$= (X_i - \bar{X})^2 + (\bar{X} - \mu)^2 + 2(X_i - \bar{X})(\bar{X} - \mu)$$

Summing over X_i in the sample gives

$$\sum_{i=1}^{N}(X_i - \mu)^2 = \sum_{i=1}^{N}(X_i - \bar{X})^2 + \sum_{i=1}^{N}(\bar{X} - \mu)^2 + 2\sum_{i=1}^{N}(X_i - \bar{X})(\bar{X} - \mu)$$

$$= \sum_{i=1}^{N}(X_i - \bar{X})^2 + N(\bar{X} - \mu)^2 + 2(\bar{X} - \mu)\sum_{i=1}^{N}(X_i - \bar{X})$$

and since

$$\sum_{i=1}^{N}(X_i - \bar{X}) = \sum_{i=1}^{N}X_i - N\bar{X}$$

$$= \sum_{i=1}^{N}X_i - N\frac{\sum_{i=1}^{N}X_i}{N}$$

$$= 0$$

$$\sum_{i=1}^{N}(X_i - \mu)^2 = \sum_{i=1}^{N}(X_i - \bar{X})^2 + N(\bar{X} - \mu)^2 \qquad (12.10)$$

This is transformed to a χ_N^2 variable by dividing by σ^2:

$$\sum_{i=1}^{N}\frac{(X_i - \mu)^2}{\sigma^2} = \sum_{i=1}^{N}\frac{(X_i - \bar{X})^2}{\sigma^2} + \frac{N(\bar{X} - \mu)^2}{\sigma^2}$$

The last term in this equation leads to a useful result. If X is normal, as per our hypothesis, then $\bar{X}$ is normal with $\sigma_{\bar{X}}^2 = \sigma^2/N$ and

$$\frac{(\bar{X} - \mu)^2}{\sigma_{\bar{X}}^2} = \frac{(\bar{X} - \mu)^2}{\sigma^2/N}$$

is χ_1^2. Notice that

$$\frac{N(\bar{X} - \mu)^2}{\sigma^2} = \frac{(\bar{X} - \mu)^2}{\sigma^2/N}$$

Hence

$$Z_N{}^2 = \sum_{i=1}^{N} \frac{(X_i - \mu)^2}{\sigma^2} = \sum_{i=1}^{N} \frac{(X_i - \bar{X})^2}{\sigma^2} + Z_1{}^2 \tag{12.11}$$

This implies that

$$\sum_{i=1}^{N} \frac{(X_i - \bar{X})^2}{\sigma^2} = Z_{N-1}^2 \tag{12.12}$$

from the additivity of χ^2 distributed variables. Recall that

$$\sum_{i=1}^{N} (X_i - \bar{X})^2 = NS^2$$

and substitute into the above expression to get

$$\frac{NS^2}{\sigma^2} = Z_{N-1}^2 \tag{12.13}$$

Thus NS^2/σ^2 is a χ_{N-1}^2 distributed random variable.

By parallel argument, using

$$\sum_{i=1}^{N} (X_i - \bar{X})^2 = (N - 1)\hat{S}^2$$

we have

$$\frac{(N - 1)\hat{S}^2}{\sigma^2} = Z_{N-1}^2 \tag{12.14}$$

Multiplying both sides by $\sigma^2/(N - 1)$ gives

$$\hat{S}^2 = \frac{Z_{N-1}^2 \sigma^2}{N - 1}$$

The expectation and variance of the sampling distribution of $\hat{S}^2$ can be found directly from these results. First, let

$$\frac{\sigma^2}{N - 1} = K$$

Then

$$\hat{S}^2 = KZ_{N-1}^2$$

and

$$E(\hat{S}^2) = E(KZ_{N-1}^2) = KE(Z_{N-1}^2)$$
$$= K(N - 1)$$
$$= \frac{\sigma^2}{N - 1}(N - 1)$$
$$= \sigma^2$$

and we have again a proof that $\hat{S}^2$ is an unbiased estimator of σ^2. The variance of $\hat{S}^2$ is obtained similarly.

$$
\begin{aligned}
\text{var}(\hat{S}^2) &= \text{var}(KZ_{N-1}^2) \\
&= K^2 \, \text{var}(Z_{N-1}^2) \\
&= K^2 2(N-1) \\
&= \frac{\sigma^4}{(N-1)^2} 2(N-1) \\
&= 2 \frac{\sigma^4}{N-1}
\end{aligned}
$$

The sampling distribution of the standard deviation can be derived easily from that of the square of the standard deviation, that is, the variance. Without proof we state that

$$
\sqrt{Z_k^2}
$$

is distributed as a chi random variable χ_k where

$$
P(Z_k^2 \leq z_k^2) = P(\sqrt{Z_k^2} \leq \sqrt{z_k^2})
$$

where Z_k^2 is χ_k^2. The distribution function of χ_k can be determined from χ_k^2 by a simple transformation that we do not develop here. Since

$$
\hat{S}^2 = \frac{Z_{N-1}^2 \sigma^2}{N-1}
$$

then

$$
\hat{S} = \frac{\sigma \sqrt{Z_{N-1}^2}}{\sqrt{N-1}}
$$

where $\sqrt{Z_{N-1}^2}$ is distributed as χ_{N-1}.

The significance of the reduction of the number of degrees of freedom from N to $(N-1)$ when $\bar{X}$ is substituted for μ cannot be overstated. This is the first example of the so-called "loss of degrees of freedom" when an estimated value is used in place of a parameter. The basic reason for this "loss" lies in the fact that the observations X_i used in calculating $\sum_{i=1}^{N} (X_i - \bar{X})^2$ have all entered into the calculation of $\bar{X}$. Hence, not all of the values in $\sum_{i=1}^{N} (X_i - \bar{X})^2$ are independent. This is formalized at the end of this chapter in a discussion relevant to all statistics based on samples.

A note of caution is called for here. For reasons we cannot go into, the sampling distributions of $\bar{X}$ and $\hat{S}^2$ have different dependencies on the normality of the distribution of the random variable sampled. Even if X is not normal, $\bar{X}$ is approximately normal if $N \geq 30$. However, if X is not normal, $\hat{S}^2$ is not χ_{N-1}^2 unless N is very large.

We now turn to a more realistic (from an applications point of view) development of the sampling distribution of $\overline{X}$. It is rare that the parameter σ^2 is known when a sample of observations is made. Hence the result that

$$\overline{X} \text{ is } N\left(\mu, \frac{\sigma^2}{N}\right)$$

is of little use when σ^2 is not known. However, we can find the distribution of $\overline{X}$ when we use $\hat{S}^2$ as an estimate of σ^2—under some special circumstances.

Distribution of $\overline{X}$ with σ^2 Unknown

The two special cases for which we have the necessary theoretical development do not cover the entire range of experimental possibilities. Hence, for some experiments where we are concerned with $\overline{X}$ we are not able to apply the models developed in this section. The developments below cover (1) large samples of independent observations on arbitrarily distributed random variables, and (2) any size sample of independent observations of normally distributed random variables. The third logical type of experiment is defined by small samples of observations on a random variable that is not normally distributed. There are extenuating circumstances relieving the gap in the models developed in this chapter. These circumstances are discussed in the context of the applications to statistical tests in the following chapters.

The Critical Ratio or Z Statistic. Regardless of the distribution of X, if a large number of independent observations are made on X, then

$$\overline{X} = \frac{1}{N}\sum_{i=1}^{N} X_i$$

is normally distributed with expectation μ and variance σ^2/N. This implies that

$$Z = \frac{\overline{X} - \mu_X}{\sigma/\sqrt{N}}$$

is distributed as $N(0, 1)$.

If σ^2 is unknown, the natural thing to try is to replace σ^2 with $\hat{S}^2$, its unbiased estimator from the sample. Now if N is large, the value $E(\hat{S}^2 - \sigma^2)^2$ is small with respect to σ^2. [See the section above for the definition of var($\hat{S}^2$).] The reader should attempt to determine from the development of the sampling distribution of

$\hat{S}^2$ as a χ^2 distributed variable just what proportion of σ^2 is involved in the standard deviation of $\hat{S}^2$.

Since for large N, $\sqrt{\text{var}(\hat{S}^2)}$ is small with respect to σ^2, and since $E(\hat{S}^2) = \sigma^2$, we may effectively consider $\hat{S}^2 \cong \sigma^2$ and hence we have

$$Z = \frac{\bar{X} - \mu}{\hat{S}/\sqrt{N}} = \frac{\bar{X} - \mu}{\hat{S}_{\bar{X}}} \tag{12.15}$$

is distributed approximately as $N(0, 1)$.

This statistic is known traditionally in psychology as the critical ratio. Henceforth we shall refer to it as the Z or normal statistic. The conventional lower limit on "large" for N is about 30. This magnitude in the context of the sampling distribution of $\bar{X}$ should insure that the distribution of the Z statistic is approximately normal. The similarity of the random variable Z above with a χ_k^2 distribution should be suggestive but not confusing.

The Z statistic has a value set (and hence a distribution) that is dependent on μ, but on no other unknown parameter. We use this fact later to make probability statements about (1) possible values for μ, and (2) hypothesized values of μ based on suggestions from psychological theory.

Student's t. In 1908 William Sealy Gosset ("Student," a brewer's mathematician) undertook to derive the distribution of the statistic

$$Q' = \frac{\bar{X} - \mu}{\hat{S}_{\bar{X}}} = \frac{\bar{X} - \mu}{S/\sqrt{N-1}}$$

when $N - 1$ is small. This case is not covered in the central limit theorem or by the argument that $\hat{S}_{\bar{X}}$ is very nearly equal to $\sigma_{\bar{X}}$ for large N. Thus, even if X is normally distributed but σ^2 is unknown, we have no basis for claiming anything about the distribution of Q'. In particular, Q' is not normally distributed even if X is normally distributed.

For the purpose of the development here we deal with

$$Q = \frac{1}{\sqrt{N-1}} Q' = \frac{1}{\sqrt{N-1}} \frac{\bar{X} - \mu}{S/\sqrt{N-1}} = \frac{\bar{X} - \mu}{S}$$

In order to develop the distribution of Q we must assume that X, the random variable sampled, is normal with $N(\mu, \sigma^2)$. As we find later, this assumption can be softened if X is not too badly nonnormal.

For the sake of simplicity we let

$$D = \bar{X} - \mu$$

Thus

$$Q = \frac{D}{S}$$

Since X is normal, then $E(\overline{X}) = \mu$ so that $E(D) = E(\overline{X} - \mu) = 0$, and

$$D: N(0, \sigma_{\overline{X}}^2)$$

The denominator of Q is distributed as the square root of a χ^2 random variable, hence as a χ random variable.

We thus have a ratio of two random variables of known distributions, a normally distributed numerator D, and a χ distributed denominator S. It remains to discover the sampling distribution of this ratio. The approach used is to work with the joint distribution of D and S. This involves defining the density function

$$f(D = d, S = s) = f(d, s)$$

for all values d and s. Since D and S are continuous random variables, the density function is defined on the real plane where $S \geq 0$. For two jointly distributed independent random variables X and Y, we have

$$f(x, y) = f(x)f(y)$$

In general, the sample mean and the sample standard deviation are not independent, and the joint distribution of $\overline{X}$ and S cannot be determined by the simple multiplication rule. However, in the case of normally distributed random variables it can be shown that $\overline{X}$ (and hence $X - \mu$) and S are independently distributed. For no other random variable is this true. The proof of this involves mathematics beyond the level of this text.

A good intuitive, geometric understanding of the joint distribution of D and S is helpful in the nonrigorous development of the distribution of Q. First we represent D as the abscissa in a pair of Cartesian coordinates and S as the ordinate, as in Figure 12.7.

The next step is to determine the set of points in the (D, S) plane that have the same values of Q. Several values of Q and the sets of points in (D, S) are represented in Figure 12.7. Here $Q = 0$ is defined by the set of points where $D = 0$; $Q = 1$ is defined by the set of points where $D > 0$ and $S = D$; $Q = -1$ is defined by the set of points where $D < 0$ and $S = D$; $Q = 2$ is defined by the set of points where $D > 0$ and $S = D/2$. Each value of Q is thus represented by a straight line radiating "upward" from $(D = 0, S = 0)$.

The density function, in Figure 12.8, is represented by contour lines like those representing altitude on geological maps. The highest point in the joint density

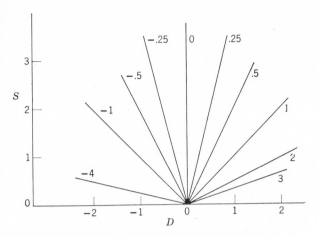

FIGURE 12.7

Sample space for the joint distribution of D and S showing outcomes, corresponding to values in the sample space of the random variable Q, as radiating lines originating at $(D = 0)$ and $(S = 0)$.

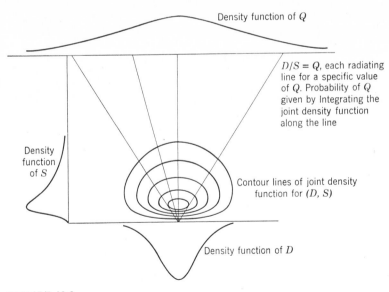

Density function of Q

$D/S = Q$, each radiating line for a specific value of Q. Probability of Q given by Integrating the joint density function along the line

Density function of S

Contour lines of joint density function for (D, S)

Density function of D

FIGURE 12.8

Illustration of density functions involved in the development of the random variable Q and its density function.

function is at the point coinciding with the maxima on the separate distributions of D and S.

Since we have represented each value of Q, say q, as a collection of points in a plane, q can be seen as an outcome in a sample space corresponding to the plane. From this it follows that $P(Q = q)$ is the "sum" of "probabilities" of the corresponding points in the outcome, that is, having a value q. However, since the joint sample space is continuous, we must use the methods of integral calculus to determine $P(Q = q)$. We do not develop here the necessary notation or mathematical apparatus to represent or perform this integration. The result is easy to grasp, however. For each value of Q we find a density function value that is the integral of (or the "total density" in) $f(D = d, S = s)$ along the line representing the value of Q. This integral gives the density function of Q:

$$f(Q = q)$$

Here Q is continuous and has as its value set the real number line. This random variable and its density function are represented at the top of Figure 12.8. This density function is Student's t distribution except for a simple transformation. It can be shown that $f(q)$ depends only on N, the sample size. The parameter σ^2 is not involved in $f(q)$ in any way. The distribution of $Q = D/S$ is free of σ^2 completely. The only parameter determining the distribution of Q is N, and N alone.

The symmetry of the D distribution insures that Q is distributed in a symmetric fashion with its expectation at .0. We state without proof here that

$$\text{var}(Q) = \frac{1}{N - 3}$$

implying that $\text{var}(Q)$ does not exist (undefined) for $N < 4$. Thus, for each value of N, the size of the sample, Q has a different standard deviation and hence the probabilities of specific values of Q vary as a function of N. The same characteristic was encountered in dealing with χ^2 distributions. The difficulty in tabling χ^2 is somewhat lessened in dealing with Q by making a very simple transformation. The result of the transformation is called Student's t distribution. Let

$$W = Q\sqrt{N - 1} = \frac{D}{S/\sqrt{N - 1}}$$

$$= \frac{\bar{X} - \mu}{\hat{S}_{\bar{X}}} \tag{12.16}$$

The parameter w is said to have Student's t distribution. The identical structure of the statistic W and the Z statistic is suggestive: When N is large, w is normally distributed. Therefore, when N is large, t is a normal distribution. However, when N is small, t is not normal. This is illustrated clearly by noting that

$$\begin{aligned}
\text{var}(W) &= \text{var}(Q\sqrt{N-1}) \\
&= (N-1)\,\text{var}(Q) \\
&= \frac{N-1}{N-3}
\end{aligned} \qquad (12.17)$$

which is always greater than 1.0. If $N = 4$, $\text{var}(W) = \frac{3}{1} = 3$. This implies that

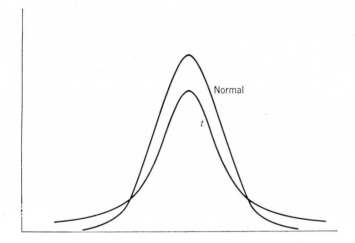

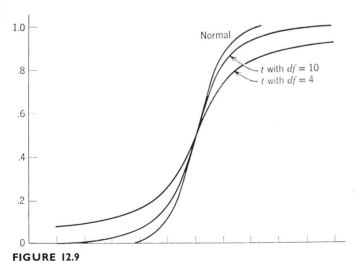

FIGURE 12.9

Density and distribution functions of Student's t and unit normal variables.

for small N, values of W as large as plus and minus 3 are observed with relatively high probability. As N is larger these values are more standard deviations from $E(W)$ and hence occur with smaller probability. When $N = 30$, $\text{var}(W) = \frac{29}{27} \cong 1.074$. When $N = 100$, $\text{var}(W) \cong 1.02060$. Thus it is clear that the t distribution is really a family of distributions. The specific members of the family depend on the size of the sample, N. For a given sample size it is conventional to deal with the number of degrees of freedom, which in the case of t distributions is $(N - 1)$. If we let

$$n = N - 1$$

we must table the distribution function

$$P(W \leq w)$$

for every value of n to cover the specific members in the family of t's. Denoting t for experiments with sample size N as t_n where $N - 1 = n$ and suitably subscripting W and w, we say then W_n is distributed as t_n. We need to table values of w_n for select values of $P(W_n \leq w_n)$. A selection of probability values and degree of freedom values are used in tabling t distributions in the table in the back of this book, Table E. The symmetry of t distributions permits us to limit the values to $P(W_n \leq w_n) \geq .5$. The tabled values w_n for $P(W_n \leq w_n) < .5$ are given by $(-1)(w_n)$ at the tabled value for $1 - P(W_n \leq w_n)$.

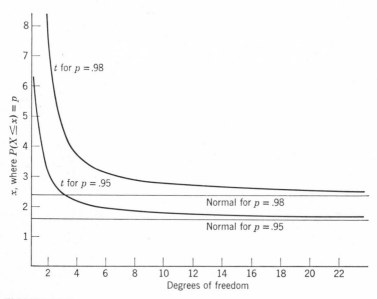

FIGURE 12.10

Graph illustrating the difference of the normal and t distributions in the upper tails.

The fact that $\text{var}(W_n) > 1.0$ where W_n is distributed as t_n implies that, in comparison to the normal distribution, more of the area in the density function of a t distributed random variable is in the "tails" of the distribution, and consequently less in the central portion of the distribution. This is illustrated graphically in Figures 12.9 and 12.10.

The normal distribution graphs are included in these two figures. The normal statistic or Z corresponds to t_∞, that is, to a t distributed statistic when $n \rightarrow \infty$.

Interval Estimates

We are now in a position to complete the discussion of estimation begun in the preceding chapter. The concept of point estimates for μ and σ^2 was discussed in detail there. However, discussion of the concept of interval estimate was delayed. This delay was necessary because the concept of sampling distributions for statistics is fundamental to interval estimation. First, we begin with the notion that an estimate is a statistic. We then reason that because a statistic is a function of observations on a random variable, the statistic is a random variable. The value of the parameter involved is but one of the values in the value set of the estimating statistic. Depending on the dispersion of the sampling distribution of the statistic, the range of "probable" values of the parameter is restricted. Only a limited range of values on either side of the observed statistic value is likely to contain the value of the parameter. The theory of interval estimation deals with models allowing probabilistic statements regarding possible values of the parameters, given the value of the statistic.

Three special cases are involved: (1) estimation of μ in large samples of observations on arbitrary random variables; (2) estimation of μ in samples on normally distributed random variables; (3) estimation of σ^2 in samples on normally distributed random variables. There is no satisfactory basis for interval estimation of μ with small samples of observations on nonnormally distributed random variables or of σ^2 except for normally distributed random variables.

The basic method for defining the interval of an estimate is to deal with the probability that, given the statistic, the parameter value is contained in the interval. We begin with the probability and find the limits of the interval.

As an example we begin with an unrealistic situation. Suppose that we sample $N = 16$ observations from a normally distributed random variable with μ unknown and $\sigma^2 = 4.0$. The sample mean, say, is $\bar{X} = 20$. We want to establish the largest and smallest "probable" values for μ from this knowledge. From developments earlier in this chapter we know that

$$Z = \frac{\bar{X} - \mu}{\sigma/\sqrt{N}}$$

is normally distributed. By the table of Φ we also know that

$$P(-1.96 < Z < 1.96) = \Phi(1.96) - \Phi(-1.96) = .95$$

and hence

$$P(-1.96 < Z < 1.96) = P\left(-1.96 < \frac{\bar{X} - \mu}{\sigma/\sqrt{N}} < 1.96\right)$$

$$= P\left(-1.96 \frac{\sigma}{\sqrt{N}} < \bar{X} - \mu < 1.96 \frac{\sigma}{\sqrt{N}}\right)$$

$$= P\left(\bar{X} - 1.96 \frac{\sigma}{\sqrt{N}} < \mu < \bar{X} + 1.96 \frac{\sigma}{\sqrt{N}}\right)$$

$$= .95$$

Rewriting this with the values of our example gives

$$P\left(20 - 1.96 \frac{2}{4} < \mu < 20 + 1.96 \frac{2}{4}\right) = P(20 - .98 < \mu < 20 + .98)$$

$$= P(19.02 < \mu < 20.98) = .95$$

The meaning of this last expression is evasive. We have indeed found two values flanking the obtained $\bar{X}$ and have attached a "probability" to the interval. However, since μ is some fixed (but unknown) value, it is either in the interval between 19.02 and 20.98 or it is not. Hence, realistically

$$P(19.02 < \mu < 20.98) = 1.0$$

or

$$P(19.02 < \mu < 20.98) = 0$$

The logical difficulty is related to the substitution of obtained values (for example, 20) for the random variable $\bar{X}$. The probability expressions used prior to substituting obtained values are legitimate.

We use a different interpretation of the .95 probability to avoid this logical problem. The interval (19.02, 20.98) is said to cover the point with the value μ with probability .95. This is sometimes interpreted as a measure of the confidence we can have that μ does indeed lie in the interval. In this sense

$$P\left(\bar{X} - 1.96 \frac{\sigma}{\sqrt{N}} < \mu < \bar{X} + 1.96 \frac{\sigma}{\sqrt{N}}\right) = .95$$

may be called the coefficient of confidence and the interval

$$\left(\bar{X} - 1.96 \frac{\sigma}{\sqrt{N}} , \bar{X} + 1.96 \frac{\sigma}{\sqrt{N}} \right)$$

is a confidence interval with confidence coefficient of .95.

By changing the constant 1.96 to 2.58, the coefficient of confidence is enhanced to .99. Using the constant 1.64 gives a coefficient of .90. Intervals with a given confidence coefficient may be defined in other ways. For example, $X: N(0, 1)$, the interval defined in

$$P(\mu < 2.58) = .995$$

is $(-\infty, 2.58)$ and has a confidence coefficient of .995 in spite of the obvious inefficiency of the estimate.

We limit ourselves here to discussing intervals symmetrical about the point estimate of the parameter μ. In the cases we consider, this model also gives the shortest (or narrowest) possible interval for the coefficient involved. For interval estimates of σ^2 the asymmetry of the χ^2 distribution is reflected in asymmetrical intervals.

Applying this theory to the estimation of μ by $\bar{X}$ when $N > 30$, we make use of the fact that $\bar{X}$ is distributed normally even where S^2 is used instead of σ^2. We showed earlier that

$$Z = \frac{\bar{X} - \mu}{S/\sqrt{N}} \qquad \text{was} \quad N(0, 1)$$

Hence we know that

$$P\left(\bar{X} - 1.96 \frac{S}{\sqrt{N}} < \mu < \bar{X} + 1.96 \frac{S}{\sqrt{N}} \right) = .95$$

In order to generalize our notation we let Z_α be the value of z such that

$$P(-z < Z_\alpha < z) = 1 - \alpha$$

If $\alpha = .05$, then

$$P(-z < Z_{.05} < z) = 1 - .05 = .95$$

and it follows that $z = 1.96$. The general expression thus becomes

$$P\left(\bar{X} - Z_\alpha \frac{S}{\sqrt{N}} < \mu < \bar{X} + Z_\alpha \frac{S}{\sqrt{N}} \right) = 1 - \alpha \qquad (12.18)$$

Since Z_α is selected by determining the degree of confidence required or acceptable, and $S/\sqrt{N}$ is given by the data, the value of μ is determined.

The logic of interval estimation may be approached in a different way. Imagine that the observed mean $\bar{X}_o$ is smaller than μ. We can specify sampling distributions from which $\bar{X}_o$ might have arisen. To do this we ask: If $\bar{X}_o$ were at the point of the sampling distribution where 5% of the distribution were below $\bar{X}_o$, what would μ have to be? Where $\bar{X}$ has a normal sampling distribution, μ is given by solving the equation

$$\Phi\left(\frac{\bar{X}_o - \mu}{S/\sqrt{N}}\right) = \alpha$$

Specifying α permits us to look up the value of z for Z_α. If $\alpha = .05$, we obtain

$$-1.64 = \frac{\bar{X}_o - \mu}{S/\sqrt{N}}$$

where $\bar{X}_o$, S, and N are given as constants from the experiment. If $\bar{X}_o = 20$, $N = 16$, and $S^2 = 4$, then

$$-1.64 = \frac{20 - \mu}{2/\sqrt{16}}$$

$$\mu = 20 + \frac{1.64}{2} = 20.82$$

Thus, if $\mu = 20.82$ we would observe $\bar{X}_o \leq 20$ with probability $p_1 = .05$. It is unlikely, $p_1 \leq .05$, that such a small value (20) could be observed if μ were 20.82 or greater. If we want more confidence (say $p_1 = .005$) for the upper limit for μ, we solve

$$-2.58 = \frac{20 - \mu}{2/\sqrt{16}}$$

giving

$$\mu = 20 + 1.29 = 21.29$$

Only five or fewer chances in 1000 will we observe an $\bar{X}_o \leq 20$ if $\mu \geq 21.29$. Thus, having observed $\bar{X}_o = 20$ it is very unlikely ($p_1 \leq .005$) that the random variable which we were observing had an expectation in excess of 21.29.

By setting the value of p_1, a normal distribution with a specific value of μ is selected, at least for the assumption that $\bar{X}_o$ was less than μ. Figure 12.11 illustrates some of the distributions specified in this fashion by letting p_1 be different values, for example, .01, .05, .10.

The calculation of upper and lower limits for the interval estimate may be achieved by extending our reasoning to a value of μ below $\bar{X}$. Call the upper value of the interval U (determined by the logic just developed) and the lower value of the interval L.

The value L is fixed by considering a value p_2 taken as the probability that a value as large or larger than $\bar{X}_o$ would be observed given $\mu = L$. Thus we want to

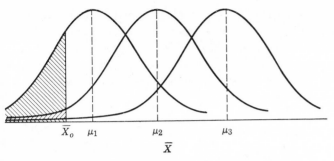

FIGURE 12.11

Illustration of selection of normal distributions with different expectations by specifying a probability in the lower tail of the distribution of $\bar{X}$.

solve the following for L:

$$\Phi\left(\frac{\bar{X}_o - L}{S/\sqrt{N}}\right) = 1 - p_2$$

Letting $p_2 = .05$ gives

$$1.64 = \frac{\bar{X}_o - L}{S/\sqrt{N}}$$

$$L = 20 - \frac{1.64}{2} = 19.18$$

Thus, if $\mu = L = 19.18$, we would observe $\bar{X}_o \geq 20$ with probability $p_2 = .05$. It is unlikely ($p_2 \leq .05$) that such a large value (20) could be observed if μ were 19.18 or smaller. If $p_2 = .005$, $L = 18.71$ and only on five or fewer samples in 1000 would we observe $\bar{X}_o \geq 20$ if $\mu \leq 18.71$.

The interval for μ is taken in this development to have L and U as its lower and upper limits, respectively. We take as the measure of confidence that μ lies in the interval the value

$$C = 1 - (p_1 + p_2)$$

Thus for $p_1 = p_2 = .05$, $C = .9$ and we have .9 confidence that

$$L \leq \mu \leq U$$

or in the numerical example

$$\bar{X}_o - 1.64 \frac{S}{\sqrt{N}} \leq \mu \leq \bar{X}_o + 1.64 \frac{S}{\sqrt{N}}$$

$$19.18 \leq \mu \leq 20.82$$

For $C = .95$,

$$\bar{X}_o - 1.96 \frac{S}{\sqrt{N}} \leq \mu \leq \bar{X}_o + 1.96 \frac{S}{\sqrt{N}}$$

$$19.02 \leq \mu \leq 20.98$$

When we compare these results with the original definitions of intervals and confidence coefficients, it is clear that they are the same. Hence the mechanics, if not the logic, leading to the intervals and their confidence coefficients are identical.

Applying the logic of confidence intervals to the experimental model when $N < 30$ but where X is normally distributed with both μ and σ^2 unknown, we obtain

$$P\left(\bar{X} - t_\alpha \frac{\hat{S}}{\sqrt{N}} < \mu < \bar{X} + t_\alpha \frac{\hat{S}}{\sqrt{N}}\right) = 1 - \alpha \qquad (12.19)$$

Here t_α is the value w_{N-1}, where w_{N-1} is distributed as Student's t with $(N - 1)$ degrees of freedom and

$$P(-w_{N-1} < W_{N-1} < w_{N-1}) = 1 - \alpha$$

For example, when $\bar{X} = 50$, $\hat{S} = 6$, and $N = 9$ and we desire a .99 confidence coefficient:

$$P\left(50 - t_\alpha \frac{6}{\sqrt{9}} < \mu < 50 + t_\alpha \frac{6}{\sqrt{9}}\right) = .99$$

The value w_8 required as t_α is obtained from the table of Student's t. Finding the row for eight degrees of freedom and the column for .005 (.5α gives the value .005—the symmetric tails of the sampling distribution involved extend equally beyond L and U) yields a $t_\alpha = 3.355$. Thus

$$P(43.29 < \mu < 56.71) = .99$$

The logic of interval estimation of σ^2 from a sample is slightly different. The lack of symmetry in the sampling distribution of the variance of a sample implies that the lower limit is somewhat nearer the sample value than is the upper limit of the interval.

The variable $Z^2 = (N - 1)\hat{S}^2/\sigma^2$ is distributed as a χ^2_{N-1} random variable. Hence we find, for a confidence coefficient $C = 1 - \alpha$, values of σ^2 to satisfy the equation

$$P\left(A \geq \frac{(N - 1)\hat{S}^2}{\sigma^2} \geq B\right) = 1 - \alpha$$

In order to express this inequality in terms of σ^2 alone, two steps are required. First, note that if $0 < a < b < c$, then $(1/a) > (1/b) > (1/c)$, which gives

$$\frac{1}{A} \leq \frac{\sigma^2}{(N - 1)\hat{S}^2} \leq \frac{1}{B}$$

Each part in the inequality may be multiplied by the same constant without changing the direction of the inequalities. Hence we deal with

$$P\left(\frac{(N-1)\hat{S}^2}{A} \leq \sigma^2 \leq \frac{(N-1)\hat{S}^2}{B}\right) = 1 - \alpha \qquad (12.20)$$

In estimation of μ we used a technique to solve for upper and lower limits symmetrical about $\bar{X}$. This gave the extremes beyond L and U each the same probability, $\alpha/2$. The symmetry of the distribution made this possible and sensible. However, in estimation of σ^2 from $\hat{S}^2$ we have the choice of an interval symmetrical with respect to $\hat{S}^2$ or with respect to the probability in the extremes. We choose here to deal with the probabilities and solve for the limits of the interval. In general, this produces an asymmetric interval about $\hat{S}^2$. For the confidence coefficient $1 - \alpha$ we want to solve for U and L so that

$$P\left(\frac{(N-1)\hat{S}^2}{B} \leq L\right) = \frac{1}{2}\alpha$$

and

$$P\left(\frac{(N-1)\hat{S}^2}{A} \geq U\right) = \frac{1}{2}\alpha$$

To solve for L and U values for N, $\hat{S}^2$ and α are required. Imagine an example where $N = 20$, $\hat{S}^2 = 12$, and $\alpha = .05$. From this, for $Z_{19}{}^2$ a $\chi_{19}{}^2$ distributed random variable, we obtain

$$P(Z_{19}{}^2 \leq B) = .025$$

and

$$P(Z_{19}{}^2 \leq A) = .975$$

which in turn yields, by inspection from a χ^2 table,

$$B = 8.91$$

$$A = 32.9$$

Substituting these values, we find

$$P\left(\frac{(19)(12)}{32.9} \leq \sigma^2 \leq \frac{(19)(12)}{8.91}\right) = .95$$

$$P(6.93 \leq \sigma^2 \leq 25.59) = .95$$

In applying this interval estimate of σ^2 from $\hat{S}^2$, it must be remembered that unless the random variable sampled is a normal random variable, $(N-1)\hat{S}^2/\sigma^2$ is not distributed as χ_{N-1}^2. The central limit theorem can be extended to cover the case of $\hat{S}^2$, and hence when N is large the methods developed for the estimate of μ may be used on the normal approximations to the χ^2 distribution to determine the interval. This is not recommended for $N < 100$. The problem here is of course in the fact that the values of A and B in the estimates are determined from a table of χ^2 distributed variables by entering the table with $\alpha/2$, $1 - \alpha/2$, and $(N - 1)$. Unless $\hat{S}^2(N - 1)/\sigma^2$ is χ^2_{N-1}, then A and B are irrelevant to the σ^2. To put it more bluntly, unless we sample from a normal random variable, the χ^2 distribution is not a model of the sampling distribution of $\hat{S}^2(N - 1)/\sigma^2$.

Degrees of Freedom as Number of Unconstrained Observations

At several points in this chapter we used the concept of degrees of freedom. The formal definition was put off in order to develop it in general and to avoid the possibility of tying it to interval estimation, t or χ^2. The number of degrees of freedom for t and χ^2, one less than the number of observations in a sample, is a special case of a general concept. We adopt the convention of symbolizing the number of degrees of freedom as df.

The value df is a parameter in the sampling distribution of a number of statistics. This parameter is always a function of the number of values entering into the calculation of the statistic. However, the calculation of df may be rather complicated in certain circumstances. It is not our intention here to broach the topic in the most general and elegant fashion. The mathematical concepts necessary are beyond the scope of this book (they involve the dimensionality of the space representing the data).

There are two roughly equivalent ways of conceptualizing the parameter df. These involve two ideas:

(1) Number of estimations of parameters used in calculating the statistic for which df is relevant.
(2) Number of constraints imposed on the data used in calculating the statistic for which df is relevant.

Briefly, when we calculate S_X^2, we have to estimate μ_X by calculating $\bar{X}$. Thus when S_X^2 is calculated from a sample of N observations, we obtain df by

df = number of independent observations on the random variable − number of independent parameter estimates based on the observations used in the calculation of the statistic

$= N - 1$

If μ_X were known and we calculated $(1/N) \sum_{i=1}^{N} (X_i - \mu_X)^2$ instead of S_X^2, no estimates of parameters are involved. Hence, in V the $df = N$. Again, if we had an estimate of μ_X, say $\bar{X}'$, which came from another collection of observations, from some theory, and so on, then

$$W = \frac{1}{N} \sum_{i=1}^{N} (X_k - \bar{X}')^2$$

would have $df = N$, because no parameter estimates based on the N observations are involved.

Turning to the number of restraints idea, we define

$df =$ number of independent observations $-$ number of constraints placed on the
    on the random variable   data used in calculating the statistic

The data observed are often transformed in some way before a statistic is calculated, the transformed data being used in the calculations. For example, S_X^2 is calculated from $(X_i - \bar{X})$ rather than from X_i. The transformation here appears so innocent that we might be tempted to conclude that it could not have an effect on our calculations. However, recall that

$$\sum_{i=1}^{N} (X_i - \bar{X}) = 0$$

regardless of the X_i and $\bar{X}$. This has the implication that the set of values $(X_i - \bar{X})$ are constrained in that their sum always must be zero. Not all of the N values $(X_i - \bar{X})$ are "free"—knowing $(N - 1)$ of the values allows us to find the Nth value arithmetically. To see this, select *any* of the $(X_i - \bar{X})$ values, say the jth, that is, $(X_j - \bar{X})$; then

$$(X_j - \bar{X}) = - \sum_{\substack{i=1 \\ i \neq j}}^{N} (X_i - \bar{X})$$

because

$$\sum_{i=1}^{N} (X_i - \bar{X}) = (X_j - \bar{X}) + \sum_{\substack{i=1 \\ i \neq j}}^{N} (X_i - \bar{X}) = 0$$

To illustrate this numerically, imagine four observed values:

$$X_1 = 20$$
$$X_2 = 16$$
$$X_3 = 3$$
$$X_4 = -3$$

The sum is 36 and the mean is 9. Set aside X_2 and find

$$\sum_{\substack{i=1 \\ i \neq 2}}^{4} (X_i - \bar{X}) = (20 - 9) + (3 - 9) + (-3 - 9) = -(X_2 - 9)$$

$$= -7 = -X_2 + 9$$

and hence

$$X_2 = 9 + 7 = 16$$

It should be clear that any value other than $\bar{X}$, for example μ_X, in $(X_i - \mu)$ does not impose a constraint: Since $\bar{X} \neq \mu$ in general (in spite of $E(\bar{X}) = \mu$), then

$$\sum_{i=1}^{N} (X_i - \mu_X) \neq 0$$

in general. Hence $V = (1/N)\sum_{i=1}^{N} (X_i - \mu_X)^2$ is calculated with no constraints on the N values $(X_i - \mu_X)$ and V has $df = N$. When $\bar{X}$ is substituted for μ_X in V to obtain S^2, a single constraint is imposed and the $df = N - 1$.

In calculating $\bar{X}$ from a sample, no prior calculations are required and no constraints are imposed. Hence, $\bar{X}$ as a statistic has $df = N$.

The three other statistics we have studied in this chapter have degrees of freedom determined in the following way. For the Z statistic the number of degrees of freedom is not relevant. We have shown that

$$Z = \frac{\bar{X} - \mu}{\hat{S}_{\bar{X}}}$$

is normally distributed, $N(0, 1)$. Since the df is not a parameter of a normally distributed variable, we know all we need to know about Z without knowing df. However, in the case of a variable W distributed as Student's t the only parameter involved is df. Although the best argument in this case is to show that the t distribution depends only on the value $(N - 1)$, we can argue the following way. The numerator of the statistic

$$\frac{\bar{X} - \mu}{\hat{S}/\sqrt{N}}$$

has no constraints. However, the denominator has a constraint. Therefore there is a total of one constraint in W, $df = N - 1$.

SUGGESTED READING

Cramér, H. (1946), *Mathematical methods of statistics*, Princeton: Princeton University Press.

GLOSSARY

Central Limit for Means. For N independent observations $X_1, \ldots, X_N$ on a random variable with expectation μ_X and variance $\sigma_X{}^2$, the sample mean,

$$\bar{X} = \frac{1}{N}(X_1 + \cdots + X_N),$$

is normally distributed in the limit with $N \to \infty$, $N(\mu_X, \sigma_X{}^2/N)$.

Chi Square Distribution. A single observation on a normally distributed variable, when standardized and squared, $Z^2 = [(X - \mu)/\sigma]^2$ has a density function called the chi square, χ^2, distribution. The distribution does not depend on μ or σ.

$$(12.1) \qquad Z_k{}^2 = \frac{(X_1 - \mu_1)^2}{\sigma_1{}^2} + \cdots + \frac{(X_k - \mu_k)^2}{\sigma_k{}^2}$$

$$(12.2) \qquad E(Z_k{}^2) = k$$

$$(12.3) \qquad \mathrm{var}(Z_k{}^2) = 2k$$

$$(12.4) \qquad Z_k{}^2 = Z_m{}^2 + Z_{k-m}^2$$

$$(12.5) \qquad P(Z_k{}^2 \le z_k{}^2) \cong \Phi(\sqrt{2z_k{}^2} - \sqrt{2k - 1})$$

$$(12.6) \qquad V = \frac{1}{N}\sum_{i=1}^{N}(X_i - \mu)^2$$

$$(12.7) \qquad P(V \le v) = P(Z_N{}^2 \le z_N{}^2)$$

$$(12.8) \qquad z_N{}^2 = \frac{N}{\sigma^2}v$$

$$(12.9) \qquad (X_i - \mu) = (X_i - \bar{X}) + (\bar{X} - \mu)$$

$$(12.10) \qquad \sum_{i=1}^{N}(X_i - \mu)^2 = \sum_{i=1}^{N}(X_i - \bar{X})^2 + N(\bar{X} - \mu)^2$$

$$(12.11) \qquad Z_N{}^2 = \sum_{i=1}^{N}\frac{(X_i - \mu)^2}{\sigma^2} = \sum_{i=1}^{N}\frac{(X_i - \bar{X})^2}{\sigma^2} + Z_1{}^2$$

$$(12.12) \qquad \sum_{i=1}^{N}\frac{(X_i - \bar{X})^2}{\sigma^2} = Z_{N-1}^2$$

$$(12.13) \qquad \frac{NS^2}{\sigma^2} = Z_{N-1}^2$$

$$(12.14) \qquad \frac{(N-1)\hat{S}^2}{\sigma^2} = Z_{N-1}^2$$

Critical Ratio or Z Statistic. Regardless of the distribution of X, if a large number of independent observations are made on X, then $\bar{X}$ is normally distributed with expectation μ and variance σ^2/N, and $Z = (\bar{X} - \mu)/(\hat{S}/\sqrt{N})$ is approximately $N(0, 1)$.

$$(12.15) \qquad Z = \frac{\bar{X} - \mu}{\hat{S}/\sqrt{N}} = \frac{\bar{X} - \mu}{\hat{S}_{\bar{X}}}$$

Student's t Distribution. When X is normally distributed and sample size is small, $W = (\bar{X} - \mu)/(S/\sqrt{N-1})$ is distributed as Student's t. The only parameter determining the t distribution is $(N - 1)$.

$$(12.16) \qquad W = \frac{\bar{X} - \mu}{\hat{S}_{\bar{X}}}$$

$$(12.17) \qquad \mathrm{var}(W) = \frac{N - 1}{N - 3}$$

$$(12.18) \qquad P\left(\bar{X} - Z_\alpha \frac{S}{\sqrt{N}} < \mu < \bar{X} + Z_\alpha \frac{S}{\sqrt{N}}\right) = 1 - \alpha$$

$$(12.19) \qquad P\left(\bar{X} - t_\alpha \frac{\hat{S}}{\sqrt{N}} < \mu < \bar{X} + t_\alpha \frac{\hat{S}}{\sqrt{N}}\right) = 1 - \alpha$$

$$(12.20) \qquad P\left(\frac{(N-1)\hat{S}^2}{A} \leq \sigma^2 \leq \frac{(N-1)\hat{S}^2}{B}\right) = 1 - \alpha$$

PROBLEMS

1. Figure 12.4 illustrated the mapping of the normal distribution into the χ^2 distribution for a single, rather wide interval on the χ^2 variable. Using intervals on the χ^2 variable of .01, construct an approximation to the graph of the χ^2 density function for one degree of freedom from the normal distribution function.

2. From the distribution function table of the χ^2 random variables, construct approximations of the density functions for the χ^2 variables with 4, 8, 16, and 30 degrees of freedom (four functions on a single graph).

3. For the density function approximations of problem 2, construct normal density functions approximating the χ^2 distributions, using the approximation equation (12.5).

4. In Chapter 11, problem 1, ten sample standard deviations were calculated from samples of 15 observations. Compare the frequency distribution of the sample

variances of these samples with the appropriate χ^2 distribution. Calculate the mean and variance of the sample variances and compare these with the expected values of the mean and variance of sample variances.

5. From the distribution function table of Student's t, construct graphs approximating the density functions for 2, 4, 8, 16, and 30 degrees of freedom. Compare these distributions with the unit standard normal density function.

6. Construct interval estimates for the expectation of X, μ, with confidence coefficients of .99, .95, .75, and .60 for the following situations. In each instance use a symmetrical interval. Represent the intervals in algebraic form and draw a simple line graph of each interval, using the same scale for all graphs.
   (a) X is $N(\mu, 14)$, the sample mean is 33.7 in a sample of 26 observations.
   (b) X is $N(\mu, \sigma^2)$, the sample mean is 33.7, and the sample variance S^2 is 11.2 in a sample of 26 observations.
   (c) X is $N(\mu, \sigma^2)$, the sample mean is 30, and the variance estimate $\hat{S}^2$ is 6.0 in a sample of 5 observations.
   (d) X is $N(\mu, 6)$, the sample mean is 30 in a sample of 5 observations.

7. Construct interval estimates for the variance with confidence coefficients of .99, .95, .75, .60 and for a normally distributed variable in the following situations. In each instance use an interval symmetrical with respect to the probability in each tail of the sampling distribution. Represent the intervals in algebraic form and draw a simple line graph of each interval, using the same scale in all of the graphs.
   (a) $N = 10$ and $\hat{S}^2 = 14.2$
   (b) $N = 10$ and $\hat{S}^2 = 26.4$
   (c) $N = 4$ and $\hat{S}^2 = 14.2$

8. Out of eight observations, one was lost after the mean of the observations was calculated. The mean value was 14.3. The other observations were 16.5, 10.2, 11.4, 19.6, 8.8, 15.2, and 16.0. What was the value of the missing observation?

9. The reaction time of a certain type of neural preparation is known to be 4.000 milliseconds under standard conditions. The reaction time is normally distributed with standard deviation of .04 milliseconds. Determine the probability that a given preparation, on a given measurement, will give a reaction time (a) equal to or in excess of 4.12 ms, (b) between 3.96 and 4.04 ms, and (c) between 3.92 and 4.06 ms.

10. Use the table of the distribution function of the unit normal random variable to find $P(X^2 \leq K) = P(-\sqrt{K} \leq X \leq \sqrt{K})$ for various values of K when X is a unit normal random variable. Use the results to construct a distribution function of values of $Y = X^2$. Compare this result with that of problem 1.

11. Using the normal approximation to the χ^2 distribution, find an approximation to the following probabilities, where X is χ_n^2:
    (a) $P(X \leq 1000)$, $n = 1000$.
    (b) $P(X \leq 210)$, $n = 200$
    (c) $P(X \leq 75)$, $n = 60$

12. If X is a χ^2 distributed random variable with n degrees of freedom, find the approximate (or exact) probabilities for the following expressions:

(a) $P(X \leq 6)$, $n = 6$

(b) $P(2 \leq X \leq 6)$, $n = 6$

(c) $P(X \leq 14)$, $n = 10$

(d) $P(X \leq 20)$, $n = 10$

(e) $P(14 \leq X \leq 22)$, $n = 18$

13. Find the probabilities in the expressions below for X having a Student's t distribution for each of the degrees of freedom, 1, 4, 10, and 30. Compare these probabilities for the same expressions under the condition that X is a unit normal random variable.

(a) $P(X \leq 5)$

(b) $P(-1 \leq X \leq 1)$

(c) $P(-1.5 \leq X \leq 1.5)$

(d) $P(-3.0 \leq X \leq 3.0)$

(e) $P(0 \leq X \leq 2.5)$

14. If X is distributed as Student's t with n degrees of freedom, find K so that the probability expressions below are approximately satisfied:

(a) $P(X \leq K) = .90$, $n = 10$

(b) $P(-K \leq X \leq K) = .95$, $n = 5$

(c) $P(K \leq X) = .95$, $n = 12$

(d) $P(-K \leq X) = .95$, $n = 6$

(e) $P(0 \leq X \leq K) = .49$, $n = 8$

CHAPTER 13

Principles of Statistical Inference: Philosophy

The preceding twelve chapters present the necessary background in probability theory and the theory of random variables in order to deal with the problem of making decisions on the basis of partial and uncertain information. It has been shown how observations are made in the real world so that certain aspects of the observations can be compared with parallel aspects of the corresponding model. At several places it has been shown that a probability model and real observations do not agree. For example, right at the outset, it was clear that the probability of an event was not necessarily equal to the relative frequency of the corresponding result. We also have seen that the average of a sample of observations on a random variable generally is not equal to the expectation of the random variable sampled. In fact, the average is a random variable itself.

Does the discrepancy between a result and the expected result, as determined by a model, indicate the inadequacy of the model? The answer is not simple! This brief chapter attempts to provide a general philosophic and logical analysis of the question in terms of models, sampling, inference, and decisions. Following this discussion, the theory of statistical inference is formalized.

Reality and Empirical Knowledge

Imagine an astronomy model dealing with individual planets and their relative positions and velocities.

> In order to bring the essential facts into prominence, a mathematical model may assume that each planet has all of its mass concentrated at one point. From the point of view of an inhabitant of a planet such an assumption is a wild distortion of the truth, but for the astronomer looking at the entire solar system it merely removes a lot of bothersome detail without adverse effect on the problem he is trying to solve. [Robert Hooke, *Introduction to scientific inference*, San Francisco: Holden-Day, 1963]

Truth and reality for the inhabitant of the planet may be well defined. The scientist, however, does not deal with Truth or Reality in the usual sense. Reality is inaccessible to the scientist, and truth is a matter of formal consistency (as in a proof within a mathematical model). With this denial of an ultimate criterion of true-false and real-unreal, the reader may wonder why the scientist goes to so much trouble in his attempts to study natural phenomena. If we cannot ever discover whether our Theory A is true or false, are we not justified in believing anything we please without bothering to gather evidence and build models and do experiments? The answer is yes, *if* you want to ignore things that can be known and the implications of these points of knowledge. If we choose the solipsistic way of answering the question, we can stop here and count what we have already done as a waste. Instead, we use what we have learned about probability and random experiments to provide *a* solution to the apparent dilemma.

The word "a" is emphasized because there are other ways of answering the question that are not based on the theory of probability and random experiments. One of the alternatives is to seek Truth through religious or mystical ways, through revelation. Another way is philosophical analysis, where logic and formal verbal analysis are applied in the search for "what must be" in the nature of things. For scientific purposes neither of these two methods is of interest.

If we cannot make any direct or intellectual appeal to reality, we must rely on what can be ascertained by observation. We must use the results of attempts to observe natural phenomena in order to ascertain whether or not our theories about the states of nature are reasonable.

If the "states of nature" involve specific things such as the number of errors a rat makes while learning a path through a maze, we might be able to observe precisely and conclusively the state of nature. Even there, we might have difficulty because of an inability to define what we mean by error or have difficulty in measuring (observing) the behavior involved. In most instances we are interested in less trivial states of nature. We would be interested, for example, in how animals of a certain species (not *a* rat) learn maze paths in terms of the sequences of correct responses and errors. Or at a higher level, we may be interested in a general law of learning which would be independent of type of animal and type of maze (or other apparatus). This last example is clearly impossible if we are forced to observe all possible organisms in all possible situations relevant to the theory. We are faced with the task of relating the specific observations of actual experiments with the general contentions of a theory.

Take, as an example, a learning theory for which a model states that the probability of a correct response on trial k following a correct response on trial $(k-1)$ is greater by some proportion θ of the probability of error on trial $(k-1)$. That is, if

$$p_j = \text{(probability of correct response on trial } j)$$
$$q_j = \text{(probability of incorrect response on trial } j)$$
$$= 1 - p_j$$

then

$$p_{j+1} = p_j + \theta q_j$$

or the increment of response probability p on trial j is

$$p_{j+1} - p_j = \theta q_j$$

Now this model pretends to be a model of a state of nature, but it does not make any assumptions about the kind of organisms, the kind of materials being learned, or the situations under which learning is to take place. Can we hope ever to say whether the model is true or false by observing specific human subjects learning specific lists of words or specific rats learning specific mazes? The answer, of course, is no! There is no way of concluding that a general statement is true or false from knowledge of particular or specific observations. We cannot say that "all crows are black" on the basis of observing crows, regardless of how many we observe.

The empirical sciences, in contrast with mathematics and philosophy, deal with highly uncertain and incomplete information. Empirical sciences do not have the "givens" or "axioms" and "definitions" of philosophy and mathematics. In classical Aristotelian logic the following ancient argument can be made without fear of logical contradiction:

> If all men are mortal,
> and Socrates is a man,
> then Socrates is mortal.

The truth of this syllogism rests in the assumption of the truth of (1) the general statement of the basic premise that "all men are mortal" and (2) the assertion that

"Socrates is a man." We are not really claiming that these are true, but if they are, then it follows in a strictly determined way that "Socrates is mortal."

Putting this into set theoretical language makes the proof obvious. Let the class (set) of all men be A and the class of all things mortal be B. "All men are mortal" then is expressed

$$A \subseteq B$$

Take a specific member of A, call it a, "Socrates":

$$a \in A$$

It follows that $a \in B$, "Socrates is mortal." This final step is carried in the definition of $\subseteq$:

$$A \subseteq B \text{ implies } A = \{a \mid a \in A \text{ and } a \in B\}$$

In empirical science we are more interested in discovering the truth or falsity of the statements $A \subseteq B$ and $a \in A$ taken to be self-evidently true. However, all men have not been observed and will not be observed in an empirical verification of this premise.

No one in his right mind would question the "validity" or "truth value" of the proposition "all men are mortal." We are confident that this is a true "theory" or model—every object satisfying the definitions of the class "man" will also satisfy the definitions of the class "mortal objects." In other terms, a statement to the contrary would be incredible, that is, have no credibility.

Degree of Credibility and Inference*

If a psychologist infers that Jones will commit suicide within a year, we want to know to what extent we can regard his inference as credible. If an economist predicts that the price of steel will decline, we are interested in the degree of credibility to attach to his statement. Degree of credibility may be stated in mathematical form, such as "The statement 'Jones will kill himself' has the probability of .1 of being confirmed," or in less precise form, such as "The statement 'Jones will kill himself' is probably false." Two questions are raised here. First, how does the prediction arise and, second, how do we check the accuracy of the prediction?

Let us begin with the second question. This has been described as the application of probability in the context of confirmation. The frequency interpretation of

* This section is taken from T. R. Sarbin, R. Taft, and D. E. Bailey, *Clinical inference and cognitive theory*, New York: Holt, Rinehart and Winston, 1960.

probability is applicable to the assessment of degrees of rational credibility through (1) knowledge of antecedent event-probabilities, and (2) knowledge of antecedent probabilities of confirmations and disconfirmations of events. For illustrative purposes let us use as a model the throw of a die. We can offer this prediction: On the next throw the die will turn up a six. If we make the same prediction a number of times, we shall be correct about one time in six. So for each throw of the die we implicitly or explicitly attach a probability-of-occurrence statement. If we have had commerce with solid, homogeneous cubes, we have built up the induction that any surface is as likely to appear as any other. The case of the die exemplifies the simple situation where all the contingencies are known and where there are only six possible response classes. Antecedent probabilities based on relative frequencies of occurrence and nonoccurrence or on theoretical deductions have already been established. Therefore the degree of rational credibility which may be attached to the statement "The next throw will not be a six" is equivalent to the probability $\frac{5}{6}$. It is important to distinguish between probability statements about events and probability statements about propositions. In the illustration above, the events are well defined. The probability statement follows from observation of relative frequencies. It is stated in the form $n/(m + n)$, in this case, $\frac{5}{6}$. The statement about the die (the prediction), however, is stated as occurrence or nonoccurrence: The next throw will not be a six. The degree of rational credibility attached to such a statement is, in this illustration, equal to the observed frequency and may be formulated probabilistically.

Among human beings (who are not always rational) the prediction does not flow mechanically from recognition of the probability statement about the event or of the rational degree of credibility that may be attached to the prediction. Subjective feelings and subjective probabilities may enter between the recognition of the actual event-probability and the assignment of degree of credibility, resulting in a distortion of the degree of belief upon which decisions are predicated. The gambler's fallacy is a case in point.

Let us now turn to a more complex and more pertinent example. On the basis of coroners' reports, hospital records, census data, and so on, we can arrive at an estimate of the probabilities of suicide for given classes of persons. For example, in the United States the base rate is 11.4 per 100,000 per year. If a behavior analyst were asked to predict whether or not a given individual would commit suicide, on such base rates and in the absence of other information he would answer "no" or "probably not." If there were no narrower classes, a high degree of credibility for the prediction could be assigned. However, if antecedent probabilities had been built up for a narrower class identified by such attributes as, let us say, "severely depressed," "feelings of worthlessness," "suicidal rumination," "intrapunitive," "urban resident," and "isolated," and the ratio of suicides to total was eight in ten and if Jones were an instance of this class, then the prediction "Jones will commit suicide" can be said to have a high credibility, expressed by the figure 80 percent.

Let us take a concrete example and show how event-probabilities enter into the assessment of degree of belief from which a decision must be made. Let us take the following sorites, where the conclusion of the first syllogism is the major premise of the second:

Major Premise. The observed frequency of association between improvement in psychotherapy and intact ego, mild anxiety, and high motivation is six in ten.
Minor Premise (usually tacit). The unobserved frequency of association between improvement in psychotherapy and intact ego, mild anxiety, and high motivation will be identical with the observed frequency of association.
Conclusion. The unobserved frequency of association between improvement in psychotherapy and intact ego, mild anxiety, and high motivation is six in ten.
Minor Premise. Jones is a member of the unobserved class of persons characterized by the attributes of intact ego, mild anxiety, and high motivation.
Conclusion. In formal logic no conclusion is valid because we have no way of knowing whether Jones belongs to the 60 percent of those who improve or to the 40 percent of those who do not.

Because we must make decisions, however, we utilize the premises by stating the conclusion in one of two ways.

The probability that Jones will improve in therapy is six in ten.

The statement "Jones will improve in therapy" has a credibility of .6. In short, for the purpose of action and decision making, we treat Jones as if he were a member of a prescribed class. The ultimate degree of belief will be formed partly on the basis of the degree of credibility that is equivalent to the event-probabilities.

Sources of Error

When we make observations in an attempt to evaluate the credibility of a theory or model of some phenomenon, two basic types of errors of observation are important (for the moment we do not consider clerical or calculational errors): errors in measurement and errors in sampling. Measurement errors are especially troubling in psychology. For example, imagine a model implying that the intelligence quotient of a given type of subject has a mean value $\bar{X}$. In order to observe the intelligence of a subject we select a given intelligence test. Even if the test result is a measure of intelligence, it is also a measure of many other things, such as motivation of the subject at time of testing, the environmental conditions of the testing, and so on. Thus the result of the test administration has some "error" as a measure of intelligence. If this error is large enough or consistent enough, it can be seriously misleading in the determination of the credibility of a model. Imagine that tests were administered under conditions that led to an average underestimate of the intelligence quotient by ten points. Unless the error of measurement were discovered and corrected (often a very difficult task), we would be likely to have a low degree of confidence in a model that claimed an $\bar{X}$ ten points greater than that observed. In this example, establishing a low degree of credibility is an erroneous conclusion to the experiment.

Another example of error in measurement illustrates its pervasiveness. Have a group of subjects measure a length of steel rod with a steel ruler under carefully controlled conditions of temperature, illumination, and so on. The rod does not change length and the rule does not change length, and yet the observed readings of length will almost certainly have a frequency distribution approximated by the distribution of a normal random variable. It seems an inherent characteristic that the specific measurements of a given quantity will differ from instance to instance—that is, be random. We cannot take time to give this subject the full discussion it deserves. However, the interested student can find an absorbing and careful analysis of probability, error, determinacy, and related topics in H. Reichenbach, *The rise of scientific philosophy*, Berkeley: University of California Press, 1951.

The second source of error can be just as misleading as the first. In the first example of this section, if we were unsuccessful in selecting appropriate subjects, nothing could be gained. If we sample the wrong population, it is obvious that our conclusions will not refer to the model we wish to study. We can arrive at a certain degree of confidence in the model or its implications, but it will be falsely based.

As it turns out, all models hinge on the assumptions of a particular sampling procedure and a particular type of population. If these assumptions are not met in the actual experiment, our conclusions will be without an adequate foundation. We discuss this point at length several places in this chapter and chapters to follow.

Categorical Versus Probabilistic Inferences

Implied in our discussion of credibility and confidence is the notion of degrees of credibility and degrees of confidence. In view of the impossibility of obtaining direct confirmation of a model by observing specific instances in nature and of the presence of measurement and possibly of sampling error, the strategy of statistical inference is to make probabilistic statements regarding the truth or falsity of the model on the basis of the observations. Categorical—*true* and *false*—statements are seldom, if ever, justified on the basis of empirical evidence. The tendency is to elevate highly probable (high degree of credibility or confidence) events and outcomes to certain ($P = 1.0$) events and outcomes. However, the history of science is replete with instances of "established" or "true" theories that have been discarded on the basis of further evidence. The careful scientist is unwilling to claim any all-or-none conclusion in the face of the uncertainty that pervades the scientific enterprise.

Inference as a Conditional Statement

We have talked as if we evaluated the model on the basis of observed results of an experiment. This is not precisely correct, although in practice it is our goal. The accepted procedure in statistics is to ask how probable it is that we would observe the particular results *if* (given that) our model were true. That is, we make statements in statistical inference like

$$P(\text{observed result} \mid \text{model is true}) \leq .001$$

This is really a fairly weak statement. It is unknown whether or not other observed results would be more probable under the model. However, the model is at least very unlikely to be true—it is very unlikely (one in one thousand) that the result would have been observed if the model were true.

Notice the difficulty in interpreting the implications of results for a model when $P(\text{observed result} \mid \text{model is true})$ is fairly large, say .25. The fact that the observed result is not in basic disagreement with (improbable under) the model does not indicate that the model is true. There may be other models under which the same results are even more probable! The most we usually can say is that the evidence does not cause us to doubt the truth value of the model.

Notice, though, that where our model has a complementary form (negation of the implication) we can apply our procedures to it and hence infer how unlikely the negation (or falseness) of our model is on the basis of the observed results. We would thus calculate $P(\text{observed result} \mid \text{model is false})$. This way is full of problems and pitfalls itself and should be used with care. For example, we have no assurance that when our "objects" are models the negation of a negation (of a model) gives the model we began with. We do know that for a real number $-(-x) = x$ and for sets $(A')' = A$, but a model is not a real number and it is not a set. If we designate a certain model as ψ and its negation as $\sim\psi$ we are not justified on the strength of $-(-x) = x$ and $(A')' = A$ in assuming that $\sim(\sim\psi) = \psi$. This also means that

$$P(\text{observed result} \mid \text{model is false}) \leq .001$$

does *not* imply that

$$P(\text{observed result} \mid \text{model is true}) \geq .999$$

or any other such number. The last probability cannot be calculated this way, and its determination is largely a nonstatistical matter.

An Example: Mathematics of a Lady Tasting Tea*
Statement of Experiment

A lady declares that by tasting a cup of tea made with milk she can discriminate whether the milk or the tea infusion was first added to the cup. We will consider the problem of designing an experiment by means of which this assertion can be tested. For this purpose let us first lay down a simple form of experiment with a view to studying its limitations and its characteristics, both those which appear to be essential to the experimental methods, when well developed, and those which are not essential but auxiliary.

Our experiment consists in mixing eight cups of tea, four in one way and four in the other, and presenting them to the subject for judgment in a random order. The subject has been told in advance of what the test will consist, namely that she will be asked to taste eight cups, that these shall be four of each kind, and that they shall be presented to her in a random order, that is in an order not determined arbitrarily by human choice, but by the actual manipulation of the physical apparatus used in games of chance, cards, dice, roulettes, etc., or, more expeditiously, from a published collection of random sampling numbers purporting to give the actual results of such manipulation. Her task is to divide the eight cups into two sets of four, agreeing, if possible, with the treatments received.

* This example is a famous paper by Sir Ronald Fisher first published as a chapter in his *Design of experiments* (Oliver and Boyd) and reprinted in Vol. III of J. R. Newman, *The world of mathematics*, New York: Simon and Shuster, 1956.

Interpretation and Its Reasoned Basis

In considering the appropriateness of any proposed experimental design, it is always needful to forecast all possible results of the experiment, and to have decided without ambiguity what interpretation shall be placed upon each one of them. Further, we must know by what argument this interpretation is to be sustained. In the present instance we may argue as follows: There are 70 ways of choosing a group of 4 objects out of 8. This may be demonstrated by an argument familiar to students of "permutations and combinations," namely, that if we were to choose the 4 objects in succession, we should have successively 8, 7, 6, 5 objects to choose from, and could make our succession of choices in $8 \times 7 \times 6 \times 5$, or 1680 ways. But in doing this we have not only chosen every possible set of 4, but every possible set in every possible order; and since 4 objects can be arranged in order in $4 \times 3 \times 2 \times 1$, or 24 ways, we may find the number of possible choices by dividing 1680 by 24. The result, 70, is essential to our interpretation of the experiment. At best the subject can judge rightly with every cup and, knowing that 4 are of each kind, this amounts to choosing, out of the 70 sets of 4 which might be chosen, that particular one which is correct. A subject without any faculty of discrimination would in fact divide the 8 cups correctly into two sets of 4 in one trial out of 70, or, more properly, with a frequency which would approach 1 in 70 more and more nearly the more often the test was repeated. Evidently this frequency, with which unfailing success would be achieved by a person lacking altogether the faculty under test, is calculable from the number of cups used. The odds could be made much higher by enlarging the experiment, while, if the experiment were much smaller even the greatest possible success would give odds so low that the result might, with considerable probability, be ascribed to chance.

The Test of Significance

It is open to the experimenter to be more or less exacting in respect of the smallness of the probability he would require before he would be willing to admit that his observations have demonstrated a positive result. It is obvious that an experiment would be useless of which no possible result would satisfy him. Thus, if he wishes to ignore results having probabilities as high as 1 in 20—the probabilities being of course reckoned from the hypothesis that the phenomenon to be demonstrated is in fact absent—then it would be useless for him to experiment with only 3 cups of tea of each kind. For 3 objects can be chosen out of 6 in only 20 ways, and therefore, complete success in the test would be achieved without sensory discrimination, i.e., by "pure chance," in an average of 5 trials out of 100. It is usual and convenient for experimenters to take 5 percent as a standard level of significance, in the sense that they are prepared to ignore all results which fail to reach this standard, and by this means, to eliminate from further discussion the greater part of the fluctuations which chance causes have introduced into their experimental results. No such selection can eliminate the whole of the possible effects of chance coincidence, and if we accept this convenient convention, and agree that an event which would occur by chance only once in 70 trials is decidedly "significant," in the statistical sense, we thereby admit that no isolated experiment, however significant in itself, can suffice for the experimental demonstration of any natural phenomenon; for the "one chance in a million" will undoubtedly occur, with no less and no more than its appropriate frequency, however surprised we may be that it should occur to us. In order to assert that a natural phenomenon is experimentally demonstrable we need, not an isolated record, but a reliable method of procedure. In relation to the test of significance, we may say that a phenomenon is experimentally demonstrable when we know how to conduct an experiment which will rarely fail to give us a statistically significant result.

Returning to the possible results of the psychophysical experiment, having decided that if every cup were rightly classified a significant positive result would be recorded, or, in other words, that we should admit that the lady had made good her claim, what should be our conclusion if, for each kind of cup, her judgments are 3 right and 1 wrong? We may take it, in the present discussion, that any error in one set of judgments will be compensated by an error in the other, since it is known to the subject that there are 4 cups of each kind. In enumerating the number of ways of choosing 4 things out of 8, such that 3 are right and 1 wrong, we may note that the 3 right may be chosen, out of the 4 available, in 4 ways and, independently of this choice, that the 1 wrong may be chosen, out of the 4 available, also in 4 ways. So that in all we could make a selection of the kind supposed in 16 different ways. A similar argument shows that, in each kind of judgment, 2 may be right and 2 wrong in 36 ways, 1 right and 3 wrong in 16 ways, and none right and 4 wrong in 1 way only. It should be noted that the frequencies of these five possible results of the experiment make up together, as it is obvious they should, the 70 cases out of 70.

It is obvious too, that 3 successes to 1 failure, although showing a bias, or deviation, in the right direction, could not be judged as statistically significant evidence of a real sensory discrimination. For its frequency of chance occurrence is 16 in 70, or more than 20 percent. Moreover, it is not the best possible result, and in judging of its significance we must take account not only of its own frequency, but also of the frequency for any better result. In the present instance "3 right and

1 wrong" occurs 16 times, and "4 right" occurs once in 70 trials, making 17 cases out of 70 as good as or better than that observed. The reason for including cases better than that observed becomes obvious on considering what our conclusions would have been had the case of 3 right and 1 wrong only 1 chance, and the case of 4 right 16 chances of occurrence out of 70. The rare case of 3 right and 1 wrong could not be judged significant merely because it was rare, seeing that a higher degree of success would frequently have been scored by mere chance.

The Null Hypothesis

Our examination of the possible results of the experiment has therefore led us to a statistical test of significance, by which these results are divided into two classes with opposed interpretations. Tests of significance are of many different kinds, which need not be considered here. Here, we are only concerned with the fact that the easy calculation in permutations which we encountered, and which gave us our test of significance, stands for something required in its interpretation. The two classes of results which are distinguished by our test of significance are, on the one hand, those which show a significant discrepancy from a certain hypothesis; namely, in this case, the hypothesis that the judgments given are in no way influenced by the order in which the ingredients have been added; and on the other hand, results which show no significant discrepancy from this hypothesis. This hypothesis, which may or may not be impugned by the result of an experiment, is again characteristic of all experimentation. Much confusion would often be avoided if it were explicitly formulated when the experiment is designed. In relation to any experiment we may speak of this hypothesis as the "null hypothesis," and it should be noted that the null hypothesis is never proved or established, but is possibly disproved, in the course of experimentation. Every experiment may be said to exist only in order to give the facts a chance of disproving the null hypothesis.

It might be argued that if an experiment can disprove the hypothesis that the subject possesses no sensory discrimination between two different sorts of object, it must therefore be able to prove the opposite hypothesis, that she can make some such discrimination. But this last hypothesis, however reasonable or true it may be, is ineligible, as a null hypothesis to be tested by experiment, because it is inexact. If it were asserted that the subject would never be wrong in her judgments we should again have an exact hypothesis, and it is easy to see that this hypothesis could be disproved by a single failure, but could never be proved by any finite amount of experimentation. It is evident that the null hypothesis must be exact, that is free from vagueness and ambiguity, because it must supply the basis of the "problem of distribution," of which the test of significance is the solution. A null hypothesis may, indeed, contain arbitrary elements, and in more complicated cases often does so: as, for example, if it should assert that the death-rates of two groups of animals are equal, without specifying what these death-rates usually are. In such cases it is evidently the equality rather than any particular values of the death-rates that the experiment is designed to test, and possibly to disprove.

In cases involving statistical "estimation" these ideas may be extended to the simultaneous consideration of a series of hypothetical possibilities. The notion of an error of the so-called "second kind," due to accepting the null hypothesis "when it is false" may then be given a meaning in reference to the quantity to be estimated. It has no meaning with respect to simple tests of significance, in which the only available expectations are those which flow from the null hypothesis being true.

Randomization; The Physical Basis of the Validity of the Test

We have spoken of the experiment as testing a certain null hypothesis, namely, in this case, that the subject possesses no sensory discrimination whatever of the kind claimed; we have, too, assigned as appropriate to this hypothesis a certain frequency distribution of occurrences, based on the equal frequency of the 70 possible ways of assigning 8 objects to two classes of 4 each; in other words, the frequency distribution appropriate to a classification by pure chance. We have now to examine the physical conditions of the experimental technique needed to justify the assumption that, if discrimination of the kind under test is absent, the result of the experiment will be wholly governed by the laws of chance. It is easy to see that it might well be otherwise. If all those cups made with the milk first had sugar added, while those made with the tea first had none, a very obvious difference in flavor would have been introduced which might well ensure that all those made with sugar should be classed alike. These groups might either be classified all right or all wrong, but in such a case the frequency of the critical event in which all cups are classified correctly would not be 1 in 70, but 35 in 70 trials, and the test of significance would be wholly vitiated. Errors equivalent in principle to this are very frequently incorporated in otherwise well-designed experiments.

It is no sufficient remedy to insist that "all the cups must be exactly alike" in every respect except that to be tested. For this is a totally impossible requirement in our example, and equally in all other forms of experimentation. In practice it is probable that the cups will differ perceptibly in the thickness or smoothness of their material, that the quantities of milk added to the different cups will not be exactly equal, that the strength of the infusion of tea may change between pouring the first and the last cup, and that the temperature also at which the tea is tasted will change during the course of the experiment. These are only examples of the differences probably present; it would be impossible to present an exhaustive list of such possible differences appropriate to any one kind of experiment, because the uncontrolled causes which may influence the result are always strictly innumerable. When any such cause is named, it is usually perceived that, by increased labor and expense, it could be largely eliminated. Too frequently it is assumed that such refinements constitute improvements to the experiment. Our view, which will be much more fully exemplified in later sections, is that it is an essential characteristic of experimentation that it is carried out with limited resources, and an essential part of the subject of experimental design to ascertain how these should be best applied; or, in particular, to which causes of disturbance care should be given, and which ought to be deliberately ignored. To ascertain, too, for those which are not to be ignored, to what extent it is worthwhile to take the trouble to diminish their magnitude. For our present purpose, however, it is only necessary to recognize that, whatever degree of care and experimental skill is expended in equalizing the conditions, other than the one under test, which are liable to affect the result, this equalization must always to be to greater or less extent incomplete, and in many important practical cases will certainly be grossly defective. We are concerned, therefore, that this inequality, whether it be great or small, shall not impugn the exactitude of the frequency distribution, on the basis of which the result of the experiment is to be appraised.

The Effectiveness of Randomization

The element in the experimental procedure which contains the essential safeguard is that the two modifications of the test beverage are to be prepared "in random order." This, in fact, is the only point in the experimental procedure in which the laws of chance, which are to be in exclusive control of our frequency distribution, have been explicitly introduced. The phrase "random order" itself, however, must be regarded as an incomplete instruction, standing as a kind of shorthand symbol for the full procedure of randomization, by which the validity of the test of significance may be guaranteed against corruption by the causes of disturbance which have not been eliminated. To demonstrate that, with satisfactory randomization, its validity is, indeed, wholly unimpaired, let us imagine all causes of disturbance—the strength of the infusion, the quantity of milk, the temperature at which it is tasted, etc.—to be predetermined for each cup; then since these, on the null hypothesis, are the only causes influencing classification, we may say that the probabilities of each of the 70 possible choices or classifications which the subject can make are also predetermined. If, now, after the disturbing causes are fixed, we assign, strictly at random, 4 out of the 8 cups to each of our experimental treatments, then every set of 4, whatever its probability of being so classified, will certainly have a probability of exactly 1 in 70 of being the 4, for example, to which the milk is added first. However important the causes of disturbance may be, even if they were to make it certain that one particular set of 4 should receive this classification, the probability that the 4 so classified and the 4 which ought to have been so classified should be the same, must be rigorously in accordance with our test of significance.

It is apparent, therefore, that the random choice of the objects to be treated in different ways would be a complete guarantee of the validity of the test of significance, if these treatments were the last in time of the stages in the physical history of the objects which might affect their experimental reaction. The circumstance that the experimental treatments cannot always be applied last, and may come relatively early in their history, causes no practical inconvenience; for subsequent causes of differentiation, if under the experimenter's control, as, for example, the choice of different pipettes to be used with different flasks, can either be predetermined before the treatments have been randomized, or, if this has not been done, can be randomized on their own account; and other causes of differentiation will be either (a) consequences of differences already randomized, or (b) natural consequences of the difference in treatment to be tested, of which on the null hypothesis there will be none, by definition, or (c) effects supervening by chance independently from the treatments applied. Apart, therefore, from the avoidable error of the experimenter himself introducing with his test treatments, or subsequently, other differences in treatment, the effects of which the experiment is not intended to study, it may be said that the simple precaution of randomization will suffice to guarantee the validity of the test of significance, by which the result of the experiment is to be judged.

The Sensitiveness of an Experiment, Effects of Enlargement and Repetition

A probable objection, which the subject might well make to the experiment so far described, is that only if every cup is classified correctly will she be judged successful. A single mistake will reduce her performance below the level of significance.

Her claim, however, might be, not that she could draw the distinction with invariable certainty, but that, though sometimes mistaken, she would be right more often than not; and that the experiment should be enlarged sufficiently, or repeated sufficiently often, for her to be able to demonstrate the predominance of correct classifications in spite of occasional errors.

An extension of the calculation upon which the test of significance was based shows that an experiment with 12 cups, six of each kind, gives, on the null hypothesis, 1 chance in 924 for complete success, and 36 chances for 5 of each kind classified right and 1 wrong. As 37 is less than a twentieth of 924, such a test could be counted as significant, although a pair of cups have been wrongly classified; and it is easy to verify that, using larger numbers still, a significant result could be obtained with a still higher proportion of errors. By increasing the size of the experiment, we can render it more sensitive, meaning by this that it will allow of the detection of a lower degree of sensory discrimination, or, in other words, of a quantitatively smaller departure from the null hypothesis. Since in every case the experiment is capable of disproving, but never of proving, this hypothesis, we may say that the value of the experiment is increased whenever it permits the null hypothesis to be more readily disproved.

The same result could be achieved by repeating the experiment, as originally designed, upon a number of different occasions, counting as a success all those occasions on which 8 cups are correctly classified. The chance of success on each occasion being 1 in 70, a simple application of the theory of probability shows that 2 or more successes in 10 trials would occur, by chance, with a frequency below the standard chosen for testing significance; so that the sensory discrimination would be demonstrated, although, in 8 attempts out of 10, the subject made one or more mistakes. This procedure may be regarded as merely a second way of enlarging the experiment and, thereby, increasing its sensitiveness, since in our final calculation we take account of the aggregate of the entire series of results, whether successful or unsuccessful. It would clearly be illegitimate, and would rob our calculation of its basis, if the unsuccessful results were not all brought into the account.

Qualitative Methods of Increasing Sensitiveness

Instead of enlarging the experiment we may attempt to increase its sensitiveness by qualitative improvements; and these are, generally speaking, of two kinds: (a) the reorganization of its structure, and (b) refinements of technique. To illustrate a change of structure we might consider that, instead of fixing in advance that 4 cups should be of each kind, determining by a random process how the subdivision should be effected, we might have allowed the treatment of each cup to be determined independently by chance, as by the toss of a coin, so that each treatment has an equal chance of being chosen. The chance of classifying correctly 8 cups randomised in this way, without the aid of sensory discrimination, is 1 in 2^8, or 1 in 256 chances, and there are only 8 chances of classifying 7 right and 1 wrong; consequently, the sensitiveness of the experiment has been increased, while still using only 8 cups, and it is possible to score a significant success, even if one is classified wrongly. In many types of experiment, therefore, the suggested change in structure would be evidently advantageous. For the special requirements of a psycho-physical experiment, however, we should probably prefer to forego this advantage, since it would occasionally occur that all the cups would be treated alike, and this, besides bewildering the subject by an unexpected occurrence, would deny her the real advantage of judging by comparison.

Another possible alteration to the structure of the experiment, which would, however, decrease its sensitiveness, would be to present determined, but unequal, numbers of the two treatments. Thus we might arrange that 5 cups should be of the one kind and 3 of the other, choosing them properly by chance, and informing the subject how many of each to expect. But since the number of ways of choosing 3 things out of 8 is only 56, there is now, on the null hypothesis, a probability of a completely correct classification of 1 in 56. It appears, in fact, that we cannot by these means do better than by presenting the two treatments in equal numbers, and the choice of this equality is now seen to be justified by its giving to the experiment its maximal sensitiveness.

With respect to the refinements of technique, we have seen above that these contribute nothing to the validity of the experiment, and of the test of significance by which we determine its result. They may, however, be important, and even essential, in permitting the phenomenon under test to manifest itself. Though the test of significance remains valid, it may be that without special precautions even a definite sensory discrimination would have little chance of scoring a significant success. If some cups were made with India and some with China tea, even though the treatments were properly randomised, the subject might not be able to discriminate the relatively small difference in flavor under investigation, when it was confused with the greater differences between leaves of different origin. Obviously, a similar difficulty could be introduced by using in some cups raw milk and in others boiled, or even condensed milk, or by adding sugar in unequal quantities. The subject has a right to claim, and it is in the interests of the sensitiveness of the experiment, that gross differences of these kinds should be excluded, and that the cups should, not as far as possible, but as far as is practically convenient, be made alike in all respects except that under test.

How far such experimental refinements should be carried is entirely a matter of judgment, based on experience. The validity of the experiment is not affected by them. Their sole purpose is to increase its sensitiveness and this object can usually be achieved in many other ways, and particularly by increasing the size of the experiment. If, therefore, it is decided that the sensitiveness of the experiment should be increased, the experimenter has the choice between different methods of obtaining equivalent results; and will be wise to choose whichever method is easiest to him, irrespective of the fact that previous experimenters may have tried, and recommended as very important, or even essential, various ingenious and troublesome precautions.

SUGGESTED READING

The readings suggested in Chapter 2 are directly relevant to the material in this chapter.

CHAPTER 14

Principles of Statistical Inference: Basic Concepts

As we have stated, empirical work in psychology begins with some guiding psychological theory or speculation. This theory or speculation generally derives from our knowledge of natural phenomena or from logical considerations. First, some body of observations and experimental data suggest generalities in nature and we wish to discover whether these generalities are in accord with the state of nature. Second, physical laws of nature, particularly biological, suggest certain forms of mathematical-logical relationships. In any event, a basic idea about the state of nature underlies every psychological research. These ideas may be very loosely formulated as, for example, "differences in child-rearing practices have effects on the children's later adult emotional adjustment." At the other extreme we find specific "theories" such as the "mathematical model" of learning mentioned in the preceding chapter.

Regardless of the level of formalization of the psychological theory we start with, the next step is to derive or deduce a formal model that is presumed to correspond in all important details to the psychological theory. This job generally is done informally in psychology and is overlooked as an actual step in much psychological work. The unfortunate aspect of this problem is that there is tacit in all research some formal model, and the lack of explicit recognition of this fact, an the model, leads to the result that the experiment performed may have only a weak bearing on the issue involved. Of course, many psychologists have fine skills at intuitive derivation of the consequences of their theories without having to be explicit and formal. The research of these gifted individuals often does not suffer in any respect except one, a very important respect, however: evaluation of the relevance and the correctness of the intuitive derivation is virtually impossible for persons not gifted with this talent. Also, it seems almost certain that the acquisition of this extraordinary skill is a matter of experience with research. In the process of acquiring the necessary experience it is not at all clear how one goes about also being fruitful and accurate.

Throughout the early chapters of this book we assumed that the models in which we are interested are probability experiments and random variables. Throughout the rest of the book we assume that this is the basic material of models in psychological research. However, we should recognize that this assumption is not necessarily accurate. There are many other useful branches of formal discipline that hold promise as models in psychology, for example, nonprobabilistic theories of formal logic, the mathematical theories of graphs, topology, and other mathematical theories of relationship and structure.

With this caution we turn our attention to models consisting of statements of (1) the existence of random variables, (2) specifications of the form of their distribution, and (3) the location and the dispersion of the random variables. The implications of these sorts of models traditionally are called statistical and are stated in terms of so-called statistical hypotheses. A statistical hypothesis is simply a statement of what we might expect to observe if our model is adequate to describe the relevant states of nature *and* if we should do a certain experiment. That is, we assume our model is an adequate statement of empirical reality and gather empirical data in an experiment to compare with the results demanded by the model.

The specific form of the hypothesis, the specific sorts of operations of the experiment, and the formal statement of the model are all heavily interrelated. The complexity of this intertwining and the limitations of space do not permit a full discussion of this topic here. However, we present several specific examples in the chapters to follow.

The next step is the execution of the experiment. This is quite simple in theory, although the practice of experimentation is more difficult. For our purposes in this chapter we assume that the experiment corresponds simply to observing one or more random variables. In later chapters we are concerned with nominal sample spaces, and we develop special techniques for statistical inference for those cases.

Finally, the data from the experiment are compared with the results specified by the model. The relevant aspects of the sample of observations (means, variances, and so on) are compared with statistical implications of the model. This comparison is done with an eye to answering the question: Are the data and the statistical implications of the model different enough to cause us to doubt the model and hence the theory on which the model is based? We must decide whether or not to "reject" the contentions of our model on the basis of the data from the experiment. It should go without saying that we never reject the data from an experiment, only the theory that led us to do the experiment.

Discrepancies of the observed data and the theoretically expected data can come about in four general ways. (1) The theory is not an adequate statement of reality. (2) The model from which the statistical implications follow is not an adequate formal expression of the theory. (3) The experiment (types of measurements, operations, experimental controls, and so on) is not a good representation of the model. (4) The results observed lead to the false conclusion that the model is not true or that the model is true when the converse is true. We are concerned here only with the first and last (1 and 4) alternatives. In addition, we consider briefly the problem of translating the model and its statistical implications into an actual experiment. A few experimental procedures are discussed in this respect because of their importance in psychological research. Also, we assume the absence of a mistaken calculation or mistranscription, and so on.

Basic Definitions

Statistical Hypotheses. A statistical hypothesis is a statement about the way a random variable is distributed. This usually takes the form of some specific numerical value for one or more parameters of the distribution of the random variable.

In the general case the model derived from the psychological theory states that the phenomenon under study is a random variable with a certain distribution. The hypothesis carries this one step further and specifies the distribution in terms of a value for one or more of the parameters determining the distribution. The hypothesis is thus a statement about reality based on a derivation from psychological theory. We act *as if* the hypothesis were true *before* the experiment and then judge whether the data make the hypothesis less credible.

In the lady-tasting-tea experiment of the last chapter, a claim was made to sensory powers that we doubt exist. The basic psychological theory involved would be the theory of the psychophysiology of taste. On the basis of our knowledge we are led to make statements that the lady cannot tell the difference between "on" and "off" tea. A complex model of psychophysiology for the discriminability of the two preparations possibly could be derived. However, we use more behavioral methods in our example. Instead of asking whether or not the subject can discriminate the "on" and the "off" tea, we ask whether or not she can correctly classify a cup of tea. Our position is that she cannot do better than chance in classification experiment. This leads directly to a formal model. If there are but two classes into which the little old lady can place each cup, then each cup-tasting directly comparable to a binomial trial. This parallel suggests the binomial random variable as a model of the situation. If we decide that we can do an experiment satisfying the properties of a binomial experiment, we have a way of deriving the statistical-probabilistic consequences we need to make our decision as to whether we are correct about the claimed skills or whether the little old lady is correct. This requires that there be more than one trial, that is, more than one cup of tea tasted; that each trial be independent of the other, that is, success with one cup does not influence success or failure with another cup; that two choices only be available; that is, that she be made to say "on" or "off" for each cup. Providing that our subject is willing to cooperate, the major problem in achieving the binomial conditions for the experiment is in achieving independence. If the subject makes choices one right after another, it is doubtful if she could say anything about the tea after the first m cups. One solution would have her come to our laboratory every morning for morning tea on n consecutive days and insure that her schedule of eating and drinking were consistent from day to day to provide for the same probability of correct discrimination each day of the experiment.

We have a model! However, there is a very large (indeed, infinite) family of special cases of the model. At this point we specify which member of the family of distributions is implicated by our theory. So far we have allowed the properties of the experiment and the properties of the behavior we shall observe to determine the structure of the model. Now, in formulating our hypothesis, we consider the specific attitude expressed in our theory. The lady states that she can do better than chance in telling the difference in "on" and "off" tea. We are skeptical, and express this skepticism by stating that we believe that she can do no better than chance—no better than if she tossed a fair coin to make her choice for each cup she tasted. For the moment, we decide to test the hypothesis of chance classification on the basis that it is a very specific hypothesis whereas our subject's hypothesis is composite in the sense that she does not say precisely how much better than chance she can do.

To formalize the situation, we have the following elements.

1. Theory. It is claimed that the lady can discriminate between "on" and "off" tea.

2. Model. On the basis of a proposed experiment, not on discrimination but on classification behavior, we formulate a model that is a binomial experiment model.

3. Hypotheses. Two hypotheses are entertained. The first is the hypothesis that we "test" in a statistical test (see below). The second is the alternative that we are considering and shall "accept" if the first is "rejected." The first is denoted as H_T and takes the form

$$H_T: p = .5 \text{ in } B(n, p)$$

The second is known as the alternative hypothesis and takes the form

$$H_A: p > .5 \text{ in } B(n, p)$$

or

$$H_A: p = .8 \text{ in } B(n, p)$$

The hypothesis H_T specifies a binomial random variable and its distribution. Once n is chosen, on the basis of decision by the experimenter, the probabilistic situation for the experiment is determined precisely. We know all the possible results in the experiment and the relative frequencies with which we would observe the results, because these parallel precisely the events in our sample space S and the probabilities in $B(n, p)$.

However, the little old lady is not so fortunate. She has an infinite number of possible distributions even when n is determined, because $p > .5$ leaves an infinite set of values for p. We call such an alternative hypothesis a composite hypothesis—in contrast with a specific hypothesis, say that $p = .8$.

Every hypothesis in a statistical test is a specific hypothesis, but we are concerned with alternative hypotheses that may be composite or specific.

Null Hypotheses. The term "null hypothesis" is widely used in the theory of statistics. A null hypothesis is the hypothesis tested or a statement of the specific parameters of the distributions of the model involved. The word "null" is perhaps misleading because of its implication that nothing occurs or that everything is absent. It derives from the situation in which there is said to be nothing to some claim or that certain events (such as experimental effects or treatments) do not have an effect on the behavior or phenomenon being observed. It also bears on the situation when we compare two groups of subjects and hypothesize no difference between the two groups. This usage for the hypothesis H_T is too restrictive for many uses of statistics. We adopt the more general meaning of the null hypothesis H_T. Any hypothesis specifying some particular distribution in advance of the experiment, which is the assumed state of nature, is the hypothesis to be tested in the statistical test. This hypothesis is called the null hypothesis or the test hypothesis and denoted by the symbol H_o or H_T. The alternatives to this hypothesis can be specific or composite. We discuss later the consequences of the difference of these two forms of alternative hypotheses.

Test Statistics. The next step is to determine the aspect of the experiment that is relevant as information regarding the parameter of the random variable and its distribution in the model, that is, the statistic that is relevant to the parameter. The considerations that enter into the choice of such test statistics are too complex to deal with here. For the applications we consider, for the moment at any rate, we need insist only that the statistic involved be an unbiased estimator of the relevant parameter.

The unbiased estimate of the parameter p is the mean number of successes in the n observations or trials, the relative frequency of success in n trials. Therefore in the present example, we observe as a test statistic the relative frequency of the result corresponding to success in the classification of "off" and "on" cups of tea. We need a symbol for this variable and we choose

$$\bar{S} = \frac{\text{number of successes in } n \text{ trials}}{n}$$

Assume that we decide to observe $n = 16$ trials in our experiment. This, with the hypothesis and the model for the experiment, specifies that if our theory is adequate the observations will be distributed as a binomial random variable with $B(16, .5)$. However, in the experiment we observe one result, namely, some one of the 17 possible numbers of successes. This is a general condition of experiments—we have a distribution of possible results but only one is observed. We can calculate the probability of observing exactly that result under the conditions of our model and the hypothesis H_T. In our example this corresponds to calculating the binomial probabilities associated with each of the 17 possible numbers of successes in the experiment. This is of little interest, since each of the 17 possible results occur with relatively small probability. We need to make one more step before we get to the goal in our experiment.

If we observed 15 successes, we would be surprised under the terms of our model and the hypothesis. We expect that the subject would get in the neighborhood of 8 successes out of the 16 possible successes, that is, that $\bar{S} = .5$ is what we expect of our subject. We would be even more surprised if she were able to get 16 successes, that is, $\bar{S} = 1.0$. The degree of our surprise and the degree of "significance" of the performance of our subject can be measured by the probability that the observed number of successes or more could have come about under the terms of the model and the hypothesis. In fact, as a matter of convention here, but with some very good reasons that we cannot go into until later, we always lump together all of the possible results of an experiment that are equal to the obtained result or more strongly inclined to support the alternative than the observed result. We now formalize this notion as the significance of the result.

Significance Probability. In this section we provide one way of answering the important question: How significant are the data we observe, in terms of the two hypotheses H_T and H_A? First, we define a significance set as the set of all possible results in the experiment that are equal to our obtained result or more extreme in the direction of H_A (with respect to H_T). Since in our example $H_A: p > .5$ as contrasted with $H_T: p = .5$, the significance set includes all possible values of $\bar{S}$ that are equal to or greater than the observed value of $\bar{S}$. If H_A had been stated with the reverse inequality, then the set would be defined by the set of possible values of $\bar{S}$ equal to or smaller than the observed $\bar{S}$.

Say that 10 of the 16 trials were successes, that is, the lady was able to classify correctly 10 of the 16 cups of tea. Under the assumptions of the model and hypothesis H_T, this is the element $\bar{S} = .625$ out of the set of possible results $R = \{.0000, .0625, .1250, .1875, .2500, .3125, .3750, .4375, .5000, .5625, .6250, .6875, .7500, .8125, .8750, .9375, 1.0000\}$. Under our definition of the significance set with $H_A: p > .5$, we have

$$\text{significance set} = \{\bar{S} \in R \mid \bar{S} \geq .6250\}$$
$$= \{.6250, .6875, .7500, .8125, .8750,$$
$$.9375, 1.0000\}$$
$$= SS_{.6250}$$

or the significance set defined by the result $\bar{S} = .6250$.

We can calculate the probability that, given the conditions of our experiment, including the model and the hypothesis, we would observe one or another element in $SS_{.6250}$. This is simply

$$P[SS_{.6250} \mid \text{success is distributed as } B(16, .5)]$$
$$= P[S \geq 10 \mid S \text{ is distributed as } B(16, .5)]$$

which is given by the sum of the probabilities of the events corresponding to the set of results making up the significance set $SS_{.6250}$. This corresponds to the probability of observing 10, 11, 12, ..., or 16 successes in 16 binomial trials where $p = .5$. This probability is easy to find. In fact, $F(9)$ for the binomial involved gives the probability of the complement of the set $SS_{.6250}$. Looking up $F(9)$ for $B(16, .5)$ gives us

$$F(9) = .7728$$

and hence the significance probability is given by

$$\text{significance probability} = P(SS_{.6250})$$
$$= 1 - F(9)$$
$$= .2272$$

The significance probability is denoted by Ψ, a function of the observed value:

$$\Psi(.6250) = P(SS) \tag{14.1}$$

This is easily interpreted: We would observe as many or more successes in a binomial experiment as we observed in this real experiment with a probability of .2272, if we were observing $B(16, .5)$. If we repeated our experiment many times, about 23 percent of the experiments would result in ten or more successes. This is not unlikely, and we probably should not be surprised at the result in our experiment. That we observed ten successes on *this* experiment does not seem remarkable. Thus, the significance of the result is not very great in light of our hypothesis.

The mechanics of this calculation were carried out in terms of the number of successes to indicate the role of the binomial distribution in the calculation. The more general procedure is to compare the parameter value specified in the null hypothesis, H_T, and the estimate of that parameter value. In the current example we would compare $p = .5$ with the obtained value of $\bar{S}$. The comparison that makes immediate intuitive sense is

$$\bar{S} - p = .6250 - .5 = .1250$$

How significant is this difference between the hypothesized parameter value and the estimated parameter value? The answer is the same as our original answer—not very. We would expect as large a difference or larger with a probability of .2272. We would have a low degree of confidence that this difference was observed because the parameter p was not equal to .5.

If we observed $\bar{S} = .8750$, our answer might be different. For this result the difference between parameter and estimate is .3750. What is the probability of observing a difference this large or larger?

The significance set for this result is given by

$$SS_{.8750} = \{.8750, .9375, 1.0000\}$$

and we have (because $\bar{S} = .8750$ corresponds to 14 successes)

$$\Psi(.8750) = P(SS_{.8750})$$
$$= P[S \geq 14 \mid S \text{ is distributed as } B(16, .5)]$$
$$= 1 - F(13) = .0020$$

On the basis of this probability we would say that it is unlikely that we would observe a difference as large or larger if our model and hypothesis were true. In about $\frac{1}{5}$ of 1 percent of a large number of replications we would observe a difference as large or larger. The importance of this is that it causes us to doubt that our result could have come about in this particular experiment if our hypothesis H_T were true. It seems more likely that the alternative hypothesis H_A is true. The degree of credibility of H_T is correspondingly low. We feel that H_T lacks credibility in view of the result observed. Since the only alternative we are considering is H_A, it gains the credibility that H_T lost.

The significance probability of every possible result can be calculated in this fashion. It conforms to the distribution function of the binomial probability distribution except that (1) values in the domain of the function are the differences between the parameter value specified in H_T and the potentially observable values of parameter estimates, and (2) the accumulation is turned "upside down," that is, complemented. Since we are dealing with the upper tail of the binomial distribution, $[1 - F(S)]$ is the significance probability. When we observe small values of $\bar{S}$, the significance probability increases to large figures, indicating little significance for the experiment in the direction of casting doubt on the null hypothesis in favor of

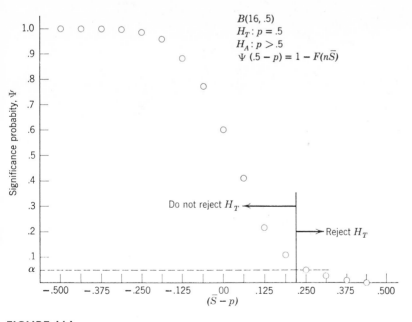

FIGURE 14.1

Difference in hypothesized parameter, $p = .5$ and observed result $\bar{S}$: $\bar{S} - p$.

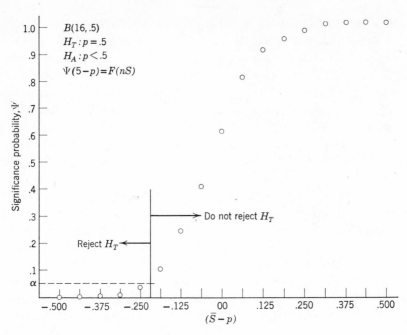

FIGURE 14.2

Difference in hypothesized parameter, $p = .5$ and observed result $\bar{S}$: $\bar{S} - p$.

the alternative. This makes intuitive sense, because we should not tend to favor a hypothesis (H_A in this case) on evidence that is closer to that required by another hypothesis (for example, $\bar{S} < .5$ is closer to .5 than any value included in H_A). Figure 14.1 shows the relationship of Ψ to ($\bar{S} - p$) in our example.

If we reverse the inequality in the alternative hypothesis, we obtain the complementary function of that plotted in Figure 14.1. That is, if $H_A: p < .5$, significance sets are formed by possible results equal to or less than the observed result and the significance probabilities correspond to the distribution function values for the lower tail of the binomial distribution. A graph of this is given in Figure 14.2.

Most of the points in the domains of the graphs in Figures 14.1 and 14.2 have such high significance probabilities that we are unwilling to attribute any significance to the alternative hypothesis being considered. This leads to an important concept. If we agree that a given significance probability is low enough to lead us to doubt the null hypothesis in favor of the alternative, we can select the set of specific results that satisfy this condition. Once we have identified this set of potential results, we need not calculate Ψ for the specific result we observe in the experiment. If the observed result is in the set, we infer that the alternative hypothesis is more likely to be true than is the null hypothesis. A rule guiding these inferences is called a decision rule or a rejection rule.

Rejection Rules. When we observe a result with a significance set that has a small probability, we reject H_T in favor of H_A. This procedure involves the risk of falsely rejecting H_T when it is true. In fact, the possibility that H_T is true in spite of extreme results (with low significance probabilities) is the probability that the result, or one more extreme, could have occurred when H_T is true. This probability is the significance probability of the result, a seeming paradox: The significance probability and the error probability are the same.

The paradox is resolved when we note that we are willing to reject H_T on the basis of small significance probabilities only. The smaller the significance probability, the greater is the significance for the rejection of H_T and the less unwilling we are to reject H_T in favor of H_A. The significance of the result increases as the risk of falsely rejecting H_T decreases.

The error that results when H_T is rejected falsely—that is, when we say that H_T is false when it is true—is called a type I error. The probability of making this sort of error for a given result is given by the significance probability Ψ.

This approach suggests that we can select, on the basis of the significance probability, the set of possible results that would cause us to reject our test hypothesis. If we wish to have strong evidence in favor of H_A as opposed to H_T, we would select a set with a small probability Ψ. That is, we would look for the largest set of possible results that still had a significance probability equal to or less than some predetermined value. Then, when the experiment was executed, we would have a rule by which to judge our result without having to make a judgment about the particular value of Ψ after the fact. If the result were more extreme than the result defining the significance set, so much the better. At any rate, we would never face the risk of a type I error with any greater probability than we are willing to face.

This procedure has an important advantage. There is a tendency to accept a significance probability of .06, for example, as "so near .05" that we would accede to the significance of a result with a significance probability of .06. This erosion is often progressive, and the scientist finds himself being impressed with significance probabilities of .10, .13, and so on, to the ultimate detriment of his science.

The accepted tradition in statistical inference is to decide, before the experiment, on a tolerable level of risk. Where it is important to have great confidence that the test hypothesis H_T is false and H_A is true, the significance probability demanded would be very small—and hence the risk small. The level of risk is specified numerically by selecting a value Ψ_{max} such that any smaller value of Ψ

associated with the obtained result would be indicative of an acceptable level of risk in rejecting H_T. This maximum significance probability is called the level of significance of the statistical test of H_T against H_A. We designate this level of significance as α:

$$\alpha = \Psi_{max} \qquad (14.2)$$

This definition leads to a rejection rule: If $\Psi \leq \alpha$, reject H_T in favor of H_A; otherwise (that is, if $\Psi > \alpha$) do not reject H_T. The rejection rule is often called an α level rejection rule or a rejection rule with an α level of significance. Widely used, but arbitrary, levels of significance in rejection rules are $\alpha = .05$, $\alpha = .01$, and $\alpha = .001$. These correspond to a risk level of falsely rejecting H_T 5 out of 100 times, 1 out of 100, and 1 in 1000, in experiments like the experiment to be performed.

Once we have decided on a level of significance, we can cast rejection rules in terms of the actual values of the random variable we observe in the experiment. To do this we ask the question: What values of the random variable, if observed, would result in a $\Psi \leq \alpha$?

The solution is simple. For H_T: $p = .5$ and H_A: $p > .5$ we want to know the smallest value $\bar{S}$ we could observe to give a value of $\Psi \leq \alpha$. This is easily translated into the distribution function of the binomial random variable. Find the critical value $\bar{S}_c$ that satisfies the equation

$$P\left[\bar{S} \geq \bar{S}_c \mid \bar{S} = \frac{S}{n} \text{ where } S \text{ is } B(n, p)\right] = \alpha$$

This equation can be rewritten in the binomial directly as

$$P[S \geq S_c \mid S \text{ is } B(n, p)] = \alpha$$

In our example with $p = .5$ and $n = 16$ we have, for $\alpha = .05$,

$$P[S \geq S_c \mid S \text{ is } B(16, .5)] = .05$$

The solution is obtained simply by finding a value S_c with distribution function value .95. Looking in the table of the binomial random variable, we find distribution function values of .9617 for $S = 12$ and .8950 for $S = 11$. That is, $\Psi = .0383$ for $S = 12$ and $\Psi = .1050$ for $S = 11$. We cannot find a critical value S_c satisfying exactly the requirement that $\Psi = \alpha$. This happens because we do not have a continuous random variable and have to take the natural values. In such cases we generally choose the more conservative of the two closest values. That is, we set S_c so that Ψ is less than the desired α. In our example, $S_c = 12$. We can now restate the rejection rule in terms of the critical value.

> *If* S $\geq$ S$_c$, *reject* H$_T$ *in favor of* H$_A$
>
> *If* S $<$ S$_c$, *do not reject* H$_T$

Translating S and S_c into $\bar{S}$ and $\bar{S}_c$, for the specific numbers in our example gives the α level rejection rule:

$$If\ \bar{S} \geq .7500,\ reject\ \mathrm{H_T}\ in\ favor\ of\ \mathrm{H_A}$$
$$If\ \bar{S} < .7500,\ do\ not\ reject\ \mathrm{H_T}$$

The Statistical Test. Several times we have referred to the test of H_T against H_A. The references have been vague to this point because of incomplete terminology. Now we can define precisely what we mean by a statistical test.

A statistical test consists of a specification of a statistic relevant to H_T and a decision rule leading to rejection of H_T or a failure to reject H_T. The rejection rule is determined by the nature of the alternative hypothesis and the significance level α.

A Second Kind of Error

The reader may have decided that if we want to avoid the risk of falsely rejecting the null hypothesis, we make our critical values very different from that specified in H_T. This would amount to making α very small. Indeed, why not set α to zero? Then the risk of a type I error would be avoided. The logic is correct: If $\alpha = 0$, the type I error is avoided. However, we would never reject H_T—not even when it was false and some other value was the true parameter. In general, as we reduce the risk of making a type I error we increase our inability to reject H_T when it is false and hence increase the risk of making a type II error.

A type II error is committed when H_T is false and we fail to reject it. Since H_T is either true or not true, and since we reject H_T or we do not, there are four possible outcomes in the test, shown in Table 14.1.

We have seen how to calculate the probability of a type I error. Now we develop the mechanics necessary to calculate the probability of a type II error. Notice that the type II error is stated in terms of failure to reject H_T when it is false. Nothing is

TABLE 14.1 Decision Table for Statistical Inference

	H_T True	H_T False
Reject H_T	Type I error	Correct decision
Do Not Reject H_T	Correct decision	Type II error

said about the alternative H_A. However, in order to calculate the probability of a type II error of the test, we must deal with specific alternative hypotheses. For example, take the tea-tasting experiment and imagine the true value of p is .8. Within the model and for our experiment H_T and H_A do not change, but the samples we draw come specifically from the binomial with $p = .8$ if the alternative to H_T is true. This hypothesis specifies a precise probability distribution. If $p = .8$ in the population (of classifications by the little old lady), we have a different but well-specified distribution from the one we would have if H_T were true.

Recall that we reject H_T when the number of successes is 12 or greater in the ($\alpha = .05$) level test. Conversely, we do not reject H_T when the number of successes is 11 or less. The probability of not rejecting H_T when $p = .8$ is

$$P[S \leq 11 \mid S \text{ is } B(16, .8)] = F(11) = .2017$$

This is the probability of making a type II error when S is $B(16, .8)$ for the ($\alpha = .05$) level test of the hypothesis $H_T: p = .5$ in $B(16, p)$. In other words, about 20 percent of the time we shall not concede the little old lady's claim even when in reality she can correctly classify 8 out of 10 cups. This probability is formally denoted β.

This calculation reveals our test to be pretty weak against making a type II error if $p = .8$ were true. This weakness is accentuated if the value of the parameter p in the real world is smaller. For example, if $p = .4$ in reality, then

$$\beta = P(\text{type II error}) \tag{14.3}$$
$$= P[S \leq 12 \mid S \text{ is } B(16, .4)] = .999$$

The probability of making the type II error in our ($\alpha = .05$) level test with $H_T: p = .5$ and $H_A: p > .5$ is nearly 1.0. If the parameter p were even smaller, this probability would be larger. Values of p far from .5 in the upper direction, that is, in the direction of H_A, produce small type II error probabilities. We have already seen that the $p = .8$ gives type II error probability $\beta = .2017$. If $p = .9$, $\beta = .0170$, and for $p = .95$, $\beta = .0009$.

An informative way of dealing with these probabilities is to turn our question around. We find the probability of rejecting H_T when the parameter has some specific value. If this parameter value is a member of the set of values designated by H_A, then

$$P(\text{reject } H_T \mid H_A \text{ is true}) = 1 - P(\text{type II error})$$
$$= 1 - \beta$$
$$= \pi \tag{14.4}$$

This can be interpreted as the probability we shall detect that H_A is true instead of

H_T. Given our test of H_T against the alternative H_A, we have two possibilities: accept H_T or reject H_T in favor of H_A. The latter course of action is a correct one if H_A is true. Thus the probability of rejecting H_T in favor of H_A when H_A is true is the probability that we would detect this fact with the test and its rejection rule. This probability π is called the power of the test against H_A.

We can apply this notion to the test of H_T: $p = .5$ in $B(16, p)$ against H_A: $p > .5$ in the tea-tasting experiment with the $\alpha = .05$ level decision rule for a specific value of the parameter:

$$If\ S \geq 12,\ reject\ \mathrm{H_T}\ in\ favor\ of\ \mathrm{H_A}$$
$$If\ S < 12,\ do\ not\ reject\ \mathrm{H_T}$$

We have already calculated the probabilities of type II error when the parameter value was variously .8, .9, and .95. Subtracting the probabilities from one, that is, $1 - \beta$, gives the power of our test against these specific alternatives, .7983, .9830, and .9991, respectively.

The alternative in the formulation of the test H_A: $p > .5$ includes all of these values as special cases. We also calculated the probability of a type II error when $p = .4$ which is not covered in the composite alternative. In general, the power for

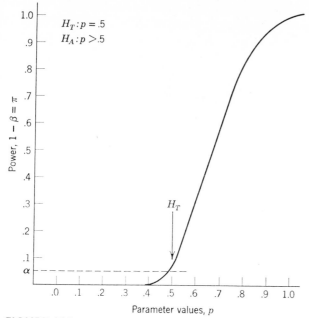

FIGURE 14.3

Power function for the test of the hypothesis H_T: $p = .5$ in $B(16, p)$ against the composite hypothesis H_A: $p > .5$.

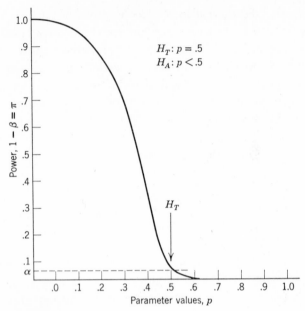

FIGURE 14.4

Power function for the test of the hypothesis H_T: $p = .5$ in $B(16, p)$ against the composite hypothesis H_A: $p < .5$.

the possible values of the parameter not included in the composite hypothesis are smaller than α. This is because the distribution, in reality, determined by the true value p, is located below the distribution specified under H_T—less of the true distribution is in the rejection region. Thus, our test would detect that H_T were false with probability less than α. This is a good thing from the standpoint that if we reject H_T under our present test, we do so in favor of H_A: $p > .5$, implying the exact opposite sort of condition in nature—that the little old lady can do better than chance when she does even more poorly.

The general relationship between the state of nature, as embodied in the parameter p, and our ability to detect this with the specific statistical test with H_T: $p = .5$ and H_A: $p > .5$ can be represented by the graph of the power of the test for all possible values of p. Letting p take all values from .00 to 1.00, and calculating the power of the test for these values, gives us the power function shown in Figure 14.3. The graph clearly indicates the lack of power to detect that the parameter value is smaller than the value specified in H_T. If we had formulated our test with H_A: $p < .5$, the power function would have been turned around, left for right, with the power at $p = .5$ remaining constant at a value α. This is shown in Figure 14.4.

Two-Sided Tests

A Problem in One-Sided Tests. Up to this point we have dealt with one or another of two composite alternative hypotheses to H_T. We assumed deviations from our hypothesis were in the direction of larger values of the parameter or smaller values than specified by H_T. That is, we had in our example

$$H_A: p > .5$$

or

$$H_A: p < .5$$

In the tea-tasting experiment we assumed that the only reasonable way the lady might perform contrary to our hypothesis was in the direction of being more accurate rather than less accurate. It would be strange if she were confused and classified all the cups incorrectly. This would imply that she could tell the difference between "on" and "off" tea perfectly well! Only her usage of the terms "on" and "off" was wrong. It is rather improbable that a person would make a claim like the one we are testing unless she were quite clear on the specifics of "on" and "off" tea. Thus, we are fairly confident that if $H_T: p = .5$ is false, then $H_A: p > .5$ is the only rational alternative.

We are led to rejection rules with H_T and H_A that would fail to reject H_T on the basis of observed extreme results. If we observed $\bar{S} = 0$ in our example where $H_A: p > .5$, we would not reject H_T. This result, $\bar{S} = 0$, has a significance probability

$$\Psi = 1.0$$

The result is as distant as possible from the rejection region in the test. We would be led to a nonrejection of H_T and an interpretation that the observed result came about simply by chance under H_T.

This situation is not satisfactory. If we had been working with $H_A: p < .5$, then the significance probability would have been

$$\Psi \cong .000$$

The difference in the decision made depends very heavily on the alternative hypothesis with which we are working. The student is encouraged to attempt to devise a number of examples, using both types of alternatives, those of the upper tail and those of the lower tail.

A revealing way of looking at the one-tail test in the light of possible deviation in nature from the hypothesized state of nature involves the power of the test. We need to make our alternative specific (single value) to calculate the power of the test. For each value of the parameter p given by the specific alternative hypothesis H_A, we calculate

$$\text{power}(p) = P(\text{rejection of } H_T \mid H_A \text{ is true})$$

The larger the power, the more likely it is that we will observe a result leading us correctly to reject H_T in favor of H_A. If the power for a specific true value p is large, we shall tend to discover the "truth" of H_A by the test. This is referred to as the power of the test to detect that the specific value given by H_A is true instead of that given by H_T.

Inspection of the power functions in Figures 14.3 and 14.4 makes it clear that we have a good deal of power for most of the specific alternatives covered by the composite alternatives of each of the single-sided tests. However, for specific alternatives not covered in the composite alternative the power is at most α. Thus, for a reasonably cautious level of significance the power for the specific values not covered is very poor. Even if the true value of the parameter were different from that given by H_T, it cannot be detected by the test unless it is covered by the composite hypothesis of the test.

The Two-Sided Alternative. A solution to this problem is possible at some cost in sensitivity of the test to specific values covered in the single-sided alternative hypothesis.

This third logical possibility in dealing with alternative hypotheses is to have rejection regions that lead to rejection for values of our test statistic sufficiently different from the hypothesized value of the parameter. No concern is directed at which *way* the test statistic is different. This is embodied in a "two-sided" or "two-tailed" test with the same null hypotheses but with an alternative H_A, stating that the parameter is different from that specified in H_T. In our example

$$H_A: p \neq .5$$

is a two-sided alternative. This may be restated

$$H_A: p > .5 \quad \text{or} \quad p < .5$$

Thus, if our observed result, say $\bar{S}$, is extreme, in either the high *or* the low direction, we are led to a rejection of H_T.

Significance Probability for Two-Sided Tests. A minor modification in the form of our rejection rules is required when we are dealing with two-sided alternatives. We reject H_T when a value on either extreme of the distribution is observed.

The significance set for a given result in a one-sided test is defined as the set of all possible results in the experiment equal to the obtained result, or more extreme, in the direction of H_A with respect to H_T. We modify this to include all possible results as extreme as the observed result, or more extreme, on both sides of H_T. In the original example we showed the significance set for a result $\bar{S} = .6250$ where $H_A: p > .5$:

$$\text{significance set} = \{\bar{S} \mid \bar{S} \geq .6250\}$$
$$= \{.6250, .6875, \ldots, 1.0000\}$$

For the two-sided hypothesis $H_A: p \neq .5$ (that is, $p > .5$ or $p < .5$)

$$\text{significance set} = \{\bar{S} \mid \bar{S} \geq .6250 \text{ or } \bar{S} \leq .3750\}$$
$$= \{.0000, .0625, \ldots, .3750, .6250, \ldots, 1.0000\}$$

The .3750 is selected as the largest value of the set of smallest values because it is as different, in an absolute sense, from $p = .5$ as is $\bar{S} = .6250$. The general way of stating this is to deal with the absolute value of the difference between the parameter value of H_T and the estimated value of the parameter in the experiment. In our example,

$$SS_{.6250} = \text{significance set}$$
$$= \{\bar{S} \mid (|.5 - \bar{S}|) \geq (|.5 - .6250|)\}$$

The significance probability Ψ is simply the probability of observing this set if H_T is true. Under $H_A: p \neq .5$, then,

$$\Psi = P[SS_{.6250} \mid \text{success is distributed as } B(16, .5)]$$
$$= P(\bar{S} \leq .3750) + P(\bar{S} \geq .6250)$$

Expressing these probabilities as terms in the distribution function of $B(16, .5)$,

$$\Psi = P(S \leq 6) + P(S \geq 10)$$
$$= F(6) + 1 - F(9) - .2272 + .2272$$
$$= .4544$$

This very high probability is reasonable: Only three of the possible values (.4375, .5000, .5625) of $\bar{S}$ are less extreme than the observed value under our two-sided alternative.

For our other example result, $\bar{S} = .8750$, we have

$$\Psi = P[\bar{S} \mid (|.5 - \bar{S}|) \geq (|.5 - .8750|)]$$
$$= P(\bar{S} \leq .1250) + P(\bar{S} \geq .8750)$$

In terms of the binomial this is

$$\Psi = F(2) + 1 - F(13)$$
$$= .002 + .002$$
$$= .004$$

A diagram of this is given in Figure 14.5.

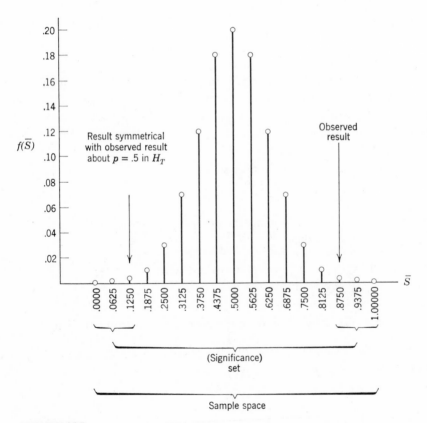

FIGURE 14.5

Density function of $B(16, .5)$ showing the significance set for the observed result $S = .8750$ under a two-tailed alternative test.

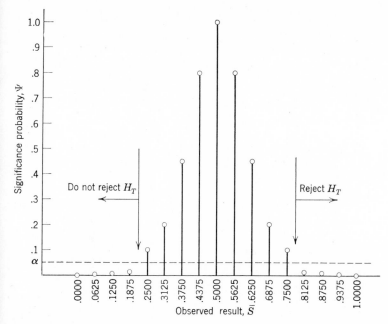

FIGURE 14.6

Significance probability function for a two-sided alternative test of H_T: $p = .5$ in $B(16, p)$.

This leads to a significance probability function of different shape from the one for the one-sided tests. The interpretation and use of Ψ' in the two-sided test is the same as in a one-sided test. If $\Psi' \leq \alpha$, reject H_T in favor of the alternative. In our example where $\bar{S} = .8750$, and $\Psi' \leq \alpha$, we would conclude that p in $B(16, p)$ was not .5. The significance probability function is shown in Figure 14.6.

One feature of this form of the test needs mentioning. The values of the test statistic, for example $\bar{S}$, must be more extreme to satisfy the condition $\Psi' \leq \alpha$ in two-sided tests than in one-sided tests. This is so because Ψ' includes the two ends of the distribution regardless of the specific value of the test statistic actually observed. This implies that we need more extreme results to be able to reject the null hypothesis under a two-sided alternative. This decreased sensitivity can be shown most easily by plotting the significance probability functions of our one- and two-sided tests in the binomial example. Figure 14.7 shows these functions. The points of the graphs of the functions are connected by lines for clarity.

The significance probability of the one-sided test is smaller at every value of the test statistic covered by the respective alternative hypothesis. The value of the test statistic necessary to have $\Psi' \leq \alpha$ is closer to the value specified in H_T for each of

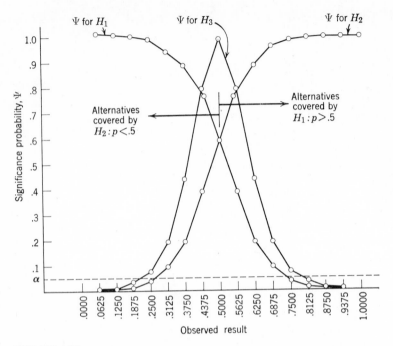

FIGURE 14.7

Significance probabilities for H_1: $p > .5$, H_2: $p < .5$, and H_3: $p \neq .5$ plotted on the same coordinate system.

the one-sided tests respectively. This means that the two-sided test requires a more extreme result on a given side of the distribution than the corresponding one-sided test does. However, it also is true that where a value of a test statistic is not included in a one-sided alternative, the two-sided test is more sensitive to the results than the one-sided test.

Critical Values for Two-Sided Tests. We wish to ascertain the values that lead to a rejection rule with an α level of significance. There are two ways of rejecting H_T in favor of a two-sided alternative: Observe a value sufficiently larger than that specified in H_T, or one sufficiently smaller. Thus, for a test of H_T: $p = .5$ against H_A: $p \neq .5$, we need critical values $\bar{S}_U$ and $\bar{S}_L$ (upper and lower). The decision rule for the two-sided test is

Reject H_T *in favor of* H_A *if*

$$\bar{S} \leq \bar{S}_L$$

or if

$$\bar{S} \geq \bar{S}_U$$

If we select a value for α, then the values of $\bar{S}_L$ and $\bar{S}_U$ are determined by the distribution of $\bar{S}$. It is conventional to equate the two rejection sets with respect to the probability associated with the sets. This is an arbitrary decision and need not be observed. For example, we may want to avoid falsely rejecting H_T in observing large values of the test statistic, but care less about the type I error if we observed small values of the test statistic. In this case we would set the upper critical value $\bar{S}_U$ to a relatively extreme value (hence less probability of a more extreme value) than the lower critical value $\bar{S}_L$. In general, however, we set the critical values to satisfy the equalities

$$P(\text{test statistic} \geq \text{upper critical value}) = .5\alpha$$
$$P(\text{test statistic} \leq \text{lower critical value}) = .5\alpha$$

When α is set and the distribution of the test statistic is known, these equalities lead to specific values for the two critical values. Taking our binomial example with $B(16, .5)$ under H_T: $p = .5$ in $B(16, p)$ and letting $\alpha = .05$, we have

$$P(\bar{S} \geq \bar{S}_U) = .025$$
$$P(\bar{S} \leq \bar{S}_L) = .025$$

By substituting the corresponding values of the binomial random variable S into these equations, we find that the closest we can come is for $S_U = 13$ or $S_U = 12$. These give

$$P(S \geq 13) = .011$$
and
$$P(S \geq 12) = .038$$

As before, we choose the more conservative critical value, giving a smaller α than desired rather than a large one. Therefore we set $S_U = 13$. The lower critical value $S_L = 3$ is found in a similar fashion. Thus the closest we can come to an ($\alpha = .05$) rejection rule is the ($\alpha = .022$) rejection rule with $\bar{S}_U = .8125$ and $\bar{S}_L = .1875$ for the upper and lower critical values of $\bar{S}$:

If $\bar{S} \geq .8125$ or if $\bar{S} \leq .1875$, reject H_T in favor of H_A
If $.1875 < \bar{S} < .8125$, do not reject H_T

Power of Two-Sided Tests. The definition of the power of a two-sided test is the same as in one-sided tests:

$$\pi = 1 - \beta$$
$$= 1 - P(\text{type II error})$$

However, we must consider the two regions of rejection for a two-sided test. For a specific alternative, for example, $p = .8$ as above,

$$\beta = P(\bar{S}_L < \bar{S} < \bar{S}_U \mid p = .8)$$
$$= P(.1875 < \bar{S} < .8125 \mid p = .8)$$

which, when we deal with S instead of $\bar{S}$, is

$$\beta = P[3 < S < 13 \mid S \text{ is } B(16, .8)]$$
$$= P[3 < S \mid S \text{ is } B(16, .8)] - P[S \geq 13 \mid S \text{ is } B(16, .8)]$$
$$= 1.0 - .598 = .402$$

Thus the power of our ($\alpha = .022$) level test of H_T: $p = .5$ against H_A: $p \neq .5$ when in reality $p = .8$ is given by $\pi = 1 - \beta = .598$. In words, we would with a probability of nearly .6 detect that $p \neq .5$ when it was actually .8 and $n = 16$.

If we plot, as before, the power of our test for all the possible values of p, we have a power function for the test. For the purposes of comparison we plot this function on the same coordinates with the power functions for the two one-sided tests in the example. Where significance probability is a function of the possible results, a discrete variable, power is a function of the possible parameter values, a continuous variable. The graphs are given in Figure 14.8.

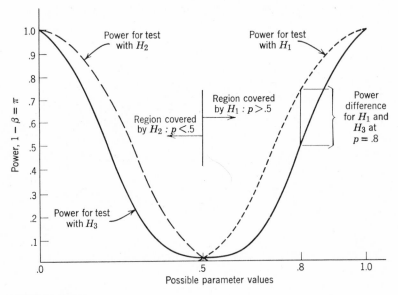

FIGURE 14.8
Power functions for one- and two-tailed tests of H_T: $p = .5$ in $B(16, p)$, H_1: $p > .5$, H_2: $p < .5$, and H_3: $p \neq .5$.

This graph indicates clearly that the one-sided tests are more powerful over the range of parameter values covered by the respective composite hypotheses. To achieve this somewhat greater power it is necessary to lose almost all sensitivity to the possibility that parameter values in the other half of the range may obtain in reality.

Level of Significance and Power

The level of significance of a test has a direct influence on the power function of the test. When we first discussed the level of significance in this chapter, it was suggested that we could reduce risk of the type I error as much as we desired by selecting an α that was sufficiently small. However, this leads to a high probability of a type II error: The more extreme our rejection regions, the more likely it is that we shall not reject H_T when it is false.

It should be quite clear why we refer to the power of the test. The test is defined by a test statistic, a specific α, and the rejection region implied by the hypotheses. Thus the power function of a test of a given H_T against a composite alternative is directly a function of the value of α.

This dependency can be demonstrated easily in the tea-tasting example. We need only do the calculations for the test where $H_T: p = .5$ and $H_A: p > .5$ to make the point.

First, choose three α values that give different rejection regions for the test with 16 observations:

$$\alpha_1 = .001, \qquad \alpha_2 = .038, \qquad \alpha_3 = .105$$

These values lead to three separate rejection rules.

A. For $\alpha_1 = .001$

$P(\bar{S}_c \leq \bar{S}) \leq .001$ implies that $\bar{S}_c = .9375$, and the rejection rule is

Reject H_T *if* $\bar{S} \geq .9375$
Do not reject H_T *if* $\bar{S} < .9375$

B. For $\alpha_2 = .038$

$P(\bar{S}_c \leq \bar{S}) = .038$ implies that $\bar{S}_c = .7500$, and the rejection rule is

Reject H_T *if* $\bar{S} \geq .7500$
Do not reject H_T *if* $\bar{S} < .7500$

C. For $\alpha_3 = .105$

$P(\bar{S}_c \leq \bar{S}) = .105$ implies that $\bar{S}_c = .6875$, and the rejection rule is

Reject H_T *if* $\bar{S} \geq .6875$
Do not reject H_T *if* $\bar{S} < .6875$

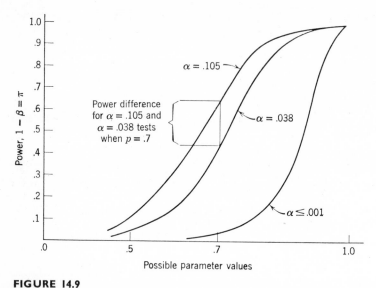

FIGURE 14.9

Power functions for three values of α of H_T: $p = .5$ in $B(16, p)$ for the alternative H_A: $p > .5$ in $B(16, p)$.

Following the mechanical steps in calculating the power functions for these three tests, we get the three curves plotted in Figure 14.9. This makes it clear that for small values of α, more difference between the value in the hypothesis H_T and in reality is required for the test to have much power to detect the difference. If α is too small, the test is insensitive to possible deviations in reality from H_T.

This appears to leave us in a very bad position: If we have a desired α, it appears that we must take whatever power we can get. Fortunately there is another alternative. We can make the power of our test, for any specific value of the parameter, as large as we like regardless of α if we can reduce the standard deviation of the test statistic. This can be done simply by having a sufficiently large sample of observations from which to calculate the test statistic.

Sample Size and Power

In order to illustrate the dependency of power on the sample size in a statistical test, we calculate the power function in our example with $n = 11$, $n = 19$, and $n = 24$. For each of these values of n we need first to establish the test involved. The limitations imposed on α by the discrete nature of the sample spaces suggest that $\alpha \simeq .03$ is convenient. For $n = 11$, an $\alpha = .033$ is possible; for $n = 19$, $\alpha = .032$; and for

$n = 24$, $\alpha = .032$. These values are close enough so that the power functions are not greatly differentiated by the differences in α. Consulting the tables of the binomial random variables for $B(11, .5)$, $B(19, .5)$, and $B(24, .5)$, we find

$$P[S \geq 9 \mid B(11, .5)] = .033$$
$$P[S \geq 14 \mid B(19, .5)] = .032$$
$$P[S \geq 17 \mid B(24, .5)] = .032$$

Thus, in terms of the binomial we set the one-tail critical values for the three tests at 9, 14, and 17, respectively. This gives the following three tests (in terms of $\bar{S}$).

A. $n = 11$, $H_T: p = .5$, $H_A: p > .5$, $\alpha = .033$, $\bar{S}_c = \frac{9}{11} = .8182$

B. $n = 19$, $H_T: p = .5$, $H_A: p > .5$, $\alpha = .032$, $\bar{S}_c = \frac{14}{19} = .7368$

C. $n = 24$, $H_T: p = .5$, $H_A: p > .5$, $\alpha = .032$, $\bar{S}_c = \frac{17}{24} = .7083$

Calculating the power function for these three tests gives rise to the graphs of Figure 14.10. The three curves show that the power of the test against all parameter values covered in H_A is greater for the larger values of n. As we increase the sample size in a test, the power of the test is increased for the parameter values covered by the alternative hypothesis.

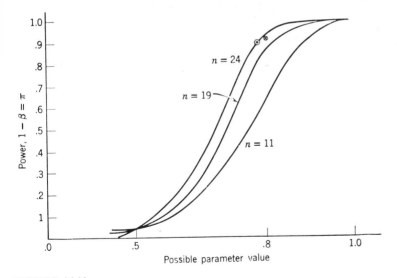

FIGURE 14.10

Power curves for tests of $H_T: p = .5$ in $B(n, p)$ against $H_A: p > .5$ with three values of n and comparable values of α.

TABLE 14.2 Values of α and β for a Selection of Values of n in the Test of H_T: $p = .5$ in $B(n, p)$ against H_A: $p > .5$

	α	β
$n = 20$	.021	.196
$n = 21$	.039	.119
$n = 22$	.026	.133
$n = 23$	.019	.073

In most experimental inference situations (not estimation) we wish to make only as many observations as necessary to satisfy the α and β requirements of our research goals. How can we discover the appropriate sample size?

We get a hint from the last graph. Suppose that we wanted $\alpha \simeq .032$ and $\beta = .10$, $\pi = .9$, and knew that $p = .8$ was the only really important competing specific hypothesis. This defines a point in our graph with $p = .8$ and $\pi = .9$. The encircled X ($\otimes$) in Figure 14.10 marks the region. This point lies between the power curves for the tests with $n = 24$ and $n = 19$. Thus the desired n is larger than 19 and smaller than 24.

In this instance, having done the work of finding the power curves for the tests with $n = 24$ and $n = 19$, it is easy to find the proper n by consulting the tables of the binomial random variables. For no value of n do we find exactly $\alpha = .032$ or $\beta = .10$. However, we can come close. The choice of n is muddled by the discrete nature of S. If we are willing to give a little on both α and β, $n = 21$ could be used. However, if we insisted that the α and β of the test must both be more conservative than $\alpha = .03$ and $\beta = .10$ then, $n = 23$ is the smallest acceptable number of observations. Table 14.2 gives the pertinent data.

We can arrive at the same conclusion using the normal approximation to the binomial.

In our discussion of the variance of a sample mean we showed that the variance is inversely proportional to the sample size. That is,

$$\text{var}(\bar{X}) = \frac{\text{var}(X)}{N}$$

where a sample of N observations is the basis for calculating $\bar{X}$. We can use this result in achieving whatever power we wish for our test and specific alternative.

In terms of our binomial example, the dispersion of $\bar{S}$, $\sigma_{\bar{S}}^2$, is given by

$$\text{var}(\bar{S}) = \frac{1}{n^2} \text{var}(S) = \frac{1}{n^2} npq = \frac{pq}{n}$$

By the central limit theorem we therefore know, for n large, that

$$\bar{S}^* = \frac{\bar{S} - p}{\sqrt{pq/n}} \quad \text{is} \quad N(0, 1)$$

Since $\sqrt{pq/n}$ decreases in value as n is increased, we need to have smaller values of $\bar{S} - p$ for $\bar{S}^*$ to maintain some given value. Suppose that we desire an ($\alpha = .05$) level test of $H_T: p = .5$ against $H_A: p > .5$. We may calculate the smallest value of n that gives us the α level test with a given power, say .9, against some specific alternative, say $p = .8$.

We work with the binomial directly and then translate back into the random variable $\bar{S}$. In the binomial, with large n,

$$S^* = \frac{S - np}{\sqrt{npq}} \quad \text{is} \quad N(0, 1)$$

Putting $p = .5$ as specified in H_T, we have

$$P(S_c \le S \mid p = .5) = .05$$

and with $p = .8$ as in the specific alternative,

$$P(S_c \le S \mid p = .8) = .9$$

The critical value of number of successes S_c is the same in each of these instances.

Taking the first equation, we have, with the correction for continuity in the normal approximation,

$$1 - \alpha = 1 - P(S_c \le S \mid p = .5)$$
$$= P(S_c > S \mid p = .5)$$
$$= \Phi(1.645) = .95$$

implying

$$\frac{S_c - \frac{1}{2} - n(.5)}{\sqrt{n(.5)(.5)}} = 1.645$$

Solving this for S_c, we have

$$S_c - \frac{1}{2} = .5n + .8225\sqrt{n}$$

Similarly, for the equation

$$P(S_c \le S \mid p = .8) = .9$$

we have

$$P(S_c \le S \mid p = .8) = 1 - \Phi\left(\frac{S_c - \frac{1}{2} - n(.8)}{\sqrt{n(.8)(.2)}}\right) = .9$$

This is also given by

$$\Phi\left(\frac{S_c - \frac{1}{2} - n(.8)}{\sqrt{n(.8)(.2)}}\right) = \Phi(-1.282) = .10$$

implying

$$\frac{S_c - \frac{1}{2} - n(.8)}{\sqrt{n(.8)(.2)}} = -1.282$$

Solving for S_c, we have

$$S_c - \tfrac{1}{2} = .8n - .5128\sqrt{n}$$

Thus, we have two expressions each equal to $S_c - \frac{1}{2}$:

$$.5n + .8225\sqrt{n} = .8n - .5128\sqrt{n}$$

Dividing each side by $n/\sqrt{n}$ gives

$$.5\sqrt{n} + .8225 = .8\sqrt{n} - .5128$$

and from there

$$.8\sqrt{n} - .5\sqrt{n} = .8225 + .5128$$

$$\sqrt{n} = \frac{1.3353}{.3}$$

$$= 4.45$$

and

$$n = 19.8$$

Substituting this value in either of the equations for $S_c - \frac{1}{2}$ gives a critical value

$$S_c = 14.06$$

However, since we are dealing with a discrete random variable, we must select the smallest integer larger than these values. Thus, we shall set

$$n = 20$$
$$S_c = 15$$

Evaluating the adequacy of this solution, we must have

$$P(S_c \leq S \,|\, p = .5) = .05$$

and

$$P(S_c \leq S \,|\, p = .8) = .9$$

Plugging $S_c = 15$ into these formulas, where S is $B(20, p)$ we have

$$P(16 \leq S \,|\, p = .5) = .021$$
$$P(16 \leq S \,|\, p = .8) = .804$$

The criterion for α is more than satisfied, but the power of the test is not high enough. This discrepancy is due to the relative inadequacy in the normal approximation to the binomial for n in the range of 30 and below.

This example suggests that we can find a test satisfying any α and β requirements we please by taking n sufficiently large. This is quite so, but it implies at least that we can *afford* to have as large a sample size as we please. This is not necessarily the case. In our example we could easily lose our subject if we should require her to attend our laboratory every day for three months ($n = 90$), where she might cooperate for three weeks ($n = 21$). In general, financial and energy resources should be spent wisely in experimentation. A well-planned experiment is based on values of α and β that are chosen in relation to the cost of making type I and type II errors. The greater precision of a test with more than the necessary number of observations is a waste of resources. The following chapter discusses these problems in depth.

The calculation of the sample size requires the availability of a specific alternative to the null hypotheses. Frequently in psychological research it is difficult to establish an appropriate specific alternative. If there are two competing theories and if the models for the theories are comparable with respect to some parameter, the alternative is given by a "new" or "challenger" theory. In this case the procedures are quite clear. In most research we are not so fortunate. If there are competing theories, the models are not developed adequately to provide specifications of values of parameters. The most usual cases are models suggesting only some sorts of magnitude comparison or order of values in the comparison of the models. In such ill-defined situations the researcher is called on to imagine how large a difference he would have to observe before he would care.

In our tea-tasting example our subject may provide us with a specific alternative to H_T by making her claim specific. If she is willing to say, for example, that she can tell "on" from "off" tea 8 out of 10 times, then we can set $p = .8$ in our calculations of power and sample size. On the other hand, if she is not willing to be so specific *we* need to establish the alternative. This is a difficult task. If we set it too high we are really being unfair to the little old lady. If we set it too low we need a very large number of observations to achieve our desired power. We might agree that 70% accuracy (50% is chance) is pretty good but less than 70% is not really very impressive. In this case we would set $p = .7$ in the absence of any other criterion.

In some experiments we are able to make judgments as to how much of a difference between the parameter value in H_T and in the alternative would be important in a practical sense. For example, we might require that a new learning procedure be 25% more effective than the old procedure before we would be willing to change. This provides the necessary value for the specific alternatives.

SUGGESTED READING

Among the references already cited those by Hays, Hodges and Lehman, and Stilson are particularly recommended.

GLOSSARY

Statistical Hypothesis. A statement of the way a random variable is distributed is a statistical hypothesis. More specifically this takes the form of some numerical value or values for one or more of the parameters of the distribution of the random variable.

Specific Hypothesis. A specific statistical hypothesis is a hypothesis specifying a single parameter value or a single distribution, as in $H: p = .5$ or $H: X$ is $N(0, 1)$.

Composite Hypothesis. A composite hypothesis is a hypothesis specifying a collection of parameter values or a family of distributions as in $H: p > .5$ or $H: X$ is $N(\mu, 1)$, $\mu > 0$.

Test Hypothesis. The hypothesis H_T to be tested is called the null hypothesis. This is always a specific hypothesis.

Alternative Hypothesis. The alternate hypothesis H_A is the hypothesis presumed to be true if H_T is false. Alternative hypotheses may be specific or composite.

Test Statistic. A sample statistic providing information about the parameter specified in H_T is a test statistic.

Significance Probability. The probability, under H_T, that the obtained test statistic or one more strongly inclined to support the alternative is observed: Ψ.

$$(14.1) \qquad\qquad \Psi = P(SS)$$

Type I Error. Rejecting H_T when H_T is true is a type I error. Probability of type I error is α.

Level of Significance. The level of significance of a statistical test is the probability of a type I error and the maximum value of the significance probability leading to a rejection of H_T: $\alpha = \Psi_{max}$.

$$(14.2) \qquad\qquad \alpha = \Psi_{max}$$

Rejection (Decision) Rule. A rejection rule specifies values of test statistics which lead to the rejection (or not) of the null hypothesis. In general, if $\Psi \leq \alpha$, then H_T is rejected in favor of H_A.

Type II Error. Rejecting H_A when H_A is true, or failing to reject H_T when H_T is false, is a type II error. Probability of a type II error is denoted by β.

$$(14.3) \qquad\qquad \beta = P(\text{type II error})$$

Power. The power of the test against H_A is the probability of rejecting H_T in favor of H_A when H_A is true: $\pi = 1 - \beta$.

$$(14.4) \qquad\qquad P(\text{reject } H_T \mid H_A \text{ is true}) = 1 - \beta = \pi$$

Two-Sided Tests. When the composite alternative of a statistical test involves values on both extremes of H_T, the test is a two-sided or two-tailed test.

Critical Values. A critical value of a test statistic is the least extreme value leading to a rejection of H_T.

PROBLEMS

1. Devise a number of experiments and models for the tea-tasting problem. For example, imagine how the idea of paired-comparison procedures might be used; confidence judgments; using a mixture of "on" and "off" as a standard; several cups per trial, some prepared "on" and some prepared "off"; and so on. Find a random variable and its distribution for each experiment.

2. For an experiment with a binomial model and 30 trials find the one-sided (both of them) significance sets for results corresponding to $\Psi = .001, \Psi = .01, \Psi = .05$, for test hypotheses with $p = .5, p = .6$, and $p = .8$.

3. Calculate the significance probability function for $H_T: p = .6$ in $B(16, p)$ against $H_A: p \neq .6$. Do the same for $H_T: p = .8$ against $H_A: p \neq .8$. Compare these with Figure 14.6 by constructing similar graphs.

4. Extend the two-part rejection rule for $H_T: p = .5$ in $B(16, p)$ against $H_A: p > .5$ so that intermediate results lead to no decision regarding H_T and H_A; that is, make a three-part rule.

5. Calculate α and β for the decision rule of problem 4.

6. Calculate the power of the test of $H_T: p = .5$ in $B(16, p)$ against $H_A: p = .5$. Explain why $\alpha = \pi$ for this case. Will $\alpha = \pi$ in any other circumstances? Explain.

7. By dividing the binomial variable S by the number of trials, S/n, we obtain the proportion of "successes" observed in n trials. Hence the binomial distribution serves as a model for proportions. This information can be used to derive a test of hypotheses about single proportions. Carry out this derivation.

8. Using the normal approximation to the binomial, calculate the sample sizes necessary to achieve tests of the hypothesis $H_T: p = P_T$ in $B(n, p)$ against the alternative $H_A: p = P_A$ in $B(n, p)$ where $P_A = P_T + .1$, for $\alpha = .01$ and $\beta = .01$ for each of the following values of $P_T: .5, .6, .7$, and $.8$. Repeat the calculations for other values of α and β. Comment on the differences in sample sizes as a function of the values of P_T.

9. All of the 10 figures in this chapter are graphical representations of the essential elements of the theory of statistical tests. The figures were all constructed from the tables of the binomial probability function and the unit standard normal distribution function at the back of this book. As an exercise designed to demonstrate your mastery of the concepts in this chapter, you should reconstruct each of the figures.

10. A physiological psychologist devises a new method for performing a delicate brain operation on mice. The previous technique resulted in a 30 percent mortality rate. The new method is applied to 10 mice and only one of them dies. At the ($\alpha = .03$) level of significance, is the psychologist justified in claiming that his new technique is reliably superior?

11. Under the hypothesis that $S: B(n, p)$, what are the significance levels possible between .005 and .1 in an upper tail test (reject H_T on large values of S) under the following special cases:

(a) $n = 8, p = .2$ (b) $n = 10, p = .1$

(c) $n = 16, p = .4$ (d) $n = 8, p = .8$

(e) $n = 12, p = .9$

12. A psychotherapist with 100 patients is instructed to select a treatment group of 50 patients at random (equal probability) to receive a special form of suggestion therapy in addition to their regular therapy during their regular therapy sessions. The remainder are to serve as controls to gauge the effectiveness of the new treatment. When the results of the experiment are available, it is found that 30 of the patients are known to be especially subject to hypnosis. Of the 50 patients selected for the special treatment, 23 of the 30 hypnosis-sensitive subjects are included. At the significance level .05, would you reject the hypothesis that the psychotherapist selected the 50 patients at random, or favored the hypnosis-sensitive subjects in the treatment group?

of H_A against H_B, we have

$$\text{When condition } A \text{ is present} \quad \begin{cases} P(\text{outcome } a) = \alpha \\ P(\text{outcome } c) = 1 - \alpha \end{cases}$$

$$\text{When condition } B \text{ is present} \quad \begin{cases} P(\text{outcome } d) = \beta \\ P(\text{outcome } b) = 1 - \beta = \pi \end{cases}$$

The expected utility of the decision rule when condition A is present is

$$E(U \mid A) = \alpha U(H_A, R_A) + (1 - \alpha)U(H_A, R_B) \tag{15.1}$$

The expected utility of a decision rule when condition B is present is

$$E(U \mid B) = \pi U(H_B, R_A) + \beta U(H_B, R_B) \tag{15.2}$$

The overall expected utility of the experiment is dependent on the probability that the respective physical conditions are present for the experiment. Denoting these probabilities $P(A)$ and $P(B)$, respectively, leads to the expression

$$E(U) = P(A)E(U \mid A) + P(B)E(U \mid B) \tag{15.3}$$

Specification of the probabilities of the physical conditions may be difficult. However, previous experience with rates of occurrence of the conditions, knowledge of some diagnostic signs, and so on, may permit estimation of these probabilities. When no basis for judging the probabilities is available, one convention is to act as if there were no difference between the probabilities, that is, to set $P(A) = P(B) = .5$. Under this latter situation,

$$E(U) = \frac{E(U \mid A) + E(U \mid B)}{2}$$

The assignment of values to U is specific to the situation. However, it is easy to contrive an example. Suppose that $4000 is the gain for the decision R_A when H_B is true and that deciding R_B when H_A is false (that is, H_B true) leads to a loss of $30,000. Deciding R_A when H_A is true leads to a loss of $10,000, whereas deciding R_B when H_A is true produces a gain of $8000 as shown in Figure 15.3.

	H_A True	H_B True
R_A	− $10,000	$4000
R_B	$8000	− $30,000

FIGURE 15.3

Example of a utility structure for a decision rule.

Setting α and β permits an evaluation of $E(U)$. Take conventional values $\alpha = .01$ and $\beta = .1$ and assume that $P(A) = P(B) = .5$. Thus

$$E(U) = \frac{(1 - \alpha)U(H_A, R_B) + \beta U(H_B, R_B) + \alpha U(H_A, R_A) + \pi U(H_B, R_A)}{2}$$

$$= \frac{.99(8000) - .1(30,000) - .01(10,000) + .9(4000)}{2}$$

$$= \frac{7920 - 3000 - 100 + 3600}{2}$$

$$= \frac{8420}{2} = 4210$$

This gives a specific value to our decision rule. We can ask if this value is the best we can do. Suppose that we test H_B against the alternative H_A with the same α and β; what would the value be? For this the outcome probabilities are different:

$$P(\text{outcome } a) = \beta$$
$$P(\text{outcome } b) = 1 - \alpha$$
$$P(\text{outcome } c) = \pi$$
$$P(\text{outcome } d) = \alpha$$

The expectation in this instance is

$$E(U) = \frac{(1 - \beta)U(H_A, R_B) + \alpha U(H_B, R_B) + \beta U(H_A, R_A) + (1 - \alpha)U(H_B, R_A)}{2}$$

$$= \frac{.9(8000) - .01(30,000) - .1(10,000) + .99(4000)}{2}$$

$$= \frac{10,860}{2} = 5430$$

This payoff is considerably better than the first evaluation. Hence, we can choose the test hypotheses intelligently; $H_T = H_B$ is the appropriate test (null) hypothesis of the two, at least when $\alpha = .01$, $\beta = .10$. By selecting larger values of α we can decrease β. Imagine that for $\alpha = .02$ we have $\beta = .05$ in the same experiment. This gives $E(U) = 5,050$ and $E(U) = 5,210$ for $H_T = H_A$ and $H_T = H_B$, respectively. Using H_B as H_T still is the best strategy, but both expectations are higher then they were with the original α, β values.

The improvement in $E(U)$ by increasing (by a factor of 2) the value of α (less significance) has a deep meaning: A higher degree of statistical significance may be less desirable than a lower degree of statistical significance. In the present example the investigator would be silly to use the criterion of statistical significance to judge the meaningfulness of his decision rule. Other utility structures might imply different decision rules and α levels, but that is not the point. The point is that α is not by itself always a measure of the meaningfulness of a statistical test. We have called α the significance level of a statistical test, but it is only one aspect of significance when significance is defined in a broader and more meaningful sense.

Two elaborations of this example are of interest. First, we should note that α and β are not independent and that their dependence is a function of the sampling distribution involved. For a given sample size an increase in α implies a decrease in β and conversely. Hence, we can find the expected utility of a decision rule as a function of α alone if we restrict our attention to a specific sampling distribution and a specific sample size. Varying both α and sample size leads to the optimal sample size with respect to α or the optimal α with respect to sample size.

The second elaboration simply is to point out that there is a certain cost to making an observation; as sample size increases, the cost of the sample increases. The cost of a sample of size N is some function of N. In some experiments this may be $C(N) = c_0 + c_1 N$ where c_0 is the cost of setting up the experiment (overhead, stimulus materials, computer costs, and so on) and c_1 is the cost per observation. Other equations can be written to imply an increasing cost per observation as N becomes large (for example, necessity of traveling greater distances to obtain additional subjects) or to imply decreasing cost per observation as N becomes large (for example, practice makes experimental observations more efficient or less wasteful of physical materials).

Decision theory and utility theory are interesting as tools for scientific and applied research. However, we cannot delve more deeply into these topics here. Several excellent sources are indicated at the end of this chapter. Our example is very limited in scope and should not be mistaken as a representation of the theory of utility or decisions.

The significance of this discussion historically has been largely lost on psychologists. It is useful to point out the possibility of optimization of a utility expectation contrary to optimization with respect to α where a utility function is defined. In most psychological work no utility function is defined—so what have we gained? Two basic facts must be pressed home: (1) where two specific hypotheses are stated the selection of a test hypothesis can bear critically on the character of the decision rule, and (2) statistical significance is a suspect concept—if it is not always an optimal concept in a utility context, it may not always be optimal in a broader scientific context.

Weak Utility. There are a number of ways in which a weak utility structure may be defined. These structures do not permit an evaluation of decision rules via consideration of maximum gain, minimum loss, and so on. However, the strategy of selecting the test hypotheses from two specific hypotheses and establishing α and β for the statistical test can be clarified by using a weak utility structure.

Imagine a study of human conditioning in which a deep narcosis is required for theoretical reasons. A drug that has the proper narcotic effects is readily available. The greater the dosage, the greater is the degree of narcosis. However, in large doses it is suspected that the drug destroys liver cells. This destruction, of course, is deleterious to the health of the subject and is therefore to be avoided. In order to determine a safe dosage an experiment is performed on pigs. This animal is chosen because of the similarity of the liver in men and pigs. A dosage is selected and the pigs treated. The liver is retrieved from the sacrificed animals after the time necessary for the liver damage to occur. The method of determining tissue damage is to ascertain the concentration of a particular enzyme per unit of liver tissue. The determination is done by biochemical methods. The concentration is a variable, for example X. In healthy tissue X is known to be normally distributed with expectation μ_h. It is known that the pathological conditions do not change the distribution form or the variance. However, a pathological condition of the liver produces an expectation of X equal to μ_p.

This gives two specific hypotheses:

$$H_h: \text{dosage } D \text{ is safe:} \qquad \mu = \mu_h \text{ in } N(\mu, \sigma^2)$$
$$H_p: \text{dosage } D \text{ is pathogenic: } \mu = \mu_p \text{ in } N(\mu, \sigma^2)$$

The value of μ may be evaluated in a sample mean of observations of X. A decision rule is established that leads to an inference that D is safe or that it is not. The errors potential in the decision process are *vital* in this example. We can decide erroneously as follows.

(1) D is a safe dose—when really it is pathogenic.

(2) D is pathogenic—when really it is safe.

The first error would lead to administration of pathogenic dosages on the false presumption that they were safe. The second error would lead to the use of a smaller dosage in the experiment, a relatively innocuous outcome. Clearly the first decision error is to be avoided with greater stringency than need be observed for the second error.

The implication is clear: H_p is the appropriate test hypothesis. By taking

$$H_T = H_p: \text{dosage } D \text{ is pathogenic, } \mu = \mu_p$$

we can determine in advance the probability of the first decision error regardless of sample size or the probability of making the other decision error. In this instance $\alpha = .001$ would be none too stringent. However, β need not be so stringent. If $\beta = .001$ were used, sample size would be very much larger than really needed (see below). A moderate value of β, for example .05, is more fitting.

Since no dollar or "hard" cost can be attributed to the decision errors, it is not possible to use utility expectations in determining the test hypotheses or values for α and β. However, the "cost" of the one decision error is so overwhelming that the strategy is very clear.

The weakest utility structure we mention here (except *no* utility structure) is that of scientific interest and conservatism. Imagine that the scientist is interested in avoiding errors leading to (1) false departures from "orthodoxy," (2) false rejection of a simple theory for a more complex theory, (3) false rejection of a theory with more experimental support for a theory less well supported, (4) false rejection of one's own theory for that of a rival, and so on. Each of these "interests" leads to using the orthodox, simple, well-supported, and "pet" hypothesis as the test hypothesis. This would make it more difficult to support departures from orthodoxy, to accept another theorist's position, and so on, if α were relatively more stringent than β. The reasonable and scientifically "fair" decision rule would be one where

$$\alpha = \beta$$

The "interests" of the theorist are satisfied by setting α as low as he wishes. However, the contending or revolutionary theory is accorded the same treatment by assigning the same value to β.

Statistical practice in psychology has a curious status in the light of this discussion. An aversion to "accepting the test hypothesis" is fairly general. This aversion probably stems from the notion that one cannot say that data in accord with a theory need to have arisen from an empirical system operating as the theory claims. Another system (requiring a different theory) may produce identical results. However, observation of results at variance with the theory tested "implies" that the theory cannot (except for α probability) be descriptive of the system generating the results. Hence the theory is rejected. This practice would be fine but for subtle "cheating." The "cheating" is accomplished by making the rival, the unorthodox, the unlikely theory the test theory. Then when the test hypothesis is rejected, the unsavory conclusion is that the championed, the orthodox, the conservative theory must be true. The logical fallacy is obvious. However, if α is small the fallacy is somewhat mitigated, or so one might argue. This argument must be provided with a bulwark, however: β must be a reasonable level in the decision rule. This requires a specified sample size—and no larger sample size is permissible. An example may be helpful.

Imagine that H_T: $\mu = \mu_T = 50$ in $N(\mu, 9)$ and H_A: $\mu = \mu_A = 53$ in $N(\mu, 9)$, and an $\alpha = .01$ decision rule (test) is established. The test hypothesis was selected by the considerations outlined in the preceding paragraph. A "good large" sample is available, such as $N = 144$. The researcher makes his observations and tests the test hypothesis with a critical value of the test statistic $\bar{X}$:

$$Z = \frac{T_c - 50}{3/12} = 2.33$$

which implies that

$$T_c = 50 + .25(2.33)$$
$$= 50.583$$

If $\bar{X} \geq 50.583$ is observed in the sample of 144 observations, H_T is rejected in favor of H_A.

To show the error of this procedure we select a reasonable value for β and find the alternative H_{A*} that is characterized by β. Imagine that $\beta = \alpha$, that is, $\beta = .01$; then

$$\frac{T_c - \mu_A{}^*}{3/12} = -2.33$$

and then

$$\mu_A{}^* = .25(-2.33) + 50.583$$
$$= 51.166$$

If $\beta = .05$ had been selected, then $\mu_{A*} = 50.994$.

The upshot is that the alternative has subtly been changed from 53 to 51.166. This means that the effective alternative is closer to H_T than to H_A. If random effects or perturbations have caused an inflation in X for the test by an amount equal to 1.166, then we shall fail to reject H_T with probability $\beta = .01$. This is far from a fair test of H_T: $\mu = \mu_T = 50$ against H_A: $\mu = \mu_A = 53$. The test would be a fair confrontation (with $\alpha = \beta = .01$) of H_T: $\mu = \mu_T = 50$ and H_{A*}: $\mu = \mu_{A*} = 51.17$.

This "unfairness" is reflected in the β of the proposed alternative H_A. This is given by

$$\frac{T_c - \mu_A}{\sigma/\sqrt{N}} = \beta$$
$$\frac{50.583 - 53}{.25} = -9.7$$

and $\beta = \Phi(-9.7) < .0000 \cdots$. Thus the test has β many times smaller than α.

The test with $N = 144$ is sensitive to small departures from H_T and extremely insensitive to large departures from H_A. Thus a statistically significant result may be scientifically meaningless. The test is so heavily biased against H_T in comparison with H_A that a rejection of H_T tells us virtually nothing about H_A. If H_A had been the test hypothesis, the observed statistic might have been even more "significant." In our example, imagine that $\bar{X} = 50.65$ had been observed. The significance probability of this result under H_T: $\mu = \mu_T = 50$ is

$$\Psi = 1 - \Phi\left(\frac{50.65 - 50}{.25}\right) = 1 - \Phi(2.6) = .0047$$

very significant indeed. However, if we tested H_T: $\mu = \mu_A = 53$,

$$\Psi = \Phi\left(\frac{50.65 - 53}{.25}\right) = \Phi(-9.4) = .00000 \ldots$$

vastly more "significant." By designating the competing hypothesis, the undesired hypothesis, the unorthodox hypothesis as the test hypothesis and using a large sample size, the researcher is going a long way toward insuring the "desired result" in terms of statistical significance. The results, though, have little or no meaning in terms of scientific significance.

The remedy is to establish sample size so that the resulting test is sensitive to both hypotheses. Differential sensitivity is appropriate under certain circumstances, but it should be intended and controlled. That is, α and β must be stated and sample size determined on the basis of α and β.

Sample size determination is dependent on the test involved. In the chapters that follow, the determination of sample size is an integral part of the development of each statistical procedure.

Two Specific Hypotheses Without Utility

In scientific work the availability of a utility structure is a rare luxury. Often the closest we come to utility is the desire to be correct, to avoid error, and to be parsimonious. The cost of a decision error may be reflected only in wasted time and effort, perhaps in laboratories acquainted with the research only through published work. Hence, methods of assigning cost values to errors of decision are very unclear. Perhaps even less clear is the assignment of payoff value for correct decisions.

Without additional information it seems reasonable to treat both specific hypotheses the same way. The choice of H_T from H_A and H_B should be random.

In order to equate the two hypotheses with respect to risk of false rejection, the two error probabilities are set to the same value, that is,

$$\alpha = \beta$$

The sample size necessary to establish this α and β level test depends on the specifics of the test. The determination of sample size is discussed in this book in the context of specific test developments. In general, it is important to avoid an inappropriate sample size. Unless the sample size is appropriate, the treatment of the test and alternative hypotheses is not balanced and statistical significance diverges from scientific significance, the latter becoming more difficult to identify.

One Specific Hypothesis Without Utility

Frequently the psychologist is faced with a statistical problem in which only one hypothesis is meaningfully specific. This comes about when theoretical considerations indicate that a variable is changed by a special treatment of the experimental subjects, that a selected subpopulation differs with respect to a parameter value, and so on. The test hypothesis in these situations is that there is no difference or no departure in the experimental condition compared with some nonspecific difference or departure implied by theoretical considerations. The problem in this situation is the lack of specificity for the terms "difference" and "departure." At best the alternative to the specific test hypothesis is a one-sided composite—at worst it is a two-sided composite.

It is difficult to imagine a reasonable articulated utility structure being defined in this vague experimental situation. In this respect we are dealing with the last discussed problem except that no specific alternative is defined. Hence, defining a specific alternative H_A transforms this special case to the case of two specific hypotheses without utility.

The following paragraphs are intended to develop a rationale for formulating a specific alternative in the absence of a natural alternative.

The scientific significance of a statistical decision is a function of (1) the sensitivity of the statistical test to alternative states of nature and (2) the scientific import of the differences among the alternative states. The sensitivity problem was discussed thoroughly, above, in connection with two given specific alternative hypotheses, H_T and H_A. However, we have not come to grips with the meaningfulness of the difference of H_T and H_A. For example, if H_T: $\mu = \mu_T = 50$ and H_A: $\mu = \mu_A = 53$, we need to be able to evaluate the meaningfulness of

$$\mu_T - \mu_A = -3$$

By itself the value -3 is meaningless. There is no connection of this value with the practical or theoretical significance of the difference of μ_T and μ_A.

A compelling unit of comparison is the standard deviation of the variable measured. If $\mu_T - \mu_A = -3$ for $\sigma^2 = 81$, the difference is one-third of one standard deviation. This means that μ_T and μ_A are relatively close to each other in terms of the variable involved. However, if $\sigma^2 = 9$, the difference -3 corresponds to one standard deviation and the expectations are relatively more distant. Since we are comparing μ_T and μ_A in terms of σ, the difference is interpretable in terms of the units of measurement of the experimental variable. This can be expressed by a single value that is independent of the particular unit of measurement and the specific difference. The value δ is the index of the difference of μ_T and μ_A with respect to the unit of measurement. Using σ as the unit of measurement gives

$$\delta = \frac{|\mu_T - \mu_A|}{\sigma} \tag{15.4}$$

The absolute value of $\mu_T - \mu_A$ is used to avoid the unnecessary distinction of upper and lower tail tests. Also, if $\alpha = \beta$, μ_T and μ_A are interchangeable, implying that the absolute value is a more general expression of the meaningful difference.

The trio of numbers α, β, and δ specify the scientific meaningfulness of a statistical test. If any one of the values of the set is not scientifically sound, the entire test and its decision rule are scientifically suspect. If δ is not large enough to have practical or theoretical meaning, the sensitivity of the test with respect to α and β is of little interest. On the other hand, if β is so small with respect to α (via large sample size) as to make the test insensitive to H_A, a meaningful δ is inconsequential.

In the chapters to follow, the determination of the meaningfulness of statistical tests is in terms of what is referred to as the alpha, beta, delta, that is, (α, β, δ), criterion. A test is referred to as an (α, β, δ) test. Procedures to evaluate a test by calculating α, β, and δ are called the (α, β, δ) procedures.

These considerations can be used to specify a reasonable specific alternative hypothesis. Once a reasonable value of δ is selected, the alternative can be calculated. Using the definition of δ and simple algebra,

$$\mu_A = \mu_T + \delta\sigma \tag{15.5}$$

or

$$\mu_A = \mu_T - \delta\sigma \tag{15.6}$$

That is, we set μ_A to a value δ standard deviation units from μ_T. For the three alternative composite hypotheses the δ method of determining specific hypotheses gives

$$H_1: \mu = \mu_A > \mu_T \qquad H_A: \mu = \mu_A = \mu_T + \delta\sigma$$
$$H_2: \mu = \mu_B < \mu_T \qquad H_B: \mu = \mu_B = \mu_T - \delta\sigma \qquad (15.7)$$
$$H_3: \mu = \mu_C \neq \mu_T \qquad H_C: \mu = \mu_C = \mu_T \pm \delta\sigma$$

where "$\pm$" is to be read "plus or minus." This procedure provides a complete statistical test once α and β are set by the researcher. Unless some utility structure is defined, the fairest procedure is to set $\alpha = \beta$. Equal values of α and β are not necessary, and the experimental circumstances may indicate a larger value for β than for α. Regardless of what values are selected, the (α, β, δ) criterion provides a clear base for evaluation of the scientific meaning of the test.

An important application of the (α, β, δ) procedure is the evaluation of the meaningfulness of statistical tests when sample size has been arbitrarily selected. As an example, take the situation in which we measure attitudes by a variable "number of items out of 50 answered in the 'scored' direction." There are 81 students just completing a special section of introductory psychology available, and they are used as subjects. We have a set of responses from 1000 sophomore students randomly selected from the student body, excluding those taking introductory psychology. Our hypothesis is that the course in psychology has not changed the mean score in the variable. The large group mean is $\bar{X}_L = 27.43$ with $S_L{}^2 = 36.21$, $S_L = 6.017$. Taking 6.017 as the standard unit of measurement, we know that one standard unit is approximately 6 items in the observed behavior (responses). We select $\alpha = .01$, $\beta = .05$, and, because of the large sample size ($N = 81$), we can apply normal distribution theory. The class mean and standard deviation are $\bar{X} = 29.15$ and $S = 5.87$. Determine Ψ by

$$\Psi = 1 - \Phi\left(\frac{29.15 - 27.43}{5.87/\sqrt{81}}\right) = 1 - \Phi(2.62)$$
$$\Psi \leq .005$$

Hence, $H_T: \mu = 27.43$ is rejected in favor of the composite alternative $H_A: \mu > 27.43$.

The critical value of $\bar{X}$ in this test is given by the following argument. Since $1 - \Phi(2.33) = .01 = \alpha$,

$$\frac{\bar{X}_c - 27.43}{5.87/9} = 2.33$$

Solving for $\bar{X}_c$ satisfying this equation gives

$$\bar{X}_c = 27.43 + 2.33\left(\frac{5.87}{9}\right)$$
$$= 28.95$$

The test hypothesis H_T: $\mu = 27.43$ is rejected with ($\alpha = .01$) risk of type I error when $\bar{X} \geq 28.95$ is observed. In order to have $\beta = .05$ the specific alternative would have $\mu = \mu_{A^*}$ such that .05 of the sampling distribution of $\bar{X}$ about μ_{A^*} is below $\bar{X}_c = 28.95$. This implies

$$\frac{\bar{X}_c - \mu_{A^*}}{5.87/9} = -1.645$$

because $\Phi(-1.645) = .05 = \beta$ when $\mu_{A^*} = \mu$ in H_A is true. Solving for μ_{A^*} gives

$$\mu_{A^*} = 28.95 + 1.645\left(\frac{5.87}{9}\right)$$
$$= 30.02$$

Thus, with $\alpha = .01$ and $\beta = .05$ in the experiment ($N = 81$) with H_T: $\mu_T = 27.43$, the effective specific alternative is H_A: $\mu_A = 30.02$. Any value less than 30.02 for μ_A would give $\beta > .05$. Any value greater than 30.02 would give $\beta < .05$. Consequently a test with the precision indicated by $\alpha = .01$ and $\beta = .05$ is discriminating H_T: $\mu_T = 27.43$ against H_A: $\mu_A = 30.02$.

The value of δ for these two hypotheses is

$$\delta = \frac{27.43 - 30.02}{5.87}$$
$$= .44$$

As a consequence the test is effectively a test with

$$(\alpha, \beta, \delta) = (.01, .05, .44)$$

The value $\delta = .44$ means that the test can discriminate expectation differences equal to .44 standard deviations in the variable measured. In the present example the standard deviation $S = 5.87$ is in terms of test items; that is, the standard deviation is equivalent to 5.87 test items. Hence the test is capable of discriminating expectations $\delta S = (.44)(5.87) = 2.59$ test items apart.

The scientific significance of the statistical result in this example is dubious. The 2.59 items of discriminability of the test is very small compared with the larger number of items of variability, in the group of 1000 general students. Chance variability, as measured by $S_L = 6.017$ in the group, indicates that discrimination of a 6 item difference in the test would not be very impressive when comparing individuals from the two groups. Since a variation of 6 items within the "naive" group is not very extreme (one standard deviation), an experiment that is sensitive to much less variation seems a bit too sensitive. Imagine that we could decide that we wished to detect expectations differing by as much or more than 1.5 standard deviations determined by the "naive" group. This difference is a magnitude of 9.025. Taking the standard deviation of the class $S = 5.87$ as the unit of relative sensitivity for the experiment, we have

$$\frac{9.025}{5.87} = 1.54$$

The desired experiment would be an $(\alpha, \beta, \delta) = (.01, .05, 1.54)$ experiment. The remaining free parameters in the experiment are the form of the sampling distribution and the sample size. For a given type of sampling distribution there is a given sample size leading to the desired (α, β, δ) test.

Any method that permits a statement regarding the relative variability that we wish to be able to discriminate serves to specify δ. If variability associated with measurement error is .5 of the total variability of a measurement variable, the experimenter probably would want to discriminate differences at least as large as that equal to the error standard deviation. In other research, where very small differences in magnitude of the observed variable are scientifically meaningful, δ should be correspondingly small. Specification of a relative magnitude by any method, by using reliability information, validity data, practical implications, and so on, leads to a scientifically meaningful test under the (α, β, δ) procedures. The statistical significance of a scientifically meaningful test is also meaningful.

The value of δ may actually have to be quite large before practical significance is achieved, regardless of the α and β of the statistical test. For example, imagine a diagnostic test used in detecting the presence of a specific type of brain tumor in human patients. We wish to be able to say with a high degree of certitude that there is a tumor or there is no tumor on the basis of the diagnostic test score for any specific individual patient. Imagine that we know that normal subjects have scores that are normally distributed with expectation 50 and standard deviation 10. Also imagine that unless the diagnostic test has fewer than 5% false positives and 5% false negatives, it does not enhance the clinical procedures already in use to detect the presence of the tumor. This restriction implies that the cutoff score (above which the clinician infers "tumor" and below which the clinician infers "no tumor") has only 5% of the normal population with higher scores, and only 5% of the persons with tumors below it. This implies that the cutoff score is 66.45 (that is, $1 - \Phi(\mu_T + 1.645\sigma) = .05$). Likewise, if the standard deviation of the score is not changed by the presence of the tumor (an unlikely circumstance), then the cutoff score is equal to 66.45, satisfying the equality $\Phi(\mu_A - 1.645\sigma) = .05$. These equations require that $\mu_A = 82.90$. These two expectations give a δ of impressive magnitude:

$$\delta = \frac{82.90 - 50.00}{10} = 3.29$$

The method for determining the sample size for the $[(\alpha, \beta, \delta) = (.05, .05, 3.29)]$ test implied by this situation is discussed in detail in the next chapter. The value of N is impressive in its own way—$N = 1$. In order to detect this magnitude of difference with more conservative error terms, for example, $\alpha = \beta = .001$ to give an $[(\alpha, \beta, \delta) = (.001, .001, 3.29)]$ test, the sample size must be at most $N = 4$. If a group of four known tumor patients randomly selected from the population of such tumor patients do not have a mean above 66.45, the diagnostic test must be rejected as insufficiently discriminatory. A larger sample with a mean value in excess of 66.45 may have sprung from a population with expectation less than the expectation required (82.90), at least within our α and β conditions.

Post-Hoc Evaluation

After the fact of an experiment and the statistical evaluation of the results of that experiment, it is too late to be rational about the design of the experiment. The least we can do, however, is to be rational about the interpretation of the results. Techniques presented above and in the chapters to follow permit the scientist reading other workers' papers to evaluate quantitatively the scientific significance of the reported research. It may not be necessary, however, to indulge in quantitative analysis of the results. Two classifications of a statistical decision made on the basis of inappropriate sample size can be established: generally safe, generally dangerous.

Even though an experiment and the consequent statistical decision are based on an inappropriate sample size, the decision may be quite supportable. If a very large sample size is used, one that would discriminate an alternative much closer to the test hypothesis than a scientifically meaningful alternative, and if the test hypothesis is not rejected, it seems clear that the conclusion is sound. If, on the other hand, a small sample (relative to the appropriate sample) is used and the test hypothesis is rejected in favor of the alternative, the conclusion also appears to be sound. In the former instance we reason that no discrimination was made even though there was superabundant power to make the discrimination. In the latter instance we reason that a discrimination was made even though there was not really enough power to expect that discrimination. Failure to catch a fish with a very fine net usually means that the fish were very small or nonexistent—catching a fish with a very loose mesh usually means that the fish were very big.

The converse results of the decision rule are worthy of doubt and additional investigation. If an experiment is executed with too many observations (relative to a balance between the competing hypotheses and the two types of error) and the test hypothesis is rejected, the reason for the rejection may be a trivial perturbation in the experiment (a small bias in the instruments used, for example). The reason for the rejection may not be clearly situated in a scientifically significant difference— the statistical test is too powerful for the job. On the other hand, when too few observations are used and the test hypothesis is not rejected, there is doubt as to whether there is just no difference or whether there is insufficient power in the test to detect the difference. Failure to reject the test hypothesis when the sample size is insufficient, and rejection of the test hypothesis when the sample size is too large, should arouse suspicion and incredulity regarding the scientific meaning of the statistical results.

SUGGESTED READING

Bakan, D. *"The test of significance in psychological research,"* Psychol. Bull., *66*, 423–437.

Chernoff, H., and Moses, L. E. (1959), *Elementary decision theory*, Wiley, New York.

Churchman, C. W. (1961), *Prediction and optimal decision*, Prentice-Hall, Englewood Cliffs, N.J.

Cohen, J. (1969), *Statistical power analysis for the behavioral sciences*, Academic Press, New York.

Fishburn, P. C. (1964), *Decision and value theory*, Wiley, New York.

Jeffrey, R. C. (1965), *The logic of decision*, McGraw-Hill, New York.

Mace, A. E. (1964), *Sample size determination*, Holt, Rineholt and Winston, New York.

Raiffa, H., and Schlaifer, R. (1961), *Applied statistical decision theory*, Harvard Business School, Boston.

Savage, L. J. (1954), *The foundations of statistics*, Wiley, New York.

Thrall, C. M., Coombs, C. H., and Davis, R. L. (eds.) (1954), *Decision processes*, Wiley, New York.

von Neuman, J., and Morgenstern, J. (1947), *Theory of games and economic behavior*, Princeton University Press, Princeton.

Wald, A. (1950), *Statistical decision functions*, Wiley, New York.

GLOSSARY

Utility Structure. The utility structure of a decision process associates values, usually economic, with all possible outcomes of the decision process. Here $U(H_A, R_B)$ is the payoff for rejecting H_B when H_A is true; $U(H_B, R_A)$ is the payoff for rejecting H_A when H_B is true; $U(H_A, R_A)$ is the cost of rejecting H_A when H_A is true; $U(H_B, R_B)$ is the cost of rejecting H_B when H_B is true.

(15.1)
$$E(U \mid A) = \alpha U(H_A, R_A) + (1 - \alpha)U(H_A, R_B)$$

(15.2)
$$E(U \mid B) = \pi U(H_B, R_A) + \beta U(H_B, R_B)$$

(15.3)
$$E(U) = P(A)E(U \mid A) + P(B)E(U \mid B)$$

Weak Utility. Attribution of nonspecific value to the possible outcomes of a decision process; usually in terms of a desire to optimize safety, orthodoxy, parsimony, and so on.

δ. The index δ is the index of the difference between the test and the alternative hypothesis.

(15.4)
$$\delta = \frac{|\mu_T - \mu_A|}{\sigma}$$

Effective Alternative. Under the circumstances in which no natural alternative is formulated but in which a δ value is definable, or after the fact of an experiment, the effective alternative is defined by equation (15.5), or (15.6).

(15.5)
$$\mu_A = \mu_T + \delta\sigma$$

(15.6)
$$\mu_A = \mu_T - \delta\sigma$$

(15.7)
$$H_1: \mu = \mu_A > \mu_T \qquad H_A: \mu = \mu_A = \mu_T + \delta\sigma$$
$$H_2: \mu = \mu_B < \mu_T \qquad H_B: \mu = \mu_B = \mu_T - \delta\sigma$$
$$H_3: \mu = \mu_C \neq \mu_T \qquad H_C: \mu = \mu_C = \mu_T \pm \delta\sigma$$

PROBLEMS

1. A patient is admitted to the university hospital with symptoms indicating that he may be suffering from a brain tumor. Neurological examination is inconclusive, so the neurologist refers the patient to the psychologist, who is to administer a performance test. The neurologist is ready to accept the psychological results to determine the course of action: brain surgery or not. If surgery is performed and needed, the patient will probably improve. If surgery is not performed even though it is needed, the patient will probably deteriorate. If surgery is performed and it is not needed, not only will the patient suffer from his current difficulties but the surgery will probably contribute to his deterioration. If surgery is not performed and not needed, the patient's disease will take its own course. If you were the psychologist charged with the responsibility of saying whether or not a tumor was indicated, what would you want to insist on with regard to the statistical properties of the test you used? What would you require in the way of preliminary validation statistics on the test? How would you perform an experiment to validate such a test, and how would you define the statistical procedures—that is, what would your sample be like, what would you use for H_T and H_A, what values would you establish for α and β, and so on?

2. For the utility structure represented in Figure 15.3, find approximately optimal values of α and β with either of the conditions A or B playing the role of H_T; maximize the expected utility. In order to do this without more mathematics than is assumed in this text you will need to calculate several expected utilities, each for a different pair of values for α and β under each of the two possible test hypotheses (A or B). Use the table of the distribution function of the unit standard normal random variable to obtain realistic values of α and β: If α is decreased it will increase β by making the critical value of the test closer to the alternative hypothesis by a precise amount, and hence a new value of β will be determined. Once you begin with an arbitrary pair of α and β values, a change in one specifies a change in the other. In order to simplify this problem, assume that initially $\alpha = \beta$.

CHAPTER 16

The One-Sample Problem

General principles of statistical inference were developed in Chapters 13 and 14 in terms of the binomial random variable. The special case of the binomial is of great importance, and we shall encounter it again. However, there is a very broad class of experiments that lead naturally to statistical models involving continuous random variables. Under certain conditions we can justify the assumption that the phenomenon of interest is normally distributed or that our sample size on the random variable is large enough to lead to a normal distribution of a statistic calculated from the sample. For these special conditions we can find statistical models employing the standard normal random variable with distribution $N(0, 1)$. These statistical models are traditional in psychology and are employed at times without sufficient justification. Often normality cannot be justified in an experimental situation or we are led to a statistical model explicitly stating that the random variable of interest is not normally distributed. In both instances we may be able to use the experimental model or some variation on it to obtain a test. If the distribution of the random variable is known, the appropriate test can be set up, just as we did with the binomial. If tables are not available, the problem becomes a calculational one—to obtain the requisite probabilities, the α, β, Ψ, and sample size for the test and the test statistic to be observed. However, the explicit distribution of the random variable of a statistical model may not be clear and hence the appropriate statistical test not clear. Almost always in these situations, some simple modification of the model leads to a new random variable with a known distribution, and consequently a test can be constructed. This last class of special cases has become known as "nonparametric," a misnomer that we discuss later.

Although these introductory remarks clearly apply to hypotheses regarding expectations, they also apply in a general sense to hypotheses regarding variances.

This chapter is restricted to a discussion of statistical models for single-sample experiments in which the main item of interest is a single characteristic of the random variable sampled, that is, the expectation or variance. The term "single-sample" implies that only one set of N observations is made. This is in contrast with the two-sample and many-sample experiments to be discussed in the next three chapters. In a multiple-sample experiment more than one sample of N observations is made and the basic item of interest is a comparison of the differences among the samples or a comparison of sample statistics with parameters prescribed by a theoretical model.

The plan of this chapter is to look first at models of experiments dealing with hypotheses about expectations. After that survey, we turn to models of experiments dealing with hypotheses about variances.

Hypotheses Regarding Expectations

We can discern eight separate cases that correspond to combinations of three binary considerations: (1) normal or nonnormal distribution of the random variable; (2) variance of the random variable known or unknown independently of

TABLE 16.1 Classification of Single Sample Problems Concerning Hypotheses About Expectations

	Distribution of the Random Variable	
	Normal	Nonnormal
σ^2 Known	Case A Large sample Small sample	Case B Large sample Small sample
σ^2 Unknown	Case C Large sample Small sample	Case D Large sample Small sample

the sample used to test the hypothesis about the expectation; (3) the size of the sample of observations, large or small.

It is convenient to deal with sample size as a "subcategory consideration" within the four special cases outlined in Table 16.1.

The plan of the next section is to discuss in detail first all the basic definitions and procedures for Case A. This discussion is parallel to the discussion in Chapter 14. Each of the other cases is taken up in turn. The special considerations needed for each of the cases are discussed in detail. However, common considerations are discussed in detail only once.

Case A: X: $N(\mu_X, \sigma_X^2)$, σ_X^2 Known

Statistical Hypothesis. The development of a statistical hypothesis in this context follows the same principles outlined in the tea-tasting experiment. Take a dart-throwing experiment as an example. Imagine presenting subjects with a target at which they are required to throw a dart. The radial distance of the point struck by the dart from the center of the target is the "error" X. Also imagine the idealistic condition where a psychologist has learned enough about sensory and motor mechanisms to construct a model predicting the expected error μ_X as a function of the quality of illumination, the intensity of the illumination, the positioning of the target, and so on. The distribution of X either is assumed or is shown to be normal. We also need imagine enough knowledge about this phenomenon to assume that σ_X^2 is a known constant. Suppose also that modifying the experiment in certain ways, for example, introducing perceptual cues distorting the visual field, change μ_X but not σ_X^2. Thus the model states that for a given set of stimulus and response conditions, X is normally distributed $N(\mu_X, \sigma_X^2)$, where σ_X^2 is constant for all experiments and μ_X is a function of the experimental conditions.

We entertain the test hypothesis that the expectation of X is given by the specific value μ_T under the condition that our model is correct: $H_T: \mu_X = \mu_T$.

In this example, assume that we are dealing with a representative psychological research problem where the alternative is composite. In addition, assume that the value structure of the problem is undefined but that there is no crucial reason to avoid either type I or type II error more than the other. Hence, we set $\alpha = \beta$ and establish the alternative by the (α, β, δ) procedure. For composite alternatives we have the associated specific alternatives

$$H_1: \mu_X = \mu_A > \mu_T \qquad H_A: \mu_X = \mu_A = \mu_T + \delta\sigma_X$$
$$H_2: \mu_X = \mu_B < \mu_T \qquad H_B: \mu_X = \mu_B = \mu_T - \delta\sigma_X$$
$$H_3: \mu_X = \mu_C \neq \mu_T \qquad H_C: \mu_X = \mu_C = \mu_T \pm \delta\sigma_X$$

Test Statistic. It was shown in the chapter on normally distributed random variables that $\bar{X}$, the mean of a sample of observations on X, is an unbiased estimator of μ_X. In addition, we have seen that for a sample of N observations on X: $N(\mu_X, \sigma_X^2)$,

$$Z = \frac{\bar{X} - \mu_X}{\sigma_X / \sqrt{N}} \tag{16.1}$$

is distributed as $N(0, 1)$. Thus, in a sample of N observations, $X_1, X_2, \ldots, X_N$, an appropriate test statistic is Z. This test traditionally is called the critical ratio or normal test. More properly, the normal test is a test of H_T: $\mu = \mu_T$ in $N(\mu, \sigma^2)$ using the critical ratio as the test statistic.

With $\bar{X}$ calculated from the sample, the Z statistic distribution is completely known because we know σ_X^2, N is given in the sample, and we are dealing with a normal random variable X.

Sample Size Determination. The definitions of α, β, H_T, H_A, and σ_X^2 must be specific in order to determine the correct sample size. In our example we use $\alpha = \beta = .05$, $\sigma_X^2 = 9$ and establish the alternative by setting $\delta = 1$, to give $(\alpha, \beta, \delta) = (.05, .05, 1.0)$.

If H_T: X is $N(50, 9)$, then

$$\text{for } H_1\colon \mu_X > \mu_T \qquad H_A\colon \mu_X = \mu_T + \delta\sigma_X = 53$$
$$\text{for } H_2\colon \mu_X < \mu_T \qquad H_B\colon \mu_X = \mu_T - \delta\sigma_X = 47$$
$$\text{for } H_3\colon \mu_X \neq \mu_T \qquad H_C\colon \mu_X = \mu_T + \delta\sigma_X = 53$$
$$\text{or} \quad H_C\colon \mu_X = \mu_T - \delta\sigma_X = 47$$

Since we are dealing with a normally distributed random variable, and hence a symmetrically distributed random variable, the N's for H_1 and H_2 are the same. Also, the same considerations hold for H_3 if we solve for N under the assumption that we are dealing with one-sided tests with $\alpha/2$ and $\beta/2$.

For the one-sided test under H_A we want to solve the two equations

$$P(T_c < T \mid \mu_T = 50 \text{ is true}) = .05$$
$$P(T_c < T \mid \mu_A = 53 \text{ is true}) = .95$$

These equations imply, respectively,

$$\frac{T_c - 50}{3/\sqrt{N}} = 1.645$$
$$\frac{T_c - 53}{3/\sqrt{N}} = -1.645$$

Solving for T_c in the two equations respectively and then for N, we find that the exact value for N is 10.824. Substituting the next larger integer and solving for the value of T_c and the significance probability of T_c, we have

$$\frac{T_c - 50}{3/\sqrt{11}} = 1.645$$

$$T_c = 51.488$$

which gives

$$\frac{51.488 - 50}{3/\sqrt{11}} = 1.645$$

$$\frac{51.488 - 53}{3/\sqrt{11}} = 1.671$$

Thus, $N = 11$ achieves an $(\alpha = .05)$ level test, but $\beta = \Phi(-1.671) = .0475$. This is produced by having $N = 11$ instead of 10.824 as prescribed. A compromise value for T_c is given by solving T_c twice, once for α and once for β, and averaging the two values of T_c. For $\alpha = .05$, T_c is

$$T_c = \frac{\sigma}{\sqrt{N}} Z_\alpha + \mu_T$$

$$= \frac{3}{\sqrt{11}} 1.645 + 50$$

$$= 51.488$$

For $\beta = .05$, T_c is

$$T_c = \frac{\sigma}{\sqrt{N}} Z_\beta + \mu_A$$

$$= \frac{3}{\sqrt{11}} (-1.645) + 53$$

$$= 51.512$$

The average is $(51.512 + 51.488)/2 = 51.500$. This T_c gives $\alpha = .049$ and $\beta = .049$, as close as possible to $\alpha = .05$, $\beta = .05$ with integer values of N. The symmetry of H_B and H_A implies the same value of N in the test with H_2 as the composite hypothesis.

For a two-sided test we solve the same equations, but with half of α and half of β associated with each of the two tails. This is equivalent to solving the equations for N with $\alpha = \beta = .025$. As above, we need solve for N in only one set of equations because the two tails are symmetrical.

The equations may be stated in general. For specific alternative hypotheses of the one-tailed variety where

$$H_T: \mu = \mu_T$$
$$H_A: \mu = \mu_A$$

For upper tail alternatives we set

$$P(Z \geq Z_\alpha \mid H_T) = \alpha$$
$$P(Z < Z_\beta \mid H_A) = \beta$$

The values of Z_α and Z_β are values from the standard normal random variables as above, for example, for $\alpha = .05$, $Z_\alpha = 1.645$, for $\beta = .05$, $Z_\beta = -1.645$. For lower tail alternatives we set

$$P(Z \leq Z_\alpha \mid H_T) = \alpha$$
$$P(Z > Z_\beta \mid H_A) = \beta$$

The values of Z_α and Z_β are substituted into the following equations, and the equations are solved for N:

$$\frac{T_c - \mu_T}{\sigma/\sqrt{N}} = Z_\alpha$$

$$\frac{T_c - \mu_A}{\sigma/\sqrt{N}} = Z_\beta$$

Again, Z_α and Z_β are values from the standard normal random variable, for example, a lower tail test, $\alpha = .01$ and $\beta = .1$, $Z_\alpha = -2.326$ and $Z_\beta = 1.29$.

A general equation for N can be derived from these equations:

$$T_c = \frac{\sigma}{\sqrt{N}} Z_\alpha + \mu_T$$

$$T_c = \frac{\sigma}{\sqrt{N}} Z_\beta + \mu_A$$

implying that

$$\frac{\sigma}{\sqrt{N}} Z_\alpha + \mu_T = \frac{\sigma}{\sqrt{N}} Z_\beta + \mu_A$$

$$\frac{\sigma}{\sqrt{N}} (Z_\alpha - Z_\beta) = \mu_A - \mu_T$$

$$\sigma(Z_\alpha - Z_\beta) = (\mu_A - \mu_T)\sqrt{N}$$

$$\left[\frac{\sigma(Z_\alpha - Z_\beta)}{\mu_A - \mu_T} \right]^2 = N \tag{16.2}$$

Letting $\beta = .05$, $\alpha = .05$ for $\mu_T = 50$, $\mu_A = 53$ with $\sigma^2 = 9$ as above, we have

$$Z_\alpha = 1.645$$
$$Z_\beta = -1.645$$
$$\left[\frac{3(3.29)}{3}\right]^2 = (3.29)^2 = 10.82$$

The Rejection Rule.   The rejection rule established by this procedure is simple. Having calculated the value of N, T_c can be calculated by solving

$$T_c = \frac{\sigma}{\sqrt{N}} Z_\alpha + \mu_T$$

and

$$T_c = \frac{\sigma}{\sqrt{N}} Z_\beta + \mu_A$$

and averaging the two values of T_c. Letting $\mu_T = 50$, $\sigma^2 = 9$, $\alpha = .01$, and $\beta = .05$ with $\mu_A = 53$, we have

$$N = \left[\frac{3(2.33 + 1.645)}{3}\right]^2 = 15.8 \simeq 16$$

$$T_c = \frac{3}{4.0}(2.33) + 50 = 51.748$$

$$T_c = \frac{3}{4.0}(-1.645) + 53 = 51.766$$

$$T_c = \frac{51.748 + 51.766}{2} = 51.757$$

giving the rejection rule:

Reject H_T *in favor of* H_A *if* $\overline{X} \geq 51.757$

Do not reject H_T *if* $\overline{X} < 51.757$

This rejection rule has, under H_T, an error probability of $\alpha = .0095$ and, under H_A, an error probability of $\beta = .049$.

The lower tail value of T_c under the alternative hypothesis $H_A: \mu_A = 47$ would be $T_c = 48.243$. The rejection rule would be stated similarly except that 51.757 would be changed to 48.243.

In the two-tailed test, where we hedge our bets that the alternative is either $\mu_A = 53$ or $\mu_B = 47$, and where we locate half of α and β (that is, $\alpha/2$ and $\beta/2$) in each tail, respectively, we can find either the upper or the lower critical value and determine from it the other critical value. Taking the upper critical value, we calculate

$$N = \left[\frac{\sigma(Z_{\alpha/2} - Z_{\beta/2})}{\mu_A - \mu_T} \right]^2$$

For $\alpha = \beta = .05$ this gives

$$N = \left[\frac{3(1.96 + 1.96)}{53 - 50} \right]^2 = 15.4$$

Let $N = 16$ and solve for the upper value of T_c:

$$T_c = \frac{\sigma}{\sqrt{N}} Z_{\alpha/2} + \mu_T$$

$$T_c = \frac{\sigma}{\sqrt{N}} Z_{\beta/2} + \mu_A$$

for $\alpha = \beta$ the average $T_c = 51.5$. The lower value consequently is 48.5 and the rejection rule is as follows.

Reject H_T *in favor of* H_A *if* $\overline{X} \geq 51.5$

or reject H_T *in favor of* H_B *if* $\overline{X} \leq 48.5$

Do not reject H_T *if* $48.5 < \overline{X} < 51.5$

The reader will have noticed that the value 51.5 has shown up as T_c in both the one- and the two-tailed tests and that 51.5 is exactly halfway between μ_T and μ_A (likewise for 48.5 and the lower tail). This is no accident. This result is a function of using $\alpha = \beta$ and of the special assumption that σ^2 does not vary with μ (that is, σ^2 is the same value under H_T and H_A), and the special convention that a two-tailed test is defined so as to split β as well as α between the two tails. If we wish to deal with any other set of conventions, we may do so. The rejection rule varies accordingly.

As β is relaxed to larger and larger values, N is decreased and T_c becomes more extreme. The effect of fixing $\beta = \alpha$ is to fix T_c halfway between μ_T and μ_A. The magnitude of $\beta = \alpha$ determines the sample size. For small $\alpha = \beta$, N is large, and conversely as either α or β are larger, N is smaller.

If we wish a two-tailed test to have a power $(1 - \beta)$ against *each* alternative (contrast with $1 - \beta/2$ in the two-tailed problem above), then the solution for N is

$$N = \left[\frac{\sigma(Z_{\alpha/2} - Z_\beta)}{\mu_A - \mu_T} \right]^2 \tag{16.3}$$

and the upper critical value is given by the average of

$$T_c = \frac{\sigma}{\sqrt{N}} Z_{\alpha/2} + \mu_T$$

$$T_c = \frac{\sigma}{\sqrt{N}} Z_\beta + \mu_A$$

and the lower critical value by the average of

$$T_c = \frac{\sigma}{\sqrt{N}} Z_{\alpha/2} + \mu_T$$

$$T_c = \frac{\sigma}{\sqrt{N}} Z_\beta + \mu_B$$

The assumption that σ^2 is the same in $N(\mu_T, \sigma^2)$ and $N(\mu_A, \sigma^2)$ need not be made. Imagine the following: If H_T is true, $\sigma^2 = \sigma_T{}^2 = 9$. If H_A is true, then $\sigma^2 = \sigma_A{}^2 = 16$. Now we solve N through T_c under H_T and H_A. Let $\alpha = \beta = .05$, $\mu_T = 50$, and $\mu_A = 53$. Then

$$T_c = \frac{\sigma_T}{\sqrt{N}} Z_\alpha + \mu_T = \frac{3}{\sqrt{N}} 1.645 + 50$$

$$T_c = \frac{\sigma_A}{\sqrt{N}} Z_\beta + \mu_A = \frac{4}{\sqrt{N}} (-1.645) + 53$$

and

$$\frac{3}{\sqrt{N}} 1.645 + 50 = \frac{4}{\sqrt{N}} (-1.645) + 53$$

so that $N = 14.73$. Letting $N = 15$ gives

$$T_c = \frac{3}{3.87} (1.645) + 50 = 51.28$$

$$T_c = \frac{4}{3.87} (-1.645) + 53 = 51.30$$

Hence a rejection rule with $T_c = 51.29$ in our example gives a test with $\alpha = \beta \cong .05$ when $\sigma^2 = 9$ under H_T and $\sigma^2 = 16$ under H_A. The general equation for sample size under these circumstances is

$$N = \left[\frac{\sigma_T Z_\alpha - \sigma_A Z_\beta}{\mu_A - \mu_T} \right]^2 \tag{16.4}$$

Significance Probability. When we have defined the sample size, the observations are made, the test statistic is calculated, and a decision is made. For our present problem $\bar{X}$ is compared with T_c in the decision rule. The decision rule leads to one of two conclusions: H_T is rejected or it is not. If this information alone were known, we would have minimum information indeed. Minimum or not, this is the standard information published—a poor practice. A more useful statement is the probability of observing the result, or one more extreme, if H_T were true. This was developed in the preceding chapter as the probability of the significance set—the significance probability.

In the context of the present case (Case A) the test statistic is

$$Z = \frac{\overline{X} - \mu_T}{\sigma/\sqrt{N}}$$

Since we know the distribution of Z under H_T, we can calculate the probability of observing extreme sets in the sample space of Z. This sample space is coincident with the real number set, and the significance sets within the sample space are given by

significance set under H_1: $SS_z = \{R \mid R \geq z\}$

significance set under H_2: $SS_z = \{R \mid R \leq z\}$

significance set under H_3: $SS_z = \{R \mid |R| \geq |z|\}$

where R is the set of real numbers.

Following the earlier notation,

significance probability $= P[SS_z \mid X: N(\mu_T, \sigma_X{}^2)]$

This equation is solved differently under H_1, H_2, and H_3. The tabled values of the $N(0, 1)$ distribution are used to find Ψ:
under H_1,

$$\Psi_1 = 1 - \Phi[z \mid X: N(\mu_T, \sigma_X{}^2)]$$

under H_2,

$$\Psi_2 = \Phi[z \mid X: N(\mu_T, \sigma_X{}^2)]$$

under H_3,

$$\Psi_3 = 1 - \Phi[|z| \mid X: N(\mu_T, \sigma_X{}^2)] + \Phi[-|z| \mid X: N(\mu_T, \sigma_X{}^2)]$$

Graphically, we have the significance sets as segments of the real lines and significance probabilities as probabilities associated with these sets ("areas under

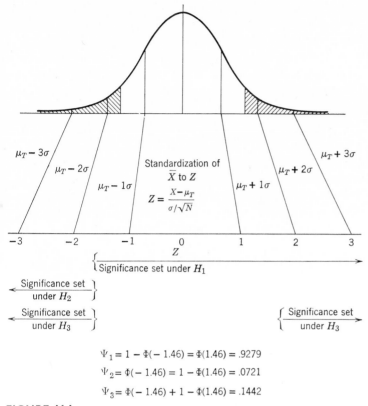

$$\Psi_1 = 1 - \Phi(-1.46) = \Phi(1.46) = .9279$$

$$\Psi_2 = \Phi(-1.46) = 1 - \Phi(1.46) = .0721$$

$$\Psi_3 = \Phi(-1.46) + 1 - \Phi(1.46) = .1442$$

FIGURE 16.1

Significance sets and significance probabilities for $Z = -1.46$.

the normal curve"). Figure 16.1 shows the significance sets for a result giving $Z = -1.46$.

      An Example. Imagine that in an attempt to classify a new psycho-therapeutic drug we are in a position to say that if it is in class A the behavioral consequence of a given dose is to produce a performance task score with a normal distribution, $N(\mu_A, \sigma_A^2)$ where $\mu_A = 100$, $\sigma_A^2 = 81$. However, if the drug is in class B, then the performance variable is $N(\mu_B, \sigma_B^2)$ where $\mu_B = 105$, $\sigma_B^2 = 49$. We wish to be able to state that the drug is an A-type drug or that it is a B-type drug. On the basis of the supposition that an error of misclassification of B as A is no more serious than classifying A as B erroneously, we elect arbitrarily to use A as the test hypothesis and B as the specific alternative. We are faced with the task of distinguishing, on the basis of a finite sample, two distributions (populations),

$N(100, 81)$ and $N(105, 49)$. We may do this either on the basis of μ or σ^2 or both. We develop here the test where

$$H_T: \mu = \mu_A = 100 \text{ in } N(\mu, 81)$$
$$H_1: \mu = \mu_B = 105 \text{ in } N(\mu, 49)$$

The relevant statistic is $\overline{X}$, and we may use

$$Z = \frac{\overline{X} - \mu_A}{\sigma_A/\sqrt{N}}$$

as the test statistic.

We wish to have a test $\alpha = \beta = .01$. Hence, we have

$$N = \left[\frac{\sigma_A Z_\alpha - \sigma_B Z_\beta}{\mu_B - \mu_A} \right]^2$$
$$= \left[\frac{9(2.326) - 7(-2.326)}{105 - 100} \right]^2 = 55.35$$

We take $N = 56$ and find

$$T_c = \frac{9}{7.483} 2.326 + 100 = 102.798$$

and

$$T_c = \frac{7}{7.483} (-2.326) + 105 = 102.824$$

for a mean value for $T_c = 102.811$.

Observing 56 subjects on the performance test after the drug has been administered results in data like that of Table 16.2.

The mean of these observations, $\overline{X} = 102.93$, leads us to reject the hypothesis that our sample came from a population in which $\mu = 100$ in favor of the alternative conclusion that we were sampling from a population in which $\mu = 105$. The significance probability or the probability that our decision is in error is

$$\Psi = 1 - \Phi\left(\frac{102.93 - 100}{9/\sqrt{56}}\right) = 1 - \Phi(2.44) = .0073$$

This specific result provides a cue for another important consideration: The significance probability of the result under the alternative hypothesis may not be very large even though the result under the test hypothesis has a smaller significance probability. The significance probability of the result observed in the data reported above, when H_1 is used as the test hypothesis, is

$$\Psi = \Phi\left(\frac{102.93 - 105}{7/\sqrt{56}}\right) = \Phi(-2.21) = 1 - \Phi(2.21) = .0136$$

TABLE 16.2 Illustrative Data and Basic Calculations

Subject	Score: X	Subject	Score: X
1	109	29	95
2	107	30	110
3	95	31	94
4	105	32	108
5	107	33	112
6	98	34	106
7	112	35	105
8	89	36	95
9	97	37	118
10	99	38	100
11	106	39	94
12	97	40	90
13	108	41	104
14	100	42	103
15	95	43	96
16	116	44	110
17	108	45	105
18	93	46	105
19	105	47	105
20	100	48	103
21	98	49	112
22	99	50	107
23	106	51	108
24	101	52	103
25	84	53	99
26	116	54	104
27	103	55	106
28	113	56	101

$$\sum_{i=1}^{56} X_i = 5764 \qquad \sum_{i=1}^{56} X_i^2 = 596028$$

$$\bar{X} = \frac{5764}{56} = 102.93$$

$$S^2 = \frac{56(596028) - (5764)^2}{(56)(56)} = \frac{97872}{3136} = 49.07$$

As a result of this comparison, the experimenter may want to consider the possibility that the drug under study derives from a class of drugs other than A or B, at least with respect to the performance variable observed.

Experimental Violations of the Model. Several strong statements are involved in the model of the experiment. First, the variable is held to be normal; the variances are very specific in the models; there is a strong assumption in writing the denominator of Z. Only if all N observations are independent do we know that $\sigma_{\bar{X}}^2 = \sigma_X^2/N$; otherwise there are covariance terms in $\sigma_{\bar{X}}^2$. We must address ourselves to the question of how seriously our inference or statistical conclusion is influenced by applying this model to an experiment that does not satisfy or match the properties of the model. Our decision rule, α, β, and N are dependent on the model. If the model is not a good model for the empirical experiment, then the decision rule leads to unrealistic and misleading decisions, with different significance probability, with type I and type II error probabilities other than those represented by Ψ, α, and β. The degree of departure of the experiment from the model determines the degree to which the values α, β, and Ψ are in error. This makes it rather difficult to be specific. However, some general statements can be made.

Departure from normality in the random variable sampled is not serious if sample size is large ($N \geq 30$) or if the distribution is not badly asymmetric. The central limit theorem operates for large samples, insuring that the sample mean is normally distributed. Where N is relatively small and a one-tailed test is used and where there is larger proportion of the total probability in that tail than in the normal distribution, α is inflated—more cases beyond T_c are observed than expected under H_T. This inflation is not reflected in Ψ. Thus, we reject H_T erroneously more often than we might desire. Increasing sample size leads to a more stringent β-level test but tends to rectify the α level. If severe departure from normality in the random variable is suspected and small samples only are available (or advised), then Case B or Case D methods (below) should be used.

If the exact variance of the observed random variable is not known within a narrow tolerance, the denominator of Z is either too large or too small and hence Z is either too small or too large, leading to poor estimates of N, Ψ, α, and β. If the value taken as σ^2 is suspect, then methods of Case C and Case D should be used.

Perhaps most serious of all problems in the application of the model of Case A is the independence of the observations. If observations covary in a sampling procedure or measurement procedure (for example, if selection of a given value makes some other values more or less likely to be observed), the standard deviation of the sampling distribution of the mean is either larger or smaller than $\sigma/\sqrt{N}$.

This has the same effect on the properties of the test as an erroneous value of σ. Great care should be taken in making independent observations, or an assessment of $\text{cov}(X_i, X_j)$ is necessary correctly to find $\sigma_{\bar{x}}$.

Sample Size: Large Versus Small. The issue of sample size in terms of large sample versus small sample is not pertinent to Case A. The sample size required to achieve a test with a specified α and β carries no additional meaning. In the paragraphs to follow, sample size, in part, determines the appropriateness of the model for the experiment.

Case B: X Nonnormal, σ^2 Known

For large samples, say $N > 30$, Case B is approximated by Case A by virtue of the central limit theorem. The departure from normality of the distribution of X determines the sample size necessary to achieve a satisfactory approximation. The more similar the distribution of X to the normal distribution, the more adequate is the approximation, even for small samples.

All of the details of Case A apply to Case B where sample size is large enough to insure approximate normality of $\bar{X}$.

For small samples in Case B, knowledge of σ^2 is not particularly useful. Case B corresponds in practice to Case D, at least so far as our hypotheses concern the expectation, where small samples are indicated and, hence, can be handled as a special instance under Case D.

Case C: X: $N(\mu_X, \sigma_X{}^2)$, $\sigma_X{}^2$ Unknown

Large Samples. In large samples this problem is equivalent to Case A when $\hat{S}^2$ is substituted for σ^2. Since N is large, $\text{var}(\hat{S}^2)$ is small and hence $\hat{S}^2$ is numerically close to σ^2 and little error is involved. In fact,

$$\lim_{N \to \infty} \frac{\bar{X} - \mu}{\hat{S}/\sqrt{N}} = \frac{\bar{X} - \mu}{\sigma/\sqrt{N}} \tag{16.5}$$

The alert reader will have spotted a difficulty with this procedure: If we do not know σ^2, how can we calculate N? Since we do not know $\hat{S}^2$ prior to the experiment and since σ^2 is unknown, the dilemma seems complete. Two ways of avoiding stalemate are available. (1) Find $\hat{S}^2$ from a pilot study in which no attempt is made to test hypotheses regarding μ. (2) Use the nature of the observations, available studies, and so on, to estimate σ^2. The latter procedure can be very good if the phenomenon under study is well known or if a great deal of data are already available. For less extensively studied phenomena the pilot study approach probably gives the best results. In any event the procedures are the same as in Case A, except that the pilot study variance, $\hat{S}_P{}^2$, is used in finding N. Also, the rejection rule should be stated in terms of the significance probability rather than the critical value: Reject H_T in favor of H_A if $\Psi \leq \alpha$; otherwise do not reject H_T. Interpretation of the results must be cautious if the initial value of $\hat{S}_P{}^2$ (for example, from a pilot study) is greatly different from the $\hat{S}^2$ obtained in the sample of N observations. Since β is a function of σ^2 and $\hat{S}^2$ is our best estimate of σ^2, a new evaluation of the test for $\hat{S}^2$ and N should be made. This may involve selecting a new value for H_A (in the light of the obtained $\hat{S}^2$) if the δ method of specifying H_A is used. The use of $\Psi \leq \alpha$ in the rejection rule determines the probability of a type I error. However, for $\hat{S}^2$, N, H_A (as redetermined), and α, the value of β may be quite different from that desired.

Small Samples. When X is $N(\mu, \sigma^2)$ and $\bar{X}$ and $\hat{S}$ are from a sample of N observations, then

$$W = \frac{\bar{X} - \mu}{\hat{S}/\sqrt{N}}$$

is distributed as Student's t. Since $\hat{S}$, $\bar{X}$, and N are known, then, if $\mu = \mu_T$, that is, if H_T is true,

$$t = \frac{\bar{X} - \mu_T}{\hat{S}/\sqrt{N}} \tag{16.6}$$

is also distributed as Student's t.

The general theory and mechanics of Case A apply to this special instance with some important distinctions. The first of these is a very fundamental difference in Z and t. In the former,

$$Z = \frac{\bar{X} - \mu_T}{\sigma/\sqrt{N}}$$

is normally distributed with expectation zero if H_T is true and expectation $(\mu_A - \mu_T)/(\sigma/\sqrt{N})$ if μ_A is true. However,

$$W = \frac{\bar{X} - \mu_T}{\hat{S}/\sqrt{N}}$$

is t-distributed if H_T is true but is not t-distributed if H_A is true. If H_A is true, W is distributed as noncentral t, with expectation and variance depending on the unknown variance σ^2 and the "noncentrality" parameter $(\mu_A - \mu_T)/(\sigma/\sqrt{N})$. In

Case A the values Z_α and Z_β needed to find N and T_c could be found by consulting the table of the unit normal distribution. Similarly, the corresponding values t_α and t_β are needed to determine N and T_c in Case C for small samples. This is not possible for t_β without knowledge of the noncentral t distribution. Calculation of t from noncentral t for H_A is difficult and tedious. However, tables of N for selected values of

$$\delta = \frac{|\mu_T - \mu_A|}{\sigma}$$

have been prepared. Such a table is given in the back of this book as Table F. The entries in the table are the number of observations needed to satisfy a selection of (α, β, δ) procedures. Also, if N has arbitrarily been set, H_A can be determined by knowing H_T, α, β, and δ for a reasonable value of δ. Power charts for a selection of α-level tests are also given at the back of the book as Chart G.

The rejection rule for this test may be stated only for the critical value(s) in terms of α, the critical value in terms of β not being calculable without the non-central t distribution. The general procedures for determining the critical values are the same in Case C as they are in Case A, except that t_α instead of Z_α is used, and $\hat{S}$ replaces σ. For a single-tailed test

$$T_c = \frac{\hat{S}}{\sqrt{N}} t_\alpha + \mu_T$$

For a two-tailed test the t_α is replaced by a $t_{\alpha/2}$ to determine the respective single-tailed critical value for each of the tails. The rejection rule is otherwise the same as in Case A.

Experimental violations of the model can be categorized in the same way the violations were classed in Case A. The comments made there generalize directly except for those regarding errors in determination of σ^2, which are not relevant.

An Example. As an example, we take an experiment in which it is extremely important to avoid a type I error and somewhat less important to avoid a type II error. A cost function cannot be applied, so that α and β must be selected arbitrarily, say $\alpha = .01$ and $\beta = .1$. Imagine H_T: $\mu_T = 50$, H_A: $\mu_A = 56$. A pilot study is performed resulting in $\hat{S}^2 = 9$, implying $\delta = 2$. The value of N is determined from the sample size table. Under the $\alpha = .01$ for one-sided tests and $\beta = .1$ with $\delta = 2$, $N = 7$.

Imagine seven observations having $\hat{S}^2 = 12$, and $\bar{X} = 54.32$. Calculating the significance probability, we find

$$t = \frac{54.32 - 50}{3.464/2.646} = 3.30$$

From the table of Student's t,

$$\Psi = P(t \geq 3.30) < .01$$

and H_T is rejected. For an exactly ($\alpha = .01$) level test,

$$\frac{T_c - 50}{3.464/2.646} = 3.143$$

The sample standard deviation $\hat{S} = 3.464$ is greater than the value used to determine δ (that is, $\hat{S} = 3.0$). This has the effect of making the initial $\delta = 2$ slightly too large. The effective δ of the test with $\hat{S} = 3.464$ is

$$\delta = \frac{\mu_T - \mu_A}{3.464} = \frac{50 - 56}{3.464} \cong 1.73$$

If N is determined for this δ, 8 observations are indicated. However, if the value of $\hat{S}$ used in determining the initial δ were based on a larger than ($N = 7$) sample, it would be more reliable (that is, narrower confidence interval than the ($N = 7$) sample $\hat{S} = 3.464$). Hence modification of the original sample size and δ is not really justified.

Case D: X Nonnormal and σ^2 Unknown

Large Samples. When a large sample of observations is indicated in an experiment, the CLT applies and we have the same situation as in Case B except that we must substitute $\hat{S}^2$ for σ^2.

Small Samples. The considerations we make here are nearly identical to the general considerations we made in our discussion of nominal and discrete quantitative sample spaces. We abandon, in most applications, the continuous quantitative characteristics of X and deal with some discrete, perhaps even nominal, function of X. The specific function we work with is determined by the nature of X, the nature of H_T and H_A and, perhaps most importantly, the availability (or derivability) of a probability distribution on the function of X. In most of our examples we deal with simple binomial functions of X. In other instances (see later chapters) we deal with other discrete variables, similar to the binomial. In this chapter we develop only one procedure. In the next few chapters we work out several procedures of this sort.

The rubrics under which these methods are generally discussed vary widely. There are many names applied: "nonparametric," "distribution-free," "exact probability," "estimation-free," "order" statistics and so on. We might surmise that the proliferation of names is associated with the breadth of specific methods, theories, and models incorporated—there are several classes of procedures involved. We can shed some light on the collection of procedures involved by looking for a moment at each of the names listed above.

"Nonparametric" is the opposite of "parametric," which refers to the fact that the parameters of a distribution are estimated and used in formulating a statistical test. For example, the parameters μ and σ^2 are estimated by $\bar{X}$ and $\hat{S}^2$ in critical ratio tests of hypotheses about expectations. The problem with the designation "nonparametric" is that the tests generally do involve some parameter—sample size if nothing else. Still wider of the mark is the designation "distribution-free." This name implies that no distribution is involved—a condition that defines the procedures right out of statistics. The application of the binomial as a function of a continuous random variable is referred to in some quarters as distribution-free. Nothing could be farther from the truth, as must be obvious from a slight acquaintance with the binomial. The reference to distribution-free is often made when no "theoretical" distribution, such as the normal, Student's t, chi square, is involved in the test. The distinction is difficult to support except (1) when the individual terms in a distribution like the binomial are easily calculated while those for, say, a normal distribution are not (excusing the advent of the electronic computer), and (2) when we can generally set up our experimental procedures so that the data do correspond to a binomial model (for example) but this is not always easily done with a normal model.

Often the term "exact probability" tests is appropriate. However, in the application of these special techniques we are often forced to use approximations (as in the normal approximation to the binomial). The same is true of the term "estimation-free." A large set of these special procedures involve the ordering of the observations according to magnitude, and hence the term "order statistics." However, many of the developments do not directly involve an ordering, as in the binomial success and failure.

Imagine that our experiment gives rise to N observations on a random variable X. Nothing is known about X with respect to its distribution or variance. However, suppose that we are able to determine that 50% of the population is above a specific value and 50% below if a certain condition (psychological, experimental, or other) is satisfied. Since a value having these properties is known as the median, we call the test developed here the one-sample median test.

The test hypothesis is easily formulated:

$$H_T: \text{Mdn} = \mu_T$$

We turn our attention to the crucial questions of sample size and alternative hypotheses after developing a function of X leading to a rejection rule. If the value μ_T is at the median of the distribution of X, then

$$P(X \le \mu_T) = .5$$
$$P(X > \mu_T) = .5$$

We define the set $A = \{X \mid X \le \mu_T\}$ and the set $B = \{X \mid X > \mu_T\}$. If any given observation is a member of set A, we define it as a "failure," and if it is in B it is a "success." Thus, each of the observations corresponds to a binomial trial. The convention of designating successes by 1 and failure by 0 gives

$$g(X) = Y = \begin{cases} 0 & \text{if } X \le \mu_T \\ 1 & \text{if } X > \mu_T \end{cases}$$

where $g(X)$ is a function of the random variable X, a binomial function. The probability associated with the values of Y is obtained easily by noting the equivalence of the set A and $y = 0$, the set B and $y = 1$. By hypothesis (that is, $P(X \le \mu_T) = .5$), $P(A) = .5$ and hence $P(y = 0) = .5$. Likewise we have $P(y = 1) = .5$. Now we repeat the observation of X N times with replacement or with N small with respect to the number of elements in the population from which we are sampling. Under these conditions, Y is a binomial variable based on N binomial trials with $p = .5$. This is equivalent to stating that $H_T : p = p_T = .5$ in $Y: B(N, p)$. This is precisely equivalent to the formal statement of H_T in the tea-tasting experiment. And as in the tea-tasting experiment, we must establish an H_A, α, β, and consequently N. The values α and β are selected by the experimenter. The alternative H_A is not so easy. It makes more sense to find H_A in terms of the variable X if that is possible. Suppose that H_A: Mdn $= \mu_A = \mu_T + K$, where K is some constant. Recalling Chebysheff's inequality, it should be clear that $P(X \le \mu_T \mid \text{Mdn} = \mu_T + K)$ is a function of the standard deviation units, expressed as K (say δS). The larger K is the smaller this probability will be (if $K > 0$, for an upper tail alternative). This solution is unsatisfactory. We must use S to find the probability and we must also make some assumption about the distribution of X or use the limit given by the Chebysheff inequality. The inequality is too gross to be of much assistance in estimating the desired value of N, and we are assuming that we have no knowledge about the distribution of X.

The upshot of this discussion is a discouraging evaluation: A treatment of power and consequently of sample size often is awkward in this class of special procedures.

The (α, β, δ) procedure can be applied on the post-facto data. Suppose that $N = 30$ observations were made. Under H_T, the critical value for number of successes out of the 30 (upper-tailed test) for an ($\alpha \simeq .01$) level test is 22 ($\alpha = .008$, more precisely). We ask the following question to get a value for p_A giving a specific value of β (for example, $\beta \simeq .05$) for $N = 30$: $P[S < 22 \mid p_A$ in $B(30, p)] = .05$. If we consult an extensive table of $B(n, p)$, this value is found to be $p_A = .86$ with the exact value of $\beta = .0497$.

Modifications of this procedure to accommodate to lower-tailed tests, and two-tailed tests are relatively straightforward. These applications are left to the reader as an exercise.

Transformations

The preceding section introduced the notion of transforming a random variable X into another random variable in such a way that distribution theory could be applied. This idea is more general than the discussion above indicates. Two general classes of transformations are useful in statistical analysis of experiments. The first type of transformation is the transformation of numerical data (scores, magnitudes, and so on) into nominal data (success and failure, and so on) or into ordinal data (ranks). We choose to call this type of transformation a discrete transformation. The second type of transformation is the transformation of numerical data into numerical data in a one-to-one continuous fashion. This type of transformation is called here a continuous transformation.

Examples of discrete transformations have occurred several times in the preceding chapters, and in particular in the single sample median test of the preceding section. Many of the "nonparametric," "order," "distribution-free," and other tests are discrete transformation tests. However, not all tests developed for the discrete or nominal case are transformation tests. The binomial, for example, is based on nominal sample spaces directly. It has a wide range of applications to discrete transformations of numerical data, but the basis of binomial tests need not be in a quantitative sample space. In the following chapters we treat statistical models of nominal or ordinal sample spaces as though they were part and parcel of the methods of discrete transformation tests that begin with a quantitative sample space. The models are for nominal or ordinal sample spaces in general, unless specifically indicated. Some of the discrete transformation test models involve the assumption that the discrete data are derivative from continuous variables. The inference in these special cases is to the continuous variables, and the application of the methods to nominal and ordinal sample spaces would not necessarily be appropriate.

Continuous transformations may be applied to obtain a variable more nearly normally distributed than the observed variable. An example of this type of transformation is the logarithmic transformation of a variable that is distributed as a logarithmo-normal random variable (see Cramér, 1946, cited in Chapter 12). Such a transformation produces a normally distributed variable. Other transformations are sometimes recommended to remove skewness in the density function of the random variables. Trigonometric transformations are sometimes applied to proportions. Most generally, transformations of these kinds are used in the context of the analysis of variance and regression analysis, topics for later chapters. A discussion of continuous transformations at this point would lead us far afield.

Hypotheses Regarding Variances

In the chapter on sampling distributions of statistics it was shown that the variance $\hat{S}^2$ has a distribution that is a function of a chi square distribution if X is normally distributed. Let V be the random variable corresponding to the variance of samples on the variable X. If W is distributed as a χ^2 random variable, then

$$V = \sigma^2 \frac{W}{N - 1}$$

Hence

$$W = \frac{(N - 1)V}{\sigma^2}$$

For example, for $\hat{S}^2 = v_{.95}$ such that

$$P(v_{.95} \leq V) = .95$$

we find the value of the χ^2 variable, w, such that

$$P(w_{.95} \leq W) = .95$$

and solve

$$v_{.95} = \sigma^2 \frac{w_{.95}}{N - 1}$$

for $N - 1 = 16$, if the value of $w_{.95}$, determined by consulting a χ^2 table, is 26.3. Thus

$$v_{.95} = \sigma^2 \frac{26.3}{16}$$

$$= \sigma^2(1.64)$$

With this development we can formulate a test of hypotheses regarding the variance of a normally distributed random variable. This normality condition is a rather hard and fast condition, even for large samples. Either a high degree of approximation to normality in X or better, strict normality, is required or V is not χ^2 distributed.

The test hypothesis has the form

$$H_T: \sigma^2 = \sigma_T{}^2$$

where $\sigma_T{}^2$ is some specific numerical value. We develop here the theory and procedures for a one-tailed test (upper) only. Lower-tailed and two-tailed tests are easily derived following the general procedures of statistical test construction and the specifics of this section. We take as our composite alternative

$$H_A: \sigma^2 = \sigma_A{}^2 > \sigma_T{}^2$$

If H_T is true, then for a sample of size N,

$$\frac{\hat{S}^2(N-1)}{\sigma_T{}^2}$$

is χ^2 distributed with $df = N - 1$. Choosing an α, we have

$$P\left[\frac{\hat{S}^2(N-1)}{\sigma_T{}^2} \geq \chi_\alpha{}^2 \,\Big|\, \sigma^2 = \sigma_T{}^2\right] = \alpha \qquad (16.7)$$

Hence the critical value of the sample variance $\hat{S}^2$ is given by

$$\frac{\hat{S}^2}{\sigma_T{}^2} = \frac{1}{N-1}\chi_\alpha{}^2$$

The power of this test is easily derived for a specific alternative

$$H_A: \sigma^2 = \sigma_A{}^2$$

where $\sigma_A{}^2$ is a specific numerical value. The power is defined by

$$P\left(\frac{\hat{S}^2}{\sigma_T{}^2} \geq \frac{1}{N-1}\chi_\alpha{}^2 \,\Big|\, \sigma^2 = \sigma_A{}^2\right) = 1 - \beta \qquad (16.8)$$

$$= P\left[\frac{(N-1)\hat{S}^2}{\sigma_A{}^2} \geq \frac{\sigma_T{}^2}{\sigma_A{}^2}\chi_\alpha{}^2 \,\Big|\, \sigma^2 = \sigma_A{}^2\right]$$

$$= P\left(\chi^2 > \frac{1}{\delta^2}\chi_\alpha{}^2\right)$$

where $\delta^2 = \sigma_A^2/\sigma_T^2$. For a given value of $(N-1)$ we can solve for a δ^2 that gives a power $(1 - \beta)$. This is given by solving

$$\frac{1}{\delta^2}\chi_\alpha^{\;2} = \chi_\beta^{\;2}$$

or

$$\delta^2 = \frac{\chi_\alpha^{\;2}}{\chi_\beta^{\;2}}$$

Sample size may be determined from this equation. Imagine that

$$H_T: \sigma^2 = \sigma_T^2 = 6.00$$
$$H_A: \sigma^2 = \sigma_A^2 = 18.00$$

Then $\delta^2 = \sigma_A^2/\sigma_T^2 = 3.0$. For $\alpha = .05$ and $\beta = .1$, find the entries in the χ^2 table in the $(.95 = 1 - \alpha)$ and $(.1 = \beta)$ columns, for a given df where the value in the .95 column is 3 times that in the β column. We find $df = 14$, $\chi_{.95}^2 = 23.66$, and $\chi_{.1}^2 = 7.79$ is the configuration of values giving the value of δ^2 nearest 3. Hence 15 observations $(df = N - 1)$ are required to test $H_T: \sigma^2 = \sigma_T^2 = 6.0$ and $H_A: \sigma^2 = \sigma_A^2 = 18.0$ with $\alpha = .05$ and $\beta = .10$.

This process can be reversed for a given sample size in order to find δ^2 for a postexperimental application of the (α, β, δ). Taking the square root of δ^2 in the χ^2 ratio gives δ in terms of the ratio of standard deviations, that is, σ_A/σ_T.

At the back of the book is a table, Table H, that can be used to obtain the number of degrees of freedom [hence sample size $(N = df + 1)$] or to evaluate δ^2 for a given df. This table is constructed for a selection of values of α and β. The entries of the table give the value of δ^2 as a function of α, β, and df. For a given α and β, specification of δ^2 leads to a df. If α, β, and df are specified, δ^2 can be determined from the table.

To use the table for $\delta^2 > 1$ (upper-tailed alternative) for a given α and β combination, the column corresponding to the α and β combination is selected. Find the two values closest to the δ^2 desired (or specified). Linear extrapolation from these values to the values of df on the left margin of the table gives a good approximation to df. For example, $\alpha = .01$, $\beta = .05$, $\delta^2 = 4$. We find $\delta^2 = 4.211$ for $df = 15$ and $\delta^2 = 3.462$ for $df = 20$. The difference in the delta squares is .749, and $\delta^2 = 4.00$ is approximately .28 of this difference below $\delta^2 = 4.211$. The corresponding difference in df is $.28(5) = 1.4$. Hence, $df = 15 + 1.4 \cong 16$ is the desired approximation.

Where a lower-tailed test is involved, that is, $\sigma_A^2/\sigma_T^2 = \delta^2 < 1$, the table is entered with $\alpha' = \beta$, $\beta' = \alpha$ and $\delta'^2 = 1/\delta^2 = \sigma_T^2/\sigma_A^2$. For example, $\alpha = .01$, $\beta = .05$, $\sigma_T^2 = 4$, $\sigma_A^2 = 1$ means $\alpha' = .05$, $\beta' = .01$ and $\delta'^2 = 4$. Entering the table through columns $\alpha' = .05$, $\beta' = .01$ and extrapolating, we find $N = 18$ is the best approximation.

Power functions of the test are given in Chart I at the end of the book. A chart is presented for each of several α level tests, giving $1 - \beta$ as a function of $\delta = \sigma_A/\sigma_T$ for $.1 \leq \delta \leq 1.0$ and for $1 \leq \delta \leq 5$. Each chart contains graphs for a variety of degrees of freedom.

SUGGESTED READING

Dixon, W., and Massey, F. (1957), *Introduction to statistical analysis* (2d ed.), McGraw-Hill, New York.

Guenther, W. C. (1965), *Concepts of statistical inference*, McGraw-Hill, New York.

Hald, A. (1952), *Statistical theory with engineering applications*, Wiley, New York.

Hays, W. (1963), *Statistics for psychologists*, Holt, Rinehart and Winston, New York.

Hoel, P. G. (1954), *Introduction to mathematical statistics* (2d ed.), Wiley, New York.

Hoel, P. G. (1960), *Elementary statistics*, Wiley, New York.

Mood, A. M. (1950), *Introduction to the theory of statistics*, McGraw-Hill, New York.

Wolf, Frank L. (1962), *Elements of probability and statistics*, McGraw-Hill, New York.

GLOSSARY

Case A. X: $N(\mu_X, \sigma_X{}^2)$, $\sigma_X{}^2$ known.

Case B. X nonnormal, σ^2 known.

Case C. X: $N(\mu_X, \sigma_X{}^2)$, $\sigma_X{}^2$ unknown.

Case D. X nonnormal and σ^2 unknown.

Alternatives. Composite alternatives and the specific alternatives under a $\delta\sigma$ rule, for H_T: $\mu = \mu_T$:

$$H_1: \mu_X = \mu_A > \mu_T \qquad H_A: \mu_X = \mu_A = \mu_T + \delta\sigma_X$$
$$H_2: \mu_X = \mu_B < \mu_T \qquad H_B: \mu_X = \mu_B = \mu_T - \delta\sigma_X$$
$$H_3: \mu_X = \mu_C \neq \mu_T \qquad H_C: \mu_X = \mu_C = \mu_T \pm \delta\sigma_X$$

(16.1) $$Z = \frac{\bar{X} - \mu_X}{\sigma_X/\sqrt{N}} \text{ is } N(0, 1) \text{ if } X \text{ is } N(\mu_X, \sigma_X{}^2)$$

Critical Ratio, or Normal Test. In (16.1) Z is referred to as the critical ratio or the normal test of hypotheses H_T: $\mu = \mu_T$.

$$(16.2) \qquad N = \left[\frac{\sigma(Z_\alpha - Z_\beta)}{\mu_A - \mu_T}\right]^2$$

Rejection Rule, Case A. Where both α and β are set, and where N is selected by a method to satisfy α and β, the rejection rule for a one-tailed test is based on the average of the critical values:

$$T_c = \frac{\sigma}{\sqrt{N}} Z_\alpha + \mu_T \qquad \text{and} \qquad T_c = \frac{\sigma}{\sqrt{N}} Z_\beta + \mu_T$$

The rejection rule is: Reject H_T in favor of H_A if $\bar{X}$ is equal to or more extreme than the average T_c. For a two-tailed rejection rule with α and β equally weighted on each tail, the upper tail critical value is calculated with $.5\alpha$ and $.5\beta$ substituted in the equations. Here H_T is rejected if $\bar{X}$ falls outside of the rejection limits, either above or below the upper or lower tail critical values, respectively. If the sample size is arbitrary, only the critical value involving α is used.

$$(16.3) \qquad N = \left[\frac{\sigma(Z_{\alpha/2} - Z_\beta)}{\mu_A - \mu_T}\right]^2$$

$$(16.4) \qquad N = \left[\frac{\sigma_T Z_\alpha - \sigma_A Z_\beta}{\mu_A - \mu_T}\right]^2$$

Significance Probability, Case A.

under H_1: $\Psi_1 = 1 - \Phi[z \mid X: N(\mu_T, \sigma_X^2)]$

under H_2: $\Psi_2 = \Phi[z \mid X: N(\mu_T, \sigma_X^2)]$

under H_3: $\Psi_3 = 1 - \Phi[|z| \mid X: N(\mu_T, \sigma_X^2)] + \Phi[-|z| \mid X: N(\mu_T, \sigma_X^2)]$

$$(16.5) \qquad \lim_{N \to \infty} \frac{\bar{X} - \mu}{\hat{S}/\sqrt{N}} = \frac{\bar{X} - \mu}{\sigma/\sqrt{N}}$$

$$(16.6) \qquad t = \frac{\bar{X} - \mu_T}{\hat{S}/\sqrt{N}} \text{ is distributed as Student's } t \text{ if } X \text{ is } N(\mu_T, \sigma^2)$$

Rejection Rule, Case C. The rejection rule is based on the critical value

$$T_c = t_\alpha(\hat{S}/\sqrt{N}) + \mu_T$$

Here H_T is rejected if the observed mean, $\bar{X}$, is equal to or more extreme than the critical value. For a two-tailed rejection rule the critical value of each tail is calculated, and if $\bar{X}$ is equal to or more extreme than either critical value, H_T is rejected.

Discrete Transformation. The transformation of numerical data into nominal data or into ordinal data.

Hypotheses about Variances. The test hypothesis H_T: $\sigma^2 = \sigma_T^2$ opposed to the alternative hypothesis H_A: $\sigma^2 = \sigma_A^2 > \sigma_T^2$.

(16.7)
$$P\left(\frac{\hat{S}^2(N-1)}{\sigma_T^2} \geq \chi_\alpha^2 \mid \sigma^2 = \sigma_T^2\right) = \alpha$$

(16.8)
$$P\left(\frac{\hat{S}^2}{\sigma_T^2} \geq \frac{1}{N-1}\chi_\alpha^2 \mid \sigma^2 = \sigma_A^2\right) = 1 - \beta$$

PROBLEMS

1. In formulating a new training regime, an educator determined that an improvement in test score of 9 raw points was needed before he could convince users of his test that the new and more complicated procedure would be of any use. The mean of the scores under the old procedure was 150, and the variance of the scores was 160. Using the (α, β, δ) procedure, formulate a specific alternative and a sample size for the test of H_T: $\mu_X = \mu_T = 150$ against the composite hypothesis H_A: $\mu_X = \mu_A$, where X is assumed to be normally distributed with variance of 144.

2. A published paper reports that the hypothesis H_T: $\mu_X = \mu_T = 0$ was rejected at the $(\alpha = .01)$ level in favor of H_A: $\mu_X > 0$ in a weight-gain experiment on surgically treated animals. What would you have to know in order to interpret the meaningfulness of this report?

3. Imagine that the experiment in problem 2 was performed on 110 animals and that normal animals of the same species varied in weight over a period of the reported experiment equivalent to a variance of 17 grams. What was the post-test value of δ and of the effective alternative hypothesis? In view of the normal, nonexperimental, weight variation, could claim to scientific importance be made (do you have enough information)?

4. Find an appropriate sample size for an $(\alpha = \beta = .01)$ test of two specific hypotheses that have $\delta = 1.4$ where the random variable being observed is normally distributed $N(\mu, 9)$.

5. Assuming that H_T: $\mu = \mu_T = 100$ in problem 4, above, formulate rejection rules for upper-tailed and lower-tailed tests. What would the value of μ_A be in two specific alternative hypotheses for these tests?

6. When the formula for selecting a sample size in the normal test gives a noninteger value, the text suggests that the two critical values for the test given by the nearest integer be averaged. The result of this is to produce α and β nonsymmetrically different than the target values for α and β. An example in the text gave $\alpha = .0089$ instead of .01, but $\beta = .043$ instead of .05, missing the mark for β by a greater

amount than for α. Devise a procedure whereby a rejection value is obtained that gives equal error in attainment of α and β, utilizing the table of the unit standard normal random variable.

7. Where the variance of the random variable being sampled is different under H_T and H_A, formula (16.4) is used for determination of an appropriate sample size for given values of α and β. No mention of δ was made in the text in the context of this problem, because of an ambiguity in the definition of δ in problems with unequal variances. Why is δ not well defined in this instance?

8. On-the-job performance in a certain task is known to have a fault rate of 36 manufactured items per hour, on the average, with a variance of 16. The faults for the entire factory are normally distributed. A psychologist claims to know how to improve the fault rate by certain manipulations of the motor skill components of the manufacturing process. The change will cause a certain decrement in the productivity of each worker, and there will be a certain savings for the factory with a reduced number of faults. Management of the factory decide that a reduction of faults on the order of 6 fewer per hour will be an improvement on the profit side. Assuming that a change in the expectation of the number of faults is not accompanied by a change in the variance, design an experiment to aid the psychologist in deciding whether or not he can stake his reputation on the claim that his technique does reduce faults 6 or more, on the average. Include in your answer all pertinent reasoning and figures.

9. In problem 8 the variance was thought to be known. If no records of this statistic were kept by the factory management and the psychologist made a quick preliminary estimate from observing a number of workers, giving $\hat{S}^2 = 24$, how would the experiment be modified? Repeat the analysis you made of problem 8 under this new circumstance.

10. The discrete transformation test developed in the text under the name of the one-sample median test can be generalized easily to cover any hypothesis that specifies how a given value splits the population into two segments. For example, a test can be devised for a hypothesis that states that 75 percent of the population will be greater than the value 101 if a certain theory is correct. Formulate the details of such a test along the lines followed in the text for the one sample median test.

11. Apply the test generated in your answer to problem 10 to the data shown in Table 16.2.

12. Generalize the discrete transformation tests involved in problem 10 so that they are lower-tailed tests, and again so that they are two-tailed tests. Apply these to the data in Table 16.2.

13. Imagine that the training regime involved in problem 1 was accepted by the school board. After it had been in use for about a year it was challenged by a parents' group, which claimed that the regime was favoring the more advanced students at the expense of the less advanced students, even though the average test score did increase. The school board ordered an investigation that was to consist of a statistical test of the

variability of the scores on the test of the students having worked under the new regime. Design a statistical test of the hypothesis H_T: $\sigma^2 = \sigma_T^2 = 144$ against H_A: $\sigma^2 = \sigma_A^2 > 144$. The school board knows that the test has a reliability limit such that three-item changes at random in a score are expected on retesting, and orders that the statistical test be able to detect a difference of variation that is at most twice this expected random change. They also order that a type I error should be risked with a probability of only .001, but that a type II error was not considered serious and could be tolerated with a probability of about .10. Design a test of the hypotheses and discuss briefly each aspect of the test. Explain why your test is pertinent to the problem. Include in your answer such details as the sample size, the (α, β, δ) specification, the critical values of the test statistic, and the rejection rule.

The Two-Sample Problem

The shift from single-sample problems and their statistical models to two-sample problems requires few new statistical concepts, but is full of experimental implications. Thus a large portion of this chapter is devoted to discussion of the principles of experimentation, of experimental design. The "design of experiments" is a term used widely in texts on the analysis of variance and other statistical tools that deal with the evaluation of more complicated experiments. We turn to that topic in the following two chapters. In the meantime, we deal with experiments in which the "design" calls for only two sets of observations on a single variable.

The plan of this chapter is first to develop the principles of design and then to turn to statistical models for experiments so designed. In discussing models for tests of hypotheses about expectations, the material of the preceding chapter is of immediate application. However, for hypotheses regarding variances we must introduce another sampling distribution: Fisher's F.

Design of Experiments

We have placed great stress on the idea that sophisticated use of statistics in scientific research involves explicit formulation of models and derivations from these models to ascertain statistical consequences that can be tested. On the basis of a handful of general models, we have developed the mechanics and interpretive concepts for statistical tests of simple hypotheses. However, we have spent relatively little effort in developing the concepts of research design and the methods of experimental manipulation. These concepts and methods are vital to the successful application of the theory of probability and statistics in scientific enterprises. This subject is extensive, and we can here do no more than briefly discuss a few of the more important varieties of research design.

Comparison of Two Treatments. In the theory of statistical tests we formulate the problem in terms of a statistical hypothesis H_T and an alternative statistical hypothesis H_A. We extend this analysis of the formation of H_T in terms of practical experimental manipulations, particularly in situations where there are competing claims, or theories, or treatments of subjects.

The term "treatment" derives from the development of much of our current statistical theory in the agricultural colleges in this country. For example, in growing wheat, it is always desired to improve the yield and quality of a planting. Various types of fertilizers can be applied to the fields in a variety of quantities. A specific treatment of the ground consists in the application of one of these fertilizers in a specific quantity. Since there are competing treatments of the ground, that is, different fertilizers in different quantities, the task is to find the most propitious treatment in terms of the quantity or quality of the resulting wheat crop.

The same basic logic pertains to many situations in psychology. Which form of psychotherapy gives the most recoveries at the least cost to the patient and society? Which method of education results in "brighter" children? Which of two methods of learning is more efficacious? The "treatments" involved in these examples are obvious: the regime imposed on patients in psychotherapy, the methods of treating children in classroom situations, and the procedures imposed on learners in a learning experiment.

Take as an example a learning theory that has as one of its consequences the "prediction" that massed practice is more efficacious in learning than practice that is distributed over a longer time period. Under massed practice, learning is supposed to be faster or to involve fewer errors. We define massed practice for a given number of practice trials as the condition where all of the trials occur one after the other with no intervening periods of time. On the other hand, distributed practice has periods of time intervening between the practice trials. The number of practice trials is the same under both treatments, and hence the total amount of practice is presumably the same. Suppose that we have a vocabulary list that is to to learned and we measure the "goodness" of learning by the number of errors made in a total of 30 practice trials. In the massed practice treatment we have the 30 trials massed into one long practice session. In the distributed practice treatment we have, for example, the list practiced five times on each of six successive days.

Ignore for the moment the considerations we must make in determining the exact character of the statistical test we use. Instead, follow three steps in the comparison of two treatments: (1) define the treatments (as above), (2) define the subjects (how are subjects selected, and how are individual subjects assigned to treatments), and (3) observe the variables involved. This last step results in N_m observations on a random variable X_m (for massed) and N_d observations of X_d (for distributed). Certain assumptions regarding the form of the distributions of X_m and X_d are necessary later.

The second step in this design problem is the most complicated of the three. At this point the elegance of the experiment and subsequent analysis of experimental data are determined. There are at least three ways of carrying out the experiment in terms of the handling of the subjects observed: (1) randomized groups of subjects (one group with massed practice, the other group with distributed practice), (2) matched subjects for both groups, and (3) subjects as their own control, repeated observations.

Experimental Versus Control Groups.   We need to digress slightly to introduce a new concept. Traditionally, when two groups are involved in a study—for example, fields in the agriculture study, batches of chemicals in a study of a chemical process, human subjects in a learning study—one group is called the "experimental group" and the other is called the "control group." These dichotomous terms are somewhat arbitrary in much of the work done in psychology, because it is not clear which group is "experimental" in comparison with the other. However, the language is based on the condition that we have one theory or one treatment that is the accepted theory or treatment and that a new theory or treatment has been proposed. The new theory or treatment is viewed as "experimental" until it is shown to be superior with respect to relevant aspects of behavior. The observations made on the control group provide a basis for comparing the efficacy of the new theory or treatment in terms of observations made on the subjects in the experimental group. If the new theory or treatment does better than the control (the current theory or treatment), we discover the difference by comparison.

In the massed versus distributed practice example, neither group is clearly the control group. However, to conform to standard usage we designate the massed practice group as the control group and the distributed practice group as the experimental group.

Random Groups. It may have occurred to the reader that if we already know a great deal about the so-called control treatment, we need not measure the control random variable. If, for example, a very large number of studies had already been reported in which massed practice of vocabulary lists was studied, we would have a very good estimate of the mean number of errors and the variance of number of errors made in such an experiment. As tempting as it seems, it is seldom a good idea to utilize data from previously executed experiments, except on one condition. Unless the experimental and control groups are observed in identical conditions and the subjects in the groups are alike in all respects except the experimental treatments, we cannot safely compare the observations. There are two parts to this stricture: The subjects must be alike, and the conditions of the experiment must be identical for both groups. This may be softened by introducing a qualification, at least insofar as the behavior being observed is or might be influenced by the conditions and subject variables.

In terms of the conditions of the experiment, we must insure that both treatments have equal opportunity to be effective. If the control group is observed under conditions not conducive to a high level of performance, whereas the opposite is true of the experimental group, part of the difference observed between the two groups is due directly to the difference in the conditions of the experiment. Thus, we would be led astray in making inferences regarding the real efficacy of the two treatments. It is always a good idea to incorporate the appropriate control group into an experiment.

In addition, we must be concerned about subject selection for the two groups. If the two groups are dissimilar in any respect that may have bearing on the performance observed in the experiment, then any difference in performance for the groups may be attributable to the subject differences rather than to the differences in the experimental and control treatments. If, for example, we selected subjects in the experimental group in such a way that they had more experience with distributed practice than the control subjects had had with massed practice, this would have an unassessed effect on the results. Even more apparent is the damage to the conclusions we draw from the experiment if we select more intelligent subjects for one of the groups than for the other.

One of the accepted ways of ensuring that the two groups are comparable in all aspects except the treatments is to select them in two steps: First, define the population of subjects acceptable in the experiment; second, select at random (equal probability) from the population each subject of each of the two groups.

In the first part, the population of potential subjects should be selected to reflect the purpose of the experiment. If we wish to infer some characteristic of a given type of subject, then the sample should be selected from the population of subjects of that type. In most human psychological work this stricture has been poorly dealt with. We see attempts to infer characteristics of "persons" based on samples of a very narrow class of persons—college sophomores who have selected a course in psychology as an elective. There is nothing inherently wrong with such a sample, but it cannot be used as a basis for inferences about a different population. In the second part, to insure that no bias is built into the experiment, in terms of the characteristics of the subjects in the two groups, each group is defined as a random sample from the population to be used. The impartiality of a table of random numbers is exploited in this selection. For example, we define the two groups by referring to a random number table—for each subject we select a number from the table of random numbers. If the number is *even* the subject is assigned to one group, if *odd* to the other group. In this way, the individual characteristics of the subjects selected are balanced in the two groups. It does not insure that the two groups are identical, but it prevents built-in biases from occurring.

A third characteristic is desirable in determining the constitution of the two groups of subjects. The sample of subjects should be as homogeneous as possible, compatible with the representativeness of the sample with respect to the population about which inferences are to be made. If the experimental treatment has an effect that is small in comparison with the variation in the subject's responses within the groups, then it is difficult to detect the difference in the sample. Thus, if all the subjects are alike with respect to the characteristics modulating their responses in the experiment, except of course the treatments, then the variation of responses from subject to subject is minimized, providing greater opportunity for the effect of the treatments to be detected. This detectability is a function of the standard deviation of the difference in the two variables, which is in turn a function of the variances of the two random variables. If the sample observations have many sources of variance other than the experimental treatment, then the estimated standard deviation of the difference between the treatment effects is inflated and obscures the difference between the treatments. In the example of massed and distributed practice there are several obvious extraneous sources of variation: intelligence, motivation, previous practice with the methods, and so on. If we select the groups at random without explicitly attending to these subject characteristics, we are in effect giving each group a chance to be as intelligent, motivated, and so forth, as the other group. If both groups are balanced in this way the observations tend to be equally variant, having the same sources of variation extraneous to the treatments. The variation may still be large, but it is not biased in favor of one group or the other.

The random groups design may be modified slightly to study individual differences. Imagine a test in differential psychology that is supposed to differentiate two subject populations. In order to test the effectiveness of the differentiation, two samples of subjects are tested. One sample is drawn from one population and the other sample from the second population. One sample is the "experimental" group and the other sample is the "control" group. Both groups are treated in identical ways: There is no experimental treatment. Hence, both groups are identical with exception of the population from which they came. If there is a difference in test scores for the two groups, it can be attributed either to sampling error or to a population difference. If the difference is significant under an appropriate (α, β, δ) procedure, the difference is attributed to a population difference.

Matched Groups.   There is a more sophisticated but more costly way of assuring equality in the groups in the comparison of two treatments. Each subject selected for one of the two groups is matched by a subject in the other group.

In matching, as many attributes of the subjects as are relevant to the behavior to be observed should be used, within limits of feasibility. In our example of the massed and distributed learning experiment the sorts of characteristics that might be used to advantage are intelligence, age, language facility, present vocabulary mastery, experience in experiments of this sort, and so on. These presumably are some of the main sources of variability from subject to subject in the two groups. If we have matched the two groups with respect to all relevant characteristics except the difference of the two treatments, we know the only way the two groups differ is the treatment they receive. This allows more confidence that any statistically significant difference in the observations for the two groups is due to the experimental treatment.

It is not necessary that each subject individually be matched with a subject in the other group. That would be one way of assuring that the two groups were similar. However, as long as the distribution of the characteristics used in the matching are the same in both groups, similar benefits accrue in general. There are other advantages to subject-by-subject matching and possible difficulties in the distribution matching procedures that cannot be dealt with here.

The analysis of the experimental results is identical with that for matched groups. If, however, the subject-by-subject matching procedures are used, each pair of subjects may be thought of as a single subject—the subject serving as a member of the experimental group and providing the idiosyncratic control comparison. The subject-by-subject matching method corresponds to a subjects-as-own-control design if the matched subject pairs are viewed as being a single subject.

Subjects as Own Controls. Under this design each subject serves in both parts of the experiment—he is observed in the experimental procedures and the control procedures. In this way, all factors except the treatments are controlled by the selection of the subjects. This is a very powerful design, as we shall see, but it has one important drawback—if one treatment can have an effect on the other treatment, the design cannot be used. If having learned one list of vocabulary items by distributed practice makes it more (or less) difficult to learn another list of equal original difficulty, this design would not be appropriate. Dependencies of this sort, mechanical or behavioral dependencies that lead to statistical dependencies, amount to confounding the effects of treatments on subjects and the effect of one treatment on another. Examples are not difficult to think of. In a test of manual dexterity, practice may have a considerable effect on the task and the subjects are more dexterous in the second of two tasks. In studies of attitude and opinion, statements of attitude may be determined in part by prior statements made by a subject.

A more general term for the subject-as-own-control designs has come into use in the theory of design of experiments. Since each subject is observed in both the experimental and the control conditions, the subject is observed repeatedly. In the analysis of variance, several treatments are involved and the subject may be observed under a number of the treatments. In this sense the design is a repeated observation design. Repetition of observations for a given subject must be distinguished from replication of the experiment on a number of subjects. In the remainder of this book "repeated observation" refers to a single subject observed under more than one experimental condition, and "replication" refers to multiple subjects in the same treatment.

The efficiency of repeated observation designs derives from a dependency of the control and experimental observations. If part of the variability of the observations in both treatments arises from personal characteristics of the subjects, then we can correct for this variability by taking a subject's performance in the control treatment as a comparison or base-line observation. If the tasks involved in both treatments are influenced by the subject's intelligence, age, personality, strength, and so on, then a subject who did well on one of the two tasks would tend to do well on the other also. Thus, if we compare the performance of a subject on one task with his performance on the other task, all factors influencing the performance except the difference in the treatments are held constant. The example of massed versus distributed practice is a good example. It seems clear that such things as the intelligence, motivation, and language skills of a given subject determine, in part, the number of errors he makes. If a subject makes few errors in one method of learning it is likely that he will make few errors in the other method of learning. Conversely, for large numbers of errors in one method there should be large numbers of errors in the other method, for a given subject. However, because all extraneous factors in the experiment-subject combination (excepting the treatments) are constant, any difference between the performances in the control situation and the experimental situation is, in the main, attributable to the different treatments.

One of the problems with the repeated measures design is the possible contamination of one treatment by the other treatment. If the second measure is dependent on the first measure, the difference between the first and second measures is due, in part, to the order of the measures. If the effect of order of treatment is the same for both orders, there is a simple way of balancing this effect. Half of the replications (that is, half of the subjects) are observed in one order and the other half in the other order. The order is counterbalanced in the design. The counter-balancing should be complete, that is, no more subjects in one order than in the other order. Also, subjects should be assigned to a given order at random, with the restriction of equal numbers.

If the order effects of the two treatments are not equal, on the average at least, the counterbalancing technique is not effective. Other techniques are available for designs of experiments with an unbalanced order effect or where the order is a part of the experimental procedures and cannot be counterbalanced. We cannot take space here to develop these techniques in detail. However, it is necessary at least to mention them. References are given at the end of the next chapter for sources of extensive discussion of these aspects of experimental design.

One important technique for evaluating the order effect of experimental treatments is to use two randomized (or matched) groups, and to let one group get both treatments whereas the other group, a control group, gets only the second treatment. A good example of this is the evaluation of the psychotherapeutic effect of a drug. An experimental group is pretested for degree of psychopathology, and then given a course of treatment with the experimental drug and post-tested for degree of psychopathology. Degree of improvement is indicated by the difference in the two scores. In order to guard against the possibility that the drug has no effect outside the "treatment," or that the degree of difference occurred spontaneously, and so on, a control group is run in the same sequence but with a placebo in place of the experimental drug. Any difference in the two groups now can be attributed to the difference in the placebo and the experimental drug, if the subjects are selected properly.

In certain experimental settings it might be that the pretesting itself was influential in the outcome of the experimental treatment. Under these circumstances four groups can be run: two experimental groups, one with pretest and one without pretest; two control groups, one with pretest and one without pretest. The comparison made is the difference of the differences within the experimental and the control groups. This design is at a level of complexity such that a more appropriate design can be found in an analysis of variance.

Statistical Procedures

The development so far in this chapter has been nonstatistical. We now apply the general theory of statistical tests to the experiments of the various designs. In doing so, the point made in earlier chapters regarding the virtual independence of statistical methodology and scientific interpretation is clarified. We apply the same general procedures to experiments of different designs. The statistical interpretation is always the same, but the scientific interpretation depends on the character of the experiment, including the design.

Only two statistical models are involved. The first deals with independent observations and the second with repeated observations. The major difference of these models is in the variance of the sampling distribution of means and differences in means. In general form, the statistical procedures are very similar to those developed for the single sample problem.

Statistical Procedures: Independent Samples

We develop this material in the same outline as that in the preceding chapter. In all of the special cases we have two sets of observations on a random variable X. We distinguish the two sets of observations by subscripts a and b. Hence for the ith observation in group a we have X_{ia}. For the jth observation in group b we have X_{jb}. Corresponding sample sizes, means, expectations, variances, and variance estimates are denoted by N_a, N_b, $\bar{X}_a$, $\bar{X}_b$, μ_a, μ_b, σ_a^2, σ_b^2, $\hat{S}_a^2$, and $\hat{S}_b^2$

where

$$\bar{X}_a = \frac{1}{N_a} \sum_{i=1}^{N_a} X_{ia}$$

$$\bar{X}_b = \frac{1}{N_b} \sum_{j=1}^{N_b} X_{jb}$$

$$\hat{S}_a{}^2 = \frac{1}{N_a - 1} \sum_{i=1}^{N_a} (X_{ia} - \bar{X}_a)^2$$

$$\hat{S}_b{}^2 = \frac{1}{N_b - 1} \sum_{j=1}^{N_b} (X_{jb} - \bar{X}_b)^2$$

Table 17.1 contains illustrative empirical data corresponding to this general experimental design. The statistics calculated from the data are also shown.

A general requirement in all of the methods developed in this section is that every observation be independent of every other observation, both within a group and between groups. This is most readily achieved by completely random assignment of subjects to groups and by rigid standards of experimental control. Each subject must be observed under identical conditions, with the exception of the treatment difference in the two groups.

The general idea of the tests is to let $(\bar{X}_a - \bar{X}_b)$ take the place of the random variable $\bar{X}$ with which we dealt in one sample problems. The difference in means has properties that make our test procedures similar to the one sample procedures.

1. If X_a and X_b, the random variables being observed, are normally distributed and independent,

$$N(\mu_a, \sigma_a{}^2) \quad \text{and} \quad N(\mu_b, \sigma_b{}^2)$$

then $(X_a - X_b)$ is normally distributed:

$$N(\mu_a - \mu_b, \sigma_a{}^2 + \sigma_b{}^2) \tag{17.1}$$

2. If X_a and X_b are normally distributed, then $(\bar{X}_a - \bar{X}_b)$ is normally distributed. Where $\bar{X}_a$ and $\bar{X}_b$ are based on sample sizes N_a and N_b, respectively, $(\bar{X}_a - \bar{X}_b)$ is

$$N\left(\mu_a - \mu_b, \frac{\sigma_a{}^2}{N_a} + \frac{\sigma_b{}^2}{N_b}\right)$$

The variance of $(\bar{X}_a - \bar{X}_b)$ is denoted by

$$\sigma^2{}_{\bar{X}_a - \bar{X}_b} = \frac{\sigma_a{}^2}{N_a} + \frac{\sigma_b{}^2}{N_b} \tag{17.2}$$

TABLE 17.1 Illustrative Data for a Two-Sample Problem for Independent Samples

Group a		Group b			
Subject	X_{ia}	Subject	X_{ib}		
1	53	1	53	$\sum_{i=1}^{20} X_{ia} = 980$	$\sum_{i=1}^{20} X_{ia}^2 = 48212$
2	48	2	47		
3	49	3	45	$\sum_{i=1}^{18} X_{ib} = 833$	$\sum_{i=1}^{18} X_{ib}^2 = 38773$
4	44	4	48		
5	49	5	44	$\bar{X}_a = \dfrac{1}{20}\sum_{i=1}^{20} X_{ia} = \dfrac{980}{20} = 49.000$	
6	45	6	44		
7	51	7	48	$\bar{X}_b = \dfrac{1}{18}\sum_{i=1}^{18} X_{ib} = \dfrac{833}{18} = 46.278$	
8	50	8	46		
9	50	9	45	$\hat{S}_a^2 = \dfrac{1}{19(20)}\left[20\sum_{i=1}^{20} X_{ia}^2 - \left(\sum_{i=1}^{20} X_{ia}\right)^2\right]$	
10	49	10	44	$= \dfrac{1}{380}(964240 - 960400)$	
11	56	11	41		
12	47	12	42	$= 10.105$	
13	44	13	54		
14	48	14	52	$\hat{S}_b^2 = \dfrac{1}{17(18)}\left[18\sum_{i=1}^{18} X_{ib}^2 - \left(\sum_{i=1}^{18} X_{ib}\right)^2\right]$	
15	46	15	43	$= \dfrac{1}{306}(697914 - 693889)$	
16	47	16	47		
17	52	17	45	$= 13.154$	
18	54	18	45		
19	50				
20	48				
$N_a = 20$		$N_b = 18$			

3. If N_a and N_b are large, then the CLT implies that $(\bar{X}_a - \bar{X}_b)$ is

$$N(\mu_a - \mu_b, \sigma^2_{\bar{X}_a - \bar{X}_b}) \tag{17.3}$$

regardless of the distribution of X_a and X_b.

Case A: X_a, X_b normal $N(\mu_a, \sigma_a^2)$ and $N(\mu_b, \sigma_b^2)$ when σ_a^2 and σ_b^2 are known. The distinction of large versus small sample size is not relevant to the use of this model except to achieve desired values of α and β. The ratio

$$Z = \frac{(\bar{X}_a - \bar{X}_b) - (\mu_a - \mu_b)}{\sqrt{(\sigma_a^2/N_a) + (\sigma_b^2/N_b)}} \tag{17.4}$$

is distributed normally, $N(0, 1)$.

The usual hypothesis tested is the hypothesis of equality of the expectations: $H_T: \mu_a = \mu_b$ and hence $H_T: \Delta_T = \mu_a - \mu_b = 0$. The alternative hypotheses take the following forms:

$$
\begin{array}{lll}
\text{composite lower-tailed} & H_1: \Delta_1 < 0 & \\
\text{composite upper-tailed} & H_2: \Delta_2 > 0 & \\
\text{composite two-tailed} & H_3: \Delta_3 \neq 0 & \quad (17.5)\\
\text{specific} & H_4: \Delta_4 = \Delta_A = \Delta_T + \delta(\sigma_{X_a - X_b}) &
\end{array}
$$

where δ is some specific value. A specific hypothesis H_A is given by $\Delta_4 = \Delta_A$. The critical value T_c in the decision rule for this test refers to the difference in the means, not to the individual means $\bar{X}_a$ and $\bar{X}_b$. The values α and β are selected in the same manner as in the one-sample problem.

Solving for the desired sample size requires a bit more thought than in the one-sample problem. Two considerations are involved: the possibility of unequal values of σ_a^2 and σ_b^2 and of unequal values of N_a and N_b. If we have $\sigma_a^2 = \sigma_b^2 = \sigma^2$,

$$\sigma^2_{\bar{X}_a - \bar{X}_b} = \sigma^2 \frac{N_a + N_b}{N_a N_b} \tag{17.6}$$

If the cost (in terms of money, effort, and so on) of observations is the same under both a and b conditions, then $N_a = N_b = N$ has certain advantages that we need not detail here. In this circumstance we have

$$\frac{N_a + N_b}{N_a N_b} = \frac{2N}{N^2} = \frac{2}{N}$$

Solving simultaneously the equations

$$T_c = \Delta_A + Z_\beta \frac{\sigma\sqrt{2}}{\sqrt{N}}$$

and

$$T_c = \Delta_T + Z_\alpha \frac{\sigma\sqrt{2}}{\sqrt{N}}$$

gives

$$N = 2\left[\frac{(Z_\alpha - Z_\beta)\sigma}{\Delta_T - \Delta_A}\right]^2 \tag{17.7}$$

If $\sigma_a{}^2 \neq \sigma_b{}^2$, we face the alternative of equal sample sizes versus unequal sample sizes. It can be shown that the test of the greatest power for a given value $N_a + N_b$ is that test where

$$\frac{N_a}{N_b} = \frac{\sigma_a}{\sigma_b}$$

Hence the two sample sizes are selected to be proportional to the standard deviations. Setting up the equations for T_c under the α and β conditions gives

$$T_c = \Delta_T + Z_\alpha \sqrt{\frac{\sigma_a{}^2}{N_a} + \frac{\sigma_b{}^2}{N_b}}$$

$$T_c = \Delta_A + Z_\beta \sqrt{\frac{\sigma_a{}^2}{N_a} + \frac{\sigma_b{}^2}{N_b}}$$

Solving these two equations, we have

$$N_a = \sigma_a(\sigma_a + \sigma_b)\left[\frac{Z_\alpha - Z_\beta}{\Delta_T - \Delta_A}\right]^2$$

$$N_b = \sigma_b(\sigma_a + \sigma_b)\left[\frac{Z_\alpha - Z_\beta}{\Delta_T - \Delta_A}\right]^2 \tag{17.8}$$

Case B: X_a and X_b Nonnormal, $\sigma_a{}^2$ and $\sigma_a{}^2$ Known. In Case B the two-sample problem does not differ in any respect from the single-sample problem except in the ways the one- and two-sample problems differ in Case A. One point is important however: Both N_a and N_b separately must be large enough for the CLT to insure approximate normality in each of $\bar{X}_a$ and $\bar{X}_b$ before the distribution of $(\bar{X}_a - \bar{X}_b)$ is sufficiently close to a normal distribution.

Case C: X_a and X_b Normal, $\sigma_a{}^2$ and $\sigma_b{}^2$ Unknown. Here, as in the single-sample problem, we have to consider two special problems: large samples and small samples. In addition, we must consider the possibility that $\sigma_a{}^2 \neq \sigma_b{}^2$, and that $N_a \neq N_b$.

In large samples $\sigma_a{}^2$ and $\sigma_b{}^2$ are estimated by the unbiased estimates $\hat{S}_a{}^2$ and $\hat{S}_b{}^2$. Substituting these values for $\sigma_a{}^2$ and $\sigma_b{}^2$ transforms Case C into Case A.

In small samples greater care must be observed in constructing the statistical test. Although we cannot develop the rationale in detail here, two procedures are necessary: one for variables where $\sigma_a^2 = \sigma_b^2 = \sigma^2$ and another for $\sigma_a^2 \neq \sigma_b^2$. Since σ_a^2 and σ_b^2 are unknown in Case C, the procedure for unequal variances is recommended for general use (contrary to current psychological practice).

Taking the instance of $\sigma_a^2 = \sigma_b^2 = \sigma^2$ first, we estimate the common variance σ^2 by

$$\hat{S}^2 = \frac{\sum_{i=1}^{N_a}(X_{ia} - \bar{X}_a)^2 + \sum_{i=1}^{N_b}(X_{ib} - \bar{X}_b)^2}{N_a + N_b - 2}$$

$$= \frac{(N_a - 1)\hat{S}_a^2 + (N_b - 1)\hat{S}_b^2}{N_a + N_b - 2} \tag{17.9}$$

and computationally by

$$\hat{S}^2 = \frac{\sum_{i=1}^{N_a} X_{ia}^2 + \sum_{i=1}^{N_b} X_{ib}^2 - (\sum_{i=1}^{N_a} X_{ia})^2/N_a - (\sum_{i=1}^{N_b} X_{ib})^2/N_b}{N_a + N_b - 2} \tag{17.10}$$

For this estimator of σ^2 the statistic

$$t = \frac{(\bar{X}_a - \bar{X}_b) - \Delta_T}{\hat{S}/\sqrt{N_a N_b/(N_a + N_b)}} \tag{17.11}$$

is distributed as Student's t with $(N_a + N_b - 2)$ degrees of freedom.

If we assume that $\sigma_a^2 \neq \sigma_b^2$, the appropriate estimate of $\hat{S}_{\bar{X}_a - \bar{X}_b}^2$ is

$$\hat{S}_{\bar{X}_a - \bar{X}_b}^2 = \frac{\hat{S}_a^2}{N_a} + \frac{\hat{S}_b^2}{N_b} \tag{17.12}$$

The ratio

$$t = \frac{(\bar{X}_a - \bar{X}_b) - \Delta_T}{\hat{S}_{\bar{X}_a - \bar{X}_b}} \tag{17.13}$$

is distributed as Student's t with $(N_a + N_b - 2)$ degrees of freedom. However, if $N_a \geq 30$ and $N_b \geq 30$, the CLT indicates that this ratio is also approximately normally distributed. Hence this is an approximation to Case A.

For $\sigma_a^2 \neq \sigma_b^2$ with small samples, the ratio (17.13) is distributed as Student's t. The degrees of freedom for this t is given approximately by

$$df = \frac{[(\hat{S}_a^2/N_a) + (\hat{S}_b^2/N_b)]^2}{\dfrac{(\hat{S}_a^2/N_a)^2}{(N_a + 1)} + \dfrac{(\hat{S}_b^2/N_a)^2}{(N_b + 1)}} - 2 \tag{17.14}$$

to the nearest integer.

Sample size determination and post-facto determination of δ in (α, β, δ) follow the principles already described in detail, if sample sizes are large. However, if sample sizes are small the determination of sample size is more complex. If $\sigma_a{}^2 = \sigma_b{}^2$ and $N_a = N_b$, the values of δ may be obtained from Table J at the back of the book, by entering the table with $N_a + N_b = N$. Conversely, for a given δ where $\sigma_a{}^2 = \sigma_b{}^2$ the value for $N = N_a + N_b$ may be determined for an (α, β, δ) test. On the other hand, where $\sigma_a{}^2 \neq \sigma_b{}^2$ the optimal sample sizes are N_a and N_b where $N_a/N_b = \sigma_a/\sigma_b$. The sample sizes presented in Table J were calculated under the assumption that $\sigma_a{}^2 = \sigma_b{}^2$. Hence, if $\sigma_a{}^2 \neq \sigma_b{}^2$, the N from the table is an approximation only. Using this value of N and finding $N_a/N_b = \sigma_a/\sigma_b$ so that $N = N_a + N_b$ gives a reasonable approximation to the sample sizes for the (α, β, δ) test, providing σ_a/σ_b is not too different from 1.0.

To illustrate these procedures we apply them to the data in Table 17.1. The sample sizes were selected arbitrarily. Imagine that the test hypothesis is that there are zero points difference between the expectations of the random variables observed in the a group and the b group. The alternative is that there is a positive difference. We select $\alpha = .05$ with $\beta = .2$, and use the significance probability in the decision rule rather than calculate the critical value of the test statistic. Applying equations (17.12) and (17.13) with the statistics calculated in Table 17.1 gives

$$\hat{S}^2{}_{\bar{X}_a - \bar{X}_b} = \frac{10.105}{20} + \frac{13.154}{18} = .5052 + .7308 = 1.2360$$

$$t = \frac{(\bar{X}_a - \bar{X}_b) - \Delta_T}{\hat{S}_{\bar{X}_a - \bar{X}_b}} = \frac{49.000 - 46.278 - 0}{1.111}$$

$$= 2.45$$

The degrees of freedom for this test is given by (17.14):

$$df = \frac{(10.105/20 + 13.154/18)^2}{(10.105/20)^2/19 + (13.154/18)^2/17} - 2 = 32.1 \cong 32$$

since we do not assume that the variances in the two groups are equal in using (17.12). Looking in the table of Student's t distribution for 32 degrees of freedom, we find that the value of t obtained gives a significance probability between .01 and .005. Consequently, we reject the test hypothesis of no difference in favor of the alternative that there is a positive difference. Using Table J at the back of the book, we determine the effective alternative under the α and β levels of the test. The entry is found in the $\alpha = .05$ section (one tailed) under the $\beta = .2$ column. Here $N = 38$ is not represented, but linear extrapolation between $N = 36$ and $N = 42$ gives a $\delta = .58$ as an approximation. Consequently, the test is a $(.05, .2, .58)$ test.

The alternative effectively playing a role in this test is given by

$$\Delta_A = \Delta_T + \delta\sqrt{\hat{S}_a^2 + \hat{S}_b^2} = .0 + .58\sqrt{10.105 + 13.154}$$
$$= 2.8$$

If we had started with a specific alternative hypothesis, such as $\Delta_A = 5.0$, and a basis for believing that the two variances were $\sigma_a^2 = 10$ and $\sigma_b^2 = 12.0$, the value of δ would have been

$$\delta = \frac{|\Delta_T - \Delta_A|}{\sqrt{\sigma_a^2 + \sigma_b^2}} = \frac{|0 - 5|}{\sqrt{10 + 12}} = 1.07$$

By consulting Table J at the back of the book, a sample size can be determined, for the test with $(\alpha, \beta, \delta) = (.05, .2, 1.07)$. The table is entered using the value of δ at the margin and then extrapolating between the sample sizes given. The value selected this way is $N = 12$. Under the circumstances of this experiment, group sizes indicated are proportional to $k = \sigma_a/\sigma_b = .833$, that is, $N_b = N/(k + 1) = 6.54 \cong 7$, and $N_a = N - N_b = 5$.

The above analysis is based on the presumption that the variances in the two groups are not equal. If there is evidence that this is not true, a slightly different picture emerges. The equations that are used now are (17.10) and (17.11):

$$\hat{S}^2 = \frac{48212 + 38773 - (980)^2/20 - (833)^2/18}{20 + 18 - 2} = 11.556$$

$$t = \frac{49.000 - 46.278 - 0}{3.399/\sqrt{(20)(18)/(20 + 18)}} = 2.47$$

The t value is only slightly larger than in the calculations for the unequal variances assumptions, largely because of the relative homogeneity of the two variances. The degrees of freedom is given by $N_a + N_b - 2 = 36$. The rejection result is the same.

Case D: X_a and X_b Nonnormal, σ_a^2 and σ_b^2 Unknown. When large samples of observation are made N_a and N_b above 30, the CLT applies and this case is the same as Case B except that $\hat{S}_a^2$ and $\hat{S}_b^2$ are substituted for σ_a^2 and σ_b^2.

Where considerations of the form or empirical nature of the variables X_a and X_b suggest that some transformation of X_a and X_b may be normally distributed, the methods described in Chapter 16 may be used. Each of the variables X_a and X_b are treated separately.

Discrete Transform Methods. When neither continuous transformation methods nor the CLT allow application of the methods developed above for nonnormal variables, discrete transformation methods can be used. These methods are similar to the methods developed in the preceding chapter for Case D. However, for the two-sample case the variety of methods is greater. We develop here only a small selection of these special methods: (1) two-sample median test, (2) Wald-Wolfowitz runs test, and (3) Wilcoxon test.

Median Test. The two-sample median test is an application of the multinomial distribution similar to the one-sample median test. The test is useful in situations where a random variable X of undetermined distribution is observed under two experimental conditions (or in two subject populations). The median test is a test of the hypothesis that X is located at the same place in both experimental conditions *if* the distributions are otherwise identical. The test is sensitive to any difference in the two distributions. Also, it must be assumed that all of the subjects in the population are represented by different values of X (the usual assumption of "continuity" is not necessary unless an unlimited number of subjects are implied by some reasoning outside the confines of the test proper). The method is as follows. The entire set of $N_1 + N_2$ values observed are combined and the grand median determined. The scores of the two groups of subjects are compared, one at a time, with the grand median. If the two groups are identically distributed, half of the subjects will fall above and half below the median. The extent that subjects are assorted above and below the median is an indicator of the degree to which the distributions are located at different points. The frequencies with which subjects are above and below the overall median for each of the groups form a 2×2 table, as illustrated in Table 17.2. The exact probability of this specific assortment is the probability of obtaining a_1 and a_2 subjects above the median out of N_1 and N_2 if observing an a (a_1 or a_2) has the same probability as observing a b (b_1 or b_2 respectively). This is given by the hypergeometric equation

$$\frac{\binom{N_1}{a_1}\binom{N_2}{a_2}}{\binom{N + N_2}{N_a}}$$

TABLE 17.2 Schematic Illustration of Frequency Table for a Two-Sample Median Test

	Group 1	Group 2	Total
Above Median	a_1	a_2	N_a
Below Median	b_1	b_2	N_b
Total	N_1	N_2	$N_a + N_b = N_1 + N_2$

The probability of observing a_2 or fewer subjects in group 2 above the median is given by

$$\frac{\binom{N_1}{a_1}\binom{N_2}{a_2}}{\binom{N_1+N_2}{N_a}} + \frac{\binom{N_1}{a_1+1}\binom{N_2}{a_2-1}}{\binom{N_1+N_2}{N_a}} + \cdots + \frac{\binom{N_1}{N_a}\binom{N_2}{0}}{\binom{N_1+N_2}{N_a}}$$

Conversely, observing a_2 or more subjects in group 2 above the median is

$$\frac{\binom{N_1}{a_1}\binom{N_2}{a_2}}{\binom{N_1+N_2}{N_a}} + \frac{\binom{N_1}{a_1-1}\binom{N_2}{a_2+1}}{\binom{N_1+N_2}{N_a}} + \cdots + \frac{\binom{N_1}{0}\binom{N_2}{N_a}}{\binom{N_1+N_2}{N_a}}$$

The symmetry of these equations indicates that alternative hypotheses about the location of the two groups are symmetric. The test of the hypothesis of equality of the treatment groups can be illustrated as follows.

Imagine a clinical psychologist working with juvenile delinquents. He postulates that a given course of therapeutic treatment and preventive counseling prevents recidivism. Two groups are formed, group 1 receiving standard treatment and group 2 receiving the special course of treatment. The groups are matched (as groups) on a number of relevant factors. At the end of the experimental period each juvenile is judged recidivist or nonrecidivist in terms of whether he is above the grand median current delinquency rating for the group of subjects in the experiment. Ten juveniles are involved, and the frequencies are shown in Table 17.3. The apparent significance of the result can be put to the test by calculating the probability that $a_2 \leq 1$ in the experiment if each juvenile has an equal chance of being recidivist or nonrecidivist:

$$P(a_2 \leq 1) = P(a_2 = 0) + P(a_2 = 1)$$

$$= \frac{\binom{5}{1}\binom{5}{4}}{\binom{10}{5}} + \frac{\binom{5}{0}\binom{5}{5}}{\binom{10}{5}} = \frac{(5)(5) + (1)(1)}{252}$$

$$= .10$$

This is the significance probability of the data. The conclusion must be that the two distributions of delinquency ratings are not significantly different, or that more data are needed to detect the difference.

TABLE 17.3 Illustrative Frequencies for a Two-Sample Median Test

	Nonrecidivist	Recidivist	Total
Experimental—Group 2	4	1	5
Control—Group 1	1	4	5
Total	5	5	10

For experiments with larger group sizes ($N \geq 20$) an approximation to the hypergeometric distribution makes this test practical. If the variable a_2 is distributed as a hypergeometric variable with

$$N = N_a + N_b$$

then

$$E(a_2) = \frac{N_a N_1}{N}$$

$$\mathrm{var}(a_2) = \frac{(N - N_a)N_a N_1(N - N_1)}{(N - 1)N^2}$$

The variable

$$Z = \frac{a_2 + .5 - E(a_2)}{\sqrt{\mathrm{var}(a_2)}}$$

is $N(0, 1)$ approximately.

This test can be applied to two-sample problems with dichotomous variables, such as "arrested" or "not arrested" during the experimental period. An approximation useful in the dichotomous variable case is discussed in the chapter on χ^2 applications.

Runs Test. The two samples of observations are put into numerical order—that is, the $N_1 + N_2$ values are ranked. This ranking is done without regard for group. If we denote a score from group 1 as a G_1 and a score from group 2 as a G_2 the resulting ordering of G's might be

$$G_1 G_1 \quad G_2 \quad G_1 G_1 \quad G_2 G_2 G_2 G_2 \quad G_1 G_1 \quad G_2$$

where contiguous G_1's and contiguous G_2's make up a run. Six runs occur in the sequence of G_1's and G_2's represented here.

If the two groups come from populations with identical distributions, we would expect the orders of observed magnitudes to be randomly assorted to the two groups—the pth largest observed value is just as likely to be in one group as in the other group. The maximum of runs $N_1 + N_2$ is obtained when the ordered values alternate between the two groups. The minimum number of runs, two, is observed when all the values of one group are larger than the other group. The greater the difference in the two groups, the smaller is the number of runs. If m runs are observed, the significance probability can be found by calculating the probability of observing m or fewer runs. This calculation is very tedious if m is large. For each odd-valued integer $2k + 1 \leq m$, the probability is

$$P(I = 2k + 1) = \frac{\binom{N_1 - 1}{k - 1}\binom{N_2 - 1}{k} + \binom{N_1 - 1}{k}\binom{N_2 - 1}{k - 1}}{\binom{N_1 + N_2}{N_1}}$$

For even-valued integers $2k \leq m$ the probability is

$$P(I = 2k) = \frac{2\binom{N_1 - 1}{k - 1}\binom{N_2 - 1}{k - 1}}{\binom{N_1 + N_2}{N_1}}$$

The sum of these probabilities is the significance probability of the observed result.

Ties in values are a problem in this test. If two subjects, one from each group, have the same score, the ordering of the subjects influences the number of runs observed. The "conservative" procedure is to break ties in a way to increase the number of runs. The more ties observed, the more arbitrary is the number of runs and consequently the more arbitrary is Ψ.

This test is questionable as a discrete transform test. The power of the test is thought to be marginal in comparison with the Wilcoxon test. However, one special use of the test is quite elegant. Imagine that a series of events crosses a gate of some kind (for example, an admissions desk, $R_1 - R_2$ in a two-choice learning experiment). Each event is of one type or another. The number of runs of types of events can be used to test the hypothesis that the two types are dependent on time of observation (seasonal variation, learning, and so on).

Wilcoxon Two-Sample Test. If a discrete transformation test is necessary, the best choice perhaps is the Wilcoxon two-sample test. This test is equivalent to the test known as the Mann-Whitney U test. The distribution of the Wilcoxon statistic is presented here along with tables sufficient for most practical applications. In addition, suggestions for evaluating the power of the test against specific alternative hypotheses are made.

TABLE 17.4 Illustrative Data for Wilcoxon Two-Sample Test

Observed Values	Experimental–Group 2	.36, 1.0, 2.5, 5.2, 6.5
	Control–Group 1	3.0, 5.8, 7.0, 7.9, 12.5
Ranks of Values	Experimental–Group 2	1, 2, 3, 5, 7 Sum = 18
	Control–Group 1	4, 6, 8, 9, 10 Sum = 37

The Wilcoxon test is based on the ranks of the observed values. Using the delinquency study as an example, imagine that the delinquency scores for the ten youngsters, together with the ranked scores, are shown in Table 17.4. A casual inspection of these data suggests that the control group has considerably higher delinquency scores. The Wilcoxon test provides a way of testing the null hypothesis of equality of the two groups. The test uses all of the order relations between the two groups.

The test statistic is the sum of the ranks in one of the two groups. The sum of ranks for both groups tends to be the same if the experimental treatment is not effective. If the experimental treatment is effective, the ranks in that group tend to be small in our example (other treatments may increase the magnitude of values observed, as in performance experiments, and large ranks might be associated with the experimental treatments). The total sum of ranks is partitioned into two complementary parts:

$$\sum \text{ranks} = W_E + W_C = 1 + 2 + \cdots + 10 = 55$$

where W_E and W_C are the sums of ranks of the experimental and control groups, respectively. It is convenient to work with the smaller value of the two. In order to determine the significance probability of W_E we may work with W_C or W_E because the probability of observing a value equal to or larger than W_C is the probability of observing a value equal to or smaller than W_E, and conversely. In our example, W_E is the smaller partial sum (18) and we develop the test procedures in terms of W_E. The first step is to determine the distribution of W_E under the hypothesis that the two groups come from the same population, with respect to the distribution of delinquency score.

If the hypothesis is correct, we might regard the experimental and control groups as randomly selected from a single group with the scores observed. That is, the scores are fixed prior to the treatments and assigned randomly to the two groups. The selection of scores for the experimental group can be made in $\binom{10}{5} =$ 252 different ways. If we dealt with the scores directly, we could form all such samples and compare the observed sum of scores with the distribution of randomly slecected scores to obtain a significance probability. However, this procedure is highly dependent on the specific values observed in the experiment. The procedures can be generalized by dealing with the ranks of the observed values—the ranks define a stable and consistent value set with which to work. The distribution of the sum of N_E out of N ranks can be derived, and it pertains to all experiments with N_E experimental subjects out of N subjects in the experiment.

The calculation of the distribution is laborious and tedious. However, it is relatively simple to find all the possible arrangements of 5 ranks of the 10 with a sum of 18 or less:

$$1 + 2 + 3 + 4 + 5 = 15$$
$$1 + 2 + 3 + 4 + 6 = 16$$
$$1 + 2 + 3 + 4 + 7 = 17$$
$$1 + 2 + 3 + 5 + 6 = 17$$
$$1 + 2 + 3 + 4 + 8 = 18$$
$$1 + 2 + 3 + 5 + 7 = 18$$
$$1 + 2 + 4 + 5 + 6 = 18$$

Thus, out of the 252 equally likely arrangements of the 5 out of 10 ranks, only 7 have sums of 18 or less:

$$\Psi = P(W_E \leq 18) = \frac{7}{252} = .028$$

This result should be compared with the result of the median test.

We now develop the distribution of W_S for the general case of $(N = N_S + N_L)$ observations where N_S is the number of observations of the smaller group and W_S is the partial sum of ranks for that group. If $N_S = N_L$, either partial sum may be used. For a random selection of N_S integers out of the first N integers $1, 2, \ldots, N$, there are $\binom{N}{N_S}$ equally likely samples. Hence

$$P(W_S \leq w) = \frac{n(W_S \leq w)}{\binom{N}{N_S}}$$

where $n(W_S \leq w)$ is the number of samples with partial sum W_S equal to or less than w. Table K at the back of the book is a table of $n(W_S \leq w)$ for $N \leq 16$ where $3 \leq N_S \leq 8$. To avoid a large table, the values of the table are not w directly.

The table is entered with the amount by which w exceeds the minimum value possible, $1 + 2 + \cdots + N_S = N_S(N_S + 1)/2$. Hence the column used in the table is headed

$$w - \frac{N_S(N_S + 1)}{2}$$

The row is chosen to conform to the values of N_S and N_L. In the example, $N_S = N_L = 5$. The observed value was $w = 18$. Hence we enter the column headed

$$w - \frac{5(5 + 1)}{2} = 18 - 15 = 3$$

The entry in the column headed 3 for the ($N_S = N_L = 5$) row is 7. This is the same figure determined by enumeration. The value $\binom{N}{N_S}$ is also given in the table.

Table K permits an evaluation of the lower tail of the distribution of the partial sum of ranks. To evaluate the upper tail, that is, to reverse the comparison, the ranks can be reversed (large values given small rank values).

Where group sizes are larger than those of Table K, an approximation using the standard normal distribution may be used. It can be proved that

$$E(W_E) = \frac{N_S(N + 1)}{2}$$

$$\operatorname{var}(W_E) = \frac{N_S N_L(N + 1)}{12}$$

The variable

$$Z = \frac{w - .5 - E(W_E)}{\sqrt{\operatorname{var}(W_E)}}$$

is approximately $N(0, 1)$ if N is large.

The determination of the power of the Wilcoxon test depends on the availability of a specific alternative in the original units of measurement. In the delinquency example, imagine that the researcher knows that unless an improvement of 6 points is achieved, the effort is not worthwhile. Hence the minimum meaningful difference in scores is 6. The test hypothesis is that the difference is 0. If the difference were in fact 6, what is the probability with $N_S = N_L = 5$ that $P(W_E \leq w) \leq .02 = \alpha$, causing us to reject the null hypothesis? This probability is the power of the test with $\alpha = .02$ and an absolute difference of 6 points. Calculation of the probability is simple but laborious. Each sample of the N_E subjects under the test hypothesis would correspond to N_E specific values of delinquency scores. If 6 were added to each of these values and the test were then calculated, the probability of rejecting the test hypothesis would be the "power" of the test. The "N_E specific scores" must be available under the test hypothesis before the artificial addition of the 6. A random sample of ($N = N_E + N_C$) scores could be the source of the necessary scores. This sample should, of course, be independent of the sample used in the experiment proper. A determination of sample size necessary to achieve sufficient power is possible by an iterative application of this method. These evaluations are not feasible without the utilization of an electronic computer if N is much greater than 5 or 6. Since the procedure suggested is based on a sample, the "power" calculated is an *estimate* of the probability of correctly rejecting H_T in the population of scores.

Statistical Procedures: Repeated Observations

Two of the designs introduced in this chapter involve observing a given subject (or matched subjects) repeatedly. The observations made in the experimental condition are linked with the observations made in the control condition. This linkage may be accompanied by a covariation of the experimental and control variables. If the variation of each of the two variables is due in part to common characteristics in the experiment, the two variables have a covariance greater than zero. This gives a special result hinted at in the chapter on variance and covariance. If we have two dependent random variables X and Y, then

$$\sigma_{X-Y}^2 = \sigma_X{}^2 + \sigma_Y{}^2 - 2\text{cov}(X, Y)$$

That is, the variance of the difference of the two random variables is smaller than the sum of the variances by an amount equal to twice the covariance. The comparison in a statistical test of expectations of random variables, discussed in the preceding section, dealt with independent groups, and the variance of the difference of the random variables was equal simply to the sum of the variances. Applying the logic of the variance of the mean to the difference between means where the random variables are dependent, we have, for two random variables X and Y,

$$\sigma_{\bar{X}-\bar{Y}}^2 = E(\bar{X} - \bar{Y} - \mu_X + \mu_Y)^2$$
$$= E(\bar{X} - \mu_X)^2 + E(\bar{Y} - \mu_Y)^2 - 2E(\bar{X} - \mu_X)(\bar{Y} - \mu_Y)$$
$$= \sigma_{\bar{X}}{}^2 + \sigma_{\bar{Y}}{}^2 - 2\text{cov}(\bar{X}, \bar{Y}) \tag{17.15}$$

We already know how to estimate $\sigma_{\bar{X}}{}^2$ and $\sigma_{\bar{Y}}{}^2$, but we have developed no way of estimating $\text{cov}(\bar{X}, \bar{Y})$. An estimate of this covariance is easily derived, but we do not need to develop it here. A much simpler procedure shortcuts the problem completely. We know that expectations and means are additive in the sense that

$$E(X + Y) = E(X) + E(Y)$$

and

$$\overline{(X + Y)} = \bar{X} + \bar{Y}$$

Thus, in the test of the hypothesis $H_T : \mu_{X_a} = \mu_{X_b}$, we can substitute the mean of the differences of the paired observations for the differences of the means. Let $D_i = X_{ia} - X_{ib}$ for the ith subject in the experiment. Then

$$\bar{D} = \bar{X}_a - \bar{X}_b \qquad (17.16)$$

is an unbiased estimate of $\mu_{X_a} - \mu_{X_b}$. Furthermore, the variance of the differences D_i is equal to the variance of $X_{ia} - X_{ib}$, including the covariance term:

$$\sigma_D^2 = \sigma_{X_a - X_b}^2$$

And the variance of the mean of the differences D_i is equal to the variance of the difference of the means

$$\sigma_{\bar{D}}^2 = \sigma_{\bar{X}_a - \bar{X}_b}^2 \qquad (17.17)$$

Where the sample estimate of the variance of the differences of the mean was complicated by the covariance factor, the sample estimate of the variance of $\bar{D}$ is simple and straightforward. We simply treat D as a random variable (instead of as a difference between random variables) and calculate the sample variance of D, from which we can calculate the variance of $\bar{D}$ with the familiar formula. Let

$$\hat{S}_D^2 = \frac{1}{N-1} \sum_{i=1}^{N} (D_i - \bar{D})^2$$

then

$$\hat{S}_{\bar{D}} = \frac{\hat{S}_D}{\sqrt{N}} \qquad (17.18)$$

The application of statistical test theory to the difference variable D is straightforward. The methods of testing hypotheses about expectations in single samples apply directly. The interpretation differs in this instance. Whereas in the one-sample problem the hypotheses were $H_T : \mu = \mu_T$ versus $H_A : \mu = \mu_A$, we now have $H_T : \mu_a - \mu_b = \Delta_T$ versus $H_A : \mu_a - \mu_b = \Delta_A$. In general, matched subjects and subjects-as-own-control experiments are designed to determine equality or inequality of treatment expectations, and $\Delta_T = 0$ while $\Delta_A \neq 0$. The (α, β, δ) procedures for a one-sample test apply directly except that δ refers to $\Delta_T - \Delta_A$.

These procedures are illustrated with data in Table 17.5. Two sets of data are presented in order to make a point regarding the reduction of the standard error of the mean difference as a function of the covariance between the two observations. Summary statistics for the two sets of data are given in Table 17.5. We first test the

TABLE 17.5 Illustrative Data for Two-Sample Repeated Observations Study

		Experiment I		
Subject	X_a	X_b	D	

Subject	X_a	X_b	D
1	17	17	0
2	15	14	1
3	17	19	−2
4	13	19	−6
5	15	19	−4
6	13	18	−5
7	12	14	−2
8	18	18	0
9	19	18	1
10	12	21	−9
11	17	19	−2
12	15	21	−6
13	13	9	4
14	17	17	0
15	16	21	−5
16	14	14	0

$$\sum_{i=1}^{16} X_{ia} = 243 \qquad \bar{X}_a = 15.188$$

$$\sum_{i=1}^{16} X_{ib} = 278 \qquad \bar{X}_b = 17.375$$

$$\sum_{i=1}^{16} X_{ia}^2 = 3763 \qquad \sum_{i=1}^{16} X_{ib}^2 = 4986$$

$$S_a^2 = \frac{1}{256} [16(3763) - (243)^2] = 4.527$$

$$S_b^2 = \frac{1}{256} [16(4986) - (278)^2] = 9.734$$

$$\sum_{i=1}^{16} X_{ia}X_{ib} = 4250$$

$$C_{ab} = \frac{1}{256} [16(4250) - (278)(243)] = 1.742$$

$$S_{X_a-X_b}^2 = S_a^2 + S_b^2 - 2C_{ab}$$

$$= 4.527 + 9.734 - 2(1.742) = 10.777$$

$$\sum_{i=1}^{16} D_i = -35 \qquad \bar{D} = -2.188$$

$$\sum_{i=1}^{16} D_i^2 = 249$$

$$S_D^2 = \frac{1}{256} [16(249) - (-35)^2] = 10.777$$

$$\hat{S}_D^2 = \frac{1}{(15)(16)} [16(249) - (-35)^2] = 11.496$$

TABLE 17.5 (continued)

				Experiment II
Subject	X_a	X_b	D	
1	11	10	1	$\sum_{i=1}^{18} X_{ia} = 389 \qquad \bar{X}_a = 21.611$
2	21	20	1	
3	30	30	0	$\sum_{i=1}^{18} X_{ib} = 328 \qquad \bar{X}_b = 18.222$
4	18	10	8	
5	14	10	4	$\sum_{i=1}^{18} X_{ia}^2 = 8917 \qquad \sum_{i=1}^{18} X_{ib}^2 = 6580$
6	20	18	2	
7	25	21	4	$S_a^2 = \dfrac{1}{324} [18(8917) - (389)^2] = 28.349$
8	20	19	1	
9	21	18	3	$S_b^2 = \dfrac{1}{324} [18(6580) - (328)^2] = 33.506$
10	24	23	1	$\sum_{i=1}^{18} X_{ia}X_{ib} = 7604$
11	36	31	5	
12	21	19	2	$C_{ab} = \dfrac{1}{324} [18(7604) - (389)(328)] = 28.642$
13	23	17	6	$S^2_{X_a-X_b} = S_a^2 + S_b^2 - 2C_{ab} = 4.571$
14	24	20	4	
15	23	20	3	$\sum_{i=1}^{18} D_i = 61 \qquad \bar{D} = 3.389$
16	20	15	5	
17	19	13	6	$\sum_{i=1}^{18} D_i^2 = 289$
18	19	14	5	
				$S_D^2 = \dfrac{1}{324} [18(289) - (61)^2] = 4.571$
				$\hat{S}_D^2 = \dfrac{1}{(17)(18)} [18(289) - (61)^2] = 4.840$

hypothesis of equal expectations by the random groups procedures, as if the two groups of scores for each experiment were independent. The hypothesis is that there is no difference between the expectations under the two treatments in an experiment. For the sake of simplicity here, we assume that the alternative is a two-tailed composite hypothesis. In experiment I the value of t is

$$t = \frac{15.19 - 17.38 - 0}{\sqrt{4.829/16 + 10.383/16}} = \frac{-2.19}{.975} = -2.25$$

The degrees of freedom by equation (17.14) is estimated at 28. The significance probability for this t is between .02 and .05. If the repeated measures procedures are used, the value of t is

$$t = \frac{-2.188 - 0}{\sqrt{11.496/16}} = -2.58$$

with 15 degrees of freedom. The significance probability for this result is between .02 and .05. The discrepancy between the two procedures is not very striking. Both procedures lead to about the same significance probability. However, since the scores in the two groups are correlated, the random groups calculation is incorrect.

In experiment II the results are not significant at the .05 level if the random groups two-sample calculations are applied:

$$t = \frac{21.611 - 18.222 - 0}{\sqrt{30.017/18 + 35.477/18}} = 1.777$$

with 34 degrees of freedom according to (17.14). On the other hand, if the subjects-as-own-control, repeated measures calculation is applied, the result is significant beyond the .01 significance probability level:

$$t = \frac{3.389 - 0}{\sqrt{4.480/18}} = 6.536$$

The degrees of freedom here is 17.

The contrast in this last example is very striking. The results are transformed from nonsignificance to a high degree of significance with the application of the correct calculation. The primary point is that when an inappropriate model is applied to a set of data, the result may be unrelated to the result that would have been obtained with the correct calculation. The pertinence of this contrast is difficult to overemphasize. The illustration presents a clear case where two models are apparently applicable, if only the data are considered. However, when the way in which the data were collected is considered, the one model is clearly inappropriate. The random groups calculations are based on the presumption that the two groups, and hence their means, are independent. The calculation of the standard error of the difference between means is carried out in terms of this assumption.

Since the means are not independent, the calculation is in error, and consequently, the significance probability attributed to the result is in error, even if all of the other assumptions of the t test are met in the phenomenon being studied.

The difference in experiments I and II, with respect to the model used, resides largely in the covariance of X_a and X_b in the two experiments. In experiment I the covariance is small in relation to the variance of the variables observed. This is seen most clearly by comparing the standard error of the mean difference of the means with the standard error of the mean of the differences. In experiment I the standard deviations are, respectively, .975 and .848, the difference being attributable to the covariance of 1.742 between X_a and X_b. In experiment II the standard deviations are, respectively, 1.906 and .518, with the covariance being 28.642.

The determination of δ in an (α, β, δ) procedure generally should be in terms of the standard deviation of the differences rather than in terms of the standard deviation of the variables. The covariance between the variables can make a great deal of difference in the interpretation of the variability of the variables and in the differences between the variables, as just demonstrated. However, if the research is intended to provide a basis for the differentiation of individuals or groups in terms of the mean of a single variable, then the value of δ and the attendant sample size probably should be determined from the standard deviations of the variables themselves, not the differences of the variables. The specific goals of the research must determine the procedures used in repeated measures experiments.

The efficiency of this design cannot be overstated. The sample size necessary to achieve a given (α, β, δ) test is a function of $\text{cov}(X, Y)$. For a given (α, β, δ) the required sample size is inversely related to $\text{cov}(X, Y)$ because of the role of $\text{cov}(X, Y)$ in $\sigma^2_{\bar{X} - \bar{Y}}$, providing that $\text{cov}(X, Y) \geq 0$. Note that a negative covariance increases the standard deviation of the mean difference. The student should attempt to think of experiments that would produce a negative covariance.

A number of discrete transformation methods are available for statistical analysis in this design. Two of these methods are presented here: (1) sign test, and (2) Wilcoxon test.

Sign Test. Each of N subjects is observed under two conditions, an experimental and a control. The score of the ith subject on the experimental task is X_{Ei}. On the control treatment the ith subject's score is X_{Ci}. If the two treatments are the same with respect to the subject population, the only sources of differences in X_{Ei} and X_{Ci} are random measurement effects. If the random measurement effects are equally pervasive in both experimental and control treatments, the difference

$$X_{Ei} - X_{Ci}$$

is just as likely to be positive as negative. The sign of the difference is recorded for all N subjects. The number of pluses is a variable S.

Under the hypothesis of equality of the treatments, S is a binomial random variable $B(N, .5)$. Therefore the sign test for repeated measurements in a two-sample problem is a simple binomial test. All of the previously developed theory of tests of hypotheses about binomial parameters pertains in this special instance.

Wilcoxon Test for Correlated Observations. This test is an extension of the Wilcoxon ranking procedures to the differences used in the sign test. The differences $X_{Ei} - X_{Ci}$ are ranked by the magnitude of the absolute values, and the sum of the ranks of the positive or of the negative differences is calculated. The sum of ranks is taken for the subjects with the least frequent sign.

There are 2^N equally likely different combinations of pluses and minuses. For each combination there is some specific value for the sum of ranks of like-signed differences. If this sum is called W, the probability of observing a value smaller than the observed partial sum w is the significance probability of w,

$$\Psi = P(W \leq w) = \frac{n(W \leq w)}{2^N}$$

where $n(W \leq w)$ is the number of the 2^N combinations having a smaller (or equally small) partial sum. Tables L, M, and N at the back of the book are necessary to easily calculate the significance probability of a given observed resu':. The partial sums of ranks of negative differences and of ranks of positive differences are distributed the same way. Hence, either sum may be used to enter the tables. Table L gives 2^N for $N \leq 20$. Table M gives $n(W \leq w)$ for $w \leq N$, and Table N gives $n(W \leq w)$ for $w \geq N$. Table 17.6 illustrates the entire test application.

TABLE 17.6 Illustration of the Wilcoxon Two-Sample Test for Repeated Measures

	X_a	X_b	D	Rank D	
1	33	44	−11	6	Sum of ranks with sign appropriate to
2	38	63	−25	8	the test of H_T against H_A: $w = 15$.
3	32	36	−4	2	
4	47	37	10	5	
5	25	28	−3	1	If $w \leq N$, use Table M to find $n(W \leq w)$.
6	57	52	5	3	
7	33	39	−6	4	
8	49	34	15	7	If $w \geq N$ use Table N to find $n(W \leq w)$.

$n(W \leq w) = 95$ $\quad 2^N = 95$

$\Psi = \frac{95}{256} = .37$

Hypotheses Concerning Variances

This section presents a very general result and its application to a special case. The general result is the distribution called Fisher's F distribution or the F ratio distribution. The special case is a test of the hypothesis that two-sample variances are from samples of observations on the same population.

Distribution of Ratios of Variance Estimates. The development of Fisher's F, the ratio of two independent estimates of a single variance, is similar to that of Student's t. We develop F here in less detail than t was dealt with in Chapter 15. Our development is limited to the special case of normally distributed random variables.

Two independent samples of observations, of size $(N_1 = n_1 + 1)$ and $(N_2 = n_2 + 1)$, are made on a random variable X that is normally distributed, $N(\mu, \sigma^2)$. The sample estimates of the variance σ^2 are distributed as modified χ^2 random variables:

$$\hat{S}_1^{\,2} = \frac{\sigma^2 W_{n_1}}{n_1}$$

$$\hat{S}_2^{\,2} = \frac{\sigma^2 W_{n_2}}{n_2}$$

where W_{n_1} is distributed as $\chi_{n_1}^{\,2}$ and W_{n_2} is distributed as $\chi_{n_2}^{\,2}$.

R. A. Fisher derived the distribution of

$$\frac{\hat{S}_1^{\,2}}{\hat{S}_2^{\,2}} = \frac{\sigma^2 W_{n_1}}{n_1} \times \frac{n_2}{\sigma^2 W_{n_2}} = \frac{n_2}{n_1} \times \frac{W_{n_1}}{W_{n_2}}$$

This ratio is the ratio of two independent χ^2 distributed random variables times the ratio of the respective degrees of freedom:

$$\frac{\hat{S}_1^{\,2}}{\hat{S}_2^{\,2}} : \frac{\chi_{n_1}^{\,2}/n_1}{\chi_{n_2}^{\,2}/n_2} \tag{17.19}$$

The substance of the derivation can be expressed in a heuristic-pictorial argument similar to that used in the discussion of Student's t.

In Chapter 12 the sampling distribution of the variance estimate $\hat{S}^2$ was developed. If there are two independent estimates of a population variance, for example $\hat{S}_1^{\,2}$ and $\hat{S}_2^{\,2}$, we know that, for W_1 and W_2 χ^2 distributed random variables,

$$P(\hat{S}_1^{\,2} \leq W_1 \quad \text{and} \quad \hat{S}_2^{\,2} \leq W_2) = P(\hat{S}_1^{\,2} \leq W_1)P(\hat{S}_1^{\,2} \leq W_2)$$

We can represent the sample spaces for $\hat{S}_1^{\,2}$ and $\hat{S}_2^{\,2}$ by $X_1 = \chi_{n_1}^{\,2}/n_1$ and $X_2 = \chi_{n_2}^{\,2}/n_2$ respectively and in turn represent the joint sample space as the positive quadrant of a plane defined by a pair of coordinates X_1 and X_2. The density function of the joint distribution of $\hat{S}_1{}^2$ and $\hat{S}_2{}^2$ is obtained by differentiation of the distribution function

$$F(\hat{S}_1{}^2 = W_1, \hat{S}_2{}^2 = W_2) = P(\hat{S}_1{}^2 \leq W_1)P(\hat{S}_2{}^2 \leq W_2)$$

This distribution is the same as the distribution of X_1 and X_2. The joint sample space and the density function on that space may be represented as a surface in two dimensions. Figure 17.1 is an illustration of such a surface represented by a contour diagram. The distributions of X_1 and X_2, that is, $\chi_{n_1}{}^2/n_1$ and $\chi_{n_2}{}^2/n_2$, are shown. Each contour represents a constant value of the density function $f(X_1, X_2)$.

All pairs of values of X_1 and X_2 in a constant ratio X_2/X_1 lie on straight lines radiating from the $(X_1 = 0, X_2 = 0)$ point. The lines in the "lower half" of the figure are for values $(X_2/X_1) < 1$, and the lines in the "upper half" are for $(X_2/X_1) > 1$. The probability of observing a given *ratio value* is given by integrating (accumulating) the density (probability) along the respective radiating line. If this integration is carried out for each value of the ratio $V = X_2/X_1$, the result is a

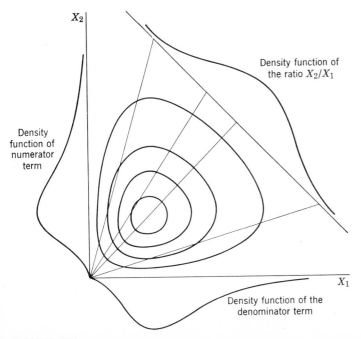

FIGURE 17.1

Schematic drawing illustrating the joint density function and marginal density functions giving rise to the Fisher F distribution. Lines radiating from (0, 0) are lines of integration producing the density function of the ratio.

density function for the variable V. Since X_1 and X_2 are the sample spaces for $\hat{S}_1{}^2$ and $\hat{S}_2{}^2$, the density function of V is the density function of the sample space for $\hat{S}_2{}^2/\hat{S}_1{}^2$. The distribution of V was derived by R. A. Fisher. The tabled version of the distribution is a slight modification of the original variable and is due to Snedecor. The tabled distribution is called the F distribution.

The two marginal variables X_1 and X_2 are dependent on the number of degrees of freedom in the estimates $\hat{S}_1{}^2$ and $\hat{S}_2{}^2$. Consequently, the F distribution is dependent on n_1 and n_2. The F distribution is independent of the variance σ^2 estimated by $\hat{S}_1{}^2$ and $\hat{S}_2{}^2$.

The fact that F is dependent on both n_1 and n_2 is a source of difficulty in tabling the distribution function of F. For each value of n_1 there is an entire F distribution for each value of n_2. Hence, to table the F distribution as finely as Z, one table for each pair of values n_1 and n_2 would be needed. Since n_1 and n_2 can take all integer values, the total F table would be unlimitedly large. Restricting the values of n_1 and n_2 to the first 100 integers and then to 200, 300, ... , 1000 would still require 108^2 tables like that of the normal. The solution to the tabling problem is much like that in the χ^2 table. A selection of values of the v such that

$$P(V \geq v) = p$$

for a selection of n_1 and n_2 values is used to construct the table of F. Thus, for $p = .05$, $n_1 = 10$, and $n_2 = 20$, the value of v satisfying

$$P(V \geq v \mid n_1 = 10, n_2 = 20) = .05$$

is tabled. This value generally is denoted

$$F_{n_2, n_1, p}$$

The interpretation of observing a ratio $\hat{S}_2{}^2/\hat{S}_1{}^2$ greater than $F_{n_2, n_1, p}$ is that $\hat{S}_2{}^2$ is so much larger than $\hat{S}_1{}^2$ that if both $\hat{S}_1{}^2$ and $\hat{S}_2{}^2$ were estimates of the same variance, the *ratio* could have occurred with probability of at most p. Table O presents values of $F_{n_2, n_1, p}$ for a selection of values of n_1 and n_2 and a number of values of p. The general availability of high-speed digital computers makes it practical to calculate probability values for any value of F for any n_1 and n_2.

Some arbitrary conventions in the use of the F distribution and some properties of the distribution are important. The equation of the density functions is complicated, and presenting it here serves no useful function. However, the only parameters determining the function are n_1 and n_2—σ^2 is not a parameter of F. Since $\hat{S}_1{}^2$ and $\hat{S}_2{}^2$ are positive real numbers, their ratio is a positive real number. Hence, V has a value set corresponding to all positive real numbers. The expectation of V is $n_1/(n_1 - 2)$, and the distribution is somewhat skewed toward the larger values. The formula for the variance of V is not particularly useful in the present context.

Test of Hypothesis that $\sigma_A{}^2 = \sigma_B{}^2$. In situations where we have two variances, one for each of two independent samples, we often wish to test

$$H_T: \sigma_A{}^2 = \sigma_B{}^2 = \sigma^2 \qquad (17.20)$$

against

$$H_1: \sigma_A{}^2 > \sigma_B{}^2$$

$$H_2: \sigma_A{}^2 < \sigma_B{}^2 \qquad (17.21)$$

$$H_3: \sigma_A{}^2 \neq \sigma_B{}^2$$

This can be done directly if the $(n_A + 1)$ and $(n_B + 1)$ observations giving $\hat{S}_A{}^2$ and $\hat{S}_B{}^2$ are independent observations on normally distributed random variables $X_A: N(\mu_A, \sigma_A)$ and $X_B: N(\mu_B, \sigma_B)$. Under the hypothesis H_T, the ratio

$$\frac{\hat{S}_A{}^2}{\hat{S}_B{}^2}$$

is distributed as Fisher's F. The critical values for the two-sided alternatives are

$$F_{n_A, n_B, .5\alpha}$$

and

$$F_{n_A, n_B, 1-.5\alpha}$$

If

$$F_{n_A, n_B, 1-.5\alpha} \geq \frac{\hat{S}_A{}^2}{\hat{S}_B{}^2}$$

or

$$F_{n_A, n_B, .5\alpha} \leq \frac{\hat{S}_A{}^2}{\hat{S}_B{}^2}$$

then the hypothesis H_T is rejected in favor of H_3. For the one-sided hypothesis the critical values and rejection rules are

$$\text{for } H_1: \text{if } F_{n_A, n_B, \alpha} \leq \frac{\hat{S}_A{}^2}{\hat{S}_B{}^2}, \text{ reject } H_T$$

$$(17.22)$$

$$\text{for } H_2: \text{if } F_{n_A, n_B, 1-\alpha} \geq \frac{\hat{S}_A{}^2}{\hat{S}_B{}^2}, \text{ reject } H_T$$

For specific alternatives the application of the (α, β, δ) procedures to determine appropriate sample size or to evaluate, post facto, the meaning of a test of the hypothesis of equality of variances is virtually identical to those of the one-sample problem. Table P at the back of the book can be used to determine the value of $\delta^2 = \sigma_A^2/\sigma_B^2$ giving a specific value of β for a specific α when $n_A = n_B$. Entering the table with $n_A = n_B = n$ permits a post-facto evaluation of δ^2. Entering the table with δ^2 permits a pre-experimental determination of $n_A = n_B$ appropriate to an (α, β, δ) test. If $\delta^2 \leq 1$, the table is entered with $1/\delta^2$. The power function of F tests are given in Chart Q at the back of the book.

It must be emphasized that the use of the F distribution to test hypotheses about variances is limited to normally distributed variables. Unless X_A and X_B are normally distributed (or very large samples are involved), $\hat{S}_A^2/\hat{S}_B^2$ is not distributed as Fisher's F. That is, the distribution function of $\hat{S}_A^2/\hat{S}_B^2$ is not the distribution function given in tables of F unless both observed variables are normally distributed. Moderate departures from normality involve the possibility of serious departures from the F distribution.

To illustrate these procedures they are applied to the data of Table 17.1. Our test hypothesis is $H_T: \sigma_a^2 = \sigma_b^2$, and the alternative is $H_A: \sigma_a^2 \leq \sigma_b^2$. Imagine that we were willing to tolerate decision errors indicated by $\alpha = .05$ and $\beta = .1$. The obtained values are $\hat{S}_a^2 = 10.105$ and $\hat{S}_b^2 = 13.154$. We compare the ratio of these two estimates with the value given by $F_{n_b, n_a, \alpha} = F_{17,19,.05}$, that is, 2.20. The obtained ratio is $13.154/10.105 = 1.30$. Clearly, the ratio is not extreme enough to cause us to reject the test. However, we must question this conclusion in the light of the sample size. Since the tables and charts giving sample sizes and powers of these tests are based on equal sample sizes, we compromise and deal with $N_a = N_b = 19$. We enter Table P at the back of the book with our risk levels, .05 and .10 respectively, and the sample size in each sample, that is, the compromise value 19. Extrapolating between sample size of 15 and 20, we find that the δ^2 value is approximately 4.0. Consequently sample size 19 in each of the two samples would permit the detection in a $[(\alpha, \beta, \delta) = (.05, .1, 2)]$ test. If we take the opposite approach and specify δ^2 by establishing specific alternative hypotheses or judging what a meaningful ratio would be, we can obtain N from Table P. Imagine that our hypothesis H_T specified that $\sigma_a^2 = \sigma_b^2 = 10$, whereas H_A specified that $\sigma_a^2 = 10$ and $\sigma_b^2 = 30$. This specification gives $\delta^2 = 3.0$, and $\delta = 1.73$. Entering Table P with δ^2, α, and β we discover that the value 3.0 is not tabled. Interpolating between 2.957 for sample size of 30 and 3.376 for sample size of 24, we obtain an approximation of the desired sample size, 29. If this is the discrimination we wish to make, the original sample size is not large enough and the failure to reject the test hypothesis is questionable. The test had sufficient power to detect a ratio of 4.0 but not one of 3.0.

SUGGESTED READING

The references given in the previous and the following chapter are pertinent to the material presented in this chapter.

GLOSSARY

Treatment. A treatment in an experimental design is a condition or manipulation of experimental procedures, under which subjects are observed in an experiment.

Experimental and Control Groups. Where two groups of observations are compared statistically in an experiment, one of the groups is called an experimental group and the other a control group, the distinction being that the experimental group receives a "nonstandard" treatment.

Random Groups. Groups that are formed by assigning subjects at random.

Matched Groups. Groups that are selected in such a way as to have certain characteristics in common, either as groups overall, or subject by subject.

Subjects as Own Controls. A single group of subjects observed under the two treatments of a two-sample problem.

Repeated Observations. In a subjects-as-own-controls design the observations made on the subjects are called repeated observations—the subjects are repeatedly observed.

(17.1) $$X_a - X_b: \quad N(\mu_a - \mu_b, \sigma_a^2 + \sigma_b^2)$$

(17.2) $$\sigma^2_{\bar{X}_a - \bar{X}_b} = \frac{\sigma_a^2}{N_a} + \frac{\sigma_b^2}{N_b}$$

(17.3) $$\bar{X}_a - \bar{X}_b: \quad N(\mu_a - \mu_b, \sigma^2_{\bar{X}_a - \bar{X}_b})$$

(17.4) $$Z = \frac{(\bar{X}_a - \bar{X}_b) - (\mu_a - \mu_b)}{\sqrt{(\sigma_a^2/N_a) + (\sigma_b^2/N_b)}}$$

(17.5)
composite lower-tailed $\quad H_1: \Delta_1 < 0$

composite upper-tailed $\quad H_2: \Delta_2 > 0$

composite two-tailed $\quad\;\; H_3: \Delta_3 \neq 0$

specific $\qquad\qquad\; H_4: \Delta_4 = \Delta_A = \Delta_T + \delta(\sigma_{X_a - X_b})$

(17.6) $$\sigma^2_{\bar{X}_a - \bar{X}_b} = \sigma^2 \frac{N_a + N_b}{N_a N_b}$$

(17.7)
$$N = 2\left[\frac{(Z_\alpha - Z_\beta)\sigma}{\Delta_T - \Delta_A}\right]^2$$

(17.8)
$$N_a = \sigma_a(\sigma_a + \sigma_b)\left[\frac{Z_\alpha - Z_\beta}{\Delta_T - \Delta_A}\right]^2$$

$$N_b = \sigma_b(\sigma_a + \sigma_b)\left[\frac{Z_\alpha - Z_\beta}{\Delta_T - \Delta_A}\right]^2$$

(17.9)
$$\hat{S}^2 = \frac{(N_a - 1)\hat{S}_a{}^2 + (N_b - 1)\hat{S}_b{}^2}{N_a + N_b - 2}$$

(17.10)
$$\hat{S}^2 = \frac{\sum_{i=1}^{N_a} X_{ia}{}^2 + \sum_{i=1}^{N_b} X_{ib}{}^2 - (\sum_{i=1}^{N_a} X_{ia})^2/N_a - (\sum_{i=1}^{N_b} X_{ib})^2/N_b}{N_a + N_b - 2}$$

(17.11)
$$t = \frac{(\bar{X}_a - \bar{X}_b) - \Delta_T}{\hat{S}/\sqrt{N_a N_b/(N_a + N_b)}}$$

(17.12)
$$\hat{S}^2_{\bar{X}_a - \bar{X}_b} = \frac{\hat{S}_a{}^2}{N_a} + \frac{\hat{S}_b{}^2}{N_b}$$

(17.13)
$$t = \frac{(\bar{X}_a - \bar{X}_b) - \Delta_T}{\hat{S}_{\bar{X}_a - \bar{X}_b}}$$

(17.14)
$$df = \frac{[(\hat{S}_a{}^2/N_a) + (\hat{S}_b{}^2/N_b)]^2}{\dfrac{(\hat{S}_a{}^2/N_a)^2}{N_a + 1} + \dfrac{(\hat{S}_b{}^2/N_b)^2}{N_b + 1}} - 2$$

(17.15)
$$\sigma^2_{\bar{X} - \bar{Y}} = \sigma_{\bar{X}}{}^2 + \sigma_{\bar{Y}}{}^2 - 2\text{cov}(\bar{X}, \bar{Y})$$

(17.16)
$$\bar{D} = \bar{X}_a - \bar{X}_b$$

(17.17)
$$\sigma_{\bar{D}}{}^2 = \sigma^2_{\bar{X}_a - \bar{X}_b}$$

(17.18)
$$\hat{S}_{\bar{D}} = \frac{\hat{S}_D}{\sqrt{N}}$$

(17.19)
$$\frac{\hat{S}_1{}^2}{\hat{S}_2{}^2} : \frac{\chi_{n_1}{}^2/n_1}{\chi_{n_2}{}^2/n_2}$$

(17.20)
$$H_T: \sigma_A{}^2 = \sigma_B{}^2 = \sigma^2$$

(17.21)
$$H_1: \sigma_A{}^2 > \sigma_B{}^2$$
$$H_2: \sigma_A{}^2 < \sigma_B{}^2$$
$$H_3: \sigma_A{}^2 \neq \sigma_B{}^2$$

(17.22)
$$\text{for } H_1: \text{if } F_{n_A, n_B, \alpha} \leq \frac{\hat{S}_A{}^2}{\hat{S}_B{}^2}, \text{ reject } H_T$$

$$\text{for } H_2: \text{if } F_{n_A, n_B, 1-\alpha} \geq \frac{\hat{S}_A{}^2}{\hat{S}_B{}^2}, \text{ reject } H_T$$

PROBLEMS

1. Design several experiments to use the various principles of design discussed in the first part of this chapter. Write down all of the characteristics of the experiment including the data collection procedures.

2. Prove that $\bar{X}_a - \bar{X}_b$ has a variance given by $\sigma_a^2/N_a + \sigma_b^2/N_b$ if X_a and X_b are independent, as claimed in equation (17.2), and that in the case when $\sigma_a^2 = \sigma_b^2 = \sigma^2$, by $\sigma(N_a + N_b)/N_aN_b$, as in equation (17.6).

3. For a diverse selection of N_a, N_b, S_a^2, and S_b^2 values, compare the estimates of equations (17.9) and (17.12). In particular, compare the estimates with equal sample sizes and with equal values of S_a^2 and S_b^2. In each set of values, find the degrees of freedom by (17.14) and compare it with $N_a + N_b - 2$.

4. For a Case C two-sample problem involving hypotheses about expectations, find sample sizes for the following (α, β, δ) tests, assuming equal variances: (.05, .2, 1.0), (.05, .1, 1.25), (.01, .05, .72), (.01, .1, 1.25).

5. Do problem 4 under the assumption that $\sigma_a^2/\sigma_b^2 = 1.4$.

6. Selecting cases at random, the data of Table 17.1 were reduced in number. Group a had subjects 1, 4, 11, and 16 removed. Group b had subjects 2, 8, and 10 removed. Recalculate the test statistics and test the hypotheses stated in the text. Calculate the effective value of δ for the $(\alpha = .05, \beta = .2)$ test.

7. For the data of Table 17.1, use the median test to test the hypothesis that the two variables have identical locations. Repeat this test on the first five observations from each group.

8. Repeat problem 7, using the runs test, and the Wilcoxon two-sample test.

9. Write an analysis of the sources and interpretation of the variation of significance probabilities obtained in the various analyses of the data of Table 17.1.

10. Remove subjects 4, 6, and 10 from the sample of experiment I of Table 17.5 and recalculate the statistics of that table and reassess the statistical tests described in the text. Make a plot of the entire sample as a joint distribution, X_a on one coordinate and X_b in the other. Write an analysis of the effect of the removal of the particular three subjects removed.

11. Apply the sign test and the Wilcoxon test for correlated observations to the data of Table 17.5, experiments I and II, and I with the subjects deleted as in problem 10.

CHAPTER 18

The Many-Sample Problem, Part I

$\mathbf{W}$e frequently wish to know the structure of the effect of several variables or treatments on a given phenomenon, or the differences among several populations with respect to a certain phenomenon. Indeed, it may be desirable to study variation of a phenomenon under several experimental conditions for several different populations. An experimental analysis method known as the analysis of variance is widely used in studies with these types of goals. This chapter and the next are an introduction to that extensively developed field of statistical theory, the analysis of variance.

The Many-Experiments Approach

One way of providing general knowledge about a phenomenon that is influenced by several factors (or that might be different in different populations) is to generalize from a number of simple, one-sample, or two-sample experiments. Imagine four meaningful treatments in an experiment on epilepsy control. Four two-sample, subjects-as-own-controls experiments are designed and run. In each of the four experiments the result is an acceptance or rejection of the superiority of the experimental treatment over the control treatment, that is, each of the experiments would be interpreted under a predefined (α, β, δ) statistical test. So far, so good. Now, we must figure some way of comparing the results of the tests in the four experiments. Several ways are apparent. For example, the significance probabilities Ψ of the four tests might be ranked, giving an indication of the relative effectiveness of the treatments. An objection to this method is seen when we ask about the differences of the four experimental treatments. They might all be insignificantly different pairwise, and yet have different significance probabilities in the comparison with a control treatment. Other objections to this procedure can be raised.

In general, the individual two-sample, or one-sample experiments-in-profusion approach to a study of a many-sample problem involves more problems than solutions. There are methods for comparing several individual samples so as to differentiate among subsets of samples that differ between subsets but not within subsets. These methods are not satisfactory as the general approach and do not provide information about the entire collection of observations. The problem in this approach is that it is difficult to draw conclusions from many separate experiments. This is particularly true under realistic circumstances of experimentation where each of the several experiments is different from the others, in comparison with the consistency of observations within an experiment. The systematic variation of chance factors from one experiment to another creates differences in the observations. Often the differences are occasioned by design, as when a new instrument, experimental location, or technology is introduced between experiments.

In this chapter and the next we develop a technology for the execution of experiments permitting the "simultaneous" evaluation of the effects of several experimental treatments or conditions on a phenomenon. The methods involved are not in themselves statistical, but depend on the concept of randomization of "errors" through the assignment of subjects to conditions and treatments in such a way as to "balance" the "error" among the conditions and treatments. The statistical methodology takes advantage of the randomization, without explicitly involving the randomization. If randomization of error is insured in an experiment, the statistical calculations developed in this chapter have desirable and powerful properties.

The statistical procedures for the "simultaneous" many-sample problem permit an evaluation of the overall effectiveness of the treatments. If the experimental treatments are all expressions of some general theme, then we are interested in the overall theme as much as in the separate treatments. Often the theme itself is the only thing of interest, the specific treatments being merely exemplars of the theme. Thus, it is crucial to have methods of stating the general effectiveness of the collection of treatments.

Imagine a complicated experiment in which two themes of experimental treatment are involved; for example, the study of learning in four types of mentally deficient children under three regimes of chemotherapy. Several things are of interest. First, the investigator is interested in evaluating the effectiveness of the therapy for each type of patient. There are twelve drug-patient combinations to be studied. However, we are interested in the effect of drugs across the deficiency types and in the effect of deficiency types across the drug treatments. At a more complex level we need to understand any differences in pattern of effect across one theme for specific cases of the other theme. For example: Does the pattern of reactions to the three drugs differ from one type of deficiency to another type?

The answers to the questions raised in the complicated experiment are difficult to obtain by pair comparison of the separate single-sample treatment-group parts of the complicated experiment. However, the methods of the analysis of variance permit an evaluation of the contribution of each of the two themes and an evaluation of the consistency of the pattern of variation in the dependent variable across the combinations of treatments and groups.

Factors and Variables and Groups

Recall the standard classification of variables as independent and dependent. An independent variable is a variable under the control of an experimenter. A dependent variable is an observation variable. The values of the dependent variable "depend" on the value of the independent variable. This conception can be generalized to experimental treatments and experimental groups. If the dependent variable is dependent on the experimental conditions under which the observations are made, then the experimental conditions have the same status as an independent variable. In order to simplify our language, the word "factor" is used to cover both types of independent variables. A factor is a collection of conditions under which the dependent variable is observed, that is, a set of values on an independent variable or experimental treatments. In general, there is a small number of special cases in a factor used in an experiment—a limited number of values of the independent variable or a few experimental treatments. These special cases of the factor are called the "levels" of the factor. This terminology has the advantage of neutrality regarding the ordinality of the special cases of a factor. Clearly the levels of an independent variable can be ordered in accordance with the magnitude of the values of the variable. However, when a factor is made up of experimental treatments, it may not be clear what order, if any, the treatments have. For example, a drug-treatment factor may be defined by an unordered set of drugs that constitute the treatments. On the other hand, an independent variable, such as the number of pretransfer training trials in a learning study, may constitute a factor with several levels defined by quantities.

If there are k levels, the experiment consists of making observations on a dependent variable under the k different experimental conditions. In terms of the preceding two chapters the experiment is a k sample experiment. If k is greater than 2, the methods of the last two chapters are not appropriate.

The most important contribution of the analysis of variance is the notion of multiple-factor experiments. Multiple-factor experiments permit the richness and complexity of behavior to emerge. In multiple-factor experiments, observations are made under all (or many) combinations of levels of factors. Information about effects of factors separately and effects of factors jointly may be isolated and quantified. The simplest experiments using multiple factors are so-called complete factorial experiments where combinations of all levels of a factor with all levels of the other are treated as separate experimental conditions. In a two-factor factorial experiment with three levels on factor A and four levels on factor B there are $4 \times 3 = 12$ combinations of factor levels. Every combination of a factor A level with a factor B level is included in the experiment. For each of these treatment combinations a given number of observations is made. A diagram of the experiment in the form of a grid of cells is a useful way of thinking about a two-factor experiment. Imagine the levels of factor A as columns and the levels of factor B as rows in a 4 by 3 grid of cells. Figure 18.1 illustrates this factorial design. Each of the 12 cells represents a combination of factor levels, that is, an experimental condition.

FIGURE 18.1

Layout of a two-way analysis of variance.

Extending this to a larger number of factors is simple. Imagine a factor C with two levels in addition to the two factors, A and B, in the above example. A treatment in the three-factor case represents a combination of a level from all three factors. For each level on the third factor the combinations of the levels of the first two factors are duplicated. Hence, there are $4 \times 3 \times 2 = 24$ treatment combinations. In constructing a diagram of the experiment, the grid is reproduced as many times as there are levels on the third factor. Letting the reproduction be on the horizontal across the page, we have super- and sub-headings for the cells in Figure 18.2. Such an experiment is called a three-way factorial experiment. The number of factors and the number of levels of each of the factors determine the number of cells, that is, the number of conditions, in the experiment. If the number of levels on the 1st, 2nd, ..., Mth factors are $n_1, n_2, ..., n_M$, then the number of treatment combinations in the experiment is given by $n_1 \times n_2 \times \cdots \times n_M$.

The levels on factors may be defined by properties of the subjects in the experiment as well as by specific values on independent variables or experimental treatments. For example, in a three-factor experiment, imagine that factor C is political affiliation, factor A age group, and factor B socio-economic status. No experimental treatments are involved. The experiment involves an assessment of the effects of subject differences on a dependent variable. Mixtures of subject factors and experimental factors are easily thought of.

The theory and practice of using the concepts of factors, levels on factors, combinations of levels, experimental treatments, subject groups, and so on, constitutes a discipline in and of itself known as the design of experiments. The discipline is untidy in the sense that there are so many facets to it that no clear-cut organization and structure is immediately apparent. Whereas there was a clear organization to the preceding chapters, this chapter and the next lack the same

	C_1			C_2		
	A_1	A_2	A_3	A_1	A_2	A_3
B_1						
B_2						
B_3						
B_4						

FIGURE 18.2

Layout of a three-way analysis of variance.

measure of completeness and structure. Only a limited number of the very many designs are presented and discussed. The entire theory of the design of experiments may not become clear in the process. The student wishing to achieve a mastery of the design and analysis of experiments must turn to the more advanced books suggested at the end of this chapter.

The material in this book deals with a special case of the general design of experiments: It is presumed that only one dependent variable is involved—each of the experimental observations is an observation of the same variable. The parameters (particularly the expectation) of the dependent variable may be a function of the experimental conditions, but it is presumed that the scale and metric of the observations is constant throughout the entire design.

The Analysis of Variance

The basic concept in the analysis of a many-sample problem is the composite nature of an observation. An observation is determined by more than one thing, such as the experimental conditions under which the observation is made, the individual characteristics of the subject (object) observed, chance fluctuations of performance and environmental conditions. If more than two experimental factors are involved, all of them contribute to the observation and perhaps even some interaction between the factors influences the observation. The analysis of variance is a general method for determining how much of the total variability of the observed values is linked to various components in the composite, and for making statistical statements regarding the components.

Imagine a single-factor experiment with four levels. When an observation is made in the experiment under one or another of the four levels, several things contribute to the value of the observation. First, there is a general level of response that is characteristic of the sort of subject studied, the sort of task involved, and the sort of situation involved in the experiment. Regardless of the experimental conditions and subjects, there is a general level or magnitude of response. This level of response may be thought of as the average, overall, score in the experiment if the experiment is exhaustive of the type of treatment involved and population of subjects. The parameter corresponding to this general level of response is denoted μ. Second, the specific experimental condition, or level of the factor under which the observation is made, has some specific effect on the value observed. This effect is presumed to be constant for every observation made under the specified level of the factor; if the effect is associated with the constant level of the factor, it must be constant. The parameter for this effect is denoted by the symbol τ_j, where the j

indicates the jth level of the factor. In the three-level single-factor experiment example, three parameters are required: τ_1, τ_2, τ_3. Third, each experimental observation is influenced by a large number of things over which the experimenter has no control. These things include individual (subject) differences, fluctuations of performance, fluctuations of experimental conditions, etc. Frequently the literature indicates these sources of variation as "error." It is clear that the use of the word "error" depends on the viewpoint taken regarding things such as uncontrolled fluctuations of illumination and individual difference. In this text the total effect of these uncontrolled sources of variation is called the residual effect. The residual effect is a random variable, and may be conceptualized as the sum of a large number of random variables, the partial sources of the variation associated with individual differences, random fluctuations of stimulation, and so on. The variable is designated by the symbol ρ. In order to identify clearly the value of the random variable observed on a given replication, it must be subscripted for the experimental condition and the replication number in that condition. Letting the subscript i stand for the observation number, the residual designation is ρ_{ij} for the ith observation in the jth level of the factor.

For the ith subject in the jth level of the factor, the dependent variable is X_{ij}. The analysis of variance is based on the concept that the dependent variable is the sum of the parameters μ, τ_j, and the random variable ρ_{ij}:

$$X_{ij} = \mu + \tau_j + \rho_{ij} \tag{18.1}$$

The statistical theory of the analysis of variance is a collection of principles and methods designed to estimate the variances in the partition (18.1) and to provide the basis for testing hypotheses about the components in (18.1). In turn, the hypotheses about the components are interpretable as hypotheses about the differences in the treatment effects τ. The theory and methods cover multiple-factor cases and the interaction of the factors.

There are two aspects to the analysis of variance. The first aspect is the analysis proper—a partitioning of the variability of observed values into parts associated with the components of the model, τ and ρ. This partitioning is a descriptive procedure, involving no assumptions about the form of the distribution of the dependent random variable. Certain assumptions are made regarding the expectation and variance of the random variable. These assumptions generally are based on experimental procedures and methods of observation rather than empirical investigation. The second aspect is the application of normative models of statistical test theory in order to make judgments regarding the confidence we can place in hypotheses about the effect of experimental treatments on the random variable observed. In this development some very restrictive assumptions are made about the distribution of the dependent variable. Actual computation in the analysis of variance involves the sums of squared mean deviations rather than variances. These sums are directly related to the variances, and the appropriate variances are estimated from the "sums of squares" by dividing by the appropriate numbers of degrees of freedom. Recall that the sample variance of N values is given by

$$S_X{}^2 = \frac{1}{N}\sum_{i=1}^{N}(X_i - \bar{X})^2$$

where $\bar{X}$ is the mean of the N values. The sum of squares is given by

$$SS = \sum_{i=1}^{N}(X_i - \bar{X})^2 = NS_X{}^2$$

Fixed Effects. The model that we give the greatest attention is the so-called fixed effects model. In this model of an experiment it is presumed that the experiment includes all levels of the factor for which implications are drawn. That is, it is assumed that the levels of the factor represented in the experiment are all the levels that one wishes to study as representative of the factor. Another model, which we introduce later and give brief attention, makes quite the opposite assumption—namely, that the levels of the factor represented in the experiment are a random sample of a population of levels to which we wish to relate our empirical observations. This second type of model is called the random effects model—if the levels are chosen at random from some population of levels, then the effects associated with those levels are random in a certain sense. In an experiment in which we select the levels on the factor because of their particular meaningfulness for the experiment, the effects $\tau_1, \tau_2, \ldots, \tau_M$ are determined by this rational selection. In this situation we intend that our experiment refer to the particular levels on the factor and not to some other, larger, collection (population) of levels. The selection of the levels may be defined by as simple logic as all of the levels that make sense for the factor. On the other hand, where there is a potentially large number of levels on a factor, a limited but representative set of the levels might be used to define the factor for the experiment. In the latter instance, the selection of the particular levels included in the factor has an influence on the result of the experiment—a different selection might have produced a different result.

Regardless of how the levels of the factor are selected, the fixed effects model leads to statistical tests, the results of which should not be generalized beyond the specific set of levels actually appearing in the experiment.

The random effects model is used in experiments in which there is a population of levels from which the levels included in the actual experiment are randomly sampled (equal probability sampling). Since the levels included in the experiment are a random sample of the possible levels, the effects included in the experiment may be considered to be a random variable. An evaluation of this variable can be used to generalize to all of the possible levels of the factor whether or not they were included in the actual experiment.

The difference in the practical aspects of the distinction of the two types of model occurs in two places: during the planning of the experiment, and during the final statistical calculations. The first difference is encountered when the levels of the factor are chosen. In the fixed effects model the levels are chosen in terms of the logic or natural features of the experimental situation. In the random effects model the levels are chosen at random from the totality of possible levels. In the statistical analysis the two types of experiment are nearly identical, right up to the formation of the F ratios used to test hypotheses about the effects of the levels of the factor. The ratios for one model are different from those in the other, for some designs.

Detailed discussion of the differences of the two types of designs will be postponed until the fixed effects model is presented in detail and the random effects model is introduced.

Between and Within

The cryptic heading of this section is an introduction to the key concept of the statistical procedures of the analysis of variance. In the two-part decomposition of the sum of squares of the dependent variable, that part associated with the experimental treatments is dependent on differences *between* the treatment effects. That part of the sum of squares associated with the residual effects is independent of the differences between the treatment effects and may be observed *within* a set of observations in a single treatment.

The crucial concept is that the within, or residual, sum of squares may be used to estimate the variation of the sampling distribution of means. This estimate provides a standard by which to judge the statistical significance of the variation between the treatments. Even if the treatments have the same effect parameters, that is, $\tau_1 = \tau_2 = \cdots = \tau_{JI}$, sample estimates of the parameters, the treatment means, have a distribution with a variance related to the variability of residual effects. In derivations in the following sections, the variability of the dependent variable is shown to be estimated by using the within-treatments observations. The within-treatments estimation is free of any influence of the differences between the treatments. The variability of the dependent variable is also shown to be estimated by using the variance of the means of the observations in the treatment conditions.

The variance of treatment means is an estimate of the variance of the sampling distribution of means, if the treatment parameters are all the same value. If the treatment parameters are different, the variance of the treatment means is an estimate of the sum of the variance of the sampling distribution of means and the variance of the treatment parameters. In order to see this, recall that the variance of the sampling distribution of the mean is

$$\sigma_{\bar{x}}{}^2 = \frac{\sigma_X{}^2}{N}$$

where N is the number of observations of X. Now if an estimate of $\sigma_{\bar{x}}{}^2$ is known, then it can be used to estimate $\sigma_X{}^2$ simply by the following relationship:

$$\sigma_X{}^2 = N\sigma_{\bar{x}}{}^2$$

It is shown below that when the M treatment parameters are different values, then

$$\sigma_{\bar{x}}{}^2 = \frac{\sigma_X{}^2}{N} + \frac{1}{M-1}\sum_{j=1}^{M}\tau_j{}^2$$

$$N\sigma_{\bar{x}}{}^2 = \sigma_X{}^2 + \frac{N}{M-1}\sum_{j=1}^{M}\tau_j{}^2$$

Also, if the treatment parameters are all the same value,

$$\sum_{j=1}^{M}\tau_j{}^2 = 0$$

giving

$$\sigma_{\bar{x}}{}^2 = \frac{\sigma_X}{N}$$

and

$$N\sigma_{\bar{x}}{}^2 = \sigma_X{}^2$$

Now, the value of $\sigma_X{}^2$ can be estimated in two independent ways. The first estimate is simply the variance of the observed values within each treatment, which we call the mean square within treatments, MS_w; the second estimate is N times the variance of the observed treatment means, which similarly is called the mean square between treatments, MS_b. It is shown below in a selection of special cases that

$$E(MS_w) = \sigma_X{}^2$$

$$E(MS_b) = \sigma_X{}^2 + \frac{N}{M-1}\sum_{j=1}^{M}\tau_j{}^2$$

Recall that the ratio of two estimates of a single variance, under certain distribution restrictions, is distributed as Fisher's F statistic. Consequently, when the treatment parameters are all the same value, implying that

$$\sum_{j=1}^{M} \tau_j^{\,2} = 0$$

then the ratio

$$\frac{MS_b}{MS_w}$$

is distributed as Fisher's F. If the treatment parameters are different, the ratio of sample variances is greater than expected—that is, the probability of observing large values of the ratio is small if the variances are estimates of the same parameter. According to the model, the only way for the ratio to be unexpectedly large, except by chance, (say with probability α), is for the treatment parameters to be variable.

Essentially, the analysis of variance is designed to test hypotheses about treatment parameters. The specific parameters are the expected differential response levels within treatment conditions.

In general, a presentation of a many-sample problem and its model has five primary parts: (1) the problem and an appropriate model, (2) the partition of the sum of squares, (3) the hypothesis structure of the problem, the expected mean squares, and the appropriate tests of the hypotheses, (4) the determination of sample size via the (α, β, δ) procedure and, finally, (5) computational forms. Derivations and detailed analyses are presented in the first problem. In successive presentations in this chapter and the following, the detail is reduced unless some basically new concept is introduced.

A Single-Factor Problem

Consider a problem in which there are several interesting variations on a theme, each variation being of some theoretical interest. The theme could be child-rearing practices, methods of presentation of stimuli in a discrimination learning task, genetic strains of mice, etc. In each case there is a single substantively defined dependent variable, such as the score achieved on an attitude questionnaire, the number of errors made in identifying tachistoscopically presented stimuli, the number of trials required to learn to a given criterion, and the amount of a given substance per ounce of urine secreted by the experimental subjects.

The experimental method is simple. Each subject is assigned to an experimental condition completely at random or on the basis of the subject characteristic. If the subject characteristic method is necessary because of the definition of the factor, the subjects are selected from the subpopulation of subjects at random. These randomization procedures insure that there is no bias in subject assignments and that the ultimate inference refers essentially to random error.

Imagine a four-level experiment with N_1, N_2, N_3, and N_4 observations respectively in the four levels. The means in the four levels are calculated. We want a method of discovering whether or not these mean values were drawn from populations with the same expectations or whether discrepancies among the means indicate that different expectations might be involved. To do this, we need to construct a statistical model of the experiment.

A Model. A model of the experiment is proposed as follows. Let the overall responsiveness be given by the average response across the M levels of the factor,

$$\mu = \frac{1}{M} \sum_{j=1}^{M} \mu_j \tag{18.2}$$

where the μ_j are the expectations of the dependent variable under the experimental conditions. The separate expectations μ_j are related to the differential response parameter so that

$$\tau_j = \mu_j - \mu \qquad \text{for} \quad j = 1, \dots, M \tag{18.3}$$

By the relationship of the μ_j and μ, it follows that

$$\sum_{j=1}^{M} \tau_j = 0 \tag{18.4}$$

In addition to the constant value ($\mu_j = \mu + \tau_j$) of the response in condition j, there is a composite quantity ρ_{ij} affecting the response, where ρ_{ij} is determined by uncontrolled factors such as environmental and organismic fluctuations and constants associated with individual differences. The dependent variable is assumed to be an additive composite of these three values:

$$X_{ij} = \mu + \tau_j + \rho_{ij}$$

The components in this equation are μ, a constant for all levels of the factor and all observations; τ_j, a constant for all observations within a level on the factor but possibly a variable across levels; and ρ_{ij}, a random variable for which

$$E(\rho_{ij}) = 0$$

$$\text{var}(\rho_{ij}) = \sigma_\rho^{\,2}$$

The expectation and variance of the dependent variable in each of the treatment levels are easily determined. Denoting the dependent variable in the jth treatment level as X_j, we have

$$E(X_j) = E(\mu) + E(\tau_j) + E(\rho_{ij})$$
$$= \mu + E(\mu_j - \mu) + 0$$
$$= \mu_j$$

On the basis of the constant value of the differential treatment effect within a treatment level,

$$\text{var}(X_j) = \text{var}(\mu + \tau_j + \rho_{ij})$$
$$= \text{var}(\rho_{ij}) = \sigma_\rho^2$$

With regard to the differential effects of the experimental conditions there are two possibilities:

$$(1) \quad \tau_1 = \tau_2 = \cdots = \tau_M$$

or

$$(2) \quad \tau_j \neq \tau_k \qquad \text{for at least one pair of } j \text{ and } k$$

A parameter indicating the degree to which the second condition holds is

$$\frac{1}{M-1} \sum_{j=1}^{M} (\mu_j - \mu)^2 = \frac{1}{M-1} \sum_{j=1}^{M} \tau_j^2$$

where

$$\frac{1}{M-1} \sum_{j=1}^{M} \tau_j^2 \begin{cases} = 0 & \text{if } \tau_1 = \tau_2 = \cdots = \tau_M \\ > 0 & \text{if it is } not \text{ true that } \tau_1 = \tau_2 = \cdots = \tau_M \end{cases} \qquad (18.5)$$

The assumptions regarding the expectation of the residual variable and variance of the dependent variable, that is, $\text{var}(\rho) = \sigma_\rho^2$ for each factor level and $E(\rho) = 0$ for each factor level, are central to the derivations. We pause here to justify the assumptions. In the first place, the residual assumptions are not easily justified without care in performing the experiment. The source of variability in a given treatment must be due strictly to the residual variables. If, for example, the experiment is poorly executed and part of the observations included in one treatment group involve a gross departure from the defined experimental procedures, part of the variance within that treatment group is due to the departure from experimental protocol. The variance within the botched set of observations would be due in part to the experimental error, and that would surely cause the variation in the botched set of observations to be different from the variance in the other experimental conditions. Second, care must be taken to assign subjects to conditions at random, in order to avoid bias in the residual effect. If there are systematic relationships between the execution of experimental conditions and the occurrence of residual factors that affect the dependent variable, the equal variance assumption is probably falsified. A particularly troublesome problem is found in studies

ent no longer can be modeled by the model proposed. Fundamentally, the model
requires independence of the treatments and the residual factor and equal variances
for all treatments. One way of insuring this is by randomization.

The assumption that the expectation of the residual variable is zero is founded
quite simply on the following argument. When a very large number of contributors
(errors, personal idiosyncrasies, and so on) make up a random variable they have
the effect of a perturbation of some constant influence with the positive and
negative perturbations equally forceful. Hence, on the whole, the negative and the
positive residual factors tend to cancel, producing an expected value of zero. The
reality of this assumption must be argued within particular experimental contexts.

The Partition of Sums of Squares. When the experiment is performed, the resulting data form a collection of M sets of values as indicated in
Table 18.1. In general, there are N_j observations in the jth level of the factor.

TABLE 18.1 General Organization of Analysis of Variance—One Way

	Level 1	Level 2	...	Level j	...	Level M
	$X_{1,1}$	$X_{1,2}$		$X_{1,j}$		$X_{1,M}$
	$X_{2,1}$	$X_{2,2}$	$\cdot$	$X_{2,j}$	$\cdot$	$X_{2,M}$
	$X_{3,1}$	$X_{3,2}$	$\cdot$	$\cdot$	$\cdot$	$\cdot$
		$X_{4,2}$	$\cdot$	$X_{i,j}$	$\cdot$	$X_{N_M,M}$
	$X_{N_1,1}$	$\cdot$		$\cdot$		
		$\cdot$		$\cdot$		
		$X_{N_2,2}$		$X_{N_j,j}$		
$\sum_{i=1}^{N_j} X_{i,j}:$	$\sum_{i=1}^{N_1} X_{i,1}$	$\sum_{i=1}^{N_2} X_{i,2}$	...	$\sum_{i=1}^{N_j} X_{i,j}$	...	$\sum_{i=1}^{N_M} X_{i,M}$
$\sum_{i=1}^{N_j} X_{i,j}^2:$	$\sum_{i=1}^{N_1} X_{i,1}^2$	$\sum_{i=1}^{N_2} X_{i,2}^2$	...	$\sum_{i=1}^{N_j} X_{i,j}^2$	...	$\sum_{i=1}^{N_M} X_{i,M}^2$

The total number of observations is $(N_1 + N_2 + \cdots + N_M = N_T)$. The mean of the observations is

$$\bar{X} = \frac{1}{N_T} \sum_{j=1}^{M} \sum_{i=1}^{N_j} X_{ij} \qquad (18.6)$$

For each score we may define a mean deviation:

$$X_{ij} - \bar{X}$$

Summing the squares of the deviations across all subjects and factor levels gives a total sum of squares, SS_T:

$$SS_T = \sum_{j=1}^{M} \sum_{i=1}^{N} (X_{ij} - \bar{X})^2 \qquad (18.7)$$

There are $(N_T - 1)$ degrees of freedom for this sum of squares, and the mean square for the total set of scores is

$$MS_T = \frac{SS_T}{N_T - 1} \qquad (18.8)$$

This mean square is the estimate of the variance of the random variable observed in the experiment.

The partition of this total variance into two parts, one associated with the differences between factor levels and the other with the variation within levels, proceeds by expressing the mean deviation of a score in two parts,

$$(X_{ij} - \bar{X}) = (X_{ij} - \bar{X}_j) + (\bar{X}_j - \bar{X})$$

where the mean of the jth treatment scores is given by

$$\bar{X}_j = \frac{1}{N_j} \sum_{j=1}^{N_j} X_{ij} \qquad j = 1, \ldots, M$$

Squaring and summing, we obtain

$$\begin{aligned}
SS_T &= \sum_{j=1}^{M} \sum_{i=1}^{N_j} (X_{ij} - \bar{X})^2 = \sum_{j=1}^{M} \sum_{i=1}^{N_j} [(X_{ij} - \bar{X}_j) + (\bar{X}_j - \bar{X})]^2 \\
&= \sum_{j=1}^{M} \sum_{i=1}^{N_j} (X_{ij} - \bar{X}_j)^2 + \sum_{j=1}^{M} \sum_{i=1}^{N_j} (\bar{X}_j - \bar{X})^2 \\
&\quad + 2 \sum_{j=1}^{M} \sum_{i=1}^{N_j} (X_{ij} - \bar{X}_j)(\bar{X}_j - \bar{X}) \\
&= SS_w + SS_b + 0
\end{aligned} \qquad (18.9)$$

The final term in the sum is zero, because of the definition of the means involved. Consequently, we have shown that

$$SS_T = SS_w + SS_b \qquad (18.10)$$

In order to calculate variances corresponding to these sums of squares, the sums of squares are divided by the respective number of free variates (degrees of freedom):

$$MS_w = \frac{SS_w}{(N_1 - 1) + (N_2 - 1) + \cdots + (N_M - 1)} = \frac{SS_w}{N_T - M} \quad (18.11)$$

$$MS_b = \frac{SS_b}{M - 1} \quad (18.12)$$

The degrees of freedom for the within mean squares is determined by the fact that each of the M treatments has an expectation that is involved in determining the sum of squares. For each sum of squares an expectation, that is, one restriction, is involved. There are M restrictions. The degrees of freedom are partitioned in an additive fashion by this procedure, so that

$$N_T - 1 = \sum_{j=1}^{M} (N_j - 1) + (M - 1) \quad (18.13)$$

The special case of equal sample sizes in all levels of the factor simplifies the notation of the equations. Equal sample sizes permit writing $N_1 = \cdots = N_M = N$ and $N_1 + \cdots + N_M = MN$. The greatest advantage to equal sample sizes is in the calculational formulas presented below; other advantages are pointed out later.

Expected Mean Squares and Sampling Distributions. We now derive the expectations of the mean squares presented above. First, we find the expectation of the mean of observations in the jth factor level. The model is assumed to be true and the expressions for the expectations are derived on the basis of that assumption:

$$E(\bar{X}_j) = E\left[\frac{1}{N_j} (X_{1j} + \cdots + X_{N_j j})\right]$$

$$= E\left[\frac{1}{N_j} (\{\mu + \tau_j + \rho_{1j}\} + \cdots + \{\mu + \tau_j + \rho_{N_j j}\})\right]$$

$$= E\left[\frac{1}{N_j} (N_j \mu + N_j \tau_j + \rho_{1j} + \cdots + \rho_{N_j j})\right]$$

$$= \mu + \tau_j + \frac{1}{N_j} [E(\rho_{1j}) + \cdots + E(\rho_{N_j j})]$$

$$= \mu + \tau_j$$

$$= \mu + (\mu_j - \mu)$$

$$= \mu_j \quad (18.14)$$

Thus the mean in each factor level has an expectation equal to the treatment effect.

It is easily shown that

$$E(\bar{X}) = \mu \tag{18.15}$$

that is, the grand mean has an expectation equal to the general level of responsiveness. The proof is left to the reader.

If the differential treatment effects are all equal, $\tau_1 = \cdots = \tau_M = \tau$, then the treatment means, $\bar{X}_1, \bar{X}_2, \ldots, \bar{X}_M$, are sample means from equivalent random variables—sample means from a single random variable X:

$$X = \tau + \mu + \rho = \mu + \rho$$

Consequently, we are in a position to estimate the variance of the sampling distribution of means of samples of observations on X by calculating the variance of the observed means. For M samples of size N, this variance estimate is

$$\hat{S}_{\bar{X}}^{\;2} = \frac{1}{M-1} \sum_{j=1}^{M} (\bar{X}_j - \bar{X})^2$$

This gives a method for estimating σ_X^2. Using the relationship

$$\sigma_X^2 = N\sigma_{\bar{X}}^2$$

applied to the estimate

$$\hat{S}_{\bar{X}}^{\;2}$$

gives

$$\hat{S}_X^{\;2} = \frac{N \sum_{j=1}^{M} (\bar{X}_j - \bar{X})^2}{M-1}$$

This is one estimate of the variability of the dependent variable, and hence of the random variable ρ, based on the assumption that the factor level means, the $\bar{X}_j$, are means of samples from a single random variable. The variance is estimated another way, and in order to avoid confusion of notation we write

$$MS_b = \frac{N \sum_{j=1}^{M} (\bar{X}_j - \bar{X})^2}{M-1} \tag{18.16}$$

For unequal sample sizes, this result must be modified to reflect the unequal contribution of the levels to the sums of squares. The term

$$\sum_{j=1}^{M} N_j (\bar{X}_j - \bar{X})^2$$

is parallel to the sums of squares

$$\sum_{j=1}^{M} \sum_{i=1}^{N_j} (\bar{X}_{ij} - \bar{X})^2$$

as suggested by the relation

$$\sigma_X{}^2 = N\sigma_{\bar{X}}{}^2$$

The weighting factor N_j in the first sum reflects the number of deviation terms given by the jth level of the factor. The mean square for the unequal sample size experiments is

$$MS_b = \frac{\sum_{j=1}^{M} N_j(\bar{X}_j - \bar{X})^2}{M - 1}$$

The second way of estimating the variance of ρ is to pool the estimates from each of the factor levels. This pooling is independent of the treatment means. The variance estimate in the jth level is

$$\hat{S}_{X_j}{}^2 = \frac{1}{N_j - 1} \sum_{i=1}^{N_j} (X_{ij} - \bar{X}_j)^2$$

Pooling the M variances by taking their weighted average (weights equal to the degrees of freedom in the respective factor levels) gives

$$\hat{S}_X{}^2 = \frac{(N_1 - 1)\hat{S}_{X_1}{}^2 + \cdots + (N_M - 1)\hat{S}_{X_M}{}^2}{N_T - M} \tag{18.17}$$

$$= \frac{\sum_{j=1}^{M} \sum_{i=1}^{N_j} (X_{ij} - \bar{X}_j)^2}{N_T - M}$$

$$= MS_w$$

Before calculating the expectations of these mean squares, the concept of mean residual must be introduced. Although ρ is a random variable with $E(\rho) = 0$ and $\text{var}(\rho) = \sigma_\rho{}^2$, the mean and variance of the sample values of the residual terms need not be equal to 0 and $\sigma_\rho{}^2$. Imagine that for the jth level, the mean sample value of ρ is $\bar{r}_j$ and that the sample mean residual of the entire experiment is $\bar{r}$. Since $E(\bar{r}_j) = E(\bar{r}) = 0$, then, for each $j = 1, \ldots, M$,

$$E(\bar{r}_j{}^2) = \sigma_{\bar{r}_j}{}^2 = \frac{\sigma_\rho{}^2}{N_j}$$

$$E(\bar{r}^2) = \sigma_{\bar{r}}{}^2 = \frac{\sigma_\rho{}^2}{N_T}$$

Using these expressions and the properties of expectation, we can derive the expectations of the mean square deviation between levels:

$$E(MS_b) = E\left[\frac{\sum_{j=1}^{M} N_j (\bar{X}_j - \bar{X})^2}{M-1}\right]$$

$$= \frac{1}{M-1} \sum_{j=1}^{M} N_j E(\bar{X}_j - \bar{X})^2$$

$$= \frac{1}{M-1} \sum_{j=1}^{M} N_j E[(\mu + \tau_j + \bar{r}_j) - (\mu + \bar{r})]^2$$

$$= \frac{1}{M-1} \sum_{j=1}^{M} N_j E[\tau_j^2 + (\bar{r}_j - \bar{r})^2]$$

$$= \frac{1}{M-1} \sum_{j=1}^{M} N_j \tau_j^2 + \frac{1}{M-1} \sum_{j=1}^{M} N_j E(\bar{r}_j - \bar{r})^2$$

$$= \frac{\sum_{j=1}^{M} N_j \tau_j^2}{M-1} + \frac{1}{M-1} \left(\sum_{j=1}^{M} \frac{N_j \sigma_\rho^2}{N_j} - \frac{\sum_{j=1}^{M} N_j \sigma_\rho^2}{N_T}\right)$$

$$= \frac{\sum_{j=1}^{M} N_j \tau_j^2}{M-1} + \frac{1}{M-1}[(M-1)\sigma_\rho^2]$$

$$= \frac{\sum_{j=1}^{M} N_j \tau_j^2}{M-1} + \sigma_\rho^2 \tag{18.18}$$

Under the condition of equal response effects, $\mu_1, \ldots, \mu_M$, the differential response effects are equal to zero, that is, $\tau_1 = \cdots = \tau_M = 0$ and

$$E(MS_b) = \sigma_\rho^2$$

Hence, under the equal effects condition, MS_b is an unbiased estimate of the residual variation.

Turning to the mean square within levels, we find a simpler derivation:

$$E(MS_w) = \frac{1}{N_T - M}\left[\sum_{j=1}^{M} E \sum_{i=1}^{N_j} (X_{ij} - \bar{X}_j)^2\right]$$

$$= \frac{1}{N_T - M} \sum_{j=1}^{M} (N_j - 1) E(\hat{S}_j^2)$$

$$= \frac{\sum_{j=1}^{M} (N_j - 1)\sigma_\rho^2}{N_T - M}$$

$$= \frac{(N_T - M)\sigma_\rho^2}{N_T - M}$$

$$= \sigma_\rho^2 \tag{18.19}$$

Hence, MS_w is an unbiased estimate of the residual variation.

A Hypothesis. The last development leads to a test of the hypothesis that all treatment (level) effects are equal against the alternative that at least some of the effects have different values. If $\mu_1 = \mu_2 = \cdots = \mu_M$, then both MS_w and MS_b have the same expectation, σ_ρ^2. We can use this fact to establish a statistical test of

$$H_T: \mu_1 = \mu_2 = \cdots = \mu_M$$

against

$$H_A: \text{for at least some } j \text{ and } k, \; \mu_j \neq \mu_k$$

Recall that if $\hat{S}_a^2$ and $\hat{S}_b^2$ are independent estimates of the variance of a normally distributed random variable with df_a and df_b, then

$$V = \frac{\hat{S}_a^2}{\hat{S}_b^2}$$

is distributed as F_{df_a, df_b}. It can be shown that for normally distributed random variables the means and variances of samples are independent; for no other random variable is this true. Since MS_w is based on variances of observations and MS_b is based on means (of factor levels), they are independent *if* the dependent variable is normally distributed. The central limit theorem does not apply to the analysis of variance ratio—the dependent variable must be normally distributed in order for the above result to hold.

The application of the F distribution to the one-way analysis is to calculate

$$V = \frac{MS_b}{MS_w} \tag{18.20}$$

When the test hypothesis is true, $E(MS_b) = \sigma_\rho^2$ and $E(MS_w) = \sigma_\rho^2$ and consequently the ratio of mean squares, both of which are estimates of the same variance, is distributed as an F-distributed random variable, if ρ is normally distributed. The numerator and denominator degrees of freedom are $df_b = M - 1$ and $df_w = N_T - M$.

If H_T is false and the level effects are different, the ratio of mean squares V is a ratio of estimates of σ_ρ^2 *plus* a sample estimate of the ratio:

$$\frac{\dfrac{1}{M-1} \sum_{j=1}^{M} N_j \tau_j^2}{\sigma_\rho^2}$$

This sample ratio is always positive. Depending on the magnitude of the numerator, which depends on M, and

$$\sum_{j=1}^{M} N_j \tau_j^2$$

TABLE 18.2 General Form of the Analysis of Variance Table

Source of Variation	SS	df	MS	EMS
Between Levels	SS_b	$M - 1$	$\dfrac{SS}{df}$	$\sigma_\rho^2 + \dfrac{1}{M-1}\displaystyle\sum_{j=1}^{M} N_j \tau_j^2$
Within Levels	SS_w	$N_T - M$	$\dfrac{SS}{df}$	σ_ρ^2
Total	SS_T	$N_T - 1$	—	—

the ratio V tends to be larger than expected under H_T; that is, large values of V are observed with larger frequency than would be the case under H_T.

The entire result of this analysis of a simple many-sample problem may be summarized in the so-called analysis of variance table, as shown in Table 18.2.

Sample Size and (α, β, δ). Determination of sample size for a many-sample problem is complicated by the very nature of the problem—there are many samples. In the simplest case of the many-sample problem, the one-way analysis of variance design, the fact that there are M samples of N_j observations each ($j = 1, \ldots, M$) complicates the issues enormously. For example, different values of M influence the value of N_j—for large values of M, each N_j can be smaller than for smaller values of M. The relation of $df_b = M - 1$ and $df_w = N_T - M$ is one of reciprocity and can have a dramatic effect on the dispersion of $V = MS_b/MS_w$. Consequently, the power of the test is affected by the balance of degrees of freedom. An even more critical problem is the difficulty in evaluating an alternative to the test hypothesis, $H_T: \mu_1 = \mu_2 = \cdots = \mu_M$. A many-sample problem is generally found in an experimental situation where specific alternatives are not well defined. In the absence of well-defined specific alternative hypotheses, we face the necessity of determining reasonable alternatives. The complexity of this task is seen when we ask: Where, among the levels of the experiment, are the departures and how is the "total" departure distributed among these levels? We might imagine one extreme in which every effect is different by an amount Δ from the effect most similar. On the other hand, all effects but two may be the same with the two distinctive effects being different by an amount Δ and differing from the other effects by an amount $\Delta/2$ or $-\Delta/2$, respectively. Between these two extremes are an almost unlimited number of other reasonable alternatives.

Looking at the problem in more detail makes it clear that we must restrict the options in order to find a solution. We deal here only with the equal sample size experiments, that is, $N_1 = N_2 = \cdots = N_M$. We disregard for the moment the question of which levels contribute to the departure from the test hypothesis by expressing the total departure in a single parameter expressed in terms of the standard deviation of the dependent variable.

The method for determining sample size is similar, in essence, to the method used in the two-sample problem. Values for α and β are selected, the value of δ is determined, and the designated sample size is looked up in Chart R at the back of the book. This look-up is based on df_b as well as the values α, β, and δ. Only $\alpha = .01$ and $\alpha = .05$ are dealt with because of the unavailability of more extensive charts or tables. The charts are constructed only for $df_b = 1, 2, 3, 4, 5, 6, 7, 8$. These cover up to 9 levels of the factor in a single factor design.

We begin with a concrete example. Imagine four levels on a factor with a test hypothesis that the level effects all are equal: $H_T: \tau_1 = \tau_2 = \tau_3 = \tau_4 = \tau$. We select $\alpha = .05$. We wish to accept this hypothesis with $\beta = .05$ if any three effects are the same but one exceeds the others by one standard deviation, σ_p, or if the effects are more widely dispersed. One of the τ's is replaced by $\tau + \sigma_p$. The restriction

$$\sum_{j=1}^{M} \tau_j = 0$$

is still in force, implying that under H_A

$$\sum_{j=1}^{M} \tau_j = \tau + \tau + \tau + (\tau + \sigma_p) = 0$$

and consequently

$$\tau = -\frac{\sigma_p}{4}$$

$$\tau + \sigma_p = \sigma_p - \frac{\sigma_p}{4} = \frac{3}{4}\sigma_p$$

Thus the τ values of the nondistinctive level effects and of the distinctive level effect are known under H_A. The mean square between levels under H_A is specifiable. These specifications require knowledge of the residual variance σ_p^2. Recall

$$E(MS_b) = \sigma_p^2 + \frac{1}{M-1}\sum_{j=1}^{M} N_j \tau_j^2$$

For equal sample sizes the last term is

$$\frac{N}{M-1}\sum_{j=1}^{M} \tau_j^2$$

The sample size N can be determined by evaluating the value of the sum. In our example

$$\sum_{j=1}^{4} \tau_j^2 = 3\left(-\frac{\sigma_\rho}{4}\right)^2 + \left(\frac{3}{4}\sigma_\rho\right)^2$$

$$= \frac{3}{16}\sigma_\rho^2 + \frac{9}{16}\sigma_\rho^2 = \frac{12}{16}\sigma_\rho^2;$$

$$= \frac{3}{4}\sigma_\rho^2$$

Hence the expected value of MS_b under H_A is greater than that under H_T by an amount

$$\frac{N}{M-1}\frac{3}{4}\sigma_\rho^2$$

In order to determine N we use Chart R at the back of the book, which allows one to read powers of the F ratio with a given value of

$$\phi = \frac{\sqrt{(N/M)\sum_{j=1}^{M}\tau_j^2}}{\sigma_\rho}$$

Notice that ϕ involves M, not $M-1$. In our example

$$\phi = \frac{\sqrt{\left(\frac{N}{4}\right)\left(\frac{3}{4}\right)\sigma_\rho^2}}{\sigma_\rho} = \sqrt{N(3/16)}$$

The value of N in this equation can be determined because of the dependency of π, N, and ϕ.

These charts were prepared by Pearson and Hartley to determine the power for F tests. The use of the charts to select N involves some crudity of estimation because of linear interpolation and inaccuracy in chart reading. However, the inaccuracy should be within one or two observations per cell.

Our alternative hypothesis is rather close to the test hypothesis, and hence the sample size possibly is large. In order to determine sample size we enter the chart for $M-1$ from the left margin with $\pi = .95$, that is, $\beta = .05$. There is a choice of entry, one for $\alpha = .05$ and another for $\alpha = .01$. The example is defined with $\alpha = .05$. The approximation to the necessary sample size is given by assuming at first that the total sample size is unlimitedly large. The point at which the horizontal line at $\pi = .95$ crosses the curve for $df_w = \infty$ corresponds to a point on the bottom of the graph, in this example $\phi = 2.04$ as nearly as can be read from the graph.

Using this value of ϕ in calculating N assures us that the power is at least .95 for the sample size determined this way. The 2.04 value is substituted into the equation for ϕ,

$$2.04 = \sqrt{N(3/16)}$$

giving

$$N = (2.04)\frac{216}{3}$$

$$= 22.2$$

$$\cong 23$$

If we set sample size for each level of the factor at 23, we are assured of power of at least .95. This sample size gives an effective total sample size of $23M = 92$ with an effective $df_w = (N - 1)M = 88$.

Since the initial $df_w = \infty$ used to obtain an initial value of ϕ is a gross overestimate of the degrees of freedom, it is instructive to use $df_w = 60$ as the initial value. This value is used because it is the largest specifically plotted real value. For $df_w = 60$, $\phi = 2.12$, implying

$$N \cong 24$$

However, since $N = 23$ guarantees a minimum power of .95, there is no sense in using a larger sample size. In general, using the $df_w = \infty$ to obtain ϕ gives the minimum sample size to obtain the specified power.

The equation for N depends on the value of ϕ looked up in the chart, but more importantly the equation for N depends on the structure of the alternative hypothesis. In the example just discussed, the alternative involved one effect with a value one standard deviation (that is, σ_ρ) larger than the other three effects, which in turn are equal. Imagine that there are six levels and that the alternative hypothesis is similar to that in the first example, one effect value one standard deviation greater than the others. This arrangement gives

$$\tau = -\frac{\sigma_\rho}{6}$$

$$\tau + \sigma_\rho = \sigma_\rho - \frac{\sigma_\rho}{6} = \frac{5}{6}\sigma_\rho$$

$$\sum_{j=1}^{M}\tau_j^{\,2} = 5\left(-\frac{\sigma_\rho}{6}\right)^2 + \left(\frac{5}{6}\sigma_\rho\right)^2$$

$$= \frac{5}{6}\sigma_\rho^{\,2}$$

Consequently

$$\phi = \frac{\sqrt{(N/M)(5/6)\sigma_\rho{}^2}}{\sigma_\rho}$$

$$= \sqrt{\frac{N}{6}\frac{5}{6}}$$

$$= .373\sqrt{N}$$

Reading from the table for $df_b = M - 1 = 5$ for $\pi = .95$ $df_w = \infty$ gives $\phi = 1.8$
Substituting this value into the equation above gives $N \cong 24$. Hence, $N = 24$ i
required for an ($\alpha = .05$, $\beta = .05$) test with the alternative specified by one leve
being one standard deviation greater than other equal valued levels. If we let th
difference of the value of the one effect be expressed in terms of the δ notation, th
difference is a ($\delta = 1$) difference, applied to one of the six effects. The applicatio
of the difference, the δ criterion, can be to as many of the effects as is called for b
the particular research problem. The spread of the effects, δ standard deviatio
units between effects, can be applied to all of the effects, or any subset of th
effects. As a consequence we need to specify with the (α, β, δ) statement the numbe

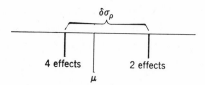

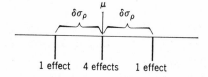

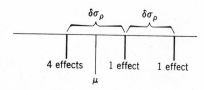

FIGURE 18.3

Illustration of some alternative hypothesis structures using δ standard deviation difference units.

of effects that the δ difference applies to. In the above examples the difference applied to one of M effects. When the difference is applied to more than one effect, the way in which the difference is distributed to the effects must be specified. For example, a two-effect application in an M-level experiment might be defined with the two different effects flanking the rest of the effects, the two different effects the same distance on either side of the rest of the effects, the two different effects spread out 2δ units from each other and δ units from the rest of the effects. Figure 18.3 illustrates this and two other alternative hypothesis structures for $M = 6$. The student should define the equation for ϕ for the alternatives indicated in the figure. The alternative with M effects, two of which are different, flanking the $(M - 2)$ equal-valued effects gives a particularly clean result. Let two of the M effects be different from the others (which are equal to each other) by an amount $-\delta\sigma_\rho$ and $+\delta\sigma_\rho$, respectively. The sum of the effects must be equal to zero, implying that all of the other $(M - 2)$ effects have values of zero. Consequently

$$\sum_{j=1}^{M} \tau_j^2 = (-\delta\sigma_\rho)^2 + (\delta\sigma_\rho)^2 = 2\delta^2\sigma_\rho^2$$

and hence

$$\phi = \frac{\sqrt{(N/M) \sum_{j=1}^{M} \tau_j^2}}{\sigma_\rho}$$

$$= \frac{\delta\sigma_\rho\sqrt{2N/M}}{\sigma_\rho}$$

$$= \delta\sqrt{\frac{2N}{M}}$$

Since δ and M are determined by the experiment, and ϕ determined by the Pearson-Hartley charts, N is the only unknown in the equation and can be solved for easily.

Take another two-effect example. The two-effect example just presented involved a 2δ standard deviation difference between the most different effects. If $\delta = 1$, the extreme effects are a full two standard deviations apart, a rather large difference. However, if we wished to look at an alternative in which the effects were at most one $\delta\sigma_\rho$ units apart, we have

$$\sum_{j=1}^{M} \tau_j^2 = \left(-\frac{\delta\sigma_\rho}{2}\right)^2 + \left(\frac{\delta\sigma_\rho}{2}\right)^2$$

$$= \frac{\delta^2\sigma_\rho^2}{2}$$

and hence

$$\phi = \frac{\delta\sigma_\rho\sqrt{N/2M}}{\sigma_\rho}$$

$$= \delta\sqrt{\frac{N}{2M}}$$

The last-presented example is a more modest departure from the null hypothesis than the previous example. It is instructive to look at the most radical departure possible. Let the alternative hypothesis be defined by a structure that strings the alternatives out, each effect $\delta\sigma_\rho$ larger than the next smaller. To simplify the notation, let the τ_j be ordered so that the order of magnitude of the effects is the same order as the magnitude of the values of j. Since the τ_j must sum to 0 the following values are implied:

$$\tau_1 = \left(1 - \frac{M+1}{2}\right)\delta\sigma_\rho$$

$$\tau_2 = \left(2 - \frac{M+1}{2}\right)\delta\sigma_\rho$$

$$\cdot$$
$$\cdot$$
$$\cdot$$

$$\tau_j = \left(j - \frac{M+1}{2}\right)\delta\sigma_\rho$$

$$\cdot$$
$$\cdot$$
$$\cdot$$

$$\tau_{M-1} = \left(M - 1 - \frac{M+1}{2}\right)\delta\sigma_\rho$$

$$\tau_M = \left(M - \frac{M+1}{2}\right)\delta\sigma_\rho$$

Evaluating the expression,

$$\frac{1}{M}\sum_{j=1}^{M}\tau_j^2$$

is equivalent to calculating the variance of a uniformly distributed random variable with value set $\{1, 2, \ldots, M\}$ multiplied by the constant $\delta\sigma_\rho$:

$$\frac{1}{M}\sum_{j=1}^{M}\tau_j^2 = \delta^2\sigma_\rho^2\left(\frac{M^2-1}{12}\right)$$

Hence

$$\phi = \delta\sqrt{\frac{N(M^2-1)}{12}}$$

Application of this result is the same as in the previous examples. Imagine four levels on a factor, that is, $M = 4$, with the alternative hypothesis stating that each effect is $\frac{1}{3}$ greater than the next smaller, $\delta = \frac{1}{3}$:

$$\phi = \frac{1}{3}\sqrt{N}\sqrt{\frac{16 - 1}{12}}$$

$$= .373\sqrt{N}$$

Reading from the Pearson-Hartley chart for $\pi = .95$, $df_w = \infty$ gives $\phi = 2.04$ and

$$2.04 = .373\sqrt{N}$$

so that

$$N = 30$$

The simplicity of the two types of alternatives just discussed makes them particularly useful. The alternative specifying that two effects are different from the other effects and that they flank the other effects and have $\delta\sigma_p$ units between them is in a compelling sense the weakest balanced alternative to the null hypothesis of all equal effects. On the other hand, the most extreme departure or the strongest alternative to the null hypothesis is the alternative with all M effects $\delta\sigma_p$ units from the nearest effect. This latter alternative gives a total of $(M - 1)\delta\sigma_p$ units between the smallest and the largest effect values. In the remainder of this chapter and in the next chapter these two alternatives are used in evaluating designs and are referred to as the 2-effect alternative and the M-effect alternative. Because the 2-effect and the M-effect alternatives define "ends" to the continuum of type of alternative being considered (uniform application of $\delta\sigma_p$ to the differing effects), they provide a framework within which the meaning of a given statistical result or planned experiment can be studied in the absence of a well-specified alternative hypothesis.

The equations for sample size in these two alternative structures are summarized as follows. For the M-effect alternative:

$$\phi = \delta\sqrt{\frac{N(M^2 - 1)}{12}}$$

$$N = \frac{\phi^2}{\delta^2(M^2 - 1)/12} \tag{18.21}$$

For the 2-effect alternative:

$$\phi = \delta\sqrt{\frac{2N}{M}}$$

$$N = \frac{\phi^2}{2\delta^2/M} = \frac{\phi^2 M}{2\delta^2} \tag{18.22}$$

In the problem where the alternative is spelled out in detail by a competing theory or other theoretical considerations, the procedures are straightforward. The sum of squares of the differential effects as specified by the competing theory is used instead of the logical alternative values indicated in the developments above. If these differential effect values are $t_1, t_2, \ldots, t_M$, the value of ϕ is obtained using

$$\phi = \frac{\sqrt{(N/M) \sum_{j=1}^{M} t_j^2}}{\sigma_\rho}$$

Notice that if σ_ρ is not the same under the theoretically interesting alternative, the analysis of variance is not an appropriate statistical procedure.

Calculational Form. The calculation of an analysis of variance in the one-factor experiment is simple. Only the sums and the sums of squares of the observations, level by level, are required. From these values, "intermediate" values are calculated from which are calculated the required sums of squares, SS_T, SS_b, SS_w.

Organization of the data tables and computational forms is basic to accurate and efficient computation. The organization of computation in the simple case is clear and trivial. However, when an experiment is large and more complicated, computational organization assumes a greater role.

In the discussion of computational procedures we develop the general form of data and equations and then apply them to a numerical example. The general layout of data is shown in Table 18.1. The observations in each level are segregated into a separate column, and the sum of the values and the sum of the squares of the values are calculated for each level.

A numerical example is given in Table 18.3, with $(N_1 = 5)$, $(N_2 = 6)$, $(N_3 = 6)$, and $(N_4 = 7)$ observations in the four treatment groups, respectively.

First we define some terms. Let the total sample size be N_T,

$$N_T = \sum_{j=1}^{M} N_j \tag{18.23}$$

let T be the total sum of scores,

$$T = \sum_{j=1}^{M} \sum_{i=1}^{N_j} X_{ij} = \left(\sum_{i=1}^{N_1} X_{i1} + \cdots + \sum_{i=1}^{N_M} X_{iM} \right) \tag{18.24}$$

and let S be the total sum of squared scores,

$$S = \sum_{j=1}^{M} \sum_{i=1}^{N_j} X_{ij}^2 = \left(\sum_{i=1}^{N_1} X_{i1}^2 + \cdots + \sum_{i=1}^{N_M} X_{iM}^2 \right) \tag{18.25}$$

TABLE 18.3 Example Data of a Four-Level One-Factor Analysis of Variance

	Treatment 1	Treatment 2	Treatment 3	Treatment 4
	53	53	49	44
	48	49	48	44
	49	50	46	42
	47	49	47	46
	49	52	50	45
		51	48	43
				42
N	5	6	6	7
$\sum_{i=1}^{N_j} X_{ij} =$	246	304	288	306
$\bar{X}_j =$	49.2	50.7	48.0	43.7
$\sum_{i=1}^{N_j} X_{ij}^2 =$	12124	15416	13834	13390
$\left(\sum_{i=1}^{N_j} X_{ij}\right)^2 =$	60516	92416	82944	93636
$N_T = 24$	$T = 1144$	$S = 54764$	$R = 54706$	$\bar{X} = 47.7$

In addition, we need the sum of the squared sums of scores, level by level, each squared sum of scores divided by the number of observations in that level:

$$R = \sum_{j=1}^{M} \frac{(\sum_{i=1}^{N_j} X_{ij})^2}{N_j} = \frac{(\sum_{i=1}^{N_1} X_{i1})^2}{N_1} + \cdots + \frac{(\sum_{i=1}^{N_M} X_{iM})^2}{N_M} \quad (18.26)$$

Note that if all N_j are equal, the N_j term factors out of the summation.

Recall the definitions of SS_T, SS_b, and SS_w. These equations lead to computational forms using N_T, T, S, and R alone. First we have

$$SS_T = \sum_{j=1}^{M}\sum_{i=1}^{N_j}(X_{ij} - \bar{X})^2$$

By squaring and distributing the summation,

$$SS_T = \sum_{j=1}^{M}\sum_{i=1}^{N_j} X_{ij}^2 + N_T\bar{X}^2 - 2\bar{X}\sum_{j=1}^{M}\sum_{i=1}^{N_j} X_{ij}$$

The last two terms in this equation, by virtue of the equality

$$\sum_{j=1}^{M}\sum_{i=1}^{N_j} X_{ij} = N_T\bar{X}$$

are equal to

$$- \frac{\left(\sum\limits_{j=1}^{M} \sum\limits_{i=1}^{N_j} X_{ij} \right)^2}{N_T}$$

Hence

$$SS_T = S - \frac{T^2}{N_T} \tag{18.27}$$

The sum of squared deviations of level means about the grand mean, summing for each subject in the entire experiment, gives

$$SS_b = \sum_{j=1}^{M} \sum_{i=1}^{N_j} (\bar{X}_j - \bar{X})^2$$

$$= \sum_{j=1}^{M} N_j (\bar{X}_j - \bar{X})^2$$

By calculations similar to those for SS_T,

$$SS_b = \sum_{j=1}^{M} \frac{\left(\sum_{i=1}^{N_j} X_{ij} \right)^2}{N_j} - \frac{\left(\sum_{j=1}^{M} \sum_{i=1}^{N_j} X_{ij} \right)^2}{N_T}$$

$$= R - \frac{T^2}{N_T} \tag{18.28}$$

Because $SS_T = SS_b + SS_w$, the calculation of SS_w is based on SS_T and SS_b:

$$SS_w = SS_T - SS_b \tag{18.29}$$

Also, by simple algebra,

$$SS_w = \sum_{j=1}^{M} \sum_{i=1}^{N_j} (X_{ij} - \bar{X}_j)^2$$

$$= \sum_{j=1}^{M} \sum_{i=1}^{N_j} X_{ij}^2 - \sum_{j=1}^{M} \frac{\left(\sum_{i=1}^{N_j} X_{ij} \right)^2}{N_j}$$

$$= S - R$$

$$= SS_T - SS_b$$

Applying these results to the numerical example gives

$$N_T = 5 + 6 + 6 + 7 = 24$$

$$T = 246 + 304 + 288 + 306 = 1144$$

$$S = 12124 + 15416 + 13834 + 13390 = 54764$$

$$R = \frac{60516}{5} + \frac{92416}{6} + \frac{82944}{6} + \frac{93636}{7} = 54706.4$$

and consequently

$$SS_T = 54764 - \frac{1308736}{24} = 54764 - 54531 = 233$$

$$SS_b = 54706 - \frac{1308736}{24} = 175$$

$$SS_w = 233 - 175 = 58$$

These results are summarized in the analysis of variance table, Table 18.4. The variance ratio

$$\frac{SS_b/df_b}{SS_w/df_w}$$

is 20.1 in the example. With these data the significance probability for the hypothesis

$$H_T: \mu_1 = \mu_2 = \mu_3 = \mu_4$$

is determined by inspection of the table of the F distribution with $df_b = df_{\text{numerator}} = 3$ and $df_w = df_{\text{denominator}} = 20$. The significance probability Ψ is less than .001, because the ratio 20.1 is larger than the ratio for a significance level of $\alpha = .001$.

TABLE 18.4 Analysis of Variance Table for Data of Table 18.3

Source of Variation	SS	df	MS	F
Between treatments	175	3	58.3	20.1
Within treatments	58	20	2.9	
Total	233	23		

Estimate of σ_ρ:

$$\hat{S}_\rho = \sqrt{2.9} = 1.703$$

Significance Probability: $\Psi < .001$.
Posttest Evaluation: Using mean sample size, $N = 6$

	Two-Effect Alternative		M-Effect Alternative	
	$\phi = \delta \sqrt{\dfrac{2N}{M}}$		$\phi = \delta \sqrt{\dfrac{N(M^2 - 1)}{12}}$	
	$\alpha = .01$	$\alpha = .05$	$\alpha = .01$	$\alpha = .05$
Power for				
$\delta = 1.0$	$\pi = .48$	$\pi = .76$	$\pi = .95$	$\pi > .99$
$\beta = .01$	$\delta = 1.83$	$\delta = 1.55$	$\delta = 1.16$	$\delta = .96$
$\beta = .05$	$\delta = 1.60$	$\delta = 1.32$	$\delta = 1.0$	$\delta = .82$

The problem analyzed was not defined initially by an (α, β, δ) procedure. Therefore a posttest evaluation of β for fixed alternative hypotheses and an evaluation of δ for the two extreme types of alternatives are needed before the test can be interpreted. Recall equations (18.21) and (18.22). These two equations permit an evaluation of β. This evaluation can be made for a given α for each of the two types of alternatives. Also, the equations can be used to evaluate δ for each type of alternative hypothesis for a selection of α and β values. Because the Pearson-Hartley charts are limited to $\alpha = .01$ and $\alpha = .05$, the evaluation also is restricted here to $\beta = .01$ and $\beta = .05$ for the purposes of finding the effective value of δ.

First we develop the explicit steps necessary to evaluate the power of the test for each of the two types of alternative hypotheses for given α and δ. Since this problem begins with values of δ, M, and N (the average of the N_j sample sizes), the value of ϕ is calculated by the equations (18.21) and (18.22). The value of ϕ is first calculated and then used to enter the chart for $df_b = M - 1$ in the section for the specified value of α. In the example, imagine that $\alpha = .01$, and that $\delta = 1.2$. For the two-effect alternative, $\phi = 2.08$. The value for N is taken as the mean value of the sample sizes in the four treatment groups. The Pearson-Hartley chart for $df = M - 1 = 3$ is entered with the value for the proper alpha level on the right portion of the chart. Finding the ordinate value of $\phi = 2.08$, we read the value of the power of the test from the abscissa for the line corresponding to $df_w = M(N - 1) = 20$. The power is difficult to read from the chart, but it is clear that the power is less than .75. If we had begun with the value $\delta = 3$, the value of $\phi = 5.2$ would have been used in entering the table and the approximate power would have been $\pi > .99$.

This procedure is reversed in order to find the effective value of δ for the test. The values of α, β, N, and M are given, the value of ϕ is determined by inspecting the Pearson-Hartley chart. Imagine that $\alpha = .05$ and $\pi = .98$ are the preselected values. Selecting the chart with $df_b = 3$ and entering from the left of the chart on the line representing .98, read across to the right to the line for $df_w = 20$ and then straight down to the value of ϕ corresponding to the intersection of these two lines. The value obtained is approximately $\phi = 2.6$. Imagine that we were applying this to the M-level alternative hypothesis. Thus, we have

$$2.6 = \delta \sqrt{\frac{6(16 - 1)}{12}}$$

$$\delta = \frac{2.6}{2.74}$$

$$= .95$$

Therefore this test is a test with $(\alpha, \beta, \delta) = (.05, .02, .95)$. The reader should calculate a number of values of δ with an assortment of values of α and β under the two types of alternative hypotheses. Some such values are shown in Table 18.4.

Empirical Violations of the Model. When an experimenter applies the methods of the analysis of variance to data, there are two sets of considerations that may be relevant. In the first instance the analysis may simply be a way for the experimenter to reduce his many observations to succinct statements about the effect of the experimental treatments on the dependent variable. Second, the experimenter may wish to make an (α, β, δ) statement about H_T versus H_A on the basis of the analysis and the variance ratio. These two approaches involve different aspects of the model. We must be concerned to justify the application of the model to the data in both approaches.

When the model is applied for descriptive analysis of the data, three assumptions must be justified. The first assumption is that the basic form of the model is appropriate—that is, an observation can be decomposed into an overall effect, a differential constant effect in each level of the factor, and a residual, random effect. The justification of this model must be made on the basis of an analysis of the scientific content of the experimental subject matter. The second assumption is that the observations are made in such a way that there is no bias in favor of some one or more of the levels. If the observations in one or more of the levels are made under conditions introducing a constant response effect, not represented in the other levels of the design, the descriptive analysis will discover this effect. The data under that circumstance, and hence the analysis, do not represent the experiment that was intended—any interpretation is in jeopardy of misleading the researcher. The third assumption is that the observations are independent throughout the experiment. If there is a covariance between observations, the partition of the sum of squares will be in error if carried out with the equations indicated above. The equations could have been developed to account for covariance in the observations, if the covariance is known. The equations given above do not take this into consideration.

When the analysis is used to make statistical inferences, there are some additional assumptions that must be accounted for. If the model is violated by the experiment, that is, the model is not a model of the experiment, the probability distribution calculations are not accurate.

Unless the probability distribution and variance assumptions of the model are true in the experiment, the distribution of the variance ratio calculated in the experiment is not the same as the Fisher F distribution. Hence the significance probability and the power probability of the experiment will not be correct—the probabilities will not be the actual state-of-nature probabilities. The degree of difference between the actual probabilities and the probabilities looked up in the tables of the F distribution and the Pearson-Hartley charts depends on the seriousness of the departure of the experiment from the assumptions. The assumptions made for descriptive application of the analysis of variance must be defended in the inferential application of the analysis, principally for the same reasons. In addition, it is assumed that the residual term in the model is distributed normally with mean of zero and variance $\sigma_\rho{}^2$ in each level of the experimental factor. The normality assumption is not automatically satisfied with large sample size, as was the case in the two-sample problem. If, and only if, the dependent random variable is normal, the variance ratio is distributed as the Fisher F distribution. However, mathematical and sampling investigation of the degree of error in the probabilities obtained from experiments in which the normality assumption was false indicates that the normality assumption is of minor importance. The assumption that the variance of the residual term is a constant across the set of levels of the factor can be violated with relative impunity if the sample size is the same in all the levels of the factor. If the sample sizes are grossly unequal, the assumption must be taken more seriously. In all circumstances, the assumption regarding independence in the observations must be regarded as important. The presence of covariance among observations seriously affects the calculation of the mean squares used in the variance ratio and, therefore, produces inaccurate values of the variance ratio.

Repeated Measures on One Factor

This section extends the statistical procedures for repeated measures from the two-sample problem to the many-sample problem. A single group of subjects is observed under M different conditions (treatments or levels on an experimental factor). All of the advantages and hazards of the two-sample repeated measures problem generalize to the many-sample problem with repeated measures.

An explanation of the strength of the design in terms of the covariance of the observations is difficult without matrix algebra because of the many possible covariances involved—$[M(M - 1)/2]$ of them. The basic idea is the same, however. Imagine that each individual subject is relatively consistent in his response to the M experimental conditions but that these subjects are quite different in their levels of response. Under this condition a great deal of the SS_T is due to individual differences that have nothing to do with the experimental treatments. This is true in the one-factor design described above; the SS_T includes the individual differences. However, in the repeated measures design there is a method of partitioning the individual-differences part of SS_T into a separate package. This partition is not possible in the design for nonrepeated observations.

Subjects have idiosyncratic responses to treatments. These responses vary from subject to subject, introducing variation within a treatment—variation that cannot be counted as coming from the treatment because the variation is within the group being treated the same way. If the groups are random groups, that is, nonrepeated measures design, then this variation is part of the residual variation and included in SS_w. The larger the individual differences within the subject population, as related to the dependent variable, the greater the MS_w is, even under very tightly controlled experimental conditions. However, if each subject is observed under all treatments, then we are able to discriminate between individual-differences contribution to the variation of the dependent variable and the other sorts of residual variation. The responsiveness of each subject, independently of the treatment effects, can be measured by the average across the treatments subject by subject. These averages, one for each subject, depend on the individual differences of the subjects in responding to the entire collection of experimental treatments. Being able to partition the individual-differences variation apart from the "within" and "between" sources of variation means that a smaller variation between experimental levels is detectable.

A Model.   The problem we wish to model is similar to the single-factor problem. However, we have but one group of subjects and several observations on each subject. Table 18.5 shows the general data layout for this experimental design. Each row corresponds to an individual subject; that is, there is an observation of X under each level of the experiment for each subject. It is clear that there are three kinds of responsiveness discriminable in this sort of design. The first is the overall responsiveness of the population of subjects to the types of experimental treatment for the dependent variable being studied. The second is the responsiveness of the subjects individually to the total experiment. The third is the responsiveness of the subjects to each of the experimental treatments. These three effects may be denoted

$$\mu = \text{overall level of response}$$
$$\mu_{i.} = \text{the level of response of subject } i$$
$$\mu_{.j} = \text{the level of response in treatment level } j$$

These values are interpreted as expectations of the dependent variable under the conditions specified. The "dot" notation, $\mu_{i.}$ and $\mu_{.j}$ is used to indicate that the expectations are taken across a set of conditions or a population of subjects respectively.

TABLE 18.5 General Organization of Analysis of Variance Single-Factor Repeated Measures Design

	Level 1	Level 2	...	Level j	...	Level M	Subject Totals
Subject 1	$X_{1,1}$	$X_{1,2}$	.	$X_{1,j}$	.	$X_{1,M}$	$\sum_{j=1}^{M} X_{1j}$
Subject 2	$X_{2,1}$	$X_{2,2}$	.	$X_{2,j}$	.	$X_{2,M}$	$\sum_{j=1}^{M} X_{2j}$
.	.	.	.	.	.	.	.
.	.	.	.	.	.	.	.
Subject i	$X_{i,1}$	$X_{i,2}$	.	$X_{i,j}$	.	$X_{i,M}$	$\sum_{j=1}^{M} X_{ij}$
.	.	.	.	.	.	.	.
Subject N	$X_{N,1}$	$X_{N,2}$	.	$X_{N,j}$	.	$X_{N,M}$	$\sum_{j=1}^{M} X_{Nj}$
$\sum_{i=1}^{N} X_{ij}:$	$\sum_{i=1}^{N} X_{i1}$	$\sum_{i=1}^{N} X_{i2}$	...	$\sum_{i=1}^{N} X_{ij}$	...	$\sum_{i=1}^{N} X_{iM}$	
$\sum_{i=1}^{N} X_{ij}{}^2:$	$\sum_{i=1}^{N} X_{i1}{}^2$	$\sum_{i=1}^{N} X_{i2}{}^2$	...	$\sum_{i=1}^{N} X_{ij}{}^2$	...	$\sum_{i=1}^{N} X_{iM}{}^2$	

For purposes of the analysis it is convenient to deal with the differential levels of response:

$$\lambda_i = \mu_{i.} - \mu \quad i = 1, \ldots, N$$
$$\xi_j = \mu_{.j} - \mu \quad j = 1, \ldots, M \tag{18.30}$$

It is necessary and reasonable to define the levels of response in such a way that both the subject levels and the treatment levels average out to the overall level of response. That is,

$$\frac{1}{N} \sum_{i=1}^{N} \mu_{i.} = \mu \quad \text{and} \quad \frac{1}{M} \sum_{j=1}^{M} \mu_{.j} = \mu$$

Consequently

$$\sum_{i=1}^{N} \lambda_i = 0 \quad \text{and} \quad \sum_{j=1}^{M} \xi_j = 0$$

In addition to the overall response level and the two differential response levels, in the observed value for any given subject in a given experimental treatment, there is some influence because of factors varying at random in the experiment and in the subject. These factors add to give a residual effect ρ. This effect is strictly random with respect to the subject and the experimental condition. It is reasonable to assume that this effect has an expectation of zero and a variance σ_ρ^2, regardless of the subject and the experimental condition. In particular, we must assume that the residual effect does not vary as a function of the differential effects ξ_j. This last assumption is important; it is sometimes referred to as the additivity assumption. The essence of this assumption is that for any two subjects the difference of the expected values on any treatment must be equal to the difference of the expected values on any other treatment. Imagine that the expected values for subjects a and b on the pth level are A_p and B_p, respectively, and that the expectations of a and b on the qth level are A_q and B_q. Additivity implies

$$A_p - B_p = A_q - B_q$$

or alternatively, if for any $p = 1, \ldots, M$

$$A_p - B_p = k$$

then

$$A_q = B_q + k$$

for all $q = 1, \ldots, M$. The value of k may be different for another pair of subjects.

Examples of departure from additivity are easy to think of. Imagine a group of subjects heterogeneous with respect to intelligence and factor levels defined by degrees of difficulty in problem sets. For levels defined by easy problems, all subjects do well. For levels defined by difficult problems, the dull subjects do poorly. Hence, for an "easy" level p, a "hard" level q, a bright subject A, and a dull subject B, we have

$$A_p - B_p = 0$$

reflecting the good performance of both subjects on the easy problems. However, A does well on q but B does poorly:

$$A_q - B_q \neq 0$$

This violates the additivity assumption. The principle of nonadditivity as used in this sense is also called interaction. That is, the subject and the levels "interacted" in such a way that subject differences depend on the experimental treatment.

In summary, the model states that each observation in the single-factor repeated observations design may be expressed as a sum of the general level of response, the two differential levels of response, and the residual effect:

$$X_{ij} = \mu + \lambda_i + \xi_j + \rho_{ij}$$

This model leads to useful results in formulating a statistical test regarding the equality of the experimental treatment effects. The general development in this design, with this model, is the same as in the previous design.

The Partition. The partition in this design is into three separate parts. Figure 18.4 is a schematic representation of the partition of the total sum of squares. The partition is obtained by expressing the deviation of an observation about the overall mean of observations as a sum of components:

$$(X_{ij} - \bar{X}) = (\bar{X}_{i.} - \bar{X}) + (\bar{X}_{.j} - \bar{X}) + (X_{ij} - \bar{X}_{i.} - \bar{X}_{.j} + \bar{X}) \quad (18.31)$$

The means in this equation are defined by

$$\bar{X} = \frac{1}{NM} \sum_{j=1}^{M} \sum_{i=1}^{N} X_{ij}$$

$$\bar{X}_{i.} = \frac{1}{M} \sum_{j=1}^{M} X_{ij} \qquad i = 1, \ldots, N$$

$$\bar{X}_{.j} = \frac{1}{N} \sum_{i=1}^{N} X_{ij} \qquad j = 1, \ldots, M$$

By the laws of summation,

$$\bar{X} = \frac{1}{N} \sum_{i=1}^{N} \bar{X}_{i.}$$

$$= \frac{1}{M} \sum_{j=1}^{M} \bar{X}_{.j}$$

Expression (18.31) makes it clear that the residual effect is dealt with in the model just as a residual: The final term is truly a residual after taking account of the mean for the subject and for the treatment. The total sum of squares is obtained by squaring and summing over i and j. All of the inner product terms on the right side of the equation can be shown to drop out because of the definition of the means involved.

We adopt a notation similar to the randomized sample design notation:

$$SS_T = \text{total sum of squares}$$
$$SS_{bs} = \text{sum of squares between subjects}$$
$$SS_{ws} = \text{sum of squares within subjects}$$
$$SS_t = \text{sum of squares between treatments}$$
$$SS_r = \text{sum of squares residual}$$

The total sum of squares, and its partition, is given by

$$SS_T = SS_{bs} + SS_t + SS_r$$

$$= \sum_{j=1}^{M} \sum_{i=1}^{N} (X_{ij} - \bar{X})^2 \qquad (18.32)$$

$$= M \sum_{i=1}^{N} (\bar{X}_{i.} - \bar{X})^2 + N \sum_{j=1}^{M} (\bar{X}_{.j} - \bar{X})^2 + \sum_{j=1}^{M} \sum_{i=1}^{N} (X_{ij} - \bar{X}_{i.} - \bar{X}_{.j} + \bar{X})^2$$

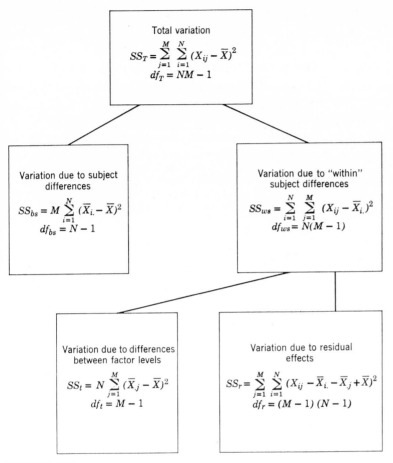

FIGURE 18.4

Partition of total sum of squares.

The degrees of freedom for these sums of squares are additive likewise:

$$df_T = df_{bs} + df_t + df_r$$
$$= (NM - 1)$$
$$= (N - 1) + (M - 1) + (N - 1)(M - 1) \qquad (18.33)$$

It can be shown that

$$SS_{ws} = SS_t + SS_r$$

In general, we are not interested in the SS_{bs}—it is simply a measure of the individual differences reflected in the experiment. Unless the subjects are selected to represent some specific subject population, there is no hypothesis about subject differences. This is not always the situation, however. For example, Winer (1964, p. 124) discusses an application of this design in the analysis of reliability of tests.

Expected Mean Squares. The mean squares corresponding to the sums of squares in the partition are obtained in the usual manner by dividing the sums of squares by the appropriate degrees of freedom value. The expected mean squares for the three sources of variation are obtained by the methods developed in the previous section, with the exception of a necessary complication in the $E(MS_r)$ occasioned by the covariability of observations under different experimental conditions. This covariability occurs because observations under every experimental condition are influenced by the same individual difference parameters, $\lambda_1, \ldots, \lambda_N$. Under the conditions imposed with the model the expectations can be shown to be

$$E(MS_{bs}) = \sigma_p^2 + \frac{M}{N-1} \sum_{i=1}^{N} \lambda_i^2 \tag{18.34}$$

$$E(MS_t) = \sigma_p^2 + \frac{N}{M-1} \sum_{j=1}^{M} \xi_j^2 \tag{18.35}$$

$$E(MS_r) = \sigma_p^2 \tag{18.36}$$

A Hypothesis. In this design the hypothesis of interest is the same as in the single-factor random observations design. If $\mu_1 = \mu_2 = \cdots = \mu_M$, then both the MS_r and the MS_t have the same expectation, σ_p^2. This is used to establish a statistical test of

$$H_T: \mu_1 = \cdots = \mu_M$$

TABLE 18.6 Analysis of Variance Table for Single Factor Repeated Measures Design

Source of Variation	SS	df	MS	EMS	F
Between Subjects	SS_{bs}	$N-1$	$\dfrac{SS}{df}$	$\sigma_p^2 + \dfrac{M}{N-1}\sum_{i=1}^{N}\lambda_i^2$	—
Between Treatments	SS_t	$M-1$	$\dfrac{SS}{df}$	$\sigma_p^2 + \dfrac{N}{M-1}\sum_{j=1}^{M}\xi_j^2$	$\dfrac{MS_t}{MS_r}$
Residual	SS_r	$(N-1)(M-1)$	$\dfrac{SS}{df}$	σ_p^2	
Total	SS_T	$NM-1$	—	—	

against
$$H_A: \text{ for at least some } j \text{ and } k, \ \mu_k \neq \mu_j$$

Under the normality condition, the ratio

$$V = \frac{MS_t}{MS_r} \qquad (18.37)$$

is distributed as Fisher's F with $(M - 1)$ and $(N - 1)(M - 1)$ degrees of freedom in the numerator and denominator, respectively, if H_T is true. Large values of the ratio V have small significance probabilities and lead to the rejection of the test hypothesis.

The entire result of the analysis is summarized in Table 18.6.

Sample Size and (α, β, δ). The problems of determining the required sample size for a given (α, β, δ) test in this design are very similar to the problems faced in the random observation single-factor design. One complication is encountered: The degrees of freedom for the residual term is less for a given sample size in this design, compared with the previous design. The procedures for sample size determination or posttest evaluation are the same except that the numerator degrees of freedom is $(N - 1)(M - 1)$ instead of $M(N - 1)$.

The implications of the decrease in the df_r are important. If partitioning the SS_{bs} out of the total variation does not reduce substantially the SS_r, the effect is to reduce the power of the test of H_T against H_A—a smaller df is available for the test without the compensating reduction in the residual variation. It is difficult to judge just how large SS_{bs} must be before one is justified in giving up the $(N - 1)$ degrees of freedom in the denominator of the F ratio. One way of judging this issue is to estimate the efficiency of the test with repeated observations with respect to the test with completely randomized design. From the analysis of variance of the repeated measures data,

$$E = \frac{(N - 1)MS_{bs} + N(M - 1)MS_r}{(NM - 1)MS_r} \qquad (18.38)$$

may be taken as a measure of the relative efficiency of the design. If E is 1.0 or greater, the repeated measures design is the more efficient and, in general, is the more powerful. If E is less than 1, the randomized observation design is the design of choice for future experiments dealing with the problem and population of subjects involved.

Calculational Form. The calculation of an analysis of variance in a one-factor repeated measures design is the same as in a one-factor random groups design for SS_T and SS_t. It may be shown by simple algebra that

$$SS_{bs} = I - \frac{T^2}{MN} \tag{18.39}$$

where

$$I = \frac{1}{M} \sum_{i=1}^{N} \left(\sum_{j=1}^{M} X_{ij} \right)^2 \tag{18.40}$$

The sum of squares for the residual factor is obtained by subtraction:

$$SS_r = SS_T - SS_{bs} - SS_t \tag{18.41}$$

A numerical example is given in Table 18.8 for the data in Table 18.7, showing a three-level design with five replications. The sums calculated for these data are the same as those calculated for the data of the random groups. In addition, the sum of scores and the square of the sum for each subject are calculated. These lead to the summary rows at the right of the data layout. The familiar equations to calculate S and T are used. The formula for R is simplified because of the equal

TABLE 18.7 Illustrative Data for a Repeated Measures Experiment

	Treatment 1	Treatment 2	Treatment 3	$\sum_{j=1}^{3} X_{ij}$	$\left(\sum_{j=1}^{3} X_{ij} \right)^2$
Subject 1	25	75	56	156	24336
Subject 2	79	100	70	249	62001
Subject 3	46	84	84	214	45796
Subject 4	27	54	40	121	14641
Subject 5	40	69	63	172	29584
$\sum_{i=1}^{5} X_{ij} =$	217	382	313		
$\bar{X} =$	43.4	76.4	62.6		
$\sum_{i=1}^{5} X_{ij}^2 =$	11311	30358	20661		
$\left(\sum_{i=1}^{5} X_{ij} \right)^2 =$	47089	145924	97969		
$N_T = 15$	$T = 912$	$S = 62330$	$R = 58196.4$	$I = 58786$	$\bar{X} = 60.8$

size of treatment samples:

$$R = \frac{1}{N} \sum_{j=1}^{M} \left(\sum_{i=1}^{N} X_{ij} \right)^2$$

Calculating these values gives

$$T = 217 + 382 + 313 = 912$$

$$S = 11311 + 30358 + 20661 = 62330$$

$$R = \frac{47089 + 145924 + 97969}{5} = 58196.4$$

$$I = \frac{24336 + \cdots + 29584}{3} = 58786$$

$$\frac{T^2}{NM} = \frac{831744}{15} = 55449.6$$

This permits an immediate evaluation of the sums of squares in the analysis of variance table:

$$SS_T = S - \frac{T^2}{NM}$$

$$SS_t = R - \frac{T^2}{NM}$$

$$SS_r = S - R - I + \frac{T^2}{NM}$$

The analysis of variance summary is given in Table 18.8.

The two alternatives give values of ϕ that are the same. A graph of the alternative structure, Figure 18.5, illustrates the reason for this relationship, which basically is due to having three levels in the experiment. The test for significance is insensitive to moderate departures from H_T because of the relatively small sample size. Even for the M-effect alternative with the least stringent α and β (.05), the best we can do is detect an alternative with the adjacent effects $\delta = 1.52$ units apart. The significance probability, in the light of this, implies a strong likelihood that the test has detected a real experimental effect. Imposing the obtained means on the graph of a three-effect alternative, Figure 18.5, likewise gives support to the notion that the experimental effects are widely separated. Although the obtained means are not as widely dispersed as the expectations in the alternative, they are more nearly like the alternative structure than the test structure.

If the repeated measures aspect of the experiment were simply forgotten and the analysis were performed in the same way a random groups analysis is performed, then $SS_r = 4134$, $df_r = 12$, $\hat{S}_\rho^2 = 344.5$, giving an estimate of σ_ρ,

TABLE 18.8 Analysis of Variance Table for Data of Table 18.7

Source of Variation	SS	df	MS	F
Between Treatments	2747	2	1373.5	13.79
Between Subjects	3337	4	834.25	
Residual	797	8	99.62	
Total	6881	14		

Estimate of σ_ρ:
$$\hat{S}_\rho = \sqrt{99.62} = 9.98$$

Significance Probability: $\Psi < .005$
Posttest Evaluation: $df_{\text{denominator}} = 8$

	Two-Effect Alternative		M-Effect Alternative	
	$\phi = \delta\sqrt{\left(\dfrac{2N}{M}\right)} = \delta(1.8256)$		$\phi = \delta\sqrt{\left[\dfrac{N(M^2-1)}{12}\right]} = \delta(1.8256)$	
	$\alpha = .01$	$\alpha = .05$	$\alpha = .01$	$\alpha = .05$
Power for				
$\delta = 1.0$	$\pi = .32$	$\pi = .65$	$\pi = .32$	$\pi = .65$
$\beta = .01$	$\delta = 2.33$	$\delta = 1.81$	$\delta = 2.33$	$\delta = 1.81$
$\beta = .05$	$\delta = 2.00$	$\delta = 1.52$	$\delta = 2.00$	$\delta = 1.52$

$\hat{S}_\rho = 18.5$. The V ratio from this analysis is 3.99 with 2 degrees of freedom in the numerator and 12 degrees of freedom in the denominator—the significance probability is between .01 and .05 but closer to the .05. The primary difference in the two analyses of the data is in the MS_r. In the correct analysis, as repeated measures data, the square root of the MS_r (the estimate of σ_ρ) is about half the value in the improper analysis in which the data are treated as random groups data

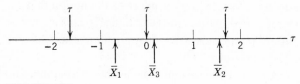

FIGURE 18.5

Differential effects (M-effect alternative) and standardized treatment means.

In the proper analysis for $\alpha = \beta = .05$ the values of the $\delta\sigma_p$ terms in the effective alternative expression of the treatment expectations are -15.2, 0, and 15.2 when $\hat{S}_p$ is used as an estimate of σ_p. On the other hand, the corresponding values in the improper random groups analysis are -28.1, 0, and 28.1. The random groups calculations make it appear that the observed means are more likely to come from a population with equal treatment effects, than from a population with treatments effects spaced with a $\delta = 1.52$. The treatment means are much more closely spaced relative to the estimate of σ_p under the random groups calculation.

These figures illustrate the powerful character of the repeated measures design. The residual variability is the standard by which we judge the variability among the treatment means. If the regularity of response on the part of individual subjects, irrelevant to the experimental treatments, can be estimated and separated from the measure of the residual variation, the residual variability is decreased, giving a finer gauge of the sampling variability of the treatment means.

Empirical Violations of the Model.   The assumptions of the repeated measures design are similar to those of the random group design. Also, we have to make the assumption of additivity. If the additivity assumption in the model is not true for the phenomenon being studied, the result of the test of H_T is not directly interpretable. If additivity does not hold, the treatment factor effect depends on the specific subject involved. Although the factor appears to have (or not) a significant effect on the dependent variable, this is not consistent unless additivity holds. It would be misleading to infer from the F ratio whether or not the factor was significant in determining the level of response for particular subjects.

A test for nonadditivity has been proposed by Tukey. Additivity holds if

$$\mu_{ij} = \mu_{i.} + \mu_{.j} - \mu$$

If this does not hold, then the inequality constant for the ijth term is

$$\mu_{ij}{}^* = \mu_{ij} - \mu_{i.} - \mu_{.j} + \mu$$

The additivity hypothesis is that $\mu_{ij}{}^* = 0$ for all i, j. The alternative is that $\mu_{ij}{}^* \neq 0$ for the same set of i, j. The Tukey test is based on the sum of squares

$$SS_A = \frac{[\sum_{j=1}^{M} \sum_{i=1}^{N} X_{ij}(\bar{X}_{i.} - \bar{X})(\bar{X}_{.j} - \bar{X})]^2}{\sum_{i=1}^{N} (\bar{X}_{i.} - \bar{X})^2 \sum_{j=1}^{M} (\bar{X}_{.j} - \bar{X})^2} \tag{18.42}$$

It can be shown that SS_r contains SS_A. Therefore, SS_r may be partitioned into a "nonadditivity" component and a "remainder" component: $SS_A + SS_{\text{remainder}} = SS_r$. The df for the residual is partitioned also: $df_A = 1$, $df_{\text{remainder}} = (M - 1)(N - 1) - 1$. Table 18.9 is the summary table for the nonadditivity

TABLE 18.9 Summary Table for Additivity Analysis

Source	SS	df	MS
Nonadditivity	SS_A	1	$MS_A = \dfrac{SS_A}{1}$
Remainder	$SS_r - SS_A$	$(N-1)(M-1) - 1$	$MS_{\text{remainder}}$ $= \dfrac{SS_r - SS_A}{(N-1)(M-1) - 1}$
Residual	SS_r	$(N-1)(M-1)$	

analysis. The appropriate F ratio is MS_A/MS_r. If the significance probability of this ratio is less than a predetermined significance level, the additivity assumption probably is justified. In the case nonadditivity is suspected, the effect of the treatment factor has to be judged inconsistent across the subjects.

By some simple algebra it can be shown that

$$SS_A = \frac{[\sum_{j=1}^{M}\sum_{i=1}^{N} X_{ij}(\sum_{j=1}^{M} X_{ij})(\sum_{i=1}^{N} X_{ij}) - (\sum_{j=1}^{M}\sum_{i=1}^{N} X_{ij})(R + I - T^2/NM)]^2}{MN(SS_{bs})(SS_t)}$$

(18.43)

For the example data of Tables 18.7 and 18.8,

$$\sum_{j=1}^{M}\sum_{i=1}^{N} X_{ij}\left(\sum_{j=1}^{M} X_{ij}\right)\left(\sum_{i=1}^{N} X_{ij}\right) = 25(156)(217) + 75(156)(382)$$

$$+ \cdots + 63(172)(313) = 56053584$$

$$\left(\sum_{j=1}^{M}\sum_{i=1}^{N} X_{ij}\right)\left(R + I - \frac{T^2}{MN}\right) = 912(61533) = 56118096$$

$$MN(SS_{bs})(SS_t) = 137501085$$

$$SS_A = \frac{(56053584 - 56118096)^2}{137501085} = \frac{(-64512)^2}{137501085} = 30.3$$

and finally

$$F = \frac{30.3}{(797 - 30.3)/7}$$

$$= \frac{30.3}{109.53} = .28$$

For an F with $df_{\text{numerator}} = 1$ and $df_{\text{denominator}} = (M-1)(N-1) - 1 = 7$, the ($\alpha = .05$) value is 5.6. Therefore, we can be confident in the additivity of the effects in these data.

The Efficiency of the Repeated Measures Design. The advantage of a repeated measures design is related directly to the separation of SS_{bs} and SS_r. If SS_{bs} is large, then the SS_r value is smaller in a repeated measures design than in a random groups design. The advantage is dissipated if the loss of degrees of freedom for the F ratio denominator offsets the reduced SS_r. The SS_r is reduced, but so also is the df_r. Consequently, unless the SS_r is reduced disproportionately, more than the df_r the MS_r tends to be increased. The loss of degrees of freedom is equal to $(N - 1)$ degrees of freedom: $df_r = M(N - 1)$ for random groups; $df_r = (N - 1)(M - 1)$ for repeated observations. Thus, if $SS_{bs} = 0$, the SS_r is the same in both designs but the MS_r is enlarged by the loss of df in the repeated measures design. A second consideration is cautionary: The smaller the df_r, the less sensitive the test of H_T is to any given H_A. That is, losing the $(N - 1)$ degrees of freedom in the denominator of F requires larger values of F before a given level of significance is reached and a consequent reduction of power for a specified alternative.

The relative efficiency of the two designs can be gauged by the ratio E defined in equation (18.38). In the example of Table 18.6 and Table 18.7, $E = 3.1$. The repeated measures design is about three times as efficient as the random measures would have been.

In general, the repeated measures design is more efficient than the random groups design in behavioral studies. Individual differences among subjects are generally major determinants of the response to tasks related to skills, experience, native capacity, personality, and so on. This might not be true for a very highly select or specific population where the dependent variable is relevant to the selection criterion. Imagine a study of maze learning under different schedules of reinforcement where the animals are a strain developed to be "maze brights." The differences between the inbred animals could quite possibly be less than adequate to outweigh the loss of df in the repeated measures design.

SUGGESTED READING

Bowker, A. H., and Lieberman, G. J. (1959), *Engineering statistics*, McGraw-Hill, New York.

Cochran, W. G., and Cox, G. M. (1957), *Experimental designs* (2d ed.), Wiley, New York.

Cox, D. R. (1958), *Planning of experiments*, Wiley, New York.

Graybill, F. A. (1961), *An introduction to linear statistical models*, McGraw-Hill, New York.

Guenther, W. C. (1964), *Analysis of variance*, Prentice-Hall, Englewood Cliffs, N.J.

Johnson, N. L., and Leone, F. C. (1964), *Statistics and experimental design*, Vol. II, Wiley, New York.

Li, C. C. (1964), *Introduction to experimental statistics*, McGraw-Hill, New York.

Mood, A. M. (1950), *Introduction to the theory of statistics*, McGraw-Hill, New York.

Myers, Jerome L. (1966), *Fundamentals of experimental design*, Allyn and Bacon, Boston.

Scheffé, H. (1959), *The analysis of variance*, Wiley, New York.

Snedecor, G. W. (1956), *Statistical methods*, Iowa State University Press, Ames, Iowa.

Winer, B. J. (1962), *Statistical principles in experimental design*, McGraw-Hill, New York.

GLOSSARY

Analysis of Variance. The analysis of variance is a general statistical and descriptive method for dealing with problems defined by many samples, either in one or in several independent classifications or variables, and one dependent variable.

Factor. A factor in the analysis of variance is a collection of conditions under which the dependent variable is observed.

General Level of Response, μ. The level of response associated with or characteristic of the collection of experimental treatments, subject populations, measuring instruments, and so on, used in an experiment. The expectation of the dependent variable over the population of subjects and factor levels and factors in the experimental universe.

Differential Level of Response, τ_j. The specific value indicating the differential degree to which a specific level of a factor influences the value of the dependent variable, for the ith level of a factor, τ_i.

Residual Effect, ρ. The residual effect is a sum of a large number of random variables, the partial sources of the variation associated with individual differences, random fluctuation of attention, and so on.

$$(18.1) \qquad X_{ij} = \mu \pm \tau_j + \rho_{ij}$$

Fixed Effects. A fixed effects analysis of variance model includes all of the levels in the experiment that the experimenter wishes to study or that are involved in inferences made on the basis of the experiment.

$$(18.2) \qquad \mu = \frac{1}{M} \sum_{j=1}^{M} \mu_j$$

$$(18.3) \qquad \tau_j = \mu_j - \mu \quad \text{for} \quad j = 1, \ldots, M$$

$$(18.4) \qquad \sum_{j=1}^{M} \tau_j = 0$$

(18.5) $\dfrac{1}{M-1} \displaystyle\sum_{j=1}^{M} \tau_j^2 \begin{cases} = 0 & \text{if } \tau_1 = \tau_2 = \cdots = \tau_M \\ > 0 & \text{if it is } not \text{ true that } \tau_1 = \tau_2 = \cdots = \tau_M \end{cases}$

(18.6) $$\bar{X} = \frac{1}{N_T} \sum_{j=1}^{M} \sum_{i=1}^{N_j} X_{ij}$$

(18.7) $$SS_T = \sum_{j=1}^{M} \sum_{i=1}^{N_j} (X_{ij} - \bar{X})^2$$

(18.8) $$MS_T = \frac{SS_T}{N_T - 1}$$

(18.9) $$SS_T = \sum_{j=1}^{M} \sum_{i=1}^{N_j} (X_{ij} - \bar{X}_j)^2 + \sum_{j=1}^{M} \sum_{i=1}^{N_j} (\bar{X}_j - \bar{X})^2 = SS_w + SS_b$$

(18.10) $$SS_T = SS_w + SS_b$$

(18.11) $$MS_w = \frac{SS_w}{N_T - M}$$

(18.12) $$MS_b = \frac{SS_b}{M - 1}$$

(18.13) $$df_T = df_w \pm df_b; \quad N_T - 1 = \sum_{j=1}^{M} (N_j - 1) + (M - 1)$$

(18.14) $$E(\bar{X}_j) = \mu_j$$

(18.15) $$E(\bar{X}) = \mu$$

(18.16) $$MS_b = \frac{N \sum_{j=1}^{M} (\bar{X}_j - \bar{X})^2}{M - 1}$$

(18.17) $$MS_w = \frac{\sum_{j=1}^{M} \sum_{i=1}^{N_j} (X_{ij} - \bar{X}_j)^2}{N_T - M} = \frac{(N_1 - 1)\hat{S}_{X_1}^2 + \cdots + (N_M - 1)\hat{S}_{X_M}^2}{N_T - M}$$

(18.18) $$E(MS_b) = \frac{\sum_{j=1}^{M} N_j \tau_j^2}{M - 1} + \sigma_\rho^2$$

(18.19) $$E(MS_w) = \sigma_\rho^2$$

(18.20) $$V = \frac{MS_b}{MS_w}$$

Two-Effect Alternative. An arbitrary alternative structure used in the text to analyze sample size. In an M-level experiment, $(M - 2)$ effects are said to have values of zero and the remaining effects are respectively $-\delta\sigma_\rho$ and $\delta\sigma_\rho$.

M-Effect Alternative. An arbitrary alternative structure used in the text to analyze sample size. In an M-level experiment all M effects are strung out $\delta\sigma_p$ units from the nearest other effect.

(18.21) $$N = \frac{\phi^2}{\delta^2(M^2 - 1)/12} \text{ for the } M\text{-effect alternative}$$

(18.22) $$N = \frac{\phi^2}{2\delta^2/M} = \frac{\phi^2 M}{2\delta^2} \text{ for the two-effect alternative}$$

(18.23) $$N_T = \sum_{j=1}^{M} N_j$$

(18.24) $$T = \sum_{j=1}^{M}\sum_{i=1}^{N_j} X_{ij} = \left(\sum_{i=1}^{N_1} X_{i1} + \cdots + \sum_{i=1}^{N_M} X_{iM} \right)$$

(18.25) $$S = \sum_{j=1}^{M}\sum_{i=1}^{N_j} X_{ij}^2 = \left(\sum_{i=1}^{N_1} X_{i1}^2 + \cdots + \sum_{i=1}^{N_M} X_{iM}^2 \right)$$

(18.26) $$R = \sum_{j=1}^{M} \frac{\left(\sum_{j=1}^{N_j} X_{ij}\right)^2}{N_j} = \frac{\left(\sum_{i=1}^{N_1} X_{i1}\right)^2}{N_1} + \cdots + \frac{\left(\sum_{i=1}^{N_M} X_{iM}\right)^2}{N_M}$$

(18.27) $$SS_T = S - \frac{T^2}{N_T}$$

(18.28) $$SS_b = R - \frac{T^2}{N_T}$$

(18.29) $$SS_w = SS_T - SS_b$$

(18.30) $$\lambda_i = \mu_{i.} - \mu \quad i = 1, \ldots, N$$
$$\xi_j = \mu_{.j} - \mu \quad j = 1, \ldots, M$$

(18.31) $$(X_{ij} - \bar{X}) - (\bar{X}_{i.} - \bar{X}) + (\bar{X}_{.j} - \bar{X}) + (\bar{X}_{ij} - \bar{X}_{i.} - \bar{X}_{.j} + \bar{X})$$

(18.32) $$SS_T = SS_{bs} + SS_t + SS_r$$
$$= \sum_{j=1}^{M}\sum_{i=1}^{N} (X_{ij} - \bar{X})^2$$
$$= M \sum_{i=1}^{N} (\bar{X}_{i.} - \bar{X})^2 + N \sum_{j=1}^{M} (\bar{X}_{.j} - \bar{X})^2 + \sum_{j=1}^{M}\sum_{i=1}^{N} (X_{ij} - \bar{X}_{i.} - \bar{X}_{.j} + \bar{X})^2$$

(18.33) $$df_T = df_{bs} + df_t + df_r$$
$$= (NM - 1)$$
$$= (N - 1) + (M - 1) + (N - 1)(M - 1)$$

(18.34) $$E(MS_{bs}) = \sigma_p^2 + \frac{M}{N - 1} \sum_{i=1}^{N} \lambda_i^2$$

(18.35)
$$E(MS_t) = \sigma_\rho^2 + \frac{N}{M-1} \sum_{j=1}^{M} \xi_j^2$$

(18.36)
$$E(MS_r) = \sigma_\rho^2$$

(18.37)
$$V = \frac{MS_t}{MS_r}$$

Additivity Assumption. In the repeated measures design it is assumed that the residual effects are not a function of the experimental treatments. For any two subjects the difference of the expected values on any treatment must be equal to the difference on any other treatment.

(18.38)
$$E = \frac{(N-1)MS_{bs} + N(M-1)MS_r}{(MN-1)MS_r}$$

(18.39)
$$SS_b = I - \frac{T^2}{MN}$$

(18.40)
$$I = \sum_{i=1}^{N} \frac{(\sum_{j=1}^{M} X_{ij})^2}{M}$$

(18.41)
$$SS_r = SS_T - SS_{bs} - SS_t$$

(18.42)
$$SS_A = \frac{[\sum_{j=1}^{M} \sum_{i=1}^{N} X_{ij}(\bar{X}_{i.} - \bar{X})(\bar{X}_{.j} - \bar{X})]^2}{\sum_{i=1}^{N} (\bar{X}_{i.} - \bar{X})^2 \sum_{j=1}^{M} (\bar{X}_{.j} - \bar{X})^2}$$

(18.43)
$$SS_A = \frac{[\sum_{j=1}^{M} \sum_{i=1}^{N} X_{ij} (\sum_{j=1}^{M} X_{ij})(\sum_{i=1}^{N} X_{ij}) - (\sum_{j=1}^{M} \sum_{i=1}^{N} X_{ij})(R + I - T^2/NM)]^2}{MN(SS_{bs})(SS_t)}$$

PROBLEMS

1. Write a description of how the cells of the designs shown schematically in Figures 18.1 and 18.2 are filled with empirical observations.

2. In an area of research of interest to you, develop a design for an experiment using the concepts of factors, groups, and variables, and draw a diagram of the design.

3. In equation (18.9) a cross-product was represented as having a value equal to zero. Prove that this is true.

4. Prove that equation (18.15) is true.

5. For the general alternative hypothesis structure where one of the effect parameters is different from all of the $(M - 1)$ other parameters by $\delta\sigma_p$ units, in an M-level randomized single-factor design, show that the general equation for sample size is

$$N = \frac{\phi^2 M^2}{(M - 1)\delta^2}$$

6. Select a variety of values of M, α, π, and δ for the analysis of variance with a variety of single-factor designs. Find the sample size specified for the designs, according to the one-effect alternative (see problem 5), the two-effect alternative, and the M-effect alternative.

7. Perform an analysis of variance on the data of Table 18.3 with only the first five listed observation values in each of the four treatment levels. Complete an analysis of variance summary table, parallel to Table 18.4, for the analysis, including a posttest evaluation. Make an analytical report of the differences in the two analyses of the data.

8. In a single-factor design with $M = 2$, show that the F-test value in an analysis of variance is identical to the square of the t-test value in the corresponding two-sample problem. Perform the calculations for both the analysis of variance and the two-sample problem using the data of treatments 2 and 3 in Table 18.3.

9. Perform an analysis of variance on the data given below under two assumptions: (a) the data are from a random groups experiment, and (b) the data are from a repeated measures experiment. Calculate the efficiency coefficient of the repeated measures model compared with the randomized design (that is, calculate E).

Replication	Treat. 1	Treat. 2	Treat. 3	Treat. 4	Treat. 5
1	5	9	8	11	15
2	3	6	7	7	11
3	4	8	9	10	14
4	1	2	4	6	8

10. Add two more replications to the experiment outlined in problem 9. Reanalyze the data and write a discussion of the comparison of the two data sets, and the differences of the analyses under the two assumptions about the experiment.

Replication	Treat. 1	Treat. 2	Treat. 3	Treat. 4	Treat. 5
5	2	1	3	3	4
6	11	11	13	13	14

CHAPTER 19

The Many-Sample Problem, Part 2

This chapter is a continuation of the preceding chapter. The topics discussed there were the simple application of the notion of the analysis of variance and the theory of statistical tests of many-sample hypotheses. In this chapter these notions are extended to more complicated experiments and models. The basic notions of the analysis and statistical applications are the same. The concept of a random effects model is introduced and compared with the fixed effects model.

A Two-Factor Design with Randomized Groups

Extension of the one-factor design to a two-factor design is simple. Imagine two factors, each independently defined by several levels, so that a subject may be observed with both factors simultaneously present. That is, the experimental conditions are meaningful and possible for each possible combination of levels across the two factors, one level from the one factor and one level from the other factor. Such a design is called a two-factor design or a two-way factorial design. The two factors are defined by the same sorts of considerations defining the single factor in a one-factor design. Hence the two-factor design essentially is a design with two independent variables (also independent in the sense that with each level of the one factor it is experimentally possible to combine any level on the other factor). However, only a single dependent variable is involved.

Experimental subjects are randomly sampled from the population and randomly assigned to a treatment or factor level combination: one level on each factor. If the levels are defined by subject characteristics the subjects are sampled at random from the populations corresponding to the factor levels.

Interaction.   A new concept must be introduced at this point. It is quite possible that the subjects in one level on a factor, say the first, have a different pattern of response to the levels on the second factor than the subjects in another level on the first factor. This can be true even if the overall mean response on the two levels of the first factor are the same. To illustrate this notion we plot the factor effects on one factor as functions of the factor levels on the other factor. Figures 19.1a through 19.1e show examples of equal and unequal factor effects with

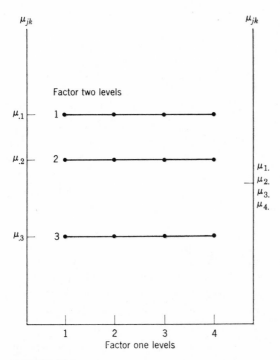

FIGURE 19.1a

Treatment effect differences on factor two, no treatment differences on factor one, no interaction.

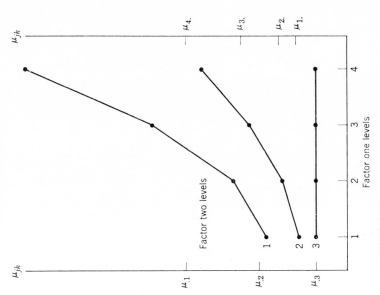

FIGURE 19.1c

Treatment differences on both factors, interaction.

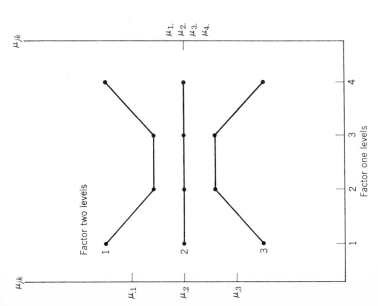

FIGURE 19.1b

Treatment differences on factor two, no treatment differences on factor one, interaction (differential patterns of expectations on levels of one factor with respect to specific levels on the other factor).

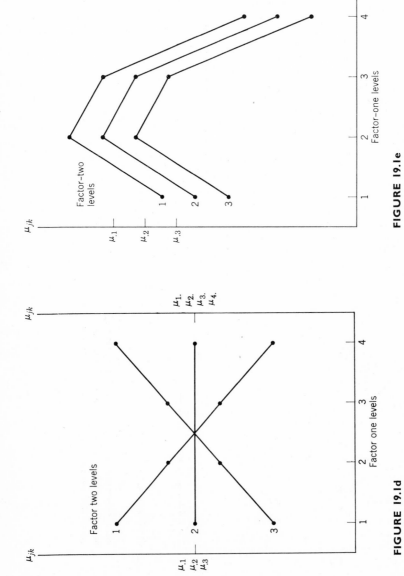

FIGURE 19.1d

Interaction but no treatment differences.

FIGURE 19.1e

Treatment effect differences on both factors but no interaction.

and without interaction. The first factor has four levels and the second factor has three factor levels. The notation for the expectations is simple:

μ_{jk} = the expectation of the dependent variable under the jth level of the first factor and the kth level of the second factor

$\mu_{j.}$ = the expectation of the dependent variable under the jth level of the first factor across all levels on the second factor

$\mu_{.k}$ = the expectation of the dependent variable under the kth level of the second factor across all levels on the first factor

A dot in an expression, as above, implies a subscript that is involved in a summation to obtain a quantity. The dot is located in the place where the subscript would appear in the full expression with the summation. The dot simply helps to keep track of the indices over which summation has taken place.

An example of interaction is useful. Imagine a learning experiment with several types of tasks and several types of training procedures. Some of the training procedures are helpful for some tasks but not for other tasks. Thus, the task-training combination itself is crucial in determining performance. The phenomenon involves an interaction of task and training. Empirical observations can lead to a detection of an "interaction" in the analysis of variance under the appropriate (α, β, δ) procedure.

Main Effects. The hypothesis structure in the two-factor design is complicated by the fact that there are two factors, levels on each of which may or may not lead to different responses, and by the fact that interaction of the two factors, when present, leads to differential performance. We follow the traditional usage and refer to the treatment effects associated with the two factors as main effects. Treatment effects associated with the specific combinations of treatments are called interaction effects.

A Model. Before developing the formal model for the two-factor experiment, it is useful to develop some notation. For each treatment combination of the design, or cell in the layout of the design, N observations are made. Imagine a levels on factor A and b levels on factor B. The data are arranged as in Table 19.1. The totals, the sums of scores, for each factor are shown in this table. These sums are taken over all of the groups of N observations for a given level on a factor; for example, for level 2 of factor A the total is the sum of the sums of observations

TABLE 19.1 Schematic Representation of the Data of a Two-Factor Design with the Totals for the Levels on Both Factors

		Factor B				Totals
		Level 1	Level 2	$\ldots$	Level b	
Factor A	Level 1	X_{111} X_{112} $\cdot$ $\cdot$ X_{11N}	X_{121} X_{122} $\cdot$ $\cdot$ X_{12N}	$\ldots$	X_{1b1} X_{1b2} $\cdot$ $\cdot$ X_{1bN}	$\sum\limits_{k=1}^{b}\sum\limits_{i=1}^{N} X_{1ki} = T_{1.}$
	Level 2	X_{211} X_{212} $\cdot$ $\cdot$ X_{21N}	X_{221} X_{222} $\cdot$ $\cdot$ X_{22N}	$\ldots$	X_{2b1} X_{2b2} $\cdot$ $\cdot$ X_{2bN}	$\sum\limits_{k=1}^{b}\sum\limits_{i=1}^{N} X_{2ki} = T_{2.}$
		$\cdot$	$\cdot$	$\ldots$	$\cdot$	$\ldots$
	Level a	X_{a11} X_{a12} $\cdot$ $\cdot$ X_{a1N}	X_{a21} X_{a22} $\cdot$ $\cdot$ X_{a2N}	$\ldots$	X_{ab1} X_{ab2} $\cdot$ $\cdot$ X_{abN}	$\sum\limits_{k=1}^{b}\sum\limits_{i=1}^{N} X_{aki} = T_{a.}$
	Totals	$\sum\limits_{j=1}^{a}\sum\limits_{i=1}^{N} X_{j1i} = T_{.1}$		$\ldots$ $\sum\limits_{j=1}^{a}\sum\limits_{i=1}^{N} X_{jbi} = T_{.b}$		$\sum\limits_{j=1}^{a}\sum\limits_{k=1}^{b}\sum\limits_{i=1}^{N} X_{jki} = T_{..}$

across all b levels on factor B. In general,

$$T_{j.} = \sum_{k=1}^{b}\sum_{i=1}^{N} X_{jki} \quad \text{for} \quad j = 1, \ldots, a \tag{19.1}$$

$$T_{.k} = \sum_{j=1}^{a}\sum_{i=1}^{N} X_{jki} \quad \text{for} \quad k = 1, \ldots, b \tag{19.2}$$

$$T_{..} = \sum_{j=1}^{a}\sum_{k=1}^{b}\sum_{i=1}^{N} X_{jki} \tag{19.3}$$

$$T_{..} = \sum_{j=1}^{a} T_{j.} = \sum_{k=1}^{b} T_{.k}$$

An important total not shown in the table is the cell total. The observations made in each level combination are summed:

$$T_{jk} = \sum_{i=1}^{N} X_{jki} \quad \text{for} \quad j = 1, \ldots, a \quad \text{and} \quad k = 1, \ldots, b \quad (19.4)$$

The model we use is very much like the single-factor model. We postulate an overall expected response to the experimental situation and an expectation for each of the levels on each of the two factors and each combination of levels:

$$\mu = \text{overall expectation}$$

$$\mu_{j.} = \text{expectation for } j\text{th level of factor } A$$

$$\mu_{.k} = \text{expectation for } k\text{th level of factor } B$$

$$\mu_{jk} = \text{expectation for the } jk\text{th level combination}$$

The expectations of factor levels are the means of the expectations of combinations of factor levels. The overall expectation is the mean of the factor level expectations.

$$\mu_{j.} = \frac{1}{b} \sum_{k=1}^{b} \mu_{jk} \qquad (19.5)$$

$$\mu_{.k} = \frac{1}{a} \sum_{j=1}^{a} \mu_{jk} \qquad (19.6)$$

$$\mu = \frac{1}{a} \sum_{j=1}^{a} \mu_{j.} = \frac{1}{b} \sum_{k=1}^{b} \mu_{.k} = \frac{1}{ab} \sum_{j=1}^{a} \sum_{k=1}^{b} \mu_{jk} \qquad (19.7)$$

If a response to the two factors is determined exclusively by the two factors independently, then the expected response in any combination of levels is the sum of the level expectations minus the overall expectation. That is,

$$\mu_{jk} = \mu + (\mu_{j.} - \mu) + (\mu_{.k} - \mu) = \mu_{j.} + \mu_{.k} - \mu \qquad (19.8)$$

However, when an interaction of the two factors exists, equality (19.8) does not hold. The difference is

$$\mu_{jk}^{*} = \mu_{jk} - \mu_{j.} - \mu_{.k} + \mu$$

The magnitude of this difference is the magnitude of the interaction effect in the jk treatment combination. Formally μ_{jk}^{*} is the same as the nonadditivity effect in the one-factor repeated measures design. However, in the two-factor design, we are able to isolate an interaction variability independent of the other sources of variation.

The differential level effects have a definition similar to that in the one-factor model:

$$\tau_j = \mu_{j.} - \mu \qquad j = 1, \ldots, a \tag{19.9}$$

$$\xi_k = \mu_{.k} - \mu \qquad k = 1, \ldots, b \tag{19.10}$$

$$(\tau\xi)_{jk} = \mu_{jk} - \mu_{j.} - \mu_{.k} + \mu \tag{19.11}$$

$$= \mu_{jk} - (\tau_j + \xi_k + \mu)$$

$$\text{for } j = 1, \ldots, a \text{ and } k = 1, \ldots, b$$

It may be shown easily that

$$\sum_{j=1}^{a} \tau_j = \sum_{k=1}^{b} \xi_k = \sum_{j=1}^{a} \sum_{k=1}^{b} (\tau\xi)_{jk} = 0 \tag{19.12}$$

All of the differential effects indicated here are constants in the model. In addition to these constant effects, the response of an individual subject, the value of any particular observation, will be influenced by the same sort of residual effects described before. These effects are pooled and are treated as a single random variable

$$\rho_{ijk} = X_{ijk} - \mu_{jk} \tag{19.13}$$

These conditions lead to a general model of the dependent variable in the two factor design:

$$X_{ijk} = \mu + \tau_j + \xi_k + (\tau\xi)_{jk} + \rho_{ijk} \tag{19.14}$$

for $i = 1, \ldots, N; j = 1, \ldots, a; k = 1, \ldots, b$.

The Partition. The partition of the deviations of observed scores from the grand mean involves four parts: one part for a residual term, one part for each factor, one for the interaction:

$$(X_{ijk} - \bar{X}) = (X_{ijk} - \bar{X}_{jk}) + (\bar{X}_{j.} - \bar{X}) + (\bar{X}_{.k} - \bar{X}) + (\bar{X}_{jk} - \bar{X}_{j.} - \bar{X}_{.k} + \bar{X}) \tag{19.15}$$

where

$$\bar{X}_{jk} = \frac{1}{N} T_{jk} \qquad \bar{X}_{j.} = \frac{1}{Nb} T_{j.}$$

$$\bar{X}_{.k} = \frac{1}{Na} T_{.k} \qquad \bar{X} = \frac{1}{Nab} T_{..}$$

Squaring and summing over i, j, k leads to

$$\sum_{j=1}^{a}\sum_{k=1}^{b}\sum_{i=1}^{N}(X_{ijk} - \bar{X})^2 = \sum_{j=1}^{a}\sum_{k=1}^{b}\sum_{i=1}^{N}(X_{ijk} - \bar{X}_{jk})^2$$

$$+ Nb\sum_{j=1}^{a}(\bar{X}_{j.} - \bar{X})^2 + Na\sum_{k=1}^{b}(\bar{X}_{.k} - \bar{X})^2$$

$$+ N\sum_{j=1}^{a}\sum_{k=1}^{b}(\bar{X}_{jk} - \bar{X}_{j.} - \bar{X}_{.k} + \bar{X})^2 + C$$

where C is the sum of three cross-product terms that can be shown to be equal to zero. As before, the terms in this expression are called sums of squares,

$$SS_T = SS_r + SS_A + SS_B + SS_{AB} \tag{19.16}$$

where SS_{AB} is used to indicate the interaction sum of squares.

The degrees of freedom associated with these sums of squares can be found by determining the number of free terms in each sum. For SS_T there are Nab terms with a single restriction, that is,

$$\sum_{j=1}^{a}\sum_{k=1}^{b}\sum_{i=1}^{c}(X_{ijk} - \bar{X}) = 0$$

and hence $df_T = Nab - 1$. For SS_r there are Nab terms in the sum but ab restrictions because

$$\sum_{i=1}^{N}(X_{ijk} - \bar{X}_{jk}) = 0 \quad \text{for} \quad j = 1, \dots, a \quad \text{and} \quad k = 1, \dots, b$$

Hence, $df_r = Nab - ab = ab(N - 1)$. The SS_A is based on terms with the single restriction that

$$\sum_{j=1}^{a}(\bar{X}_{j.} - \bar{X}) = 0$$

so that $df_A = a - 1$. Similarly for factor B, $df_B = b - 1$. The interaction sum of squares involves ab terms for which there are b restrictions that

$$\sum_{j=1}^{a}(\bar{X}_{jk} - \bar{X}_{.k}) = 0 \quad \text{for} \quad k = 1, \dots, b$$

and a restrictions that

$$\sum_{k=1}^{b}(\bar{X}_{jk} - \bar{X}_{j.}) = 0 \quad \text{for} \quad j = 1, \dots, a$$

However, the $(a + b)$ restrictions count twice the restriction that

$$\sum_{j=1}^{a}(\bar{X}_{j.} - \bar{X}) = \sum_{k=1}^{b}(\bar{X}_{.k} - \bar{X}) = 0$$

Thus, $df_{AB} = ab - a - b + 1 = (a - 1)(b - 1)$.

Dividing each sum of squares by the appropriate degrees of freedom gives the respective mean square.

Expected Mean Squares. By making essentially the same assumptions for this model that are made in the random groups, single-factor model, we can derive the expected mean squares. First, it is assumed that the residual terms are mutually independent and are not dependent on the treatment effects or the interaction effects. And it is assumed that $E(\rho_{ijk}) = 0$ and $\text{var}(\rho_{ijk}) = \sigma_\rho^2$. The expectations are

$$E(MS_r) = \sigma_\rho{}^2 \tag{19.17}$$

$$E(MS_A) = \sigma_\rho{}^2 + Nb\frac{\sum_{j=1}^{a} \tau_j{}^2}{a-1} \tag{19.18}$$

$$E(MS_B) = \sigma_\rho{}^2 + Na\frac{\sum_{k=1}^{b} \xi_k{}^2}{b-1} \tag{19.19}$$

$$E(MS_{AB}) = \sigma_\rho{}^2 + N\frac{\sum_{j=1}^{a} \sum_{k=1}^{b} (\xi\tau)_{jk}{}^2}{(a-1)(b-1)} \tag{19.20}$$

Hypotheses. There are three test hypotheses usually examined in this design; one for each main effect and one for the interaction. The test hypothesis in each case is the hypothesis of equal treatment or interaction effects. The alternatives specify inequalities among the effects.

For factor A,

$$H_{T,A}: \mu_{1.} = \mu_{2.} = \cdots = \mu_{a.}$$
$$H_{A,A}: \text{not all } \mu_{j.} \text{ are the same}$$

For factor B,

$$H_{T,B}: \mu_{.1} = \mu_{.2} = \cdots = \mu_{.b}$$
$$H_{A,B}: \text{not all } \mu_{.k} \text{ are the same}$$

For the interaction,

$$H_{T,AB}: (\mu_{jk} - \mu_{j.} - \mu_{.k} + \mu) = 0 \qquad \text{for all } j \text{ and } k$$
$$H_{A,AB}: (\mu_{jk} - \mu_{j.} - \mu_{.k} + \mu) \neq 0 \qquad \text{for some } jk$$

Tests of these three hypotheses can be constructed using the mean squares of the analysis of variance if the dependent variable is normally distributed. This normality requirement is expressible as the familiar requirement that ρ_{ijk} be distributed normally: $N(0, \sigma_\rho^2)$.

It can be shown that the ratio

$$V_A = \frac{MS_A}{MS_r}$$

is distributed as Fisher's F with $(a - 1)$ and $[ab(N - 1)]$ degrees of freedom in the numerator and denominator, respectively, if $H_{T,A}$ is true and if ρ_{ijk}: $N(0, \sigma_\rho^2)$. Large values of V_A are attributed to the $\sum_{j=1}^{a} \tau_j^2$ term in the $E(MS_A)$ and lead to rejection of $H_{T,A}$ in favor of $H_{A,A}$.

Also, by the same reasoning,

$$V_B = \frac{MS_B}{MS_r}$$

is distributed as Fisher's F with $(b - 1)$ and $[ab(N - 1)]$ degrees of freedom in the numerator and denominator, respectively, if $H_{T,B}$ is true and if ρ_{ijk}: $N(0, \sigma_\rho^2)$. Large values of V_B are attributed to the $\sum_{k=1}^{b} \xi_k^2$ term in the $E(MS_B)$ and lead to rejection of $H_{T,B}$ for $H_{A,B}$.

By similar arguments,

$$V_{AB} = \frac{MS_{AB}}{MS_r}$$

is distributed as Fisher's F with $[(a - 1)(b - 1)]$ and $[ab(N - 1)]$ degrees of freedom in the numerator and denominator, respectively, if $H_{T,AB}$ is true and if ρ_{ijk}: $N(0, \sigma_\rho^2)$. Large values of V_{AB} are attributed to the $\sum_{j=1}^{a} \sum_{k=1}^{b} (\xi\tau)_{jk}^2$ term in $E(MS_{AB})$ and lead to rejection of $H_{T,AB}$ in favor of $H_{A,AB}$.

The analysis of variance is summarized in Table 19.2.

TABLE 19.2 Analysis of Variance for a Two-Factor Randomized Groups Design

Source	SS	df	MS	EMS	F
Treatments					
A	SS_A	$a - 1$	MS_A	$\sigma_\rho^2 + \dfrac{Nb \sum_{j=1}^{a} \tau_j^2}{a - 1}$	$\dfrac{MS_A}{MS_r}$
B	SS_B	$b - 1$	MS_B	$\sigma_\rho^2 + \dfrac{Na \sum_{k=1}^{b} \xi_k^2}{b - 1}$	$\dfrac{MS_B}{MS_r}$
AB	SS_{AB}	$(a - 1)(b - 1)$	MS_{AB}	$\sigma_\rho^2 + \dfrac{N \sum_{j=1}^{a} \sum_{k=1}^{b} (\xi\tau)_{jk}^2}{(a - 1)(b - 1)}$	$\dfrac{MS_{AB}}{MS_r}$
Residual	SS_r	$ab(N - 1)$	MS_r	σ_ρ^2	
Total	SS_T	$ab(N - 1)$			

The significance probabilities of the F ratios in the analysis are determined one at a time with the respective degrees of freedom. Since several tests are involved, we need to be concerned with the probability that *one or more* of them will lead to a type I error. If all three test hypotheses are true, the three F ratios are sample values from the "same" distribution. With three tests we are more likely to observe an extreme value, leading to a type I error, than with but a single test. Let the level of significance for the three tests be α_A, α_B, and α_{AB}. The probability of rejecting one or more hypotheses when all hypotheses are true is

$$\alpha^* = 1 - (1 - \alpha_A)(1 - \alpha_B)(1 - \alpha_{AB})$$

For example, where $\alpha_A = \alpha_B = \alpha_{AB} = .05$, $\alpha^* = .143$. This leads to a change of meaning of the rejection rule and should be taken into account when interpreting results of the tests.

A commonly advised procedure is the "pooling" of sums of squares and degrees of freedom when making tests of main effects. The logic, which is erroneous, is that if $H_{T,AB}$ is not rejected, then no interaction exists and SS_{AB} is attributable only to residual variation and may be pooled with SS_r to obtain a new estimate of σ_ρ^2. The impropriety of this is easily seen when we recall that the failure to reject $H_{T,AB}$ is based on sample data and does not insure that $H_{T,AB}$ is *true*. In order for a pooling to be justified, we must be able to justify the equality

$$E(MS_{AB}) = \sigma_\rho^2$$

This cannot be done on the basis of sample data.

Also, the pooled degrees of freedom, when sums of squares are pooled, leads to a distortion of the meaningfulness of the tests when the tests are defined under an (α, β, δ) procedure.

One final comment is called for in a discussion of the hypotheses and their tests. If the interaction is found to be significant, that is, the sample data lead to a rejection of $H_{T,AB}$, the interpretation of main effects is made more difficult. The actual focus of significant effects in a given factor depends on the levels of the other factor. Thus, simple statements regarding the influence of the treatments cannot be made. A specific example of this is given in the following paragraphs.

Sample Size and (α, β, δ). The determination of sample size for a two-factor experiment requires no new concepts or procedures. The alternative hypothesis for the main effects and interaction must be specified. If, as before, we specify them in terms of $\delta\sigma_\rho$ units, evaluation of the noncentrality parameter is straightforward. The Pearson-Hartley charts can be used to determine the sample sizes necessary to achieve the desired (α, β, δ) status for each of the three tests.

The noncentrality parameters and sample sizes for the three tests are as follows.
For factor A:

$$\phi_A = \frac{\sqrt{(bN/a) \sum_{j=1}^{a} \tau_j^2}}{\sigma_\rho} \tag{19.21}$$

$$\phi^2 = N \frac{b}{a} \frac{1}{\sigma_\rho^2} \sum_{j=1}^{a} \tau_j^2$$

$$F = \frac{1}{\sigma_\rho^2} \sum_{j=1}^{a} \tau_j^2$$

$$N = \phi^2 \frac{a}{(b/\sigma_\rho^2) \sum_{j=1}^{a} \tau_j^2}$$

$$= \phi^2 \frac{a}{bF}$$

with $df_{\text{denominator}} = (a-1)$, $df_{\text{numerator}} = ab(N-1)$.
For factor B:

$$\phi_B = \frac{\sqrt{(aN/b) \sum_{k=1}^{b} \xi_k^2}}{\sigma_\rho} \tag{19.22}$$

with $df_{\text{denominator}} = b-1$, $df_{\text{numerator}} = ab(N-1)$.
For interaction AB:

$$\phi_{AB} = \frac{\frac{N}{(a-1)(b-1)+1} \sum_{j=1}^{a} \sum_{k=1}^{b} (\xi\tau)_{jk}^2}{\sigma_\rho} \tag{19.23}$$

with $df_{\text{denominator}} = (a-1)(b-1)$, $df_{\text{numerator}} = ab(N-1)$.
For the two types of alternatives developed in the preceding chapter, the 2-effect alternative and the M-effect alternative, we have the following special expressions for ϕ, and N.
For factor A:

2-Effect Alternative	a-Effect Alternative	
$\phi_A = \delta\sqrt{\dfrac{2bN}{a}}$	$\phi_A = \delta\sqrt{\dfrac{bN(a^2-1)}{12}}$	(19.24)
$\phi^2 = \delta^2 \dfrac{2b}{a} N$	$\phi^2 = \delta^2 \dfrac{b(a^2-1)}{12} N$	
$N = \phi^2 \dfrac{a}{2b\delta^2}$	$N = \phi^2 \dfrac{12}{\delta^2 b(a^2-1)}$	

For factor B:

2-Effect Alternative   b-Effect Alternative

$$\phi_B = \delta\sqrt{\frac{2aN}{b}} \qquad \phi_B = \delta\sqrt{\frac{aN(b^2 - 1)}{12}} \qquad (19.25)$$

$$\phi^2 = \delta^2\frac{2a}{b}N \qquad \phi^2 = \delta^2\frac{a(b^2 - 1)}{12}N$$

$$N = \phi^2\frac{b}{2a\delta^2} \qquad N = \phi^2\frac{12}{\delta^2 a(b^2 - 1)}$$

For the interaction we must take into account all ab parameters $(\xi\tau)_{11}, \ldots,$ $(\xi\tau)_{ab}$. Hence, we have the following equations.

2-Effect Alternative   ab-Effect Alternative

$$\phi_{AB} = \delta\sqrt{\frac{2N}{ab}} \qquad \phi_{AB} = \delta\sqrt{\frac{N[(ab)^2 - 1]}{12}} \qquad (19.26)$$

$$\phi^2 = \delta^2\frac{2}{ab}N \qquad \phi^2 = \delta^2\frac{[(ab)^2 - 1]}{12}N$$

$$N = \phi^2\frac{ab}{2\delta^2} \qquad N = \phi^2\frac{12}{\delta^2[(ab)^2 - 1]}$$

Evaluating these equations for N leads to a disturbing finding—no single sample size will permit a consistent (α, β, δ) for the three tests, except where $a = b$, for main effects tests. For a given (α, β, δ), different sample sizes are required for each of the three tests. Take a specific example. Let $a = 3$, $b = 5$. Let $\alpha = \beta = .01$ and $\delta = 1.0$ so that the $(\alpha, \beta, \delta) = (.01, .01, 1.0)$ for all tests. We enter the Pearson-Hartley tables with $df_{\text{numerator}} = 2$ to find ϕ_A. Entering from the left at the $(\pi = .99)$ line, a value for ϕ is read from the bottom of the $(\alpha = .01)$ portion of the chart below the point of the intersection of the $df_{\text{denominator}} = \infty$ curve and the $(\pi = .99)$ line. The value of ϕ is 3.05 approximately. Hence, for the 2-effect alternative,

$$\phi_A = \sqrt{\frac{10N}{3}}$$

$$(3.05)^2 = N\frac{10}{3}$$

$$N \cong 3$$

Entering the table with $df_{\text{numerator}} = 4$ leads to $\phi = 2.51$. Hence, for the 2-effect alternative,

$$\phi_B = \sqrt{\frac{6N}{5}}$$

$$(2.51)^2 = N\frac{6}{5}$$

$$N \cong 6$$

Entering the table with $df_{\text{numerator}} = 8$ leads to $\phi = 2.04$. Therefore, for the 2-effect alternative,

$$\phi_{AB} = \sqrt{\frac{2N}{15}}$$

$$(2.04)^2 = N\frac{2}{15}$$

$$N \cong 32$$

Clearly, an experiment with all three sample sizes is impossible. The sample sizes are so different that not even a compromise is possible. The source of this problem for the test of interaction is the proliferation of "levels" in the interaction term as a function of the number of levels in the factors. The source for the different sample sizes for the tests of A and B is the different numbers of levels in the factors themselves.

The problem takes on a different complexion under the second type of alternative. Applying the all-level alternative with the same (α, β, δ) as above, we find

for factor A $N = 2.79$

for factor B $N = 1.05$

for interaction $N = .23$

Clearly, this type of alternative also has its disadvantages.

In the 2-level alternatives the parameters, the τ's, the ξ's, and the $(\xi\tau)$'s were bunched up, providing little variability to detect. The all-level alternative spreads the parameters far afield, providing a superabundance of detectable variability.

The lesson of these calculations must not be mistaken. We are not trying to determine a type of alternative for which a single sample size is appropriate. The type of alternative natural to a given experiment is the alternative with which the experimenter must be concerned. However, the example calculations dramatically point out the problems of achieving consistency of the (α, β, δ) character of a multifactor experiment in general. A design might have to have excessive power for some tests to achieve sufficient power for other tests or vice versa. The balance of power depends on the number of factors, the number of levels on the factors, the type of alternatives for the tests and the values in (α, β, δ). The values in the (α, β, δ) might be different for the various tests, reflecting the various degrees of concern for the information sought in the tests. If this condition is present, a consistency of sample size might be possible.

At this point it becomes clear what is meant by the word "design." In order to provide tests of hypotheses about several (two or more) sources of variation in an empirical phenomenon, a many-sample experiment can be designed. If the design is a good design, the experiment will include an appropriate number of factors, an appropriate number of levels on each factor, and, of course, an appropriate number of observations. Great care must be taken to select the levels on the factors to avoid needless proliferation of levels. What is desired is the necessary but sufficient information regarding the scientific questions.

Designing an experiment often requires a good deal of ingenuity and hard work. A well-designed experiment will provide appropriate (α, β, δ) tests of all useful or relevant hypotheses at reasonable cost. An experiment that does not meet these conditions is a poorly defined experiment; it probably involves a waste of time and money, and possibly will be misleading. One is tempted to say that a poorly designed experiment is best not executed—the effort and expense is saved, and misleading or false conclusions are avoided.

Some Comments on Sample Size.   In the foregoing we presumed a fixed number of observations in each treatment combination—the equal sample size condition. This turns out to be an odious restriction in many experimental situations: Good observations in some cells may be hard to obtain; data are "lost" and cannot be replaced; subjects are "lost" (for example, animals dying during the course of an experiment) and are not replaceable.

A detailed discussion of the problems encountered in the case of unequal sample sizes is beyond the scope of this book. In general, however, the problems are of two sorts: calculational and statistical-inferential. The calculational problems are simply a nuisance, although for more complicated designs they become more severe. The statistical problems are more complex. The effect on power of tests of the inequality of sample sizes is unknown. Also, the effect of unequal variances (σ_ρ^2) in the cells of the design on the distribution of the variance ratios is accentuated in the unequal sample size case. The analysis is relatively insensitive to empirical violations of the model where sample sizes are equal. However, the analysis is relatively more sensitive to departures from the equal variance condition of the model when sample sizes are not equal: The true significance probabilities of obtained results will not correspond to the values obtained from the tables of the F distribution.

There are mechanical ways of dealing with this problem at the planning stage, at least for some types of experimental situations. The first is to define ways of establishing a list of "surrogate subjects" that will provide the necessary back-up data. The second is to routinely observe more subjects than required for the design and select those to be retained at random after making all the observations. Other such schemes are "thinkable." Objections to these schemes on practical and theoretical grounds can be found readily.

A special problem arises when the sample size is *one*, that is for a single observation in each cell of the design. Recall that $df_r = ab(N - 1)$. If $N = 1$, then $N - 1 = 0$, and $df_r = 0$, implying that we cannot estimate $\sigma_p{}^2$ directly. However, if $H_{T,AB}$ were known to be true, then MS_{AB} could be used as an estimate of $\sigma_p{}^2$. In general, the complete lack of interaction is hard to defend and no adequate tests of $H_{T,A}$ and $H_{T,B}$ are well defined. Certain nonfactorial designs are capable of dealing with the single-observation case, but they are beyond the scope of this book.

If the number of observations in each cell is not a constant, three considerations are generated. In the first instance the calculational formulas given in the sections above are not appropriate. The influence of the unequal sample sizes can have an unwarranted effect on the conclusions drawn from the experiment. Where the inequality of sample sizes is due to chance effects and is not systematically a function of the treatments, the method of unweighted means is the method of choice. A more complicated but possibly more powerful means of estimation of the mean squares is the general least squares solution. These two techniques are described in Winer (1962)—cited in the preceding chapter. Where the sample sizes in the cells of the design are proportional to the marginal sample sizes, the calculational equations described above may be modified to weight each of the means entering into the sums of squares according to the sample sizes actually observed. The modified equations are easier to use than the unweighted means equations or the least squares equations. The equations are developed in Myers (1966)—preceding chapter.

In addition to calculational differences, the presence of unequal sample sizes in a factorial design implies the possibility of bias introduced by over-representation of certain aspects of the experiment. Also, if there is a lack of equality in treatment combination variances, the failure to have equal sample sizes heightens the possibility of inappropriate application of the distribution theory to the ratios of mean squares.

Calculational Form. Calculations in the two-factor random groups design are an extension of the one-factor random groups design calculations. The extension involves keeping track of the sums of observations for each level on each factor and for each level combination. Table 19.3 gives example data and the sums and sums of squares of observations for a 3×4 experiment with $N = 3$. The sums of scores and square scores for each level on both factors are found. The means for all factor levels are also calculated at this stage. Note that the sums of scores and square scores can be checked by summing across the level totals on factor A and across the level totals on factor B. These double summations give the overall sum of scores and squared scores; both summations should have the same value.

The computing formulas developed in the single-factor design are adapted easily to the two-factor design. These formulas are sums and differences of five basic expressions (the bracketed numbers are to be treated as algebraic quantities):

$$[1] = \frac{(\sum_{j=1}^{a} \sum_{k=1}^{b} \sum_{i=1}^{N} X_{ijk})^2}{Nab}$$

$$[2] = \sum_{j=1}^{a} \sum_{k=1}^{b} \sum_{i=1}^{N} X_{ijk}^2$$

$$[3] = \frac{\sum_{j=1}^{a} \sum_{k=1}^{b} (\sum_{i=1}^{N} X_{ijk})^2}{N}$$

$$[4] = \frac{\sum_{j=1}^{a} (\sum_{k=1}^{b} \sum_{i=1}^{N} X_{ijk})^2}{Nb}$$

$$[5] = \frac{\sum_{k=1}^{b} (\sum_{j=1}^{a} \sum_{i=1}^{N} X_{ijk})^2}{Na}$$

It can be shown by simple algebra from the partition that

$$SS_A = [4] - [1]$$
$$SS_B = [5] - [1]$$
$$SS_{AB} = [3] - [5] - [4] + [1]$$
$$SS_r = [2] - [3]$$
$$SS_T = [2] - [1]$$

The sums and sums of squared sums, and so forth, can best be obtained by making up a table of cell totals, as illustrated in Table 19.4 for the example data of Table 19.3. Each entry in the main body of this summary table is the sum of scores in the respective cell of the data layout. The row and column sums of the summary table

TABLE 19.3 An Example of Data and Basic Sums in a Two-Factor Design

		Factor B				$\sum_{k=1}^{b}\sum_{i=1}^{N} X_{ijk}$	$\sum_{k=1}^{b}\sum_{i=1}^{N} X_{ijk}^{2}$	$\bar{X}_{j.}$
		$k=1$	$k=2$	$k=3$	$k=4$			
Factor A	$j=1$	54 55 50	48 50 52	50 48 51	53 54 55	620	32104	51.7
	$j=2$	50 47 53	47 53 51	48 53 54	46 54 48	604	30502	50.3
	$j=3$	46 46 45	48 50 52	50 49 53	46 48 44	577	27831	48.1
$\sum_{j=1}^{a}\sum_{i=1}^{N} X_{ijk}$		446	451	456	448	$\sum_{j=1}^{a}\sum_{k=1}^{b}\sum_{i=1}^{N} X_{ijk} = 1801$		
$\sum_{j=1}^{a}\sum_{i=1}^{N} X_{ijk}^{2}$		22216	22635	23144	22442	$\sum_{j=1}^{a}\sum_{k=1}^{b}\sum_{i=1}^{N} X_{ijk}^{2} = 90437$		
$\bar{X}_{.k}$		49.56	50.11	50.67	49.78	$\bar{X} = 50.03$		

must check against the factor level sums of the basic data table. The basic sums are determined directly from Table 19.4.

The analysis of variance table is constructed like Table 19.2. For the example data the analysis of variance table is Table 19.5. The significance probabilities are determined by the usual table look-up. For example, by inspecting the tabled values of F with $df_{\text{numerator}} = 2$ and $df_{\text{denominator}} = 24$, we find that the ($\alpha = .005$) value of F is larger than the observed value of 6.50 and that the ($\alpha = .01$) value of F is smaller than 6.50. Therefore the significance probability of the test of $H_{T.A}$ is between .01 and .005.

A posttest evaluation of the (α, β, δ) character of the analysis is complicated by the fact that we must deal with three tests. The details of the calculations, for each of the tests, are identical with procedures developed earlier. The results of the calculations are shown in Table 19.6. For the evaluation of the test for interaction a special alternative is devised to illustrate the development of such special cases.

Imagine that the factors represent qualitative continua with factor A levels 1 and 3 and factor B levels 1 and 4 being extremes. Further, assume that, according to a theory under investigation, the effect on factor B of level 1 of factor A is to make the expectations higher in the extremes of B; the effect on factor B of level 3 of factor

TABLE 19.4 Summary Table of Cell Totals and Other Quantities for the Analysis of Variance

		Factor B				$\sum\limits_{k=1}^{4}\sum\limits_{i=1}^{N} X_{ijk}$	$\sum\limits_{k=1}^{4}\left(\sum\limits_{i=1}^{4} X_{ijk}\right)^2$
		$k=1$	$k=2$	$k=3$	$k=4$		
Factor A	$j=1$	159	150	149	162	620	96226
	$j=2$	150	151	155	148	604	91230
	$j=3$	137	150	152	138	577	83417

$$\sum\limits_{j=1}^{3}\sum\limits_{i=1}^{N} X_{ijk} \qquad 446 \qquad 451 \qquad 456 \qquad 448$$

$$\sum\limits_{j=1}^{3}\left(\sum\limits_{i=1}^{N} X_{ik}\right)^2 \qquad 66550 \qquad 67801 \qquad 69330 \qquad 67192$$

$$\sum_{j=1}^{a}\left(\sum_{k=1}^{b}\sum_{i=1}^{N} X_{ijk}\right)$$
$$=\sum_{k=1}^{b}\left(\sum_{j=1}^{a}\sum_{i=1}^{N} X_{ijk}\right)=1801$$

$$\sum_{k=1}^{b}\sum_{j=1}^{a}\left(\sum_{i=1}^{N} X_{ijk}\right)^2$$
$$=\sum_{j=1}^{a}\sum_{k=1}^{b}\left(\sum_{i=1}^{N} X_{ijk}\right)^2=270873$$

$$[1]=\frac{\left(\sum\limits_{j=1}^{a}\sum\limits_{k=1}^{b}\sum\limits_{i=1}^{N} X_{ijk}\right)^2}{Nab}=\frac{3243601}{36}=90100 \qquad\qquad SS_A=[4]-[1]=79$$

$$[2]=\sum\limits_{j=1}^{a}\sum\limits_{k=1}^{b}\sum\limits_{i=1}^{N} X_{ijk}^2=90437 \qquad\qquad SS_B=[5]-[1]=6$$

$$[3]=\frac{\sum\limits_{j=1}^{a}\sum\limits_{k=1}^{b}\left(\sum\limits_{i=1}^{N} X_{ijk}\right)^2}{N}=\frac{270873}{3}=90291 \qquad SS_{AB}=[3]-[5]-[4]+[1]=106$$

$$[4]=\frac{\sum\limits_{j=1}^{a}\left(\sum\limits_{k=1}^{b}\sum\limits_{i=1}^{N} X_{ijk}\right)^2}{Nb}=\frac{1082145}{12}=90179 \qquad\qquad SS_r=[2]-[3]=146$$

$$[5]=\frac{\sum\limits_{k=1}^{b}\left(\sum\limits_{j=1}^{a}\sum\limits_{i=1}^{N} X_{ijk}\right)^2}{Na}=\frac{810957}{9}=90106 \qquad\qquad SS_T=[2]-[1]=337$$

TABLE 19.5 Analysis of Variance for the Data of Table 19.3

Source	SS	df	MS	F
A	79	2	39.5	6.50
B	6	3	2.0	.33
AB	106	6	17.7	2.91
Residual	146	24	6.08	
Total	337	35		

Significance Probabilities:

Factor A $.005 < \Psi < .01$

Factor B $.1 < \Psi$

Interaction $.025 < \Psi < .05$

TABLE 19.6 Posttest Evaluation of Test Regarding Factor A

2-Effect Alternative

$$\phi = \delta\sqrt{bN2/a} = \delta(2.828)$$

	$\alpha = .05$	$\alpha = .01$
Power for		
$\delta = 1.0$	$\pi = .98$	$\pi = .93$
$\beta = .01$	$\delta = 1.004$	$\delta = 1.170$
$\beta = .05$	$\delta = .856$	$\delta = 1.011$

3-Effect Alternative

$$\phi = \delta\sqrt{bN(a^2 - 1)/12} = \delta(2.828)$$

	$\alpha = .05$	$\alpha = .01$
Power for		
$\delta = 1.0$	$\pi = .98$	$\pi = .93$
$\beta = .01$	$\delta = 1.004$	$\delta = 1.170$
$\beta = .05$	$\delta = .856$	$\delta = 1.011$

TABLE 19.6 (continued) Posttest Evaluation of Test Regarding Factor B

2-Effect Alternative
$$\phi = \delta\sqrt{aN2/b} = \delta(2.122)$$

	$\alpha = .05$	$\alpha = .01$
Power for		
$\delta = 1.0$	$\pi = .92$	$\pi = .76$
$\beta = .01$	$\delta = 1.239$	$\delta = 1.460$
$\beta = .05$	$\delta = 1.051$	$\delta = 1.254$

4-Effect Alternative
$$\phi = \delta\sqrt{aN(b^2 - 1)/12} = \delta(3.354)$$

	$\alpha = .05$	$\alpha = .01$
Power for		
$\delta = 1.0$	$\pi > .99$	$\pi > .99$
$\beta = .01$	$\delta = .787$	$\delta = .924$
$\beta = .05$	$\delta = .671$	$\delta = .799$

TABLE 19.6 (continued) Posttest Evaluation of Test Regarding Interaction

Specific Alternative Described in Text
$$\phi = \delta\sqrt{12/7} = \delta(1.309)$$

	$\alpha = .05$	$\alpha = .01$
Power for		
$\delta = 1.0$	$\pi = .60$	$\pi = .32$
$\beta = .01$	$\delta = 1.742$	$\delta = 2.017$
$\beta = .05$	$\delta = 1.505$	$\delta = 1.780$

A is to make the expectations lower in the extremes of B. Imagine the effect may be stated in terms of $\delta\sigma_\rho$ units at each of the four affected parts. The values of $(\xi\tau)$ are zero for all jk except for four-level combinations that take on relative values given by

$$(\xi\tau)_{11} = (\xi\tau)_{14} = -(\xi\tau)_{31} = -(\xi\tau)_{34} = \delta\sigma_\rho$$

The sum of these values, when squared, is

$$\sum_{j=1}^{a}\sum_{k=1}^{b}(\xi\tau)_{jk}^{2} = 4\delta^2\sigma_\rho^{2}$$

and hence

$$\phi^2 = \frac{N}{(a-1)(b-1)+1} \, 4\delta^2$$

$$= \frac{3}{(3-1)(4-1)+1} \, 4\delta^2$$

$$= \frac{12}{7} \, \delta^2$$

From this the value of δ for a given value of π or the value of π for a given δ can be calculated for the α and β values tabled.

This specific alternative may be meaningful in any number of applications. However, in this instance, the procedure used to establish this as a meaningful alternative interaction pattern was to look at the graph of the treatment means. The means corresponding to the table entries of Table 19.3 are plotted in the fashion of Figure 19.1. The example data are plotted as Figure 19.2. This figure makes it seem that our alternative is a reasonable one—if we had to select an alternative to the rejected test hypothesis ($\alpha = .05$), the alternative suggested by the data seems an appropriate choice in the lack of a more compelling basis.

The results of the analysis are reasonably clear. The main effect of factor A is a significant effect with a very small significance probability. The experiment was designed without concern for achieving a given power against a specific alternative. Thus, we are thrown back on the necessity of developing a posttest evaluation of the meaning of the "significance." The power of the test is good, $\pi = .93$, against the two-level type alternative with $\delta = 1$ at the ($\alpha = .01$) level of significance. Since there are three levels, the two- and three-level alternatives are the same. Imagine that we are able to state, on the basis of our knowledge of the variability of the phenomenon under study, that a spacing of one standard deviation between the adjacent three effects is a meaningful and important difference. This statement implies that the experiment has detected (with error probability .01) just such a difference. The power against this alternative is only .93 for the ($\alpha = .01$) test and yet the test rejected the test hypothesis. A type II error cannot be made when the test hypothesis is rejected, and yet the probability of the type II error for this two-level ($\delta = 1$) alternative is .07. The implication is that the effective alternative, and quite likely the state of nature, is a set of effects more widely separated than the one standard deviation spacing. If we had failed to reject the test hypothesis, the conclusion would have been quite different. The power against the ($\delta = 1$) two-alternative hypothesis is insufficient, and failure to reject the test hypothesis would not permit a clear interpretation that the state of nature corresponding to the test hypothesis was more likely than the state of nature of the alternative.

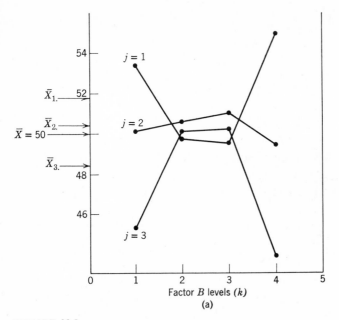

FIGURE 19.2

Graph of means from data of Table 19.3: levels on factor A plotted as a function of levels on factor B.

Turning to the main effect of factor B, it is apparent that we have insufficient power to detect a ($\delta = 1$) 2-effect alternative (three effects zero, one effect -1 standard deviation, and the remaining effect $+1$ standard deviation from zero). However, there is more than sufficient power to detect an alternative with the effects having values of -1.5, $-.5$, $.5$, and 1.5 standard deviations. If the four-level alternative is the meaningful alternative, then the conclusion from the failure to reject is that the alternative is not the state of nature and that the test hypothesis describes the state of nature, at least so far as it is practical to pin down the precise state of nature. Inspection of the observed means lends support to the notion that the test hypothesis is very much more nearly the picture reflected in the data than the alternative hypothesis.

The significance of the interaction provides a good deal of meat for discussion. First, everything we have said about the main effects must be highly qualified. In order most easily to see this, imagine that the same data had been observed but only for levels 2 and 3 of factor B. The design would have been a 3 by 2 design in this case with ($k = 1$) and ($k = 4$) not involved. The result of the analysis of variance would have been quite different. Neither factor A nor factor B would have been significant. This clearly isolates the source of the significance of the A factor within

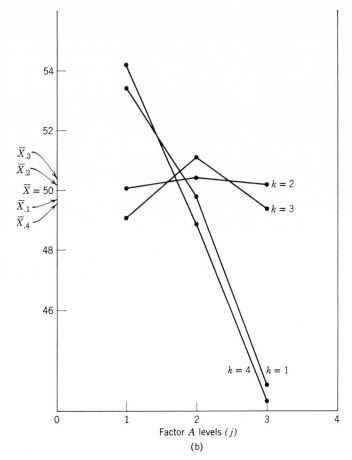

Figure 19.2 (Continued.)

the differences among the four cells in the full design representing factor B levels $(k = 1)$ and $(k = 4)$. The significance of the effects of factor A does not extend to all of the levels of factor B. Therefore the statement that factor A significantly influences the dependent variable is true only in a weak sense. The effect of the levels of factor A depends in a very strong way on the presence of certain levels of the B factor. There is a decrease in the average response from $(j = 1)$ to $(j = 3)$ for the $(k = 1)$ and $(k = 4)$ factor B levels and a level average response for the $(k = 2)$ and $(k = 3)$ factor B levels.

The level of significance of the interaction is somewhat less extreme than that of factor A. This tends to lead us to be less certain of the reality of the effect of the interaction. The graphs of the means certainly show a very clear pattern. However, this pattern is for the means—the individual response patterns, subject by subject, are not as neat. That is, the variation within cells is large enough to soften the interpretation of the differential patterns of mean response to the design as a whole. On the other hand, the power of the test against the one standard deviation separation of the four extreme points in the graph is quite high, indicating that a rejection even at a low level of statistical significance probably indicates a reasonably large real difference in the phenomenon. The effect dispersion for the special alternative with $\alpha = .05$ and $\beta = .05$ is $\delta = 1.505$, a rather large difference—and the analysis leads to a rejection. It would appear that, although the risk of a type I error is rather large, the difference that is detectable is of such a magnitude as to be of practical importance.

Empirical Violations of the Model. The model is essentially the same as the model of the single-factor random groups design developed in the preceding chapter. The assumptions in deriving the expected mean squares are the same with all of the components of the model. Also, the assumptions leading to the tests of all of the hypotheses are the same as before. Consequently, the catalog of possible violations is essentially the same as in the one-factor analysis of variance.

Higher Order Factorial Designs

An extension of the analysis of variance from two independent variables or factors to many independent variables or factors is conceptually easy but tedious in detail. A three-factor design was outlined early in the preceding chapter.

This section outlines briefly the character of the model and the calculational form of a three-factor design. Assume that each subject is randomly assigned to one of the cells of the design and that all cells have the same number of subjects. Only one dependent variable is involved.

Name the three factors A, B, and C and let the number of levels on the respective factors be a, b, and c. For the N observations in each of the combinations of the three factors there is a symbolic designation with four subscripts:

$$X_{jkmi} \text{ where } j = 1, 2, \ldots, a; k = 1, 2, \ldots, b$$
$$m = 1, 2, \ldots, c; \ i = 1, 2, \ldots, N$$

The data layout can be seen as an extension of the data layout of the two-factor problem shown in Table 19.1, each level on factor C duplicating the cells of Table 18.1. A three-factor example is shown in Table 19.7. The summary data necessary

TABLE 19.7 Data Layout for a Three-Factor Factorial Experiment

		Factor C						
		Level 1			...	Level c		
Factor B:		Level 1	...	Level b	...	Level 1	...	Level b
Factor A	Level 1	X_{1111} X_{1112} . . . X_{111N}	...	X_{1b11} X_{1b12} . . . X_{1b1N}		X_{11c1} X_{11c2} . . . X_{11cN}	...	X_{1bc1} X_{1bc2} . . . X_{1bcN}
	. . .							
	Level a	X_{a111} X_{a112} . . . X_{a11N}	...	X_{ab11} X_{ab12} . . . X_{ab1N}		X_{a1c1} X_{a1c2} . . . X_{a1cN}	...	X_{abc1} X_{abc2} . . . X_{abcN}

are (1) the sums of the observations in each of the $a \times b \times c$ cells of the design; (2) the sums of the observations made under each of the a levels of A, the b levels of B, and the c levels of C; (3) the total sum of the observations; (4) the sums of observations in each of the $a \times b$ combinations of levels on A and B, across all of the levels on C; (5) the sums of observations in each of the $a \times c$ combinations of levels on A and C, across all of the levels on B; (6) the sums of observations in each of the $b \times c$ combinations of levels on B and C across all of the levels on A; and (7) the sum of squares of all of the observations made in the entire experiment. Tables 19.8 through 19.11 are the general forms of the summary tables for these data. Some of the sums are repeated in more than one table. The repeated sums provide checks in calculation. A mean for each of the treatment combinations and factor levels can be calculated by dividing the appropriate sum by the corresponding number of individual scores in the sum.

A Model. A model similar to the model for a two-way analysis of variance is readily constructed. An expectation for each of the cells in the design is

TABLE 19.8 First Table of Summary Data: A by B by C (Cells) Sums, Sums of B by C Categories, Sums of Factor A Levels, and Total Sum

Factor B:	Factor C					Factor A Totals
	Level 1		$\cdots$	Level c		
	Level 1	Level b	$\cdots$	Level 1	Level b	
Factor A — Level 1	$T_{111} = \sum\limits_{i=1}^{N} X_{111i}$	$T_{1b1} = \sum\limits_{i=1}^{N} X_{1b1i}$		$T_{11c} = \sum\limits_{i=1}^{N} X_{11ci}$	$T_{1bc} = \sum\limits_{i=1}^{N} X_{1bci}$	$T_{1..} = \sum\limits_{k=1}^{b}\sum\limits_{m=1}^{c} T_{1km}$
$\cdots$						
Level a	$T_{a11} = \sum\limits_{i=1}^{N} X_{a11i}$	$T_{ab1} = \sum\limits_{i=1}^{N} X_{ab1i}$		$T_{a1c} = \sum\limits_{i=1}^{N} X_{a1ci}$	$T_{abc} = \sum\limits_{i=1}^{N} X_{abci}$	$T_{a..} = \sum\limits_{k=1}^{b}\sum\limits_{m=1}^{c} T_{akm}$
	$T_{.11} = \sum\limits_{j=1}^{a} T_{j11}$	$T_{.b1} = \sum\limits_{j=1}^{a} T_{jb1}$		$T_{.1c} = \sum\limits_{j=1}^{a} T_{j1c}$	$T_{.bc} = \sum\limits_{j=1}^{a} T_{jbc}$	$T_{...} = \sum\limits_{j=1}^{a} T_{j..}$ $= \sum\limits_{k=1}^{b}\sum\limits_{m=1}^{c} T_{.km}$

TABLE 19.9 Second Table of Summary Data: A by B Sums, Sums of Levels on Factors A and B

		Factor B			Factor A Totals
		Level 1	...	Level b	
Factor A	Level 1	$T_{11.} = \sum\limits_{m=1}^{c} T_{11m}$		$T_{1b.} = \sum\limits_{m=1}^{c} T_{1bm}$	$T_{1..} = \sum\limits_{k=1}^{b} T_{1k.}$
	...				
	Level a	$T_{a1.} = \sum\limits_{m=1}^{c} T_{a1m}$		$T_{ab.} = \sum\limits_{m=1}^{c} T_{abm}$	$T_{a..} = \sum\limits_{k=1}^{b} T_{ak.}$
	Factor B Totals	$T_{.1.} = \sum\limits_{j=1}^{a} T_{j1.}$		$T_{.b.} = \sum\limits_{j=1}^{a} T_{jb.}$	$T_{...} = \sum\limits_{j=1}^{a} T_{j..}$ $= \sum\limits_{k=1}^{b} T_{.k.}$

TABLE 19.10 Third Table of Summary Data: A by C Sums, Sums of Levels on Factors A and C

		Factor C			Factor A Totals
		Level 1	...	Level c	
Factor A	Level 1	$T_{1.1} = \sum\limits_{k=1}^{b} T_{1k1}$		$T_{1.c} = \sum\limits_{k=1}^{b} T_{1kc}$	$T_{1..} = \sum\limits_{m=1}^{c} T_{1.m}$
	...				
	Level a	$T_{a.1} = \sum\limits_{k=1}^{b} T_{ak1}$		$T_{a.c} = \sum\limits_{k=1}^{b} T_{akc}$	$T_{a..} = \sum\limits_{m=1}^{c} T_{a.m}$
	Factor C Totals	$T_{..1} = \sum\limits_{j=1}^{a} T_{j.1}$		$T_{..c} = \sum\limits_{j=1}^{a} T_{j.c}$	$T_{...} = \sum\limits_{j=1}^{a} T_{j..}$ $= \sum\limits_{m=1}^{c} T_{..m}$

TABLE 19.11 Fourth Table of Summary Data: B by C Sums, Sums of Levels on Factors B and C

		Factor C			Factor B Totals
		Level 1	...	Level c	
Factor A	Level 1	$T_{.11} = \sum_{j=1}^{a} T_{j11}$		$T_{.1c} = \sum_{j=1}^{a} T_{j1c}$	$T_{.1.} = \sum_{m=1}^{c} T_{.1m}$
	...				
	Level b	$T_{.b1} = \sum_{j=1}^{a} T_{jb1}$		$T_{.bc} = \sum_{j=1}^{a} T_{jbc}$	$T_{.b.} = \sum_{m=1}^{c} T_{.bm}$
	Factor C Totals	$T_{..1} = \sum_{k=1}^{b} T_{.k1}$		$T_{..c} = \sum_{k=1}^{b} T_{.kc}$	$T_{...} = \sum_{k=1}^{b} T_{.k.}$ $= \sum_{m=1}^{c} T_{..m}$

defined along with expectations for each of the three two-way layouts cutting across the three-way design. In addition, each level of each factor is characterized by an expectation. In this way the following expectations are defined:

μ_{jkm} = level of response in the cell defined by the jth level of factor A, the kth level of factor B, the mth level of factor C

$\mu_{jk.} = \dfrac{1}{c} \sum_{m=1}^{c} \mu_{jkm}$ = level of response in the treatment combination defined by the jth level on factor A and the kth level on factor B, across the c levels on factor C

$\mu_{j.m} = \dfrac{1}{b} \sum_{k=1}^{b} \mu_{jkm}$ = level of response in the treatment combination defined by the jth level on factor A and the mth level on factor C, across the b levels on factor B

$\mu_{.km} = \dfrac{1}{a} \sum_{j=1}^{a} \mu_{jkm}$ = level of response in the treatment combination defined by the kth level of factor B and the mth level of factor C, across the a levels on factor A

$$\mu_{j..} = \frac{1}{bc}\sum_{k=1}^{b}\sum_{m=1}^{c}\mu_{jkm} = \text{level of response in the } j\text{th level on factor } A, \text{ across the}$$

bc treatment combinations on factors B and C

$$\mu_{.k.} = \frac{1}{ac}\sum_{j=1}^{a}\sum_{m=1}^{c}\mu_{jkm} = \text{level of response in the } k\text{th level on factor } B, \text{ across the}$$

ac treatment combinations on factors A and C

$$\mu_{..m} = \frac{1}{ab}\sum_{j=1}^{a}\sum_{k=1}^{b}\mu_{jkm} = \text{level of response in the } m\text{th level on factor } C, \text{ across the}$$

ab treatment combinations on factors A and B

$$\mu = \frac{1}{abc}\sum_{j=1}^{a}\sum_{k=1}^{b}\sum_{m=1}^{c}\mu_{jkm} = \text{level of response across all } abc \text{ level combinations}$$

of factors A, B, and C.

As before, it is postulated that the response of an individual subject, X_{jkmi}, is determined by the general experimental conditions, the class of subject involved, and the combination of treatments involved. Any deviation from the cell expectation in an individual score is attributed to a residual term, ρ.

It is convenient to work with the differential levels of response instead of the expected level of response values. These are defined in the same general format as they were in the two-factor problem:

$$\tau_j = \mu_{j..} - \mu$$
$$\xi_k = \mu_{.k.} - \mu$$
$$\lambda_m = \mu_{..m} - \mu$$
$$(\tau\xi)_{jk} = \mu_{jk.} - \mu_{j..} - \mu_{.k.} + \mu$$
$$(\tau\lambda)_{jm} = \mu_{j.m} - \mu_{j..} - \mu_{..m} + \mu$$
$$(\xi\lambda)_{km} = \mu_{.km} - \mu_{.k.} - \mu_{..m} + \mu$$
$$(\tau\xi\lambda)_{jkm} = \mu_{jkm} - \mu_{jk.} - \mu_{j.m} - \mu_{.km} + \mu_{j..} + \mu_{.k.} + \mu_{..m} - \mu$$

Because of the properties of expectations, it can be shown that the sums of each of the differential parameters are zero. Consequently, the fixed effect model for the three-way design can be written

$$X_{jkmi} = \mu + \tau_j + \xi_k + \lambda_m + (\tau\xi)_{jk} + (\tau\lambda)_{jm} + (\xi\lambda)_{km} + (\tau\xi\lambda)_{jkm} + \rho_{jkmi}$$

$$(19.27)$$

The statistical assumption is that the residual terms are normally distributed and independent: ρ_{jkmi} are independently $N(0, \sigma_\rho^2)$.

The Partition. The deviation of a score from the grand mean is similar to that in the two-factor problem:

$$
\begin{aligned}
(X_{jkmi} - \bar{X}) = {}& (\bar{X}_{j..} - \bar{X}) + (\bar{X}_{.k.} - \bar{X}) + (\bar{X}_{..m} - \bar{X}) \\
& + (\bar{X}_{jk.} - \bar{X}_{j..} - \bar{X}_{.k.} + \bar{X}) + (\bar{X}_{j.m} - \bar{X}_{j..} - \bar{X}_{..m} + \bar{X}) \\
& + (\bar{X}_{.km} - \bar{X}_{.k.} - \bar{X}_{..m} + \bar{X}) \\
& + (\bar{X}_{jkm} - \bar{X}_{jk.} - \bar{X}_{j.m} - \bar{X}_{.km} + \bar{X}_{j..} + \bar{X}_{.k.} + \bar{X}_{..m} - \bar{X}) \\
& + (X_{jkmi} - \bar{X}_{jkm})
\end{aligned}
\tag{19.28}
$$

Squaring and summing over j, k, m, and i gives the sums of squares associated with the several factors, their interactions, and the residual term:

$$
SS_T = SS_A + SS_B + SS_C + SS_{AB} + SS_{AC} + SS_{BC} + SS_{ABC} + SS_r
\tag{19.29}
$$

The degrees of freedom and the expected mean squares associated with each of these elements of the partition are given in Table 19.12, the analysis of variance table for a three-factor, fixed effects, random group experiment.

Hypothesis Structure. The hypotheses about main effects and all four interaction parameter sets are

$H_{T.A}$:	$\tau_j = 0$	for $j = 1, 2, \ldots, a$
$H_{A.A}$:	$\tau_j \neq 0$	for some $j = 1, 2, \ldots, a$
$H_{T.B}$:	$\xi_k = 0$	for $k = 1, 2, \ldots, b$
$H_{A.B}$:	$\xi_k \neq 0$	for some $k = 1, 2, \ldots, b$
$H_{T.C}$:	$\lambda_m = 0$	for $m = 1, 2, \ldots, c$
$H_{A.C}$:	$\lambda_m \neq 0$	for some $m = 1, 2, \ldots, c$
$H_{T.AB}$:	$(\tau\xi)_{jk} = 0$	for $j = 1, 2, \ldots, a$ and $k = 1, 2, \ldots, b$
$H_{A.AB}$:	$(\tau\xi)_{jk} \neq 0$	for some j, k
$H_{T.AC}$:	$(\tau\lambda)_{jm} = 0$	for $j = 1, 2, \ldots, a$ and $m = 1, 2, \ldots, c$
$H_{A.AC}$:	$(\tau\lambda)_{jm} \neq 0$	for some j, m
$H_{T.BC}$:	$(\xi\lambda)_{km} = 0$	for $k = 1, 2, \ldots, b$ and $m = 1, 2, \ldots, c$
$H_{A.BC}$:	$(\xi\lambda)_{km} \neq 0$	for some k, m
$H_{T.ABC}$:	$(\tau\xi\lambda)_{jkm} = 0$	for $j = 1, 2, \ldots, a, k = 1, 2, \ldots, b$, and $m = 1, 2, \ldots, c$
$H_{A.ABC}$:	$(\tau\xi\lambda)_{jkm} \neq 0$	for some j, k, m

TABLE 19.12 Analysis of Variance Table for a Three-Factor, Fixed Effect, Random Groups Experiment

Source	SS	Degrees of Freedom	MS	EMS	F
A	SS_A	$a - 1$	MS_A	$\sigma^2 + \dfrac{bcN}{a-1}\sum\limits_{j=1}^{a}\tau_j^2$	$\dfrac{MS_A}{MS_r}$
B	SS_B	$b - 1$	MS_B	$\sigma^2 + \dfrac{acN}{b-1}\sum\limits_{k=1}^{b}\xi_k^2$	$\dfrac{MS_B}{MS_r}$
C	SS_C	$c - 1$	MS_C	$\sigma^2 + \dfrac{abN}{c-1}\sum\limits_{m=1}^{c}\lambda_m^2$	$\dfrac{MS_C}{MS_r}$
AB	SS_{AB}	$(a-1)(b-1)$	MS_{AB}	$\sigma^2 + \dfrac{cN}{(a-1)(b-1)}\sum\limits_{j=1}^{a}\sum\limits_{k=1}^{b}(\tau\xi)_{jk}^2$	$\dfrac{MS_{AB}}{MS_r}$
AC	SS_{AC}	$(a-1)(c-1)$	MS_{AC}	$\sigma^2 + \dfrac{bN}{(a-1)(c-1)}\sum\limits_{j=1}^{a}\sum\limits_{m=1}^{c}(\tau\lambda)_{jm}^2$	$\dfrac{MS_{AC}}{MS_r}$
BC	SS_{BC}	$(b-1)(c-1)$	MS_{BC}	$\sigma^2 + \dfrac{aN}{(b-1)(c-1)}\sum\limits_{k=1}^{b}\sum\limits_{m=1}^{c}(\xi\lambda)_{km}^2$	$\dfrac{MS_{BC}}{MS_r}$
ABC	SS_{ABC}	$(a-1)(b-1)(c-1)$	MS_{ABC}	$\sigma^2 + \dfrac{N}{(a-1)(b-1)(c-1)}\sum\limits_{j=1}^{a}\sum\limits_{k=1}^{b}\sum\limits_{m=1}^{c}(\tau\xi\lambda)_{jkm}^2$	$\dfrac{MS_{ABC}}{MS_r}$
Residual	SS_r	$abc(N-1)$	MS_r	σ^2	—
Total	SS_T	$abcN - 1$	—	—	—

Test ratios for these hypotheses are indicated in the analysis of variance table, Table 19.12. The considerations made regarding degrees of freedom and significance levels, and so on, in dealing with the two-factor problem are all to be made in a three-factor problem. One of the particularly bothersome difficulties in higher order factorial designs is the probability that one or more of the test hypotheses will be rejected when all of the test hypotheses are true. Since there are seven test hypotheses, there are seven levels of significance, α_A, α_B, α_C, α_{AB}, α_{AC}, α_{BC}, α_{ABC}. The probability of falsely rejecting one or more of the hypotheses is given by

$$\alpha^* = 1 - (1 - \alpha_A)(1 - \alpha_B)(1 - \alpha_C)(1 - \alpha_{AB})(1 - \alpha_{AC})(1 - \alpha_{BC})(1 - \alpha_{ABC})$$

For a significance level of .05 for each hypothesis, $\alpha^* = .302$. The student should calculate the value of α^* when the significance level is .01 for all tests. As the number of factors is increased, this problem becomes more acute. For four factors there are 15 separate hypotheses and tests. Consequently, for a common significance level of .05 the probability of rejecting one or more of the hypotheses when all of them are true is $\alpha^* = .54$.

Sample Size. Determination of sample size by the (α, β, δ) procedure and the evaluation of the tests by a posttest (α, β, δ) statement proceeds exactly as in the two-factor problem. There are seven separate tests and consequently seven potentially different sample sizes or seven posttest (α, β, δ) statements to be made. The numerator term in the equation for the noncentrality parameter is the same as the second term in the equation for the expected mean square, with the simple difference that a one is added to the denominator of the expected mean square. Thus, for the AB interaction term the equation is

$$\phi = \frac{\dfrac{cN}{(a - 1)(b - 1) + 1} \sum_{j=1}^{a} \sum_{k=1}^{b} (\tau\xi)_{jk}^2}{\sigma_\rho}$$

The other equations are similarly formed. Evaluating these equations for the two types of extreme alternatives considered in detail above gives specific equations for the parameters.

For factor A:

2-Effect Alternative

$$\phi_A = \delta\sqrt{\frac{bcN2}{a}}$$

a-Effect Alternative

$$\phi_A = \delta\sqrt{\frac{bcN(a^2 - 1)}{12}} \tag{19.30}$$

For factor B:

2-*Effect Alternative*

b-*Effect Alternative*

$$\phi_B = \delta\sqrt{\frac{acN2}{b}} \qquad\qquad \phi_B = \delta\sqrt{\frac{acN(b^2 - 1)}{12}} \qquad (19.31)$$

For factor C:

2-*Effect Alternative*

c-*Effect Alternative*

$$\phi_C = \delta\sqrt{\frac{abN2}{c}} \qquad\qquad \phi_C = \delta\sqrt{\frac{abN(c^2 - 1)}{12}} \qquad (19.32)$$

For the AB interaction:

2-*Effect Alternative*

ab-*Effect Alternative*

$$\phi_{AB} = \delta\sqrt{\frac{cN2}{ab}} \qquad\qquad \phi_{AB} = \delta\sqrt{\frac{cN[(ab)^2 - 1]}{12}} \qquad (19.33)$$

For the AC interaction:

2-*Effect Alternative*

ac-*Effect Alternative*

$$\phi_{AC} = \delta\sqrt{\frac{bN2}{ac}} \qquad\qquad \phi_{AC} = \delta\sqrt{\frac{cN[(ac)^2 - 1]}{12}} \qquad (19.34)$$

For the BC interaction:

2-*Effect Alternative*

bc-*Effect Alternative*

$$\phi_{BC} = \delta\sqrt{\frac{aN2}{bc}} \qquad\qquad \phi_{BC} = \delta\sqrt{\frac{aN[(bc)^2 - 1]}{12}} \qquad (19.35)$$

For the ABC interaction:

2-*Effect Alternative*

abc-*Effect Alternative*

$$\phi_{ABC} = \delta\sqrt{\frac{N2}{abc}} \qquad\qquad \phi_{ABC} = \delta\sqrt{\frac{N[(abc)^2 - 1]}{12}} \qquad (19.36)$$

Calculational Form. The calculation of the sums of squares indicated in the analysis of variance table, Table 19.12, are found from sums of the squares of the totals indicated in Tables 19.8 through 19.11 and the sum of squares of the entire collection of observations. In a manner similar to that in dealing with the two-factor problem, let

$$[1] = \frac{T^2_{...}}{abcN}$$

$$[2] = \sum_{j=1}^{a} \sum_{k=1}^{b} \sum_{m=1}^{c} \sum_{i=1}^{N} X^2_{jkmi}$$

$$[3] = \frac{\sum_{j=1}^{a} T^2_{j..}}{bcN}$$

$$[4] = \frac{\sum_{k=1}^{b} T^2_{.k.}}{acN}$$

$$[5] = \frac{\sum_{m=1}^{c} T^2_{..m}}{abN}$$

$$[6] = \frac{\sum_{j=1}^{a} \sum_{k=1}^{b} T^2_{jk.}}{cN}$$

$$[7] = \frac{\sum_{j=1}^{a} \sum_{m=1}^{c} T^2_{j.m}}{bN}$$

$$[8] = \frac{\sum_{k=1}^{b} \sum_{m=1}^{c} T^2_{.km}}{aN}$$

$$[9] = \frac{\sum_{j=1}^{a} \sum_{k=1}^{b} \sum_{m=1}^{c} T^2_{jkm}}{N}$$

Using the notation on the left of the equal sign in the above list to stand for the algebraic-arithmetic quantities on the right, the sums of squares of the partition can be expressed as follows:

$$SS_A = [3] - [1]$$
$$SS_B = [4] - [1]$$
$$SS_C = [5] - [1]$$
$$SS_{AB} = [6] - [3] - [4] + [1]$$
$$SS_{AC} = [7] - [3] - [5] + [1]$$
$$SS_{BC} = [8] - [4] - [5] + [1]$$
$$SS_{ABC} = [9] - [6] - [7] - [8] + [3] + [4] + [5] - [1]$$
$$SS_r = [2] - [9]$$
$$SS_T = [2] - [1]$$

A brief remark about significant three-factor interactions is in order before going on to another topic. Two-factor interactions were interpreted in the two-factor problem as an indication that the level of response in a given treatment combination (cell) in the design depended not on the separate treatments (levels on the separate factors) but on idiosyncratic characteristics of the combination. If a three-factor interaction is present in a given experiment, the interpretation is much more complicated. The presence of a three-factor interaction is interpretable as an equation in expectations. When the following holds, a three-factor or three-way interaction is said to exist:

$$(\tau \xi \lambda)_{jkm} = \mu_{jkm} - \mu_{jk} - \mu_{j.m} - \mu_{.km} + \mu_{j..} + \mu_{.k.} + \mu_{..m} - \mu \neq 0$$

for some combination of j, k, and m

The specific part of this long combination of expectations that causes the three-factor effect to be nonzero is not indicated in the finding that the interaction is present. The interaction effect is not attributable to differences in A, B, or C, or to A and B, or to A and C, or to B and C. Interpretation of a three-factor interaction, consequently, is very difficult. Also, a significant three-factor interaction makes an interpretation of other significant effects difficult and hazardous.

Repeated Measures in Two-Way and Higher Order Factorials

In the last two chapters we have been concerned with designs in some of which more than one observation is made on each experimental unit (subject). These designs, designated repeated measures designs as contrasted with the randomized designs, have the possibility of offering a great deal of efficiency from a statistical point of view. Also, they involve a smaller total number of physical subjects. This type of design can be extended to multiple-factor experiments. Furthermore, the randomized design and the repeated measures design can be combined into a "mixed methods" design in which the repeated observations refer to some factors but not to other factors.

Imagine that we are interested in the effect of a set of experimental conditions on a process that may be observed over a series of trials or time intervals. On the first time interval the N subjects in each of the experimental conditions are observed—providing a data set that may be modeled by the one-way analysis of variance with randomized groups. On the second time interval the same subjects are all observed a second time. Conceptualizing the two time intervals as levels on a second factor, the data layout now is the layout for a two-way analysis of variance with the special condition that the data on the second factor are repeated measures. Imagine that the treatment conditions are designated as the A factor with a levels and that there are b time intervals defining the second factor B. Subjects are assigned randomly to the levels of factor A, and each subject is observed under each of the levels of factor B. The experiment is a mixture of a repeated measures design and a randomized design. The A factor is referred to as a between-subjects factor and the B factor is referred to as a within-subjects factor.

Examples are easy to think of. The A factor might be indoctrination groups, each group of randomly selected subjects receiving a different course of indoctrination; factor B might be several scenes to be rated for social value. Factor A might be motivational conditions imposed on groups randomly formed by selection from an animal colony; factor B might be a number of choice points in a "T" maze. Factor A might be genetic strains of mice; factor B might be different periods during the life of the animals (age of one month, six months, one year, two years).

A specific example of a two-factor design with one within-subjects factor (repeated measures) and one between-subjects factor will be developed here in some detail. The reader is encouraged to refer to Winer (1962) for a discussion of a large number of other designs using repeated measures.

Imagine b levels on the repeated measures factor (factor B) and a levels on the random subjects factor (factor A). Each of the a groups of subjects observed in the respective levels of the A factor contain N subjects. As a consequence there are aN different subjects and abN separate observations. The data layout is illustrated in Table 19.1 with the proviso that all of the observations represented in a given row in the table are from a single subject.

Several different models can be written to deal with this situation. The models differ fundamentally in terms of whether the additivity assumption is made or not. In the model that we work with here, the assumption is made and no attempt is made to deal with it directly. Other procedures do not make this assumption and take advantage of the presence of nonadditivity in the data. The complete model is more complex and more difficult to work with. Where serious departures from additivity are suspected, the more rigorous tests should be used. The reader is referred to more advanced texts for description of the problem and methods.

The observations are interpreted as arising from multiple components,

$$X_{jki} = \mu + \tau_j + \xi_{i(j)} + \lambda_k + \tau\lambda_{jk} + \xi\lambda_{ki(j)} + \rho_{i(jk)} \qquad (19.37)$$

where

$\mu =$ the overall level of response of the population of subjects to the collection of levels on A and B and the general type of task involved in measuring X; a constant

$\tau_j =$ the differential level of response to the jth level of the A factor, $j = 1, 2, \ldots, a$; a constant

$\xi_{i(j)}$ = the differential level of response of the ith subject in the jth level of the A factor. Since the subscript i is used for a different subject in one group and in another group, the parenthetic (j) is used to specify the group, $i = 1, 2, \ldots, N$ for $j = 1, 2, \ldots, a$; a random variable defined in the population of subjects from which the sample of subjects is chosen

λ_k = the differential level of response to the kth level of the B factor, $k = 1, 2, \ldots, b$; a constant

$\tau\lambda_{jk}$ = the interactive differential level of response of factors A and B, $j = 1, 2, \ldots, a$ and $k = 1, 2, \ldots, b$; a constant

$\xi\lambda_{ki(j)}$ = the interactive differential level of response of individual subjects and factor B, $k = 1, 2, \ldots, b$ and $i = 1, 2, \ldots, N$ for each $j = 1, 2, \ldots, a$; a random variable defined in the population of subjects from which the sample of subjects is chosen

$\rho_{i(jk)}$ = the residual term associated with each observation for each subject under the j, k conditions of factor A and factor B; a random variable

The reader should notice that the model does not include a term to represent any carry-over effect from level to level under the B factor. The presumption is made that there is no carry-over effect. If there does tend to be a carry-over effect in any given order of application in the factor B treatments, the order of the treatments in factor B should be randomized for each subject. If this randomization is not possible because of the sequential nature of factor B levels, the experimenter must turn to another model or redesign his experiment. We do not develop methods of dealing with this problem in this book.

There is no term in the model indicating an interaction between the levels of factor A and the subjects. This comes about because each subject is "nested" within a level of factor A. Consequently, all of the responses of a subject are within a single level and there is no opportunity for a subject's response to vary with factor A levels.

The primary statistical assumption is that the random variable terms are independent and normally distributed, ρ: $N(0, \sigma_\rho^2)$, ξ: $N(0, \sigma_\xi^2)$, and $\xi\lambda$: $N(0, \sigma_{\xi\lambda}^2)$. The additional assumption in the model is that of additivity of observations within the repeated measures groups of observations (within A levels across the B levels in the example being used here).

The entire set of abN observations may be partitioned into sums of squares associated with each of the differential response terms in the general model. The derviation of these sums of squares is beyond the scope of this presentation. However, it is not possible to obtain a sum of squares associated simply with the residual term—the residual term is confounded with the subjects-within-groups variation. This fact does not provide a serious difficulty in formulating tests of hypotheses about the main effects and their interaction. The unavailability of a simple estimate of σ_ρ is a nuisance in posttest evaluation of (α, β, δ) for an experiment. If the experimenter is willing to include individual differences in the residual term in order to evaluate the discriminations made in the data, then there is no difficulty.

The summary table of the analysis of variance is given as Table 19.13. There are two major divisions of the sum of squares, between subjects and within subjects. The total sum of squares is the sum of these two major parts. Within each of the two major divisions the sum of squares is partitioned further. For the between-subjects division only two subparts are defined: SS_A, the sum of squares associated with the A factor, and $SS_{S|A}$, the sum of squares between subjects within the levels of factor A. The SS_A is interpretable in the way that the main effects sum of squares in random groups designs are interpretable. The $SS_{S|A}$ may be interpreted as the within-groups sum of squares taking the overall performance (across the B factor) for each subject within an A level as the measure of subject response. Both SS_A and $SS_{S|A}$ include sums of squares because of individual differences over and above the residual term, that is, because of the sample of observations on the random variable ξ. For the within-subjects division of the sums of squares there are three subdivisions: SS_B, the sum of squares associated with the B factor; SS_{AB}, the sum of squares associated with the interaction of the two factors, A and B; $SS_{SB|A}$, the sum of squares associated with the interaction of factor B and subjects within the groups on levels of the A factor.

The hypotheses that we wish to test involve only the main effects and the interaction of the main effects. In general we have no interest in the parameters $\xi\lambda$ or ξ because they depend on individual-difference random variables that are a nuisance value only in evaluating the effect of the experimental treatments. Thus, we test

$$H_{T,A}: \quad \tau_j = 0 \qquad \text{for all } j = 1, 2, \ldots, a$$
$$H_{A,A}: \quad \tau_j \neq 0 \qquad \text{for some } j = 1, 2, \ldots, a$$
$$H_{T,B}: \quad \lambda_k = 0 \qquad \text{for all } k = 1, 2, \ldots, b$$
$$H_{A,B}: \quad \lambda_k \neq 0 \qquad \text{for some } k = 1, 2, \ldots, b$$
$$H_{T,AB}: \tau\lambda_{jk} = 0 \qquad \text{for all } j = 1, 2, \ldots, a \text{ and } k = 1, 2, \ldots, b$$
$$H_{A,AB}: \tau\lambda_{jk} \neq 0 \qquad \text{for some } j = 1, \ldots, a \text{ and } k = 1, 2, \ldots, b$$

The appropriate variance ratios to test the hypotheses are indicated in Table 19.13. This is the first instance of a situation calling for general comment. The

TABLE 19.13 Analysis of Variance for a Two-Factor Experiment, Repeated Measures on Factor B

Source	SS	df	MS	EMS	V
Between Subjects	—	$Na - 1$	—	—	—
A	SS_A	$a - 1$	MS_A	$\sigma_\rho^2 + b\sigma_\xi^2 + \dfrac{Nb}{a-1}\sum\limits_{j=1}^{a}\tau_j^2$	$\dfrac{MS_A}{MS_{S\mid A}}$
Subjects Within Groups	$SS_{S\mid A}$	$a(N-1)$	$MS_{S\mid A}$	$\sigma_\rho^2 + b\sigma_\xi^2$	—
Within Subjects	—	$aN(b-1)$	—	—	—
B	SS_B	$b-1$	MS_B	$\sigma_\rho^2 + \sigma_{\xi\lambda}^2 + \dfrac{Na}{b-1}\sum\limits_{k=1}^{b}\lambda_k^2$	$\dfrac{MS_B}{MS_{SB\mid A}}$
AB	SS_{AB}	$(a-1)(b-1)$	MS_{AB}	$\sigma_\rho^2 + \sigma_{\xi\lambda}^2 + \dfrac{N}{(a-1)(b-1)}\sum\limits_{j=1}^{a}\sum\limits_{k=1}^{b}\tau\lambda_{jk}^2$	$\dfrac{MS_{AB}}{MS_{SB\mid A}}$
B by Subjects Within Groups	$SS_{SB\mid A}$	$a(N-1)(b-1)$	$MS_{SB\mid A}$	$\sigma_\rho^2 + \sigma_{\xi\lambda}^2$	—
Total	—	$abN - 1$	—	—	—

principle in formulating variance ratios to test specific hypotheses has up to this time been very simple. The mean square associated with an effect and the mean square associated with the residual (within-cells) source of variation were estimates of the same variances of normally distributed random variables, as indicated by the *EMS*, whenever the test hypothesis was presumed to be true. Consequently, the ratio of the mean squares were distributed as Fisher's *F* when the test hypothesis was true. The expected mean squares in Table 19.13 do not suggest any way of forming the variance ratios under such a simple rule. None of the expected mean squares is a simple composite of the residual variance and the noncentrality parameter. The residual variance is always confounded with certain individual-difference variances. The solution to the seeming dilemma is really rather simple. Recall that the sum of two independent normal random variables is a normal random variable with variance equal to the sum of the variances of the summed variables. Therefore, if the null hypothesis is true, the mean squares in the variance ratios indicated in Table 19.13 will be distributed as Fisher's *F* with the respective degrees of freedom.

Sample size for an experiment with this design is determined in the manner outlined in the sections above. The basic difference is that the standard deviation used is not the standard deviation of the residual term but the standard deviation of the composite of the residual and the individual-differences terms. In the specific design being discussed, the increase in sensitivity of discrimination associated with the repeated measures is primarily restricted to the within-subjects variable, factor *B*, and the interaction of the other factor with the within-subjects variable (that is, the *AB* interaction). As a consequence, smaller sample sizes may be needed to make the same absolute (as contrast to relative) discriminations on the within-subjects factors and interactions. Because of this, the design of experiments with repeated measures and random groups admixed among the factors of the experiment generally are centered on the repeated measures factors. The advantages of the design are clearly located in these factors. The risks the experimenter runs in introducing nonnormal individual-difference variables, nonadditivity across the repeated measures factors, the differential sensitivity, and the possibility of subtle carry-over effects make the design unattractive unless it is useful in making discriminations not otherwise possible with practical sample sizes. The random groups factor in this type of experiment is, by contrast, generally a factor of little interest other than as a nuisance factor that the experimenter wishes to partition out of the sums of squares in evaluating the effects on the repeated measures variable.

For posttest evaluations of the (α, β, δ) status of an analysis, the mean squares used as the estimate of the variance of the observed variable are the subjects-within-groups mean square for the *A* factor, and the *B* by subjects-within-groups mean square for both the *B* factor and the *AB* interaction.

Computational procedures are simple. From the summary tables for a two-way analysis of variance the following computational elements are calculated:

$$[1] = \frac{T_{..}^{\;2}}{abN}$$

$$[2] = \sum_{j=1}^{a} \sum_{k=1}^{b} \sum_{i=1}^{N} X_{jki}^2$$

$$[3] = \frac{\sum_{j=1}^{a} T_{j.}^{\;2}}{bN}$$

$$[4] = \frac{\sum_{k=1}^{b} T_{.k}^{\;2}}{aN}$$

$$[5] = \frac{\sum_{j=1}^{a} \sum_{k=1}^{b} T_{jk}^{\;2}}{N}$$

$$[6] = \frac{\sum_{j=1}^{a} \sum_{i=1}^{N} T_{j.i}^{2}}{b}$$

The sums of squares are computed from these computational elements.

between-subjects total sum $= [6] - [1]$

$$SS_A = [3] - [1]$$

$$SS_{S|A} = [6] - [3]$$

within-subjects total sum $= [2] - [6]$

$$SS_B = [4] - [1]$$

$$SS_{AB} = [5] - [3] - [4] + [1]$$

$$SS_{SB|A} = [2] - [5] - [6] + [3]$$

Higher order factorial designs with repeated measures on one or more factors and random groups on the remaining factors are common in the psychological research literature. These designs are of the same general nature as the two-factor design just presented. The problems and the advantages are essentially the same. The reader is referred to other sources for a more complete and detailed exposition (particularly Winer and Myers).

Detailed Analyses Within a Many-Sample Problem

The information gained by an investigator from the results of an analysis of variance is rather uninformative. The investigator knows whether or not the factor "makes a difference," but if it does make a difference, the investigator does not know how or where. Also, the analysis of variance neglects an important field of methodology in which several samples are involved but no single set of factors or interactions is directly of interest. This section is a brief review of two methods for dealing with statistical questions coming out of these considerations. First, we look at the experiment in which there are several groups but no single unifying theme to the entire experiment in the same sense as in the designs already discussed. The totality of the experiment can be used to safeguard against the dangers inherent in the tests-in-profusion approach to an analysis of the data. The basic principles of the analysis of variance can be applied to unify the analysis. Second, we examine the methods of "digging in" in detail to study the location of the significant effects discovered in the analysis of variance. The information already at hand from the calculations of the analysis of variance is used to give a sound statistical basis to the analysis of the detail data of the experiment.

Planned Comparisons. In designing an experiment involving many samples, the experimenter may have several specific questions he wants to answer—he wants to compare some samples with others, pairwise and in larger combinations. The comparisons he wishes to make are all stated in advance and the statistical tests planned before the experiment. Instead of making overall tests as in the analysis of variance, the researcher carries out the planned comparisons.

Consider an experiment designed to study the influence of age and instructions on the use of relationships in learning in a two-choice discrimination problem. Two ages are of critical importance, preschool and older. In order to keep conditions as similar as possible the experimenter takes his sample during the first month of the school year, using kindergarten and first-grade students. One set of instructions includes a detailed description of the stimuli, and the other simply indicates the goal behavior, discriminative response to each stimulus complex. The design forms a two-by-two factorial, but the hypothesis of greatest interest is not in age as such, but in one type of interaction which can be best stated as a comparison of select means in the design. As an ancillary interest the experimenter wants to compare the two instructions. The design is illustrated in Figure 19.3. The comparisons of interest are the following.

(1) The mean response in cell 2 is the same as the means in the other cells.

(2) The mean response in cell 1 is the same as the means in cells 3 and 4.

(3) The mean response level in age I is the same as the mean response level in age II.

An overall analysis of variance would answer the third question but not, in any direct sense, the first two. The interaction term, if significant, might possibly imply

Age of Subject		
	Kindergarten: I	First Grade: II
Simple Instructions	1	2
Elaborated Instructions	3	4

(Row label spanning both instruction rows: Instructions)

FIGURE 19.3

Illustration design for planned comparison example.

the first statement. However, other arrangements could lead to the interaction being significant, and the analysis of variance is thus not an appropriate tool for answering the specific questions asked.

In an earlier chapter it was indicated that the sums, even the weighted sums, of normally distributed variables were normally distributed with expectations and variances dependent on the expectations, variances, and covariances of the summed variables. The approach to making tests of planned comparisons is to express the comparison as a sum of variables and structure this sum under a model given by the test hypothesis.

A comparison in the model of an experiment is defined as a weighted sum of the expectations in the model. The simplest way of expressing this is to denote the model as a collection of M expectations defined for the experimental conditions (sample categories) in the experiment. Thus, in our example, there are four ($M = 4$) expectations μ_1, μ_2, μ_3, and μ_4. A comparison, denoted κ, is given by

$$\kappa = c_1\mu_1 + c_2\mu_2 + \cdots + c_M\mu_M = \sum_{j=1}^{M} c_j\mu_j \tag{19.38}$$

where the constants $c_1, \ldots, c_M$ are real numbers, not all zero, and where

$$\sum_{j=1}^{M} c_j = 0 \tag{19.39}$$

Applying this model directly to data gives an estimate of the comparison:

$$K = c_1\bar{X}_1 + c_2\bar{X}_2 + \cdots + c_M\bar{X}_M = \sum_{j=1}^{M} c_j\bar{X}_j \tag{19.40}$$

Since $\bar{X}_j$ is an unbiased estimate of μ_j, it can be shown easily that K is an unbiased estimate of κ.

The variance of the estimate of a comparison is simply defined. It is assumed that the observations in each treatment group are decomposable into the expectation of that treatment and a residual term:

$$X_{ji} = \mu_j + \rho_{ji}$$

By assuming that the residual term is independently normally distributed for all treatment groups with $N(0, \sigma_\rho^2)$, it can be shown easily that the sampling distribution of the mean in the jth treatment group is distributed normally, $N(\mu_j, \sigma_\rho^2/N_j)$. From this it follows directly by the rule of variances of sums that

$$\text{var}(K) = \sigma_\rho^2 \sum_{j=1}^{M} \frac{c_j^2}{N_j} \tag{19.41}$$

The variance of the residual term may be estimated as it was in the one-way analysis of variance (random groups and repeated measures design, depending on how the experiment being analyzed is designed). Thus the estimate of the variance of K in a random groups design is given by

$$\hat{S}_K^2 = MS_r \sum_{j=1}^{M} \frac{c_j^2}{N_j} \tag{19.42}$$

This completes the repertoire of parameters and estimates necessary to devise a statistical test of hypotheses regarding the parameters. The appropriate test is the t test for a single-sample hypothesis about an expectation (only one comparison is involved—it is treated as a single-sample mean).

Going back to the example with which we began this section, the hypotheses for the three comparisons are developed as follows. The contrast is first defined as the coefficients $c_1, \ldots, c_M$ expressing the character of the hypothesis. In the instance of comparison (1) listed above, we want to test the hypothesis that the expected response in cell 2 is the same as the expectations in the other cells, 1, 3, and 4. The average of the expectations in 1, 3, and 4 is equal to the expectation in 2 under the test hypothesis. Consequently, if we set

$$c = -\tfrac{1}{3} \qquad c_2 = 1 \qquad c_3 = -\tfrac{1}{3} \qquad c_4 = -\tfrac{1}{3}$$

the contrast is

$$\kappa = \mu_2 - \tfrac{1}{3}(\mu_1 + \mu_3 + \mu_4)$$

The test hypothesis is that this contrast is zero. An alternative to the test hypothesis is that the expectation in cell 2 is $\delta\sigma_\rho$ units larger than the average of the other expectations. Thus this contrast leads to

$$H_T: \kappa = \kappa_T = 0$$
$$H_A: \kappa = \kappa_A = \kappa_T + \delta\sigma_\rho$$

The sample comparison is formed as indicated above. For the four-group example the sample comparison is

$$K = -\tfrac{1}{3}\bar{X}_1 + \bar{X}_2 - \tfrac{1}{3}\bar{X}_3 - \tfrac{1}{3}\bar{X}_4$$

The general form of the test is

$$t = \frac{K - \kappa_T}{\hat{S}_K} \tag{19.43}$$

which, if H_T is true, is distributed as student's t with $\sum_{j=1}^{M}(N_j - 1)$ degrees of freedom. Sample size determination, interpretation of the results of the calculation of t in the framework of the (α, β, δ) analysis, and so on, are identical to the same considerations in the one-sample problem regarding hypotheses about expectations. Some complications in the assumptions are made: equal variances in all groups, no carry-over effects if the repeated measures approach is used, additivity in repeated measures.

The variance, $\hat{S}_K{}^2$, is given by multiplying the MS_r by $\sum_{j=1}^{M}(c_j{}^2/N_j)$. Presuming $N_1 = 12$, $N_2 = 10$, $N_3 = 9$, and $N_4 = 10$, this term is given by

$$\frac{1/9}{12} + \frac{1}{10} + \frac{1/9}{9} + \frac{1/9}{10} = .1327$$

Imagine that the data were such that $MS_r = 33.91$. The net result is that $\hat{S}_K{}^2 = (.1327)(33.91) = 4.5$. Imagine that the means observed in the experiment were

$$\bar{X}_1 = 51 \qquad \bar{X}_2 = 54 \qquad \bar{X}_3 = 49 \qquad \bar{X}_4 = 51$$

The observed contrast is

$$K = -\tfrac{1}{3}(51) + (54) - \tfrac{1}{3}(49) - \tfrac{1}{3}(51) = 3.667$$

The test statistic is $3.667/2.12 = 1.73$. The $(\alpha = .05)$ value of t for $df = 37$ is 2.03. Consequently, the significance probability of the test statistic for the test hypothesis is greater than .05 and very slightly less than .10. In order to evaluate the meaning of this statement, imagine that $\delta = .9$ was judged to be a significant departure of K under the alternative with respect to the test hypothesis. If $\alpha = \beta = .05$ is to be used in the (α, β, δ) for the test, then the value of $\delta = .54$ is discriminable with $df = 37$. A confusion is likely to be made between the standard deviation of the comparison and the standard deviation of the residual variable. The MS_r is to be used in specifying the value of $\delta\sigma_p$ of the alternate hypothesis. In this example, the appropriate value is $\sqrt{33.91} = 5.82$, not the 2.12 used in the equation for t. Since the discrimination power of the test is rather more than is necessary to make a meaningful statement about the results of the test, the failure to reject H_T is not in any way a dubious result, within the α and β risk levels. The power against this happening when the alternative is true exceeds .95 by a good margin.

Turning to the second and third comparisons defined in the statement of the study, the question of the definition of multiple comparisons becomes important. First, denote a comparison by the letter C. Since the comparison is defined by the constants $c_1, \ldots, c_M$, there are several sets of constants where there are several comparisons. Subscripting the comparisons $C_1, \ldots, C_R$ for R comparisons suggests carrying the subscript over into the specifications of the constants. Let the first subscript on a constant for a comparison designate the comparison, and the second subscript designate the treatment condition to which the constant is applied. Thus, for the jth comparison and the kth treatment, the constant is c_{jk}. The whole set of comparisons is denoted:

$$C_1: c_{11}, c_{12}, \ldots, c_{1M}$$

$$\vdots \qquad \vdots$$

$$C_j: c_{j1}, c_{j2}, \ldots, c_{jM}$$

$$\vdots \qquad \vdots$$

$$C_R: c_{R1}, c_{R2}, \ldots, c_{RM}$$

A question of independence of the comparisons arises immediately. Just as the tests-in-profusion approach was questionable because of the dependency among a collection of two-sample tests, the R comparisons among the M treatment groups are also questionable. There are at most $(M-1)$ degrees of freedom in an

M-sample problem with respect to hypotheses about the expectations of the M samples. Consequently, if there are more than $(M-1)$ comparisons, some of them must be dependent on the others. It is simple to determine whether or not two comparisons are independent: if two comparisons C_a and C_b are related in such a way that

$$\omega_{ab} = \sum_{k=1}^{M} \frac{C_{ak}C_{bk}}{N_k} = 0 \tag{19.44}$$

the comparisons are said to be independent (the term "orthogonal" is often used). If the comparisons are independent, the test results are also independent. Consequently, significance in one of two independent comparisons has no implication for the significance in the other.

The three comparisons proposed for the example experiment are given by the coefficients

$$C_1: -\tfrac{1}{3}, 1, -\tfrac{1}{3}, -\tfrac{1}{3}$$
$$C_2: 1, 0, -\tfrac{1}{2}, -\tfrac{1}{2}$$
$$C_3: \tfrac{1}{2}, -\tfrac{1}{2}, \tfrac{1}{2}, -\tfrac{1}{2}$$

Recall $N_1 = 12$, $N_2 = 10$, $N_3 = 9$, and $N_4 = 10$. Calculating ω for all three combinations of contrasts gives

$$\omega_{12} = \frac{1}{135} \qquad \omega_{13} = \frac{67}{1080} \qquad \omega_{23} = \frac{14}{360}$$

indicating that the comparisons are all dependent. Comparisons C_1 and C_2 are not as severely dependent as C_1 and C_2 are with C_3. However, the caution still holds: Do not interpret significance or lack of significance of any of the comparisons independently of consideration of the significance or lack of significance in the other comparisons.

The example just worked out is illustrative of the difficulties we can introduce by permitting unequal sample sizes in the various treatment conditions of an experiment. If the same set of comparisons are defined in an experiment in which the sample sizes were all the same, say $N = 10$, the values of ω are quite different:

$$\omega_{12} = 0 \qquad \omega_{13} = \frac{4}{60} \qquad \omega_{23} = \frac{1}{20}$$

With equal sample sizes in the treatment groups, the first two comparisons may be interpreted independently. However, the third comparison is redundant to some extent on each of the first two comparisons. A little thought reveals that the third comparison is equivalent to the test of the age factor in the two-way analysis of variance that could be performed on the design. Then C_1 and C_2 involve parts of the data used to test the age factor hypotheses, and thus C_3 is partially captured in C_1 and C_2.

Defining other comparisons to "use up" the three degrees of freedom in the example may not be of any interest to the experimenter. However, the reader is encouraged to explore the comparisons that are equivalent with the three hypotheses in the analysis of variance for the two-way factorial design implied in the study. If C_3 had been the principle comparison of interest, the other two, C_1 and C_2, might have been replaced with other comparisons to form an independent set of comparisons.

The reader is also encouraged to express the $\binom{4}{2}$ two-sample t tests arising out of the tests-in-profusion approach to studying a four-sample problem in terms of the corresponding comparisons, and to investigate the dependency properties of this set of comparisons. Assume equal sample sizes for simplicity.

Post-hoc Comparisons in the Analysis of Variance. The planned comparisons just discussed are intended to replace the analysis of variance as the method of determining the significance of tests of statistical hypotheses when the main effects and interaction effects approach of the analysis of variance is not appropriate. The many sample problem confronted after the analysis of variance has located significant main effects and (or) interactions is to find the location of the source of the significance in the effects. Whereas the planned comparison approach set out to test hypotheses not well stated in the main and interaction effects of the analysis of variance, the post-hoc comparison approach is designed to explain the source of rejection of test hypotheses in the analysis of variance. In this sense the post-hoc comparison is an exploratory device designed to facilitate an understanding of the significant results in an analysis of variance. It is not appropriate to attempt to "salvage" an experiment with disappointing results (for example, no significant effects) by comparing the most extreme means in the study and attempting to draw conclusions about the phenomenon under study from comparisons of these extremes. In general, it might be expected that the more extreme (opposite directions from the overall mean) means will be "significantly" different when analyzed by the method of comparisons without implying anything about the influence of the independent variable or treatments involved in the study. The experimenter is justified in trying to understand a contribution of an experimental variable or treatment condition *only* when it is shown that that variable (factor, interaction) has made an overall significant difference in the dependent variable.

The definition of a comparison is the same in the post-hoc comparison as it was in the planned comparison. A post-hoc comparison is defined as a weighted sum of treatment-category expectations (means in the sample) where the weights are not all zero and where the sum of the weights is zero. A post-hoc comparison is defined for one of the sets of means involved in a main effect or an interaction. The summary tables for the totals used in calculating sums of squares for the main effects and interactions can be used as the basis for defining the treatment-category means. For example, in a three-factor factorial design the summary tables, Table 19.8 through Table 19.11, contain the data necessary to define the means for the A factor, the B factor, the C factor, the AB interaction cells, the AC interaction cells, and the BC interaction cells. Table 19.8 contains the ABC interaction cells data. Each of the totals in these tables must be divided by the number of observations entering into the totals—the mean for each relevant cell is calculated.

Imagine that the AB interaction is significant. Table 19.9 is used to obtain the means of the AB treatment combinations. There are ab means, and some comparison among these means will illuminate the meaning of the interaction in terms of the levels on A and B.

The statistical method for testing post-hoc comparisons is the same regardless of the effect involved, first-order or interaction. Assume that there are R separate treatment levels (or combinations of levels) defining the effect. Thus, if there are a levels on factor A and b levels on factor B and c levels on factor C, there are $(R = a)$ separate treatment conditions to be considered in a post-hoc comparison within the A factor and $(R = abc)$ separate treatment combinations to be considered in a post-hoc comparison within the ABC interaction. The statistical test of a comparison depends on the value of R. It also depends on the values of the degrees of freedom used in the variance ratio in testing the hypothesis concerning the factor or interaction being investigated—for example, df_n for the numerator degrees of freedom and df_d for the denominator degrees of freedom.

Let

$$F^* = \sqrt{(R - 1)F} \qquad (19.45)$$

where F is the α level critical value of the variance ratio distributed as Fisher's F with df_n and df_d. Also, let

$$\hat{S}_K = \frac{MS_r}{m} \sum_{j=1}^{R} c_i^2$$

where MS_r is the denominator mean square of the effect or interaction involved, m is the number of observations entering into each of the means in the summary table involved, and c_i are the comparison constants applied to the R means.

Scheffé has proved that the probability is $(1 - \alpha)$ that all imaginable comparisons will be "captured" or enclosed by the set of intervals given by

$$P(\kappa - F^*\hat{S}_K \leq K \leq \kappa + F^*\hat{S}_K) = 1 - \alpha \qquad (19.46)$$

where κ is any one of all possible comparisons and K is the corresponding sample comparison. Thus, if a single-factor experiment is involved, the numerator degrees of freedom for the F is $(R - 1)$ for R levels, and the denominator degrees of freedom is df_r.

In the example, above, of a two-factor factorial experiment (see Table 19.5), conjectures regarding the source of the significance of the A factor and the AB interaction were formulated. These conjectures play the role of informal hypotheses that can now be specified more formally and tested by the method of post-hoc comparisons. There are twelve cells in the design with which the conjectures may be tested. Using the double subscript notation for the comparison constants, we have

$$
\begin{array}{llll}
c_{11} = 1 & c_{12} = 0 & c_{13} = 0 & c_{14} = 1 \\
c_{21} = 0 & c_{22} = 0 & c_{23} = 0 & c_{24} = 0 \\
c_{31} = -1 & c_{32} = 0 & c_{33} = 0 & c_{34} = -1
\end{array}
$$

as the contrast for the alternate hypothesis formulated for factor A. Taking $\alpha = .05$, we find F by looking up the critical value of Fisher's F with $df_n = 3$ and $df_r = 24$; $F = 3.01$. Hence

$$F^* = \sqrt{11(3.01)} = 5.75$$

For this comparison, $m = 3$ observations enter into each of the interaction level (cell) means. The $MS_r = 6.08$, from Table 19.5. Hence

$$\hat{S}_K{}^2 = \frac{6.08}{3} \sum_{i=1}^{3} \sum_{j=1}^{4} c_{ij}{}^2 = 8.107$$

$$\hat{S}_K = 2.85$$

and the value that the obtained comparison is compared with is

$$F^*\hat{S}_K = 16.39$$

The obtained value of the comparison is given (using data of Table 19.4) by

$$
\begin{aligned}
K = {} & 1(\tfrac{15\,9}{3}) + 0(\tfrac{15\,0}{3}) + 0(\tfrac{14\,9}{3}) + 1(\tfrac{16\,2}{3}) \\
& + 0(\tfrac{15\,0}{3}) + \quad \cdots \quad + 0(\tfrac{14\,8}{3}) \\
& - 1(\tfrac{13\,7}{3}) + 0(\tfrac{15\,0}{3}) + 0(\tfrac{15\,2}{3}) - 1(\tfrac{13\,8}{3}) \\
= {} & 15.3
\end{aligned}
$$

Since the value of K, that is, 15.3, is smaller than the sampling limits about κ for a $(1 - \alpha = .95)$ interval, the conjecture is not supported by the comparison. The test hypothesis that this comparison does not play a role in the factor A significance is not rejected. Other comparisons may be found to be significant. The interpretation of multiple comparisons with statistical significance depends on the nature of the content of the experiment.

The Random Effects Model

Thus far we have been concerned exclusively with the fixed effects model of the analysis of variance. In most circumstances the fixed effects model is satisfactory for psychological research. However, there are a number of important applications in psychology where the fixed effects model is not sufficient to the need. Generally these applications involve large numbers of treatment-level possibilities or large numbers of subpopulations of experimental units or subjects. If it is not practical or useful to design an experiment embodying all of the treatment levels or sub-populations, a sample of treatments and subpopulations is used in defining the experiment. The extreme case is where there are an unlimited number of possible levels on an experimental factor—for example, the dose of a certain drug used in treating mental retardates, the number of footcandles of illumination used in a detection experiment, the intertrial delay time. Clearly, no experiment can be defined with an exhaustive representation of any of these sorts of independent variables. If the experimenter is content to select a few levels of the factor and restrict his inferences to those levels, the fixed effects model is perfectly reasonable. However, if the experimenter wishes to infer that the variable, not just the levels included in the experiment, is effective in determining the level of response in the dependent variable, then the fixed effects model is inappropriate. Instead of the fixed effects model the experimenter can use the random effects model.

Single-Factor Designs. The structural form of the random effects model is the same as the fixed effects model. For a single-factor experiment the structural form of the model is

$$X_{ij} = \mu + \tau_j + \rho_{ij} \qquad i = 1, \ldots, N; j = 1, \ldots, M$$

(where the experiment has arbitrarily been restricted to the equal sample size case).

The assumptions regarding the components of the structural model are different. In the fixed effects model the differential treatment effect τ_j is considered to be a constant defined strictly by the jth level of the factor. In the random effects model no assumption about the specific value of the effect can be made prior to the experiment (and, in practice, only sample data are available), and the value of τ_j depends on the specific level sampled from the population of factor levels. Consequently, the value of τ_j is a random variable. The governing assumption in the random effects model is that the random variable τ_j is normally distributed, $N(0, \sigma_\tau^2)$. It is also necessary to assume that the τ_j's are independent, that is, that the selection of a given value of the parameter for one level does not influence the selection of a given value of the parameter for another level. In summary, the τ_j's are variable quantities that have normal distributions, whereas in the fixed effects model they are a set of fixed but unknown constants. The random effects model makes the same assumption as that made in the fixed effects model regarding the residual variable, that is, that ρ_{ij} are independently $N(0, \sigma_\rho^2)$ for all ij.

The test hypothesis of equal-treatment effects and the alternative hypothesis of differential-treatment effects are stated in terms of the variance of the treatment variable:

$$H_T: \sigma_\tau^2 = 0$$
$$H_A: \sigma_\tau^2 \neq 0$$

(19.47)

Although the assumptions appear to be radically different, the test of this hypothesis structure is the same as in the fixed effect model, when the sample of levels is small in comparison with the population of levels (see below). The expected mean square for the treatment factor is $\sigma_\rho^2 + N\sigma_\tau^2$. The calculational form is the same in both models, and the possible empirical violations are similar; plus the added necessity of meeting the independently normal distribution assumption about the treatment effect parameters in the random effects model.

A more interesting class of test hypotheses and alternatives is available. Imagine that we wished to test the hypothesis that the variability of response to a treatment variable was equal to or less than one half of the variability of response as a function of residual factors. This leads to the following hypothesis structure:

$$H_T: \sigma_\tau^2 \leq \frac{\sigma_\rho^2}{2}$$

$$H_A: \sigma_\tau^2 > \frac{\sigma_\rho^2}{2}$$

The test for this structure is defined only for the equal sample size condition. In general, where $\delta = \sigma_\tau^2/\sigma_\rho^2$ ($\delta = \frac{1}{2}$, above) the hypothesis structure

$$H_T: \sigma_\tau^2 \leq \delta\sigma_\rho^2$$
$$H_A: \sigma_\tau^2 > \delta\sigma_\rho^2$$

(19.48)

leads to a test involving the ratio

$$\frac{MS_t}{MS_r} \frac{1}{1 + N\delta}$$

(19.49)

which, when the test hypothesis is true and the assumptions of the model are met, is distributed as Fisher's F with $(M - 1)$ and $[M(N - 1)]$ degrees of freedom in the numerator and denominator, respectively.

Sample size determination for experiments of this type is very similar to the sample size determination in the fixed effects model. The major difference is in the different structure of the alternative hypotheses. The determination of (α, β, δ) in a posttest evaluation follows the general outline of the evaluation in the fixed effects model with the difference of the new definition of δ. The power function for the hypothesis structure of equations (19.47) is available in the form of charts similar to the Pearson-Hartley charts for the power function in fixed effects models. These charts may be found in the book by Bowker and Lieberman referenced at the end of the previous chapter. Where the second type of hypotheses structure, that is, equations (19.48), is involved the power function may be determined with the availability of extensive tables of Fisher's F (see the book by Graybill referenced in the list at the end of the previous chapter). The power of the test depends on two values of the ratio—δ, specified in the hypothesis structure, and another, δ', which may be specified as the alternative value in H_A, or calculated when the power is stated specifically. The power of the test is given by

$$P\left(F \geq F^* \frac{1 + N\delta}{1 + N\delta'}\right) \tag{19.50}$$

where F is Fisher's F-distributed random variable, and F^* is the critical value (α level of significance), both with $(M - 1)$ degrees of freedom in the numerator and $[M(N - 1)]$ degrees of freedom in the denominator. When the values of δ and δ' are specified, along with N and F^*, the probability can be looked up in the F table. When the probability, the values of N, F^*, and δ are specified, the value of δ' can be calculated.

Two-Factor Factorial, Both Factors Random Effects Factors.

Imagine a two-factor experiment in which the levels on both factors are selected at random from populations of levels. The general form of this experiment is exactly like the two-factor fixed effect experiment, in terms of organization, calculation, and structural model. The assumptions, the hypothesis structure, and the tests are different, however. The structural model is familiar:

$$X_{jki} = \mu + \tau_j + \xi_k + (\tau\xi)_{jk} + \rho_{jki}$$

$$j = 1, \ldots, a; k = 1, \ldots, b; i = 1, \ldots, N$$

where it is assumed that

$$\tau_j \quad \text{are independently } N(0, \sigma_\tau^2)$$
$$\xi_k \quad \text{are independently } N(0, \sigma_\xi^2)$$
$$(\tau\xi)_{jk} \text{ are independently } N(0, \sigma_{\tau\xi}^2) \tag{19.51}$$
$$\rho_{jki} \quad \text{are independently } N(0, \sigma_\rho^2)$$

The computation of the sums of squares and the mean squares from the observed data is the same as in the fixed effects factorial design. The hypothesis structure is limited to statements regarding the variances of the components in the structural model:

$$H_{T.A}: \quad \sigma_\tau^2 = 0$$
$$H_{A.A}: \quad \sigma_\tau^2 \neq 0$$
$$H_{T.B}: \quad \sigma_\xi^2 = 0$$
$$H_{A.B}: \quad \sigma_\xi^2 \neq 0 \tag{19.52}$$
$$H_{T.AB}: \sigma_{\tau\xi}^2 = 0$$
$$H_{A.AB}: \sigma_{\tau\xi}^2 \neq 0$$

The expected mean squares for each of the sources of variability (each factor and their interaction) and the appropriate variance ratio for each of the respective tests of hypotheses in the hypothesis structure are indicated in Table 19.14. Note that the denominator in the variance ratios of the tests for the main effects is the interaction mean square. This procedure is required in order to have (1) the ratio distributed as F under the test hypothesis and (2) departures from the F distribution solely a function of the main effect variance involved in the test. The expected mean

TABLE 19.14 Expected Mean Squares and Variance Ratios for a Two-Factor Random Groups Design, Both Factors Random Effects

Source	EMS	V
Treatments A	$\sigma_\rho^2 + N\sigma_{\tau\xi}^2 + bN\sigma_\tau^2$	$\dfrac{MS_A}{MS_{AB}}$
B	$\sigma_\rho^2 + N\sigma_{\tau\xi}^2 + aN\sigma_\xi^2$	$\dfrac{MS_B}{MS_{AB}}$
AB	$\sigma_\rho^2 + N\sigma_{\tau\xi}^2$	$\dfrac{MS_{AB}}{MS_r}$
Residual	σ_ρ^2	

squares depend on the equal sample size assumption and on the assumption that the levels defining the factors are selected at random from a very large population of levels.

Two-Factor Factorial, Factor A Fixed Effects, Factor B Random Effects. Imagine that the two factors in a factorial are mixed, one (for example, factor A) with a fixed effects model, and the other (factor B) with a random effects model. Such a design might arise out of a study in which one factor represents arbitrary instruction sets, the other some time-defined variable; one factor might be strain of animal and the other factor defined by a dosage of a given drug; one factor might be qualitative properties of stimulus materials (a fixed set) and the other factor the amount of illumination used in displaying the stimuli; and so on.

The assumptions for the model are slightly more complicated but essentially are like those of the previous example of a design. The calculations and data layout are the same as in the two-factor fixed effect design with the exception of the variance ratios used to test the hypotheses. The hypotheses are the same as those of the previous example, detailed in equations (19.52). The expected mean squares and the appropriate variance ratios to test the respective hypotheses are indicated in Tables 19.15 and 19.16.

TABLE 19.15 Expected Mean Squares for Three-Factor Experiments Having N Observations in Each Cell

$$EMS_A = \sigma_\rho^2 + N\left(1 - \frac{b}{G}\right)\left(1 - \frac{c}{H}\right)\sigma_{\tau\xi\lambda}^2 + Nb\left(1 - \frac{c}{H}\right)\sigma_{\tau\lambda}^2 + Nc\left(1 - \frac{b}{G}\right)\sigma_{\tau\xi}^2 + Nbc\sigma_\tau^2$$

$$EMS_B = \sigma_\rho^2 + N\left(1 - \frac{a}{E}\right)\left(1 - \frac{c}{H}\right)\sigma_{\tau\xi\lambda}^2 + Na\left(1 - \frac{c}{H}\right)\sigma_{\xi\lambda}^2 + Nc\left(1 - \frac{a}{E}\right)\sigma_{\tau\xi}^2 + Nab\sigma_\xi^2$$

$$EMS_C = \sigma_\rho^2 + N\left(1 - \frac{a}{E}\right)\left(1 - \frac{b}{G}\right)\sigma_{\tau\xi\lambda}^2 + Na\left(1 - \frac{b}{G}\right)\sigma_{\xi\lambda}^2 + Nb\left(1 - \frac{a}{E}\right)\sigma_{\tau\lambda}^2 + Nab\sigma_\lambda^2$$

$$EMS_{AB} = \sigma_\rho^2 + N\left(1 - \frac{c}{H}\right)\sigma_{\tau\xi\lambda}^2 + Nc\sigma_{\tau\xi}^2$$

$$EMS_{AC} = \sigma_\rho^2 + N\left(1 - \frac{b}{G}\right)\sigma_{\tau\xi\lambda}^2 + Nb\sigma_{\tau\lambda}^2$$

$$EMS_{BC} = \sigma_\rho^2 + N\left(1 - \frac{a}{E}\right)\sigma_{\tau\xi\lambda}^2 + Na\sigma_{\xi\lambda}^2$$

$$EMS_{ABC} = \sigma_\rho^2 + N\sigma_{\tau\xi\lambda}^2$$

$$EMS_r = \sigma_\rho^2$$

TABLE 19.16 Expected Mean Squares for Three Special Cases of Experiments Covered in Table 19.15

Fixed Effects Model: All Effects Fixed

$$EMS_A = \sigma_\rho{}^2 + Nbc\sigma_\tau{}^2$$
$$EMS_B = \sigma_\rho{}^2 + Nac\sigma_\xi{}^2$$
$$EMS_C = \sigma_\rho{}^2 + Nab\sigma_\lambda{}^2$$
$$EMS_{AB} = \sigma_\rho{}^2 + Nc\sigma_{\tau\xi}{}^2$$
$$EMS_{AC} = \sigma_\rho{}^2 + Nb\sigma_{\tau\lambda}{}^2$$
$$EMS_{BC} = \sigma_\rho{}^2 + Na\sigma_{\xi\lambda}{}^2$$
$$EMS_{ABC} = \sigma_\rho{}^2 + N\sigma_{\tau\xi\lambda}^2$$
$$EMS_r = \sigma_\rho{}^2$$

Random Effects Model: All Effects Random

$$EMS_A = \sigma_\rho{}^2 + N\sigma_{\tau\xi\lambda}^2 + Nb\sigma_{\tau\lambda}{}^2 + Nc\sigma_{\tau\xi}{}^2 + Nbc\sigma_\tau{}^2$$
$$EMS_B = \sigma_\rho{}^2 + N\sigma_{\tau\xi\lambda}^2 + Na\sigma_{\xi\lambda}{}^2 + Nc\sigma_{\tau\xi}{}^2 + Nac\sigma_\xi{}^2$$
$$EMS_C = \sigma_\rho{}^2 + N\sigma_{\tau\xi\lambda}^2 + Na\sigma_{\xi\lambda}{}^2 + Nb\sigma_{\tau\lambda}{}^2 + Nab\sigma_\lambda{}^2$$
$$EMS_{AB} = \sigma_\rho{}^2 + N\sigma_{\tau\xi\lambda}^2 + Nc\sigma_{\tau\xi}{}^2$$
$$EMS_{AC} = \sigma_\rho{}^2 + N\sigma_{\tau\xi\lambda}^2 + Nb\sigma_{\tau\lambda}{}^2$$
$$EMS_{BC} = \sigma_\rho{}^2 + N\sigma_{\tau\xi\lambda}^2 + Na\sigma_{\xi\lambda}{}^2$$
$$EMS_{ABC} = \sigma_\rho{}^2 + N\sigma^2$$
$$EMS_r = \sigma_\rho{}^2$$

Mixed Effects Model: Factor A Random, B and C Fixed

$$EMS_A = \sigma_\rho{}^2 + Nbc\sigma_\tau{}^2$$
$$EMS_B = \sigma_\rho{}^2 + Nc\sigma_{\tau\xi}{}^2 + Nac\sigma_\xi{}^2$$
$$EMS_C = \sigma_\rho{}^2 + Nb\sigma_{\tau\lambda}{}^2 + Nab\sigma_\lambda{}^2$$
$$EMS_{AB} = \sigma_\rho{}^2 + Nc\sigma_{\tau\xi}{}^2$$
$$EMS_{AC} = \sigma_\rho{}^2 + Nb\sigma_{\tau\lambda}{}^2$$
$$EMS_{BC} = \sigma_\rho{}^2 + N\sigma_{\tau\xi\lambda}^2 + Na\sigma_{\xi\lambda}{}^2$$
$$EMS_{ABC} = \sigma_\rho{}^2 + N\sigma_{\tau\xi\lambda}^2$$
$$EMS_r = \sigma_\rho{}^2$$

Rules for Forming EMS's and Variance Ratios

This section deals with a method for determining expected mean squares (EMS) for three-factor (or smaller) factorial randomized designs for fixed effects, random effects, and mixed models when equal sample sizes are involved. The two models are special cases of a general rule: Out of R treatment (group) levels, r are sampled; if $r = R$, the model is a fixed effects model; if $r < R$, the model is a random effects model. In the random effects models discussed above it is assumed that R is very much larger than r so that $r/R = 0$. Mixed models are mixtures of exhaustive and partial sampling of factor levels. Assume that factor A represents a population of E levels of which a are actually sampled for the experiment. Likewise for factors B and C with populations of levels with G and H levels, respectively, from which b and c levels are sampled. The EMS for all of the mean squares for a three-factor factorial are indicated in Table 19.15. In the interest of notational simplicity, the variance symbol σ^2 is used throughout this table, even where the sum of squares of the treatment parameters divided by the appropriate degrees of freedom was used earlier (fixed effect parameters). Hence, in Table 19.15 σ_τ^2 is substituted for

$$\frac{1}{a-1} \sum_{j=1}^{a} \tau_j^{2}$$

when A is a fixed effect factor (that is, $a = E$), and similarly for other factors.

The equations in Table 19.15 may be used to determine EMS's for any factorial experiment with up to three factors. The first step in using the equations for fewer than three factors is to cross out all terms involving parameters not included in the experiment, for example, λ in a two-factor experiment. For the remaining terms, evaluate the coefficients

$$\left(1 - \frac{a}{E}\right) \qquad \left(1 - \frac{b}{G}\right) \qquad \left(1 - \frac{c}{H}\right)$$

If the number of levels used in the experiment is equal to the number of levels in the population, the coefficient is 0. If the number of levels used is much smaller than the number of levels in the population, the coefficient is essentially 1.0. The intermediate condition will not be discussed here. The terms with coefficients of zero can be crossed out. The remaining terms (substituting the value 1.0 for the coefficients) define the EMS.

Table 19.16 illustrates three applications of this logic. Three special cases are defined: three fixed effect factors; three random effects factors; a mixed model with factor A a random effects factor and factors B and C fixed effects factors.

The analysis of variance for all of the applications of this general scheme is the same as that developed for the fixed effects models discussed in earlier sections of this chapter and in the previous chapter. The sums of squares are calculated by the equations developed for the fixed effects designs. The degrees of freedom are also the same.

It is clear from the three illustrated special cases that the variance ratios necessary to test some hypotheses may not be defined by the mean squares arising in the experiment. For example, in the experiment with three random effects there is no mean square that can be used as a denominator to test hypotheses about factor A.

The general form of a variance ratio to test a hypothesis, for example, about an effect ω (main or interaction), is

$$\frac{\text{(function of mean squares in design)} + MS_\omega}{\text{(function of mean squares in design)}}$$

where the functions of mean squares in the denominator and numerator are the same. Two types of functions lead to two types of variance ratios. Where the function is given by a mean square defined in Table 19.16, the ratio is distributed as Fisher's F if the model and assumptions are accurate. However, where the function is not a mean square defined in Table 19.16, the ratio is distributed *quasi* the Fisher's F. The first type of ratio has been discussed in detail. The second type (quasi F) needs explication.

In attempting to test

$$H_{T.A}: \sigma_A{}^2 = 0$$

in a three-factor factorial design with random effect factors, the natural denominator term in the ratio would be

$$\sigma_\rho{}^2 + N\sigma_{\tau\xi\lambda}^2 + Nb\sigma_{\tau\lambda}{}^2 + Nc\sigma_{\tau\xi}{}^2$$

If there were a mean square having this expression as its expectation, for example, MS_D, with df_D, then under the model and $H_{T.A}$ the ratio

$$\frac{MS_A}{MS_D}$$

would be distributed as Fisher's F with df_A and df_D in the numerator and denominator, respectively. However, MS_D is not defined in the experiment. Thus, no natural variance ratio is defined. The "solution" is to devise a "mean square," MS_Q, out of the mean squares available. By searching through the mean squares, we see that it is possible to define

$$MS_Q = MS_{AC} + MS_{AB} - MS_{ABC}$$
$$EMS_Q = \sigma_\rho{}^2 + N\sigma_{\tau\xi\lambda}^2 + Nb\sigma_{\tau\lambda}{}^2 + Nc\sigma_{\tau\xi}{}^2$$

Using this form, we can calculate a mean square ratio

$$V_Q = \frac{MS_A}{MS_{AC} + MS_{AB} - MS_{ABC}}$$

This form is not altogether a good solution, because the denominator term can be negative if MS_{ABC} is larger than the sum of MS_{AC} and MS_{AB}. This is not consistent with the general definition of a mean square, and another "solution" is desirable. The problem is avoided if MS_{ABC} is added to the denominator and numerator of V_Q:

$$V_Q = \frac{MS_A + MS_{ABC}}{MS_{AC} + MS_{AB}}$$

This expression has the correct form if EMS's are substituted for mean squares:

$$\frac{2\sigma_\rho{}^2 + 2N\sigma_{\tau\xi\lambda}^2 + Nb\sigma_{\tau\lambda}{}^2 + Nc\sigma_{\tau\xi}{}^2 + Nb\sigma_\tau{}^2}{2\sigma_\rho{}^2 + 2N\sigma_{\tau\xi\lambda}^2 + Nb\sigma_{\tau\lambda}{}^2 + Nc\sigma_{\tau\xi}{}^2}$$

If the test hypothesis is true and the model is correct, V_Q is distributed in a way that is approximately Fisher's F with df_n in the numerator and df_d in the denominator where

$$df_n = \frac{(MS_A + MS_{ABC})^2}{(MS_A)^2/df_A + (MS_{ABC})^2/df_{ABC}}$$

$$df_d = \frac{(MS_{AC} + MS_{AB})^2}{(MS_{AC})^2/df_{AC} + (MS_{AB})^2/df_{AB}}$$

This "trumped-up" test depends on mean squares used elsewhere in the design for other tests and hence is not independent of tests using those mean squares.

SUGGESTED READING

References indicated at the end of the previous chapter are relevant to the material of this chapter.

GLOSSARY

Multifactor Design. A multifactor design is defined by multiple factors that are independently defined so that a subject may be observed with the several factors simultaneously present, with all combinations of levels across the several factors.

Interaction. Interaction between factors is defined by a pattern of response on one factor at a given level of the other factor (or factors) that is not the same as a pattern of response at another level of the second factor.

Main Effects. Main effects are the treatment effects in the analysis of variance associated with the factors taken singly.

Interaction Effects. Interaction effects are the treatment effects in the analysis of variance associated with the factors taken in combinations.

$$(19.1) \quad T_{j.} = \sum_{k=1}^{b} \sum_{i=1}^{N} X_{ijk} \quad \text{for } j = 1, \ldots, a$$

$$(19.2) \quad T_{.k} = \sum_{j=1}^{a} \sum_{i=1}^{N} X_{ijk} \quad \text{for } k = 1, \ldots, b$$

$$(19.3) \quad T_{..} = \sum_{j=1}^{a} \sum_{k=1}^{b} \sum_{i=1}^{N} X_{ijk}$$

$$T_{..} = \sum_{j=1}^{a} T_{j.} = \sum_{k=1}^{b} T_{.k}$$

$$(19.4) \quad T_{jk} = \sum_{i=1}^{N} X_{ijk} \quad \text{for } j = 1, \ldots, a \text{ and } k = 1, \ldots, b$$

$$(19.5) \quad \mu_{j.} = \frac{1}{b} \sum_{k=1}^{b} \mu_{jk}$$

$$(19.6) \quad \mu_{.k} = \frac{1}{a} \sum_{j=1}^{a} \mu_{jk}$$

$$(19.7) \quad \mu = \frac{1}{a} \sum_{j=1}^{a} \mu_{j.} = \frac{1}{b} \sum_{k=1}^{b} \mu_{.k} = \frac{1}{ab} \sum_{j=1}^{a} \sum_{k=1}^{b} \mu_{jk}$$

$$(19.8) \quad \mu_{jk} = \mu + (\mu_{j.} - \mu) + (\mu_{.k} - \mu) = \mu_{j.} + \mu_{.k} - \mu$$

$$(19.9) \quad \tau_{j} = \mu_{j.} - \mu \quad j = 1, \ldots, a$$

$$(19.10) \quad \xi_{k} = \mu_{.k} - \mu \quad k = 1, \ldots, b$$

$$(19.11) \quad (\tau\xi)_{jk} = \mu_{jk} - \mu_{j.} - \mu_{.k} + \mu$$

$$= \mu_{jk} - (\tau_{j} + \xi_{k} + \mu)$$

$$\text{for } j = 1, \ldots, a \text{ and } k = 1, \ldots, b$$

$$(19.12) \quad \sum_{j=1}^{a} \tau_{j} = \sum_{k=1}^{b} \xi_{k} = \sum_{j=1}^{a} \sum_{k=1}^{b} (\tau\xi)_{jk} = 0$$

$$(19.13) \quad \rho_{ijk} = X_{ijk} - \mu_{jk}$$

$$(19.14) \quad X_{ijk} = \mu + \tau_{j} + \xi_{k} + (\tau\xi)_{jk} + \rho_{ijk}$$

$$\text{for } i = 1, \ldots, N; j = 1, \ldots, a; k = 1, \ldots, b$$

$$(19.15) \quad (X_{ijk} - \bar{X}) = (X_{ijk} - \bar{X}_{jk}) + (\bar{X}_{j.} - \bar{X}) + (\bar{X}_{.k} - \bar{X})$$

$$+ (\bar{X}_{jk} - \bar{X}_{j.} - \bar{X}_{.k} + \bar{X})$$

(19.16) $\quad SS_T = SS_r + SS_A + SS_B + SS_{AB}$

(19.17) $\quad E(MS_r) = \sigma_\rho^{\ 2}$

(19.18) $\quad E(MS_A) = \sigma_\rho^{\ 2} + Nb\,\dfrac{\sum_{a=1}^{j} \tau_j^{\ 2}}{a-1}$

(19.19) $\quad E(MS_B) = \sigma_\rho^{\ 2} + Na\,\dfrac{\sum_{k=1}^{b} \xi_k^{\ 2}}{b-1}$

(19.20) $\quad E(MS_{AB}) = \sigma_\rho^{\ 2} + N\,\dfrac{\sum_{j=1}^{a} \sum_{k=1}^{b} (\xi\tau)_{jk}^{\ 2}}{(a-1)(b-1)}$

(19.21) $\quad \phi_A = \dfrac{\sqrt{(bN/a)\,\sum_{j=1}^{a} \tau_j^{\ 2}}}{\sigma_\rho}$

(19.22) $\quad \phi_B = \dfrac{\sqrt{(aN/b)\,\sum_{k=1}^{b} \xi_k^{\ 2}}}{\sigma_\rho}$

(19.23) $\quad \phi_{AB} = \dfrac{\sqrt{\dfrac{N}{(a-1)(b-1)+1}\,\sum_{j=1}^{a} \sum_{k=1}^{b} (\xi\tau)_{jk}^{\ 2}}}{\sigma_\rho}$

(19.24) $\qquad$ *2-Effect Alternative* $\qquad\qquad\qquad\qquad$ a-*Effect Alternative*

$$\phi_A = \delta\sqrt{\dfrac{2bN}{a}} \qquad\qquad\qquad \phi_A = \delta\sqrt{\dfrac{bN(a^2-1)}{12}}$$

$$\phi^2 = \delta^2\,\dfrac{2b}{a}\,N \qquad\qquad\qquad \phi^2 = \delta^2\,\dfrac{b(a^2-1)}{12}\,N$$

$$N = \phi^2\,\dfrac{a}{2b\delta^2} \qquad\qquad\qquad N = \phi^2\,\dfrac{12}{\delta^2 b(a^2-1)}$$

(19.25) $\qquad$ *2-Effect Alternative* $\qquad\qquad\qquad\qquad$ b-*Effect Alternative*

$$\phi_B = \delta\sqrt{\dfrac{2aN}{b}} \qquad\qquad\qquad \phi_B = \delta\sqrt{\dfrac{aN(b^2-1)}{12}}$$

$$\phi^2 = \delta^2\,\dfrac{2a}{b}\,N \qquad\qquad\qquad \phi^2 = \delta^2\,\dfrac{a(b^2-1)}{12}\,N$$

$$N = \phi^2\,\dfrac{b}{2a\delta^2} \qquad\qquad\qquad N = \phi^2\,\dfrac{12}{\delta^2 a(b^2-1)}$$

(19.26)   2-*Effect Alternative*   ab-*Effect Alternative*

$$\phi_{AB} = \delta \sqrt{\frac{2N}{ab}} \qquad\qquad \phi_{AB} = \delta \sqrt{\frac{N[(ab)^2 - 1]}{12}}$$

$$\phi^2 = \delta^2 \frac{2}{ab} N \qquad\qquad \phi^2 = \delta^2 \frac{[(ab)^2 - 1]}{12} N$$

$$N = \phi^2 \frac{ab}{2\delta^2} \qquad\qquad N = \phi^2 \frac{12}{\delta^2[(ab)^2 - 1]}$$

(19.27) $X_{jkmi} = \mu + \tau_j + \xi_k + \lambda_m + (\tau\xi)_{jk} + (\tau\lambda)_{jm} + (\xi\lambda)_{km}$
$$+ (\tau\xi\lambda)_{jkm} + \rho_{jkmi}$$

(19.28) $(X_{jkmi} - \bar{X}) = (\bar{X}_{j..} - \bar{X}) + (\bar{X}_{.k.} - \bar{X}) + (\bar{X}_{..m} - \bar{X})$
$$+ (\bar{X}_{jk.} - \bar{X}_{j..} - \bar{X}_{.k.} + \bar{X}) + (\bar{X}_{j.m} - \bar{X}_{j..} - \bar{X}_{..m} + \bar{X})$$
$$+ (\bar{X}_{.km} - \bar{X}_{.k.} - \bar{X}_{..m} + \bar{X})$$
$$+ (\bar{X}_{jkm} - \bar{X}_{jk.} - \bar{X}_{j.m} - \bar{X}_{.km} + \bar{X}_{j..} + \bar{X}_{.k.} + \bar{X}_{..m} - \bar{X})$$
$$+ (X_{jkmi} - \bar{X}_{jkm})$$

(19.29) $SS_T = SS_A + SS_B + SS_C + SS_{AB} + SS_{AC} + SS_{BC} + SS_{ABC} + SS_r$

(19.30)   2-*Effect Alternative*   a-*Effect Alternative*

$$\phi_A = \delta \sqrt{\frac{bcN2}{a}} \qquad\qquad \phi_A = \delta \sqrt{\frac{bcN(a^2 - 1)}{12}}$$

(19.31)   2-*Effect Alternative*   b-*Effect Alternative*

$$\phi_B = \delta \sqrt{\frac{acN2}{b}} \qquad\qquad \phi_B = \delta \sqrt{\frac{acN(b^2 - 1)}{12}}$$

(19.32)   2-*Effect Alternative*   c-*Effect Alternative*

$$\phi_C = \delta \sqrt{\frac{abN2}{c}} \qquad\qquad \phi_C = \delta \sqrt{\frac{abN(c^2 - 1)}{12}}$$

(19.33) 2-*Effect Alternative*   ab-*Effect Alternative*

$$\phi_{AB} = \delta \sqrt{\frac{cN2}{ab}} \qquad\qquad \phi_{AB} = \delta \sqrt{\frac{cN[(ab)^2 - 1]}{12}}$$

(19.34) 2-*Effect Alternative*   ac-*Effect Alternative*

$$\phi_{AC} = \delta \sqrt{\frac{bN2}{ac}} \qquad\qquad \phi_{AC} = \delta \sqrt{\frac{bN[(ac)^2 - 1]}{12}}$$

(19.35) 2-*Effect Alternative*   bc-*Effect Alternative*

$$\phi_{BC} = \delta \sqrt{\frac{aN2}{bc}} \qquad\qquad \phi_{BC} = \delta \sqrt{\frac{aN[(bc)^2 - 1]}{12}}$$

(19.36) 2-*Effect Alternative*   abc-*Effect Alternative*

$$\phi_{ABC} = \delta \sqrt{\frac{N2}{abc}} \qquad\qquad \phi_{ABC} = \delta \sqrt{\frac{N[(abc)^2 - 1]}{12}}$$

(19.37) $X_{jki} = \mu + \tau_j + \xi_{i(j)} + \lambda_k + \tau\lambda_{jk} + \xi\lambda_{ki(j)} + \rho_{i(jk)}$

Planned Comparisons. A method of comparing specified portions and combinations of portions of the data in an analysis of variance design. The comparisons are all stated in advance of the experiment, as specified hypotheses.

(19.38) $\kappa = c_1\mu_1 + c_2\mu_2 + \cdots + c_M\mu_M = \sum\limits_{j=1}^{M} c_j\mu_j$

(19.39) $\sum\limits_{j=1}^{M} c_j = 0$

(19.40) $K = c_1\bar{X}_1 + c_2\bar{X}_2 + \cdots + c_M\bar{X}_M = \sum\limits_{j=1}^{M} c_j\bar{X}_j$

(19.41) $\text{var}(K) = \sigma_\rho^2 \sum\limits_{j=1}^{M} \dfrac{c_j^2}{N_j}$

(19.42) $\hat{S}_K^2 = MS_r \sum\limits_{j=1}^{M} \dfrac{c_j^2}{N_j}$

(19.43) $t = \dfrac{K - \kappa_T}{\hat{S}_K}$

(19.44) $\omega_{ab} = \sum\limits_{k=1}^{M} \dfrac{c_{ak}c_{bk}}{N_k} = 0$

Post-Hoc Comparisons. A method of exploration in an analysis of variance that has revealed significant effects that permits the discovery of the sources of the significance in specific comparisons.

(19.45) $F^* = \sqrt{(R - 1)F}$

(19.46) $P(\kappa - F^*\hat{S}_K \leq K \leq \kappa + F^*\hat{S}_K) = 1 - \alpha$

Random Effects Model. The random effects model is an analysis of variance model based on the assumption that the levels on one or more factors in the design are defined by a random variable.

(19.47) $\begin{aligned} H_T &: \sigma_\tau^2 = 0 \\ H_A &: \sigma_\tau^2 \neq 0 \end{aligned}$

(19.48)
$$H_T: \sigma_\tau^2 \leq \delta\sigma_\rho^2$$
$$H_A: \sigma_\tau^2 > \delta\sigma_\rho^2$$

(19.49)
$$\frac{MS_t}{MS_r} \frac{1}{1 + N\delta}$$

(19.50)
$$P\left(F \geq F^* \frac{1 + N\delta}{1 + N\delta'}\right)$$

(19.51) τ_j are independently $N(0, \sigma_\tau^2)$

           ξ_k are independently $N(0, \sigma_\xi^2)$

           $(\tau\xi)_{jk}$ are independently $N(0, \sigma_{\tau\xi}^2)$

           ρ_{jki} are independently $N(0, \sigma_\rho^2)$

(19.52) $H_{T,A}: \sigma_\tau^2 = 0$

           $H_{A,A}: \sigma_\tau^2 \neq 0$

           $H_{T,B}: \sigma_\xi^2 = 0$

           $H_{A,B}: \sigma_\xi^2 \neq 0$

           $H_{T,AB}: \sigma_{\tau\xi}^2 = 0$

           $H_{A,AB}: \sigma_{\tau\xi}^2 \neq 0$

PROBLEMS

1. Prove equation (19.12).

2. For a two-way analysis of variance design with four levels on factor A and five levels on factor B, calculate sample size for a variety of values of α, β, and δ. Attempt to discover a compromise within the set of values for α and β for a fixed value of δ. Do these calculations for both the 2-effect and the M-effect alternatives. If the interaction effect is of no immediate concern, except as a safety element in the experimental design, it might be ignored in the selection of sample size. Find a compromise value for the two main-effect tests.

3. In problem 2, what is the effect of removing one level from factor B?

4. Perform the calculations of the analysis of variance for data of Table 19.3 with one of the three observations in each cell deleted. Use a random number table to select the data value to delete in each cell. Compare the results of your calculations with those of Table 19.5.

5. Evaluate the effect of the reduction of sample size specified in problem 4 by applying the posttest procedures illustrated in Table 19.6.

TABLE P-19.1

		Factor B			
		$k = 1$	$k = 2$	$k = 3$	$k = 4$
Factor A	$j = 1$	49 51 46	43 45 47	46 44 47	49 49 50
	$j = 2$	46 43 48	42 48 45	44 48 53	41 50 44
	$j = 3$	41 42 40	43 50 47	45 43 50	41 44 40

6. Imagine that the experiment giving rise to the data of Table 19.3 were replicated under a second set of environmental conditions and the results were as shown in Table P-19.1. Perform the analysis of variance calculations on these data, in the full three-way design, using the data of Table 19.3 and the data of Table P-19.1. Do a posttest analysis of the scientific significance of all of the results and compare this information with the information presented about the two factors of the original data.

7. Assuming that the data of Table 19.3 came from an experiment in which observations on factor B were repeated measures, that is, for each row of scores in the table the scores for the four levels of B were obtained from a single subject. There would then be three subjects in each level of factor A. Perform the calculations of the repeated measures analysis of variance on the data, and compare the results with the results of the analysis presented in the text.

8. Repeat problem 7, reversing the role of factors A and B in the repeated measures aspect of the problem. Make an analysis of the consequences of the design mode on the outcome of the analysis. If the inappropriate model is used, along with the attendant erroneous calculational form, what are the consequences for the statistical and scientific inferences one makes?

9. Prove that equation (19.41) is true.

10. Using the method of planned comparisons, make a comparison of level 1 and level 4 of factor B in the data of Table 19.3. Make a comparison between the cell defined by level 1 on A with level 2 on B and the cell defined by level 4 on A and level 3 on A. Are these two comparisons independent? Find a comparison that is independent of the first comparison specified in this exercise, and check to determine if it is independent of the second comparison specified.

11. Using the method of post-hoc comparisons, discover the sources of major differences affecting the results of the analysis of variance performed on the data of problem 6.

CHAPTER 20

Regression and Correlation

T his chapter is a brief excursion into a topic deserving a volume in its own right, both from the point of view of the interest and usefulness of the topic and from the point of view of its complexity. In general, this chapter is a discussion of the analysis of experiments in which each event in the sample space of the experiment is characterized by multiple values, that is, by multiple random variables. For each of the events in the sample space there are defined values on M random variables $X_1, \ldots, X_M$. Each of these random variables is fully formed, having a separate value set and probability distribution. In addition, it is possible to speak of the *joint* value set and the joint distribution of the M random variables. Many of the problems we discuss impose a special status on one of the M variables, for one reason or another (for example, to indicate it as a variable to be predicted or as a dependent variable in situations where the other variables are directly measured or controlled). In this case, the multiple variables are indicated by the set $(Y, X_1, \ldots, X_M)$, where Y is the distinctive variable.

The types of problems addressed in these situations are many and varied. In all of the problems, however, there is a central theme which holds this topic together— What can be learned about Y by having knowledge about $X_1, \ldots, X_M$, or what information do the X's convey regarding one another? The answer to this question, indeed the specific form of the general question, depends on the context in which it is asked. For example, we might be interested in predicting Y from the X's, or discovering if there is a common source of determination of the events in the sample space with regard to two or more X's, or we may wish to correct (as best we can) observations on Y for contamination by the presence of effects that are measured by the X's, and so forth. In order to provide a conceptual foundation for the topics to be discussed, some of the basic concepts must be presented first.

Imagine that X and Y are two random variables defined on a sample space in such a way that they are indicators of the characteristics of the events to which the values are assigned. For example, take two random variables defined on the events in the sample space of an experiment consisting of tossing a coin three times and observing the head-tail sequence. Let the random variable Y be defined as the number of heads on the first two tosses and the variable X be defined as the number of heads on the last two tosses. Clearly, the two variables are related by the involvement of the middle toss in each of the variables. The relationship between X and Y may be stated in a number of ways, for example, if Y is 2 then X must be 1 or more, and values tend to be the same in both variables, and so on. In this simple example, the relationship of X and Y is clear and the source of the relationship is unmistakable. However, when we turn to more interesting random variables, such as those encountered in psychology, the relationships among the variables may not be as transparent. Nevertheless, we are interested in the "relationship" of the variables. Since there is no clear meaning to the word "relationship," we must examine the possible meanings and discover whether any of the meanings are useful in understanding the phenomena giving rise to the variables, in the sense that the coin-tossing basis of X and Y in the example leads to an understanding of the relationship of X and Y.

In the face of a lack of clear indicators of the relationship of two variables, we develop formal models of relationship that, in turn, become the basis for interpretation of the relationship between two variables. The most frequent model used is that of some general function such as a linear or a quadratic function connecting X and Y through a well-defined rule, giving a value Y for each value of X. Since for each value of X there is a single value of Y defined by such a function, it is clear that the relationship of random variables cannot, in general, be very well described by the function. For example, in the coin-tossing experiment the relationship is not a single-valued relationship. For $X = 1$ all three values of Y are possible although the probability that $Y = 1$ is greater than the probability that $Y = 0$. If $X = 2$ it is not possible that $Y = 0$, but both $Y = 1$ and $Y = 2$ are possible. Hence, any single-valued function, giving a single value Y as a function of the value of X, could not possibly describe the relationship. However, it is possible that an appropriate function may capture the essence of the relationship between two variables although the detail and scatter of the values in the value sets of the variables are not represented in the function. We can devise measures of the degree to which the function represents the major aspects of the relationship of the variables and make a judgment regarding whether the function is a satisfactory description of the relationship. Several properties of the "fit" of the function to the relationship might be proposed. The properties and measures of the goodness of the fit must be discussed in the context of the specific form of the function and the purposes of studying the variables. These considerations are developed in some detail in the sections to follow in this chapter.

Another, and related, approach is to attempt to discover some summary or descriptive quantity representing some aspect of the relationship in quantitative terms. The most familiar of such concepts are the coefficients of regression and correlation. Often such an approach involves statements regarding the variance of Y without knowledge of X and the variance of Y conditional on knowledge of X.

In the sections to follow in this chapter we are concerned with three practical problems. First, imagining that Y is a probabilistic function of X, we address the problem of determining the best estimate of the value of Y for an event when X is known but Y is not known. Second, when Y is composed of effects of several characteristics of the events, including those determining, in part, the variable X, we address the problem of estimating that part of Y that is independent of the effects determining X. Third, we wish to determine the extent that the two variables, X and Y, are influenced by common sources or characteristics of the events.

Whereas in the general case we may be interested in M variables, $X_1, \ldots, X_M$, plus a special variable Y, we attend in this chapter primarily to the study of two variables. We are concerned generally with a variable X and a variable Y. The character of X and Y with respect to the experimental or observational procedures determines whether we treat X and Y differentially. In some instances we treat Y as a dependent variable and X as an independent variable, under the control of the investigator or presumed to be known. On the other hand, we have occasion to treat X and Y identically, without distinction in terms of conceptual meaning.

Our attention is restricted along another dimension. We deal exclusively in this chapter with single-sample or two-sample problems. In the first instance we are concerned with problems in regression and correlation when we have a single sample of N observations on the joint distribution of X and Y. In the second instance, we are concerned with the differences (or lack of differences) in the characteristics of the joint distribution of X and Y in the two samples of observations.

Jointly Distributed Random Variables

In order to provide the probability basis for the analysis of regression and correlation of two random variables, and eventually for an analysis of sampling problems regarding regression and correlation, we need to recall the basic definitions of a random variable. For each event in the sample space of an experiment there is defined a value from the value set of X, depending on the characteristics of the event. Similarly a random variable Y may be defined on the event space.

Take a simple example of a sample space and two random variables. Let the experiment be defined by a series of three tosses of a coin. The event set is defined by the sequence of heads and tails: HHH, HHT, HTH, THH, HTT, THT, TTH, TTT. Define two random variables on this sample space: (1) X, the number of heads in the three tosses, and (2) Y, the number of tails prior to the last head. These two random variables are defined over the same eight events in the sample space of the experiment, as illustrated in Table 20.1. The value sets of both variables are the same: $\{0, 1, 2, 3\}$.

The distributions of the two variables are given in Table 20.2. Clearly these

TABLE 20.1 Two Random Variables Defined on the Sample Space of Three Tosses of a Coin

Event	X	Y
HHH	3	0
HHT	2	0
HTH	2	1
THH	2	1
HTT	1	0
THT	1	1
TTH	1	2
TTT	0	3

TABLE 20.2 Probability Distributions of the Variables Defined in Table 20.1

Value	$P(X = x)$	$P(Y = y)$
0	$\frac{1}{8}$	$\frac{3}{8}$
1	$\frac{3}{8}$	$\frac{3}{8}$
2	$\frac{3}{8}$	$\frac{1}{8}$
3	$\frac{1}{8}$	$\frac{1}{8}$

TABLE 20.3 An Illustration of a Joint Distribution Showing the Events in the Sample Space and the Attendant Probabilities

| | | \multicolumn{4}{c}{X} | |
		0	1	2	3	$P(Y = y)$
	0	None $P = \frac{0}{8}$	HTT $P = \frac{1}{8}$	HHT $P = \frac{1}{8}$	HHH $P = \frac{1}{8}$	$\frac{3}{8}$
	1	None $P = \frac{0}{8}$	THT $P = \frac{1}{8}$	HTH, THH $P = \frac{2}{8}$	None $P = \frac{0}{8}$	$\frac{3}{8}$
Y	2	None $P = \frac{0}{8}$	TTH $P = \frac{1}{8}$	None $P = \frac{0}{8}$	None $P = \frac{0}{8}$	$\frac{1}{8}$
	3	TTT $P = \frac{1}{8}$	None $P = \frac{0}{8}$	None $P = \frac{0}{8}$	None $P = \frac{0}{8}$	$\frac{1}{8}$
$P(X = x)$		$\frac{1}{8}$	$\frac{3}{8}$	$\frac{3}{8}$	$\frac{1}{8}$	

two random variables are different, although they are defined on a single sample space and have the same value set.

In order to investigate the relationship between the two random variables we can look at the two-way layout or the joint distribution of the variables. A table, Table 20.3, of the joint occurrence of the values of X and Y is constructed, and the events associated with the paired values are indicated in the cells of the table. The probability in a cell of the layout is the probability of observing, on a given execution of the experiment, the values represented in the joint distribution of the two variables. That is, the probability of observing $X = 2$ and $Y = 1$ is 2/8; the events HTH and THH are in the respective cell. The probabilities in this table are called the probabilities of the joint distribution of the two random variables. Note that the basis of the probabilities of a given pair of values, for example, $(X = x, Y = y)$, is the occurrence of events in the sample space to which the pair of values (x, y) are assigned. Only if the two random variables are related through a common sample space is it possible to define a joint distribution of the random variables.

The probabilities in the joint distribution of the two random variables display a characteristic that is useful in the next few paragraphs. For each value in the value set of one of the variables, for example, X, we find that

$$P(X = x) = \sum_{y \in Y} P(X = x, Y = y) \tag{20.1}$$

That is, the sum of probabilities over one variable for a given value of the other variable is the probability of the given value, without respect for the first variable. The definition of the probabilities of the joint distribution and the summation rule just stated indicate that the probabilities in the joint distribution are proper probabilities: They sum to 1.0 and are bounded by zero and one.

The probabilities of the values in the value set of one of the jointly distributed variables, for example, $(X = x)$ where X and Y are jointly distributed, are called the marginal probabilities of the joint distribution. Equation (20.1) indicates that the marginal probabilities of a given variable are the summations across the value set of the other variable.

Independence of Random Variables. The concept of independence of outcomes in sample spaces may be generalized to random variables. The arguments and general meaning of independence in random variables are similar to those in our discussion of outcomes in sample spaces. One difference is apparent immediately. In order to speak of the independence of the random variables, we must refer to the entire value sets of both variables. Since each value in the value set of each variable and each combination of values in the joint of the variables must be treated as outcomes, the concept of independence in random variables is parallel to the concept of pairwise independence of a collection of outcomes in a sample space.

First, we define the conditional probability of a value on one variable, given a particular value on the other variable. Imagine that the variable Y is taken as the conditioning variable, and that the specific value is y. Then we define the conditional probability of $X = x$, given $Y = y$, by

$$P(X = x \mid Y = y) = \frac{P(X = x, Y = y)}{P(Y = y)} \tag{20.2}$$

In the example given above,

$$P(X = 2 \mid Y = 1) = \frac{P(X = 2, Y = 1)}{P(Y = 1)} = \frac{2/8}{3/8} = \frac{2}{3}$$

The definition of independence of random variables is that if for all x and y in the value sets of the variables

$$P(X = x \mid Y = y) = P(X = x) \tag{20.3}$$

then the variables are independent. If for any value in the joint distribution of X and Y this relationship does not hold, the variables are dependent. By replacing the conditional probability $P(X = x \mid Y = y)$ with the marginal probability $P(X = x)$ and multiplying both sides of the last equation by the conditioning probability $P(Y = y)$, the multiplication rule for independence is expressed:

$$P(X = x, Y = y) = P(X = x)P(Y = y) \tag{20.4}$$

The example given above is indicative of dependence of the two variables, X and Y: $P(X = 2 \mid Y = 1) = \frac{2}{3} \neq P(X = 2) = \frac{3}{8}$. If the example variables had been independent, with the marginals defining $P(X = x)$ and $P(Y = y)$, the joint distribution, given by applying the multiplication rule, would have been that displayed in Table 20.4. The values in the cells of this table were obtained by multiplying the respective row and column marginal probabilities. The values in this table should be compared with the probabilities of the joint distribution as defined by the experiment, that is, Table 20.3.

As an example of independent random variables, the reader is encouraged to construct the joint distribution of two random variables W and Z defined on the sample space of a series of three tosses of a coin. Indicate W to be the number of heads on the first toss and Z to be the number of heads on the second and third tosses. In this example, $P(W = w)P(Z = z) = P(W = w, Z = z)$ for all combinations of w and z.

Deviations of joint distributions from independence may be slight or they may be relatively extreme. The definition of independence and the rules for deciding whether jointly distributed variables are independent give no hint at a method of quantifying the degree of dependency. Since there are many ways in which we could define a departure from independence, there are many ways in which we can quantify the degree of departure. Essentially, all of the topics of the remainder of this chapter are directed at determining the degree of departure from independence of jointly distributed variables and "ways" in which the joint distribution departs from independence.

TABLE 20.4 Example of Two Independent, Jointly Distributed, Random Variables, Indicating the Joint Distribution

			X			
		0	1	2	3	$P(Y = y)$
Y	0	$\frac{3}{64}$	$\frac{9}{64}$	$\frac{9}{64}$	$\frac{3}{64}$	$\frac{3}{8}$
	1	$\frac{3}{64}$	$\frac{9}{64}$	$\frac{9}{64}$	$\frac{3}{64}$	$\frac{3}{8}$
	2	$\frac{1}{64}$	$\frac{3}{64}$	$\frac{3}{64}$	$\frac{1}{64}$	$\frac{1}{8}$
	3	$\frac{1}{64}$	$\frac{3}{64}$	$\frac{3}{64}$	$\frac{1}{64}$	$\frac{1}{8}$
	$P(X = x)$	$\frac{1}{8}$	$\frac{3}{8}$	$\frac{3}{8}$	$\frac{1}{8}$	

The product rule for independently distributed variables indicates that knowledge of the value of one variable for an event in the sample space provides literally no information about the value of the other variable for the event. The distribution of the X values for any given $Y = y$ is the same as the marginal distribution, that is, the X distribution is not dependent on the particular value of Y. On the other hand, when the variables are dependent, the distribution of the X values may be different for different outcomes involving the various values of Y. In our original example of dependent variables, knowledge that $Y = 1$ leads to the knowledge that X cannot be equal to 3, whereas knowledge that $Y = 0$ leads to the knowledge that 1, 2, and 3 are all equally likely. Thus, when the variables are dependent, the information we might have about one variable is informative about the other variable.

Part of the information about Y given by knowledge of the value of X can be captured by describing the distribution of Y given a specific value of X. Such a distribution is called a conditional distribution. Certain properties of these distributions lead to quantifications of the degree of dependency of jointly distributed variables.

Conditional Distributions. The conditional probability of a value in the value set of Y, given a value from the value set of X, was defined earlier, in (20.2). Improving the notation to permit some simple algebra, we define the jth value in the value set of X to be the value x_j, and the kth value in the value set of Y to be the value y_k. Also, let M_X and M_Y be the number of values in the value sets of X and Y, respectively. Then we define the $(M_X \times M_Y)$ conditional probabilities, with respect to conditioning on X by equation (20.2). These probabilities define the conditional distribution of Y, given $X = x$. It must be presumed that $P(X = x) \neq 0$.

The conditional distribution may be viewed simply as a distribution of the variable Y under the restricted sample space of the experiment corresponding to that part of the sample space composed of events to which the value x is assigned by X. It is easy to prove that

$$\sum_y P(y \mid x) = 1$$

$$0 \leq P(y \mid x) \leq 1.0 \qquad \text{for all} \quad x \text{ and } y$$

The properties of conditional distributions that interest us most are the expectations and variances of the distributions. In order to distinguish these from the expectation and variance of the variable Y, they are called the conditional expectations and conditional variances. The definition of the expectation of a conditional distribution is formally the same as the definition of the expectation of a random variable, except that the conditional probabilities are involved:

$$\mu_{Y|x} = E(Y \mid x) = \sum_y P(y \mid x)y \qquad (20.5)$$

The notation in this expectation is complicated, so we will at times represent the expectation of Y given $X = x_j$ by the symbol $\mu_{j.}$. Do not confuse this with the expectation of the jth value of the X variable—it is the expectation of the Y variable. This expectation is a function of the index of the X variable, that is, j, because a different conditional expectation is defined for each $j = 1, \ldots, M_X$.

The variances of the conditional distributions are given by

$$\sigma^2_{Y|x} = \text{var}(Y \mid x) = \sum_y P(y \mid x)(y - \mu_{Y|x})^2 \qquad (20.6)$$

If the variables are independent, the conditional probabilities of Y, given X, are the same as the marginal probabilities of Y. Therefore, in the expression for $\mu_{j.}$ the conditional probabilities are the same as the marginal probabilities and $\mu_{j.} = \mu_Y$, for all $j = 1, \ldots, M_X$. For the same reason, the conditional variances are all equal to the variance of Y, that is, $\sigma_Y^2 = \sigma^2_{Y|x_j}$ for $j = 1, \ldots, M_X$. As we see shortly, deviations of conditional means about the mean of Y give us some leverage in quantifying the degree of departure from independence in the joint distribution. In addition, the conditional variances permit an assessment of the relative extremity of the departure from independence. In particular, we are able to demonstrate that the variance of Y can be partitioned into a part associated with dependency on X and a part not related to X. This partition, in turn, leads to a well-behaved measure of the degree of dependency of the two variables.

At this point we develop an example that we carry through, up to the point of an introduction to sampling in jointly distributed random variables. Imagine (X, Y) with the value sets of X and Y

$$X = \{1, 2, 3, 4, 5, 6\}$$
$$Y = \{1, 2, 3, 4, 5, 6, 7, 8\}$$

and probability distributions, marginals and joint, as indicated in Table 20.5. The table includes all of the conditional expectations and variances. The conditional distributions of Y given X are similar to those of X given Y, except that the roles of X and Y are reversed. The quantities involved are as follows:

$$P(x \mid y) = \frac{P(X = x, Y = y)}{P(Y = y)}$$
$$\mu_{X|y} = \sum_x P(x \mid y)x$$
$$\sigma^2_{X|y} = \sum_x P(x \mid y)(x - \mu_{X|y})^2$$

TABLE 20.5 Example of Jointly Distributed Random Variables—The Joint Probabilities, the Marginal Probabilities, the Conditional Expectations and Variances, and the Marginal Expectations and Variances

		\multicolumn{6}{c}{X}								
		1	2	3	4	5	6	$P(Y=y)$	$\mu_{.k}$	$\sigma_{.k}^2$
	8	.00	.00	.00	.01	.05	.01	.07	5.00	.29
	7	.00	.00	.00	.04	.06	.02	.12	4.83	.47
	6	.00	.00	.02	.05	.03	.04	.14	4.64	1.09
	5	.01	.00	.04	.09	.01	.02	.17	3.88	1.28
Y	4	.02	.01	.09	.04	.00	.01	.17	3.12	1.28
	3	.04	.03	.05	.02	.00	.00	.14	2.36	1.09
	2	.02	.06	.04	.00	.00	.00	.12	2.17	.47
	1	.01	.05	.01	.00	.00	.00	.07	2.00	.29
$P(X=x)$		.10	.15	.25	.25	.15	.10			
$\mu_{j.}$		3.00	2.00	3.68	5.32	7.00	6.00			
$\sigma_{j.}^2$		1.20	.80	1.58	1.58	.80	1.20			

$$\mu_X = 3.5 \qquad\qquad \sigma_X^2 = 2.05$$
$$\mu_Y = 4.50 \qquad\qquad \sigma_Y^2 = 3.93$$

It is sometimes useful to think of the conditional distribution of a random variable, conditional on a value of a variable with which it is jointly distributed, in terms of the graphical representation of conditional probabilities defining the distribution. Figure 20.1 shows the graphs of the probability function of (X, Y) tabled in Table 20.5. The marginal probabilities $P(X = x)$ and $P(Y = y)$ are shown as well as the joint probabilities, $P(X = x, Y = y)$. Figure 20.2 shows the graphs of the conditional probability function $(Y \mid X)$, that is, $P(Y = y \mid X = x)$ for all (X, Y). These conditional distributions should be compared by the student with the distribution $P(Y = y)$ of Figure 20.1. The student is urged to construct a graph of the conditional distribution of $(X \mid Y)$ corresponding to Figure 20.2, that is, reversing the roles of X and Y.

Generalized Regression Functions. For each value of X in the example above we have a number of values of Y with nonzero probabilities. These values and their probabilities are a complete description of the information we have about the variable Y conditional on the value $X = x$. It is useful in a number of situations (for example, prediction, simple description) to be able to assign a value to each of the conditional distributions that captures the essence of that distribution

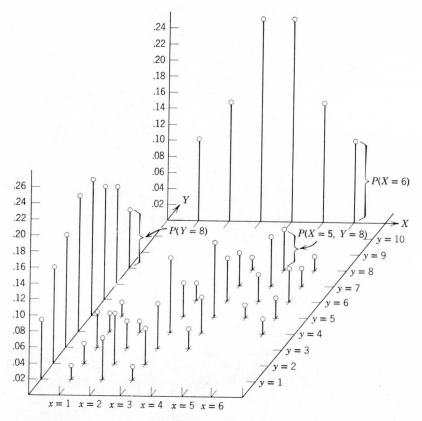

FIGURE 20.1

Graph of joint distribution of (X, Y) from Table 20.5.

with respect to some particular application or with respect to some particular probabilistic property. A single value for each of the conditional distributions is the most desirable. Such an assignment is a function of the value of the conditioning variable, or some function of the conditioning variable, or at least some function of the conditional distribution for each value of the conditioning variable. Let $g(x)$ be the function. This leads to the definition of a set of pairs, $[x, g(x)]$. This set of pairs is called the generalized regression function. We are interested in some special cases of the generalized regression function. In particular, we are interested in the regression function when

$$g(x) = E(Y \mid X = x) = \mu_{Y|x} \qquad (20.7)$$

and when

$$g(x) = \xi + \theta x \qquad (20.8)$$

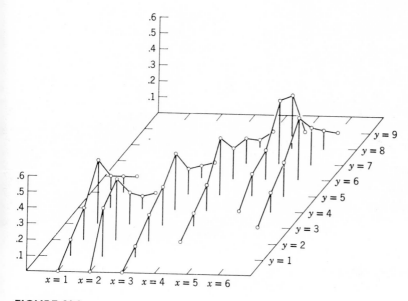

FIGURE 20.2

Graph of conditional distributions of $(Y \mid X)$ from Table 20.5.

Other functions in which we might be interested are, for example, the function giving $g(x)$ as the largest value of Y with a nonzero probability in the conditional distribution defined on x, or the function giving $g(x)$ as the value of the median of the conditional distribution. These two examples might be used as the regression function of Y on X if we were interested in stating with certainty that the value of Y were no larger than the specified value (that is, a maximum y given x) or if we were interested in stating that half of the values (with respect to probability) lay above the regression function and half below.

The set of pairs, $[x, g(x)]$, defines a set of points in the Cartesian coordinates, of the plane, defined by (X, Y). If these points fall on a line, the line is said to be the regression line. If the points fall on a curve, the curve is said to be the regression curve. In general, when dealing with a limited set of values of X, the set of pairs defines a limited set of points in the plane (X, Y) and the points fall on no clearly defined line or curve. However, it is common practice to connect by "interpolation" the points by line segments (straight or curved) to define more clearly the locus of the entire set of points in the regression line or curve. Figure 20.3 is a graph of the connected points of the regression functions defined by the conditional expectations of the conditional distributions of X on Y and of Y on X. In specifying the regression of Y on X, the arguments in the regression function are the values in the value

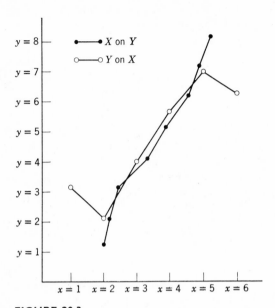

FIGURE 20.3

Regression of Y on X, and X on Y, for data of Table 20.5.

set of X. Conversely, the arguments in the regression function of X on Y are the values of Y. Care must be taken to differentiate the direction of the implication—regression of Y on X and regression of X on Y are different, as indicated in Figure 20.3.

Of all of the regression functions we might devise, the function defined by the conditional expectations is optimal in a particular and useful sense. The variability about the regression function is a minimum, as measured by the expected deviation of the values of the variable Y, with respect to the conditional probabilities, for each and every conditional distribution defined by values of X. For a given conditional distribution we can depend on an old result,

$$E(X - a)^2 > E(X - \mu_X)^2 \qquad \text{for any} \quad a \neq \mu_X$$

to obtain the result that the variation about the conditional distributions of Y given X is a minimum for all regression functions:

$$\sum_x P(X = x)E[Y - h(x)]^2 < \sum_x P(X = x)E[Y - g(x)]^2$$

$$g(x) \neq h(x) = \mu_{Y|x}$$

$$\sigma^2_{y|h(X)} < \sigma^2_{y|g(X)} \tag{20.9}$$

In other words, the regression defined by the conditional expectations is a least squares regression, with respect to the conditional distributions.

It seems quite natural at this point to ask, "So what?" Thus far we have developed simple descriptive properties of the joint distribution of two random variables defined on the same sample space. If we have complete knowledge of the random variables, the only benefit gained from applying the rules and manipulations we have developed is that of convenience in describing the joint distribution. On the other hand, when we come to the problem of sampling the jointly distributed random variables, several problems arise. Imagine that we know the entire joint distribution of the random variables (either from some theoretical consideration or from very large sampling experiments in which both variables were observed jointly) but that for practical purposes, observation of Y was not possible—how best can we "estimate" the value of Y when we have a value of X for a result observed? The development above indicates that the expected value of Y, conditional on the observed value of X, is the "best" estimate in the sense that it has the smallest expected deviation from the natural value of Y (which is unobserved). Other basic reasons for being interested in the joint distribution of random variables become apparent later in the chapter.

The use of the conditional expectations to develop a regression function is a case of an algorithmic development. The function is not defined in terms of some equation or analytic expression but rather in terms of a step-by-step procedure. On the other hand, there are certain analytic functions that can be defined that have the character of regression functions. We develop two such special, analytic functions here—a linear function and a polynomial function.

A linear function of a variable X is a set of values of Y defined by

$$\hat{Y} = \xi + \theta X$$

which has a graph that is a straight line, in the (X, Y) Cartesian coordinates. The linear function is specified by two parameters, ξ and θ. The two parameters are interpretable in terms of the geometry of the line in the coordinate system. The parameter ξ is the value of $\hat{Y}$ when $X = 0$, and consequently is called the intercept of the linear function (the value of $\hat{Y}$ at which the line intercepts the Y axis). The parameter θ is interpretable as the slope of the line. The values of $\hat{Y}$ at a pair of points on X one unit apart have a certain difference. The difference of the $\hat{Y}$ values indicates degree of change in $\hat{Y}$ for a unit difference on X—the rate of change per unit of X. The difference of $\hat{Y}$ values expressed in the unit of X, the rate of change, is the slope of the line, indicated by the parameter θ. Clearly the model of a linear function is too broad to allow its use as a regression function without additional restrictions. A general purpose of defining a regression function is that of determining a best value of Y in the case that X alone is known. Consequently, we want to select the pair of parameter values in the equation that come closest in describing the location of all of the conditional distributions of Y given X. The criterion is the same as that used to evaluate the regression function defined by the conditional expectations,

$$\sigma^2_{Y|g(x)} = \sum_x P(X = x)E[Y - g(x)]^2 \tag{20.10}$$

where $g(x)$ is the regression function value at the value of the X variable. For each possible pair of values of the two parameters ξ and θ there is a value of this evaluative criterion. The values of ξ and θ that simultaneously give the smallest value of

$$\sum_x P(X = x)E[Y - (\xi + \theta x)]^2 \tag{20.11}$$

are called the least squares values of the parameters. The equation

$$\hat{Y} = \xi + \theta X$$

is called the least squares linear regression equation when ξ and θ are the least squares values of the parameters. From now on in this book the coefficients ξ and θ in the linear regression equation refer to the least squares values. By simple methods of the differential calculus, it can be shown that the least squares values are given by

$$\theta = \frac{E(XY) - E(X)E(Y)}{E(X^2) - E(X)^2} \tag{20.12}$$

$$\xi = E(Y) - \theta E(X) \tag{20.13}$$

A special case of the regression through the conditional expectations of Y given X is that case where

$$\mu_{Y|X} = \xi + \theta X$$

The conditional expectations all fall on the line defined by

$$\hat{Y} = \xi + \theta X$$

and the linear regression is also the optimal model of the regression because

$$\sum_x P(X = x)E(Y - \mu_{Y|x})^2 = \sum_x P(X = x)E[Y - (\xi + \theta x)]^2$$

Since the conditional expectation regression is the best (least squares criterion) of all possible regression functions, it stands as a criterion regression form. If the linear regression model gives

$$\sum_x P(X = x)E(Y - \mu_{Y|x})^2 < \sum_x P(X = x)E[Y - (\xi + \theta x)]^2$$

then the linear model is inadequate, to some degree, perhaps measured by the magnitude of the inequality. When the linear regression model is not adequate, the regression is said to be nonlinear, a rather nondescriptive characterization. Many possible regression functions are "nonlinear," for example, $\hat{Y} = \log X$, $\hat{Y} = e^{-X}$, $\hat{Y} = \xi + \theta_1 X + \theta_2 X^2 + \theta_3 X^3 + \cdots$. The last example just given is a widely used nonlinear regression model called the polynomial regression. The coefficients $\xi, \theta_1, \theta_2, \ldots$ in the polynomial can be determined by a least squares procedure. A polynomial

$$\hat{Y} = \xi + \theta_1 X + \theta_2 X^2 + \cdots + \theta_n X^n \qquad (20.14)$$

is called a polynomial of the nth degree. The adaptability of the polynomial to specific forms of nonlinearity is illustrated by the following. If there are n values in the value set of X, then the least squares solution to the polynomial with degree $(n - 1)$ will pass through each of the n conditional expectations $\mu_{Y|X}$. That is, where $f(x)$ is the least squares polynomial of degree one less than the values in the value set of X,

$$\sum_x P(X = x)E(Y - \mu_{Y|x})^2 = \sum_x P(X = x)E[Y - f(x)]^2 \qquad (20.15)$$

Thus the polynomial is always a "good" model, if its degree can be arbitrarily selected. However, this statement should not be generalized to the substantive meaning of the relationship between X and Y. It is possible that nonpolynomial equations describe more meaningfully the "process" or "mechanism" by which Y is a function of X.

Some Properties of Jointly Distributed Random Variables.

We use the notion of regression to study the properties of the joint distributions of random variables. Out of this come several useful descriptive and practical (applicable) concepts. First, we inspect the deviations about a regression line or curve. Second, we develop several standard measures of degree of relationship of two random variables. It is the parameters that are developed in the second part that are most frequently encountered in tests of hypotheses about correlations of variables.

The method of approach to studying the properties of joint distributions is to examine the variances and other variance-like quantities connected with regression. First, we find the conditional distributions in (X, Y), that is, $(Y | X)$. These distributions have expectations $\mu_{Y|X}$ defined in equation (20.5) and a regression value $\hat{Y} = g(X)$. If $g(X)$ is a completely adequate regression function, $\hat{Y} = \mu_{Y|X}$. However, if $g(X)$ is not a perfect model of the regression through the conditional expectations, there is a regression bias term $\lambda_X = \hat{Y}_X - \mu_{Y|X}$. The expectation and variance of each of the conditional distributions free of this bias term are defined in equations (20.5) and (20.6). However, if $\lambda \neq 0$ for the conditional distributions, then

$$\sigma^2_{Y|g(x)} = \sum_Y P(Y \mid x)(Y - \lambda_x - \mu_{Y|x})^2 \qquad \text{for each } x \text{ in } X$$

$$= \sigma^2_{Y|X} + \sigma_\lambda^{\;2}.$$

(20.16)

The two parameters $\sigma^2_{Y|X}$ and σ_λ^2 are indicative of important properties of the joint distribution. If $\sigma_\lambda^2 = 0$, the implication is that the regression function is the optimal function, passing through the conditional expectation. Otherwise $g(x)$ is to some extent, indicated by the magnitude of σ_λ^2, an inappropriate model of the regression of Y on X. The parameter $\sigma^2_{Y|X}$ is an indication of the degree to which the variance of Y is "due" to or associated with the variable X. If $\sigma^2_{Y|X} = \sigma_Y^2$, clearly the two variables are independent because this can occur only when the distribution of $(Y \mid X)$ is the same for all x's and is the same as the distribution of Y in a univariate sense. When $\sigma^2_{Y|X} \leq \sigma_Y^2$, the difference of the inequality is said to be due to the regression of Y on X. Figure 20.4 illustrates some of these relationships.

This basic idea, that part of σ_Y^2 is due to σ_X^2, is expressed in the partition of variance

$$\sigma_Y^{\;2} = \sigma_{\hat Y}^{\;2} + \sigma^2_{Y|g(X)}$$

(20.17)

This is a statement simply that the variance of Y is partitioned into a part associated with $\hat Y$ values (predicted) and values varying about $\hat Y$. As a consequence of this relationship there is a trade-off between $\sigma_{\hat Y}$ and $\sigma^2_{Y|g(X)}$ as the values $\hat Y$ become more variable. Hence, as $\sigma_{\hat Y}^2$ increases, indicating increasing differentiation of the predicted values of Y, $\sigma^2_{Y|g(X)}$ is reduced. If each of the values of $\hat Y$ are the same, they are also equal to the expectation of the dependent variable, indicating that the variables are independent and that the regression of Y on X is a constant, and hence indicative of no relationship.

A measure of the degree of dependency of Y on X is suggested by the inequality $\sigma^2_{Y|X} \leq \sigma_Y^2$. Clearly, if each of the conditional distributions are degenerate with all of their density located at $\mu_{Y|x}$, then $\sigma^2_{Y|X} = 0$, and when the regression of Y on X is known, the value of Y for any x is also known. In the latter case, Y is a determinate (in contrast with a probabilistic) function of X, and in that sense completely dependent on X. A useful way of summarizing the degree of dependency is given by the ratio

$$\eta^2_{Y|X} = \frac{\sigma_Y^{\;2} - \sigma^2_{Y|X}}{\sigma_Y^{\;2}}$$

(20.18)

The behavior of this ratio is regular and conforms to an intuitive notion of degree of dependency. As $\sigma^2_{Y|X} \rightarrow \sigma_Y^2$ then $\eta^2_{Y|X} \rightarrow 0$, and as $\sigma^2_{Y|X} \rightarrow 0$ then $\eta^2_{Y|X} \rightarrow 1.0$. It can be shown (see below), for the special case of linear regression, that the

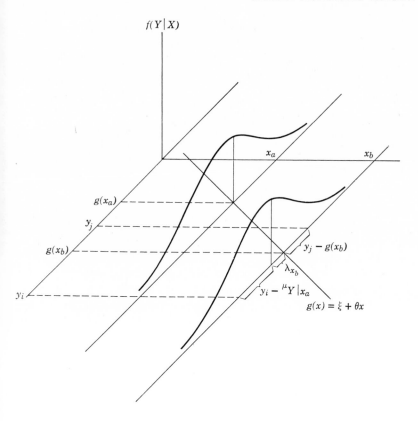

$f(Y|X)$

$g(x_a)$

y_j

$g(x_b)$

x_a

x_b

$y_j - g(x_b)$

λx_b

$y_i - {}^{\mu}Y|x_a$

$g(x) = \xi + \theta x$

y_i

FIGURE 20.4

An example of a regression $g(x) = \xi + \theta x = y$ where $g(x) \neq \mu_{y|x}$, showing two conditional distributions.

ratio $\eta^2_{Y|X}$ is the percent of variance of Y due to variation in X. For the general regression case this is not as easily demonstrable but seems intuitively obvious once pointed out. The parameter $\eta^2_{Y|X}$ is known as the correlation ratio. It is known in much of the psychological literature as the coefficient of curvilinear correlation. This terminology is imprecise. The ratio is applicable to all regression equations if they are such that $\sigma_\lambda^2 = 0$. If $\sigma_\lambda^2 \neq 0$, $\eta^2_{Y|X}$ is an ambiguous measure of dependency, the degree of ambiguity depending on σ_λ^2.

In addition to the regression of Y on X defined by some function $[g(X) = Y]$, it is possible to define the regression of X on Y to be some function $[h(Y) = X]$. In general, the function is different, and the ratio

$$\eta^2_{X|Y} = \frac{\sigma_X^2 - \sigma^2_{X|Y}}{\sigma_X^2}$$

will not be the same as the ratio $\eta^2_{Y|X}$. We do not extend this discussion, because when regression is of primary concern the problem usually involved is the prediction or "correction" of one variable from another. Hence the variable Y is a dependent variable and it is the value of $\eta^2_{Y|X}$ or $\sigma^2_{Y|X}$ that is of interest, or the regression estimate values $\hat{Y}$, or the residual terms $(Y - \hat{Y})$ that are of interest.

These characteristics of (X, Y) can be specialized with some useful results for a linear regression function. Where $\hat{Y} = \xi_{Y|X} + \theta_{Y|X}X$ is a model of the regression of Y on X and $\hat{X} = \xi_{X|Y} + \theta_{X|Y}Y$ is a model of the regression of X on Y, a useful measure of relationship between X and Y is the covariance of X and Y defined by

$$\sigma_{XY} = E(XY) - E(X)E(Y) \qquad (20.19)$$

which is equal and symmetric to $\sigma_{YX} = E(YX) - E(Y)E(X)$.

One problem in dealing with σ_{XY} is that it is a function of σ_X^2 and σ_Y^2 so that the value of σ_{XY} is dependent on the absolute scaling of the value sets of X and Y. It would be more desirable to have a measure of relationship that had a uniform value set, as for example η^2 where $0 \le \eta^2 \le 1.0$. For an illustration of the problem, take an arbitrary transformation of random variables without changing the conditional distributions relative to the original joint distribution. Let $R = 3W$ so that $\sigma_R = 3\sigma_W$ and $E(R) = 3E(W)$. Under these conditions

$$\begin{aligned}
\sigma_{RZ} &= E(RZ) - E(R)E(Z) \\
&= 3E(WZ) - 3E(W)E(Z) \\
&= 3\sigma_{WZ}^2
\end{aligned}$$

Thus, by changing the scale factor on W to give R, the covariance is changed by the same scaling factor. Although (R, Z) and (W, Z) are basically the same joint distributions, the covariances are quite different. For purposes of comparing degrees of relationship without consideration of variances, the standardized random variables may be used. Standardizing X and Y by

$$X^* = \frac{X - E(X)}{\sigma_X}$$

$$Y^* = \frac{Y - E(Y)}{\sigma_Y}$$

gives $\sigma_{Y^*}^2 = \sigma_{X^*}^2 = 1$. The covariance of any arbitrary pair of jointly distributed random variables is scaled in the same way as any other arbitrary pair, when the variables are standardized first. The values of the covariance of standardized variables are limited to -1.0 up to 1.0. The covariance of standardized random variables is given by

$$\sigma_{X^*Y^*} = E(X^*Y^*) - E(X^*)E(Y^*)$$

$$= \frac{\sigma_{XY}^2}{\sigma_X \sigma_Y} \tag{20.20}$$

For the sake of notational conventionality,

$$\rho_{XY} = \sigma_{X^*Y^*} \tag{20.21}$$

This parameter, the covariance of standardized random variables, is known as the Pearson product-moment correlation coefficient (do not confuse the use of ρ_{XY} with the use of ρ in the analysis of variance). If X and Y are related in a linear fashion, we say that X and Y are correlated if $\rho_{XY} \neq 0$. If $\rho_{XY} = 0$, X and Y are said to be uncorrelated. Note that if $\sigma_X = 0$ or $\sigma_Y = 0$, ρ is not defined. The limits of the correlation coefficient are

$$-1.0 \leq \rho_{XY} \leq 1.0$$

The correlation coefficient is related to the regression coefficient θ in linear models of regression. The regression coefficients as defined in (20.12) can be reexpressed in terms of covariance and variance:

$$\theta_{Y|X} = \frac{\sigma_{XY}}{\sigma_X^2}$$

By substitution, this gives

$$\theta_{Y|X} = \rho_{XY} \frac{\sigma_Y}{\sigma_X} \tag{20.22}$$

The regression of X and Y can be shown to be

$$\theta_{X|Y} = \rho_{XY} \frac{\sigma_X}{\sigma_Y}$$

The relationship between the regression coefficient and the correlation coefficient is illuminated when we are concerned with variables with unit variances, that is, $\sigma_X^2 = \sigma_Y^2 = 1.0$:

$$\theta_{Y|X} = \rho_{XY} \frac{\sigma_Y}{\sigma_X} = \rho_{XY}$$

$$\theta_{X|Y} = \rho_{XY} \frac{\sigma_X}{\sigma_Y} = \rho_{XY}$$

This implies that

$$\pm \sqrt{\rho_{XY}^2} = \pm \sqrt{\theta_{X|Y} \theta_{Y|X}} \tag{20.23}$$

That is, the correlation coefficient is the square root of the product of the regression coefficients, the geometric mean of the coefficients. The sign is the sign of ρ.

The deviation from regression, that is, $(Y - \hat{Y})$, can be studied in terms of variance and correlation. The values of ξ and $\theta_{Y|X}$ in

$$\hat{Y} = \xi + \theta_{Y|X} X$$

are the least square values in the sense that

$$E(Y - \hat{Y})^2$$

is minimized by setting

$$\theta_{Y|X} = \rho_{XY} \frac{\sigma_Y}{\sigma_X} \quad \text{and} \quad \xi = E(Y) - \rho_{XY} \frac{\sigma_Y}{\sigma_X} E(X)$$

The expectation of the squared "error" or deviation about the regression line has the value

$$E\left[Y - E(Y) + \rho_{XY} \frac{\sigma_Y}{\sigma_X} E(X) - \rho_{XY} \frac{\sigma_Y}{\sigma_X} X \right]^2$$

$$= E\left\{ [Y - E(Y)] - \rho_{XY} \frac{\sigma_Y}{\sigma_X} [X - E(X)] \right\}^2$$

$$= E\left\{ [Y - E(Y)]^2 - 2\rho_{XY} \frac{\sigma_Y}{\sigma_X} [Y - E(Y)][X - E(X)] + \rho_{XY}^2 \frac{\sigma_Y^2}{\sigma_X^2} [X - E(X)] \right\}^2$$

$$= \sigma_Y^2 - 2\rho_{XY} \frac{\sigma_Y}{\sigma_X} \sigma_{XY} + \rho_{XY}^2 \frac{\sigma_Y^2}{\sigma_X^2} \sigma_X^2$$

$$= \sigma_Y^2 - \frac{2\sigma_{XY}}{\sigma_X^2} + \frac{\sigma_{XY}}{\sigma_X^2} = \frac{\sigma_X^2 \sigma_Y^2}{\sigma_X^2} - \frac{\sigma_{XY}}{\sigma_X^2}$$

$$= \sigma_Y^2 - \sigma_Y^2 \rho_{XY}^2 = \sigma_Y^2 (1 - \rho_{XY}^2)$$

Recall that $E(Y^2 - \hat{Y})^2 = \sigma_{Y|X}^2$, the variability about the regression of Y on X, where the model is correct. The last derived expression gives

$$\sigma_{Y|X}^2 = \sigma_Y^2 (1 - \rho_{XY}^2) \tag{20.24}$$

implying that the square of the correlation coefficient is the percentage of the variance of Y, dependent on variation in X. By simple algebra

$$\sigma_Y^2 = \sigma_{Y|X}^2 + \sigma_Y^2 \rho_{XY}^2$$

$$= \sigma_Y^2 (1 - \rho_{XY}^2) + \sigma_Y^2 \rho_{XY}^2 \tag{20.25}$$

which states that σ_Y^2 is partitioned into a part independent of X, that is, $\sigma_{Y|X}^2$, and a part determined by the relationship between X and Y, that is, $\sigma_Y^2 \rho^2$. The value ρ^2 is sometimes called the coefficient of determination and the value

$(1 - \rho^2)$ the coefficient of nondetermination. The student should substitute the extreme values of ρ, that is, $+1.0$ and -1.0, into these last equations to gain some additional insight into the meaning of the correlation coefficient.

Sampling Jointly Distributed Random Variables

A sample of N observations on the joint distribution of X and Y is the collection of N pairs of values

$$(X_i, Y_i) \; i = 1, \ldots, N$$

Clearly the sample statistics of interest on each of the two variables are the same in this experiment as in single-variable experiments:

$$\bar{X} = \frac{1}{N} \sum_{i=1}^{N} X_i \qquad\qquad \bar{Y} = \frac{1}{N} \sum_{i=1}^{N} Y_i$$

$$S_X^2 = \frac{1}{N} \sum_{i=1}^{N} X_i^2 - \bar{X}^2 \qquad S_Y^2 = \frac{1}{N} \sum_{i=1}^{N} Y_i^2 - \bar{Y}^2$$

$$\hat{S}_X^2 = \frac{N}{N-1} S_X^2 \qquad\qquad \hat{S}_Y^2 = \frac{N}{N-1} S_Y^2$$

Because of its wider use we discuss first the linear regression case and only then some of the statistics on nonlinear regression.

Imagine that $N = 30$ pairs of observations are made. They can be tabulated as those shown in Table 20.6 along with the sums and sums of squares. For an analysis of the data in terms of linear regression and correlation, the sum of products of the X and Y values in pairs is also found. The respective sample means and sample variances are also shown in Table 20.6.

We restrict our attention here to the basic features of the sample. Some of the properties of joint distributions of random variables are not dealt with in this section. Of particular importance in practice are the regression equations, the correlation coefficient, and the variance of the dependent variable about the regression line. Also of interest in practical contexts are the regression values and the differences between the observed values and the regression values.

Fundamental to the description of the sample is a graph of the pairs of values observed. Such a graph of the pairs is called a scatter plot. It corresponds roughly to a graph of the frequencies of the values observed in a univariate sample. Each observation, that is, (X_i, Y_i) is represented in Cartesian (X, Y) coordinates by a dot or point indicating simultaneously both the value X_i and the value Y_i. Thus the observation $(24, 46)$ for observation number 10 is plotted as a point directly

TABLE 20.6 Example Data: Thirty Observations on Jointly Distributed Variables X and Y

Observation (i)	X_i	Y_i
1	25	48
2	23	45
3	26	49
4	27	49
5	25	45
6	24	46
7	23	46
8	21	44
9	23	47
10	24	46
11	24	47
12	23	46
13	25	46
14	22	44
15	27	47
16	25	46
17	24	48
18	21	43
19	22	45
20	24	45
21	24	47
22	25	47
23	23	45
24	27	48
25	24	46
26	26	47
27	22	43
28	23	44
29	24	45
30	24	46

$$\sum_{i=1}^{N} X_i = 720 \qquad \sum_{i=1}^{N} Y_i = 1380$$

$$\sum_{i=1}^{N} X_i^2 = 17356 \qquad \sum_{i=1}^{N} Y_i^2 = 63552$$

$$\sum_{i=1}^{N} X_i Y_i = 33178$$

$$\overline{X} = 24.0 \qquad \overline{Y} = 46.0$$
$$S_X^2 = 2.53 \qquad S_Y^2 = 2.40$$
$$S_X = 1.59 \qquad S_Y = 1.55$$

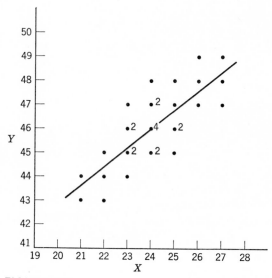

FIGURE 20.5

Scatter plot and line of regression of Y on X for data of Table 20.6.

above 24 and to the right of 46 on the X and Y axes, respectively, in Figure 20.5. All N observations are plotted. Multiple dots at a given (X, Y) point are indicated by a number to the right of the point.

The basic criterion of least squares used in finding the linear regression in the sections above is applicable in the sample. Instead of dealing directly with the sample values, it is convenient at this point to work with the deviations of X_i from $\bar{X}$, that is, $(X_i - \bar{X})$. The equation equivalent to

$$\hat{Y} = \xi_{Y|X} + \theta_{Y|X}X$$

is expressible in terms of $\bar{Y}$ instead of ξ, and $b_{Y|X}$ instead of $\theta_{Y|X}$:

$$\tilde{Y} = \bar{Y} + b_{Y|X}(X_i - \bar{X}) \tag{20.26}$$

Equation (20.26) is solved for $b_{Y|X}$ so that

$$\sum_{i=1}^{N}(\tilde{Y}_i - Y_i)^2$$

is a minimum. It is not difficult to prove that

$$b_{Y|X} = \frac{1/N \sum_{i=1}^{N} X_i Y_i - \bar{X}\bar{Y}}{1/N \sum_{i=1}^{N} X_i^2 - \bar{X}^2}$$

$$= \frac{N \sum_{i=1}^{N} X_i Y_i - (\sum_{i=1}^{N} X_i)(\sum_{i=1}^{N} Y_i)}{N \sum_{i=1}^{N} X_i^2 - (\sum_{i=1}^{N} X_i)^2} \tag{20.27}$$

satisfies the least squares criterion. The statistic $b_{Y|X}$ permits plotting the line on which all of the values Y_i fall. In the example data of Table 20.6 and Figure 20.5,

$$b_{Y|X} = \frac{1740}{2280}$$
$$= .763$$

Consequently, for $X_i = 27$ the value of $\tilde{Y}_i$ is given by

$$\tilde{Y}_i = \bar{Y} + b_{Y|X}(X_i - \bar{X})$$
$$= 46.0 + .763(27 - 24)$$
$$= 48.289$$

If this calculation is repeated for each value of X observed and plotted in Figure 20.5, they fall on the line shown in Figure 20.5. A simple procedure is used to define the regression line. Select any two values of X_i. It is convenient to select $\bar{X}$ and some relatively large value, say 27, in the numerical example. The two corresponding values of $\tilde{Y}$ are $\bar{Y}$ and $\bar{Y} + b_{Y|X}(X_i - \bar{X})$, respectively. These two points are plotted on the scatter plot and a line drawn through them. All of the other regression values fall on this line.

Table 20.7 shows the regression scores $\tilde{Y}$ for the least squares linear regression of Y on X for the data of Table 20.6 and Figure 20.5. In addition, the differences of the regression values and the observed values of Y are tabled:

$$D_i = \tilde{Y}_i - Y_i \tag{20.28}$$

The values of D may be interpreted in two interesting ways. In the first way, D is a residual score, that part of an observation on Y not associated with X. In the second way of interpreting D, the values are seen as errors of prediction. In both interpretations, the variance of D is useful in interpreting the degree of independence in the two variables observed. In the residual interpretation, the variance of D is the variance of Y not due to variability in X. In the error interpretation, the variance of D is the variance of Y due to inaccuracy in predicting Y from X. The source of this error is not ascertainable from the data themselves. If we wish to use X to select subjects having particular Y values in order to control Y, when Y is not observable, the variance of D is a measure of the degree to which our control fails.

The variability of the $\tilde{Y}$ values is an index of the degree to which Y is a function of X, either in a prediction sense or in the sense of common sources of variation. In the prediction interpretation the variance of $\tilde{Y}$ is a measure of the goodness of the linear prediction. If the variance of the regression values is large with respect to the variance of Y, the prediction is good. If the variance of the regression values

TABLE 20.7 Regression Values and Differences of the Regression Values and Observed Values

Individual (i)	$\tilde{Y}_i$	$D_i = \tilde{Y}_i - Y_i$	Rescaled Y_i: Y^*
1	46.76	−1.24	44.76
2	45.24	.24	46.24
3	47.53	−1.47	44.53
4	48.29	−.71	45.29
5	46.76	1.76	47.76
6	46.00	.00	46.00
7	45.24	−.76	45.24
8	43.71	−.29	45.71
9	45.24	−1.76	44.24
10	46.00	.00	46.00
11	46.00	−1.00	45.00
12	45.24	−.76	45.24
13	46.76	.76	47.76
14	44.47	.47	46.47
15	48.29	1.29	47.29
16	46.76	.76	46.76
17	46.00	−2.00	44.00
18	43.71	.71	46.71
19	44.47	−.53	45.47
20	46.00	1.00	47.00
21	46.00	−1.00	45.00
22	46.76	−.24	45.76
23	45.24	.24	47.24
24	48.29	.29	46.29
25	46.00	.00	46.00
26	47.53	.53	46.53
27	44.47	1.47	47.47
28	45.24	1.24	47.24
29	46.00	1.00	47.00
30	46.00	.00	46.00

$$\sum_{i=1}^{N} \tilde{Y}_i = 1380 \qquad \sum_{i=1}^{N} D_i = .000 \qquad \sum_{i=1}^{N} Y_i^* = 1380$$

$$\sum_{i=1}^{N} \tilde{Y}_i^2 = 63524.28 \qquad \sum_{i=1}^{N} D_i^2 = 27.699 \qquad \sum_{i=1}^{N} Y_i^{*2} = 63507.70$$

$$\bar{Y} = 46.0 \qquad \bar{D} = .0 \qquad \bar{Y}^* = 46.00$$

$$S_{\tilde{Y}}^2 = 1.476 \qquad S_D^2 = .923 \qquad S_{Y^*}^2 = .923$$

is small, with respect to the variance of Y, the prediction is relatively poor. In terms of the interpretation of the variance of the regression values, the variables are said to have more in common with regard to variability when the variance of the regression values is large with respect to the variance of Y.

The interpretations both of D and of $\tilde{Y}$ in terms of their variability derive from the relationship of these variances to the variance of Y:

$$S_Y^2 = S_D^2 + S_{\tilde{Y}}^2 \tag{20.29}$$

The relationship of variances in (20.29) is the sample equivalent to the relationship of the random variables expressed in equation (20.17). Using the variances in (20.29) in a way parallel to the developments outlined in the previous section, equations (20.24) and (20.25), the correlation coefficient of the sample can be defined by

$$r_{XY}^2 = \frac{S_Y^2 - S_D^2}{S_Y^2} \tag{20.30}$$

Reversing the logic, it can be shown that

$$S_Y^2 = S_Y^2 r_{XY}^2 + S_Y^2 (1 - r_{XY}^2) \tag{20.31}$$

The data of Table 20.7 provide illustrations of these results. From Table 20.6 we find $S_Y^2 = 2.40$. Comparing this with data of Table 20.7 by way of equation (20.29) gives

$$S_Y^2 = S_D^2 + S_{\tilde{Y}}^2$$
$$2.40 = .923 + 1.476$$

As a consequence,

$$r_{XY}^2 = \frac{2.40 - .923}{2.40}$$
$$= .615$$
$$r_{XY} = .784$$

The variation in Y ranges from 43 to 49, a span of 6 points. When the variation of Y is restricted by gauging it relative to $\tilde{Y}$, that is, the variation of D, the values range from -2.0 to 1.76, a span of 3.76 points. The restriction reduces the range by less than a factor of 2.

Several other relationships of statistics in the sample of observations parallel the structure of the random variables observed. Take the regression of X on Y:

$$\tilde{X} = \bar{X} + b_{X|Y}(Y_i - \bar{Y})$$

The values of $\tilde{X}$ are linear regression least squares values when

$$b_{X|Y} = \frac{1/N \sum_{i=1}^{N} X_i Y_i - \bar{X}\bar{Y}}{1/N \sum_{i=1}^{N} Y_i^2 - \bar{Y}^2} \tag{20.32}$$

The correlation coefficient can be defined by

$$\pm\sqrt{r_{XY}^2} = \pm\sqrt{b_{Y|X}b_{X|Y}} \tag{20.33}$$

Evaluating these statistics for the data of Table 20.6 gives

$$b_{X|Y} = \frac{1740}{2160}$$

$$= .806$$

and

$$r_{XY} = \sqrt{(.763)(.806)}$$

$$= \sqrt{.615}$$

$$= .784$$

The correlation coefficient in a sample of observations is also the covariance of the standardized scores. Let

$$X_i^* = \frac{X_i - \bar{X}}{S_X} \qquad Y_i^* = \frac{Y_i - \bar{Y}}{S_Y}$$

Then

$$r_{XY} = \frac{1}{N}\sum_{i=1}^{N} X_i^* Y_i^* \tag{20.34}$$

This result can be derived easily from (20.27), (20.32), and (20.33).

The most appropriate calculational form for the sample correlation coefficient is

$$r_{XY} = \frac{N\sum_{i=1}^{N} X_i Y_i - (\sum_{i=1}^{N} X_i)(\sum_{i=1}^{N} Y_i)}{\sqrt{[N\sum_{i=1}^{N} X_i^2 - (\sum_{i=1}^{N} X_i)^2][N\sum_{i=1}^{N} Y_i^2 - (\sum_{i=1}^{N} Y_i)^2]}} \tag{20.35}$$

Care should be taken in using this equation on a digital computer when N is very large or when the magnitudes of the X and Y values are large. When either or both of these conditions are present, calculation should be based on the values $(X_i - \bar{X})$ and $(Y_i - \bar{Y})$. Then

$$r_{XY} = \frac{\sum_{i=1}^{N} (X_i - \bar{X})(Y_i - \bar{Y})}{\sqrt{\sum_{i=1}^{N} (X_i - \bar{X})^2 \sum_{i=1}^{N} (Y_i - \bar{Y})^2}}$$

Application of formula (20.35) in digital computers may result in error from inadequately represented sums of very large magnitudes because of the limited significant arithmetic in digital computers.

An analysis of the residual values $D_i = (\tilde{Y}_i - Y_i)$ is instructive. If the linear regression model is appropriate *for the sample*, the values of D should be independently distributed with respect to Y and also with respect to X. Plotting D as a

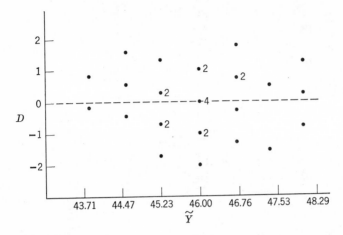

FIGURE 20.6

Scatter plot of D against $\tilde{Y}$.

function of X and then as a function of Y from Tables 20.6 and 20.7 gives Figures 20.6 and 20.7. The D values should be centered about the horizontal line at $D = 0$ without any systematic variability of the conditional mean $\bar{D}$ as a function of $\tilde{Y}$. A departure from a mean of zero or a linear, nonconstant, regression of D on Y indicates an error in analysis. If the mean residuals were some nonlinear function of $\tilde{Y}$, the linear model is inadequate. Corresponding comments apply to the scatter

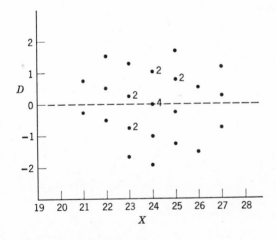

FIGURE 20.7

Scatter plot of D against X.

plot of D as a function of X. Departures from these regression conditions in the sample do not necessarily imply corresponding departures in the random variables being observed. Means of detecting "significant" departures are discussed in sections to follow.

Table 20.7 includes the Y scores rescaled to represent Y free of variability related to X. That is, the location of the sample is held constant but the variability of the scores is reduced to be equal to $S_D{}^2 = S_Y{}^2 - S_{\bar{Y}}{}^2$. These scores are of interest in part because they permit testing hypotheses about μ_Y with the estimate of $\sigma_{\bar{Y}}{}^2$ free of variability associated with X. Imagine a single-sample problem dealing with hypotheses about expectations. Let Y be the dependent variable specified in the hypothesis. If Y is very heavily influenced by the variable X, to the extent, for example, that makes σ_Y a great deal larger than $|\mu_T - \mu_A|$ in the test structure, we would like to control that part of the variability resulting from subject differences on X. This can be done by using the scores

$$Y_i{}^* = \bar{Y} + (\tilde{Y}_i - Y_i)$$

This transformation gives scores such that

$$\bar{Y}^* = \bar{Y} \qquad S_{Y^*} = S_D$$

As a consequence, under a Case C one-sample problem the statistics

$$Z = \frac{\bar{Y}^* - \mu_T}{S_{Y^*}/\sqrt{N}} = \frac{\bar{Y} - \mu_T}{S_D/\sqrt{N}} \tag{20.36}$$

$$t = \frac{\bar{Y}^* - \mu_T}{S_{Y^*}/\sqrt{N-1}} = \frac{\bar{Y} - \mu_T}{S_D/\sqrt{N-1}} \tag{20.37}$$

are distributed normally and as Student's t $(df = N - 1)$, respectively. The value $S_{Y^*} = S_D$ may be calculated without finding the individual values $Y_i{}^*$ or the values D_i. Since $S_D = S_Y\sqrt{1 - r^2}$, the equivalent expressions are

$$Z = \frac{\bar{Y} - \mu_T}{S_Y\sqrt{1 - r^2}/\sqrt{N}} \tag{20.38}$$

$$t = \frac{\bar{Y} - \mu_T}{S_Y\sqrt{1 - r^2}/\sqrt{N-1}} \tag{20.39}$$

The sensitivity of a test of a hypothesis about μ_Y is enhanced proportionally to $1/\sqrt{1 - r_{XY}{}^2}$ when information about X is used in the manner described here. For example, if $\sqrt{1 - r_{XY}{}^2} = .5$, the sensitivity of a test to differences in μ_T and μ_A doubles.

TABLE 20.8 Example Data Set, Nonlinear

Observation (i)	X_i	Y_i
1	10	50
2	30	40
3	30	48
4	20	54
5	10	42
6	15	50
7	15	46
8	10	48
9	10	48
10	25	52
11	20	54
12	25	54
13	20	52
14	15	48
15	10	46
16	15	50
17	20	52
18	25	46
19	30	42
20	25	50
21	30	46
22	20	54
23	10	46
24	15	54
25	20	50
26	25	52
27	25	50
28	30	44
29	20	56
30	20	52

$$\sum_{i=1}^{N} X_i = 595 \qquad \sum_{i=1}^{N} Y_i = 1476$$

$$\sum_{i=1}^{N} X_i^2 = 13175 \qquad \sum_{i=1}^{N} Y_i^2 = 73096$$

$$\sum_{i=1}^{N} X_i Y_i = 29200$$

$$\bar{X} = 19.83 \qquad \bar{Y} = 49.20$$
$$S_X^2 = 45.81 \qquad S_Y^2 = 15.89$$
$$S_X = 6.768 \qquad S_Y = 3.986$$

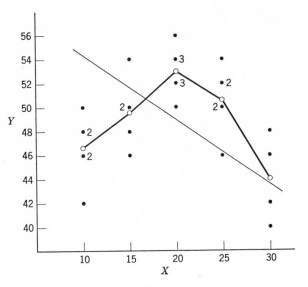

FIGURE 20.8

Scatter plot for data of Table 20.8, showing linear regression line and regression on means.

Turning to the problem of nonlinear regression, imagine that the data of Table 20.8 were observed in an experiment. The scatter plot of X and Y, Figure 20.8, reveals a nonlinearity. Although it might be possible to postulate some specific regression function, we limit our discussion here to the regression function defined by conditional means of the Y values for each value of X. One reason for this is the limitation of this book to nonmatrix methods and restriction on time and space. The reader is referred to more advanced texts listed at the end of this chapter. A second reason is that the regression function defined by the conditional means is optimal in the sense that the variability about the regression function is equal to or smaller than any other function.

The analysis of the data, when they are reorganized as shown in Table 20.9, is similar to the analysis of variance. The variability about the observed regression of Y on X is simply the variation about the means in the X-value categories. The variation of the regression values is the variation of the means of Y across the X-value categories. From the analysis of variance we have

$$SS_{total} = SS_{between} + SS_{within}$$

This partition parallels, in data, the parametric relationship

$$\sigma_Y{}^2 = \sigma_{\tilde{Y}}{}^2 + \sigma_{Y|X}^2$$

Thus, all of the procedures of a one-way analysis of variance can be used in determining the regression properties of the sample of observations on (X, Y). The correlation ratio, defined in equation (20.18), is estimated by

$$E^2_{Y|X} = \frac{SS_{\text{between}}}{SS_{\text{total}}} \tag{20.40}$$

where SS_{between} and SS_{total} are calculated by the procedures of a one-way analysis of variance.

For the data of Table 20.9 the sums of squares are

$$SS_{\text{total}} = 73096 - \frac{(1476)^2}{30} = 476.8$$

$$SS_{\text{between}} = 72922.2 - \frac{(1476)^2}{30} = 303$$

Consequently

$$E^2_{Y|X} = \frac{303}{476.8} = .635$$

In discussing conditional variances, we introduced the notion that the variance of Y about any regression function value could be partitioned into two parts,

TABLE 20.9 Reorganized Data of Table 20.8

	X					
	10	15	20	25	30	
	42	46	50	46	40	
	46	48	52	50	42	
	46	50	52	50	44	
	48	50	52	52	46	
	48	54	54	52	48	
	50		54	54		
			54			
			56			
N	6	5	8	6	5	Totals
ΣY	280	248	424	304	220	1476
ΣY^2	13104	12336	22496	15440	9720	73096
$\bar{Y}_x$	46.67	49.60	53.00	50.67	44.00	49.20
$\dfrac{(\Sigma Y)^2}{N}$	13066.7	12300.8	22472.0	15402.7	9680.0	72922.2

$\sigma^2_{Y|X}$ and $\sigma_\lambda{}^2$ (see equation [20.16]). The variance $\sigma_\lambda{}^2$ indicates the degree to which the regression function $g(X)$ deviates from the regression function defined by the conditional expectations. If the regression function is

$$g(X) = \xi + \theta X$$

as defined in (20.8), then $\sigma_\lambda{}^2$ represents the degree to which the regression departs from the linear model. Applying the same principles to sample data leads to a partition of the variance of Y in the sample:

$$SS_{\text{total}} = SS_{\text{residual}} + SS_{\text{nonlinear}} + SS_{\text{linear}} \qquad (20.41)$$

The SS_{linear} is the sum of squares associated with the differences among the linear regression values. The $SS_{\text{nonlinear}}$ is the sum of squares of the differences between the conditional means and the linear regression values. The SS_{residual} is the sum of squares of the observed values about the observed means—the quantity familiar to the reader from the analysis of variance. It can be shown that under conditions of the least square solution in (20.26), the SS_{between} of the one-way analysis of variance can be partitioned so that

$$SS_{\text{between}} = SS_{\text{linear}} + SS_{\text{nonlinear}} \qquad (20.42)$$

Hence, in order to evaluate the relative departure of the regression of means from a linear function, only the SS_{linear} and SS_{between} need be calculated.

It is convenient to calculate SS_{between}, SS_{total}, and SS_{residual} by the analysis of variance methods. Also it is convenient to calculate

$$SS_{\text{linear}} = \frac{N[\sum_{i=1}^{N} X_i Y_i - (\sum_{i=1}^{N} X_i)(\sum_{i=1}^{N} Y_i)/N]^2}{N(\sum_{i=1}^{N} X_i{}^2) - (\sum_{i=1}^{N} X_i)^2} \qquad (20.43)$$

The partition of SS_{between} leads to

$$E^2_{Y|X} = \frac{SS_{\text{between}}}{SS_{\text{total}}} = \frac{SS_{\text{linear}} + SS_{\text{nonlinear}}}{SS_{\text{total}}}$$

$$= \frac{SS_{\text{linear}}}{SS_{\text{total}}} + \frac{SS_{\text{nonlinear}}}{SS_{\text{total}}} \qquad (20.44)$$

That is, the correlation ratio is partitioned into two components, one indicating the degree that Y is linearly dependent on X and the other indicating the degree to which the linear model is inadequate.

In the data of Tables 20.8 and 20.9 the SS_{linear} value is

$$SS_{\text{linear}} = \frac{30[29200 - (595)(1476)/30]^2}{30(13175) - (595)^2}$$

$$= 3.98$$

TABLE 20.10 Regression and Difference Values for Least Squares Linear Regression and Mean Regression Functions of X

			Linear		Mean Regression	
Observation (i)	X_i	Y_i	$\tilde{Y}_i$	D_i	$\tilde{Y}_i$	D_i
1	10	50	49.7	−.3	46.7	−3.3
2	30	40	48.7	8.7	44.0	4.0
3	30	48	48.7	.7	44.0	−4.0
4	20	54	49.2	−4.8	53.0	−1.0
5	10	42	49.7	7.7	46.7	4.7
6	15	50	49.5	−.5	49.6	−.4
7	15	46	49.5	3.5	49.6	3.6
8	10	48	49.7	1.7	46.7	−1.3
9	10	48	49.7	1.7	46.7	−1.3
10	25	52	48.9	−3.1	50.7	−1.3
11	20	54	49.2	−4.8	53.0	−1.0
12	25	54	48.9	−5.1	50.7	−3.3
13	20	52	49.2	−2.8	53.0	1.0
14	15	48	49.5	1.5	49.6	1.6
15	10	46	49.7	3.7	46.7	.7
16	15	50	49.5	−.5	49.6	−.4
17	20	52	49.2	−2.8	53.0	1.0
18	25	46	48.9	2.9	50.7	4.7
19	30	42	48.7	6.7	44.0	2.0
20	25	50	48.9	−1.1	50.7	.7
21	30	46	48.7	2.7	44.0	−2.0
22	20	54	49.2	−4.8	53.0	−1.0
23	10	46	49.7	3.7	46.7	.7
24	15	54	49.5	−4.5	49.6	−4.4
25	20	50	49.2	−.8	53.0	3.0
26	25	52	48.9	−3.1	50.7	−1.3
27	25	50	48.9	−1.1	50.7	.7
28	30	44	48.7	4.7	44.0	.0
29	20	56	49.2	6.8	53.0	−3.0
30	20	52	49.2	−2.8	53.0	1.0

Linear		Mean Regression	
$\sum_{i=1}^{N} \tilde{Y}_i = 1476.2$	$\sum_{i=1}^{N} D_i = .2$	$\sum_{i=1}^{N} \tilde{Y}_i = 1476.4$	$\sum_{i=1}^{N} D_i = .4$
$\sum_{i=1}^{N} \tilde{Y}_i^2 = 72642.62$	$\sum_{i=1}^{N} D_i^2 = 473.82$	$\sum_{i=1}^{N} \tilde{Y}_i^2 = 72961.1$	$\sum_{i=1}^{N} D_i^2 = 173.9$
$S_{\tilde{Y}}^2 = .78$	$S_D^2 = 15.8$	$S_{\tilde{Y}}^2 = 10.4$	$S_D^2 = 5.78$
$S_{\tilde{Y}} = .88$	$S_D = 3.92$	$S_{\tilde{Y}} = 3.23$	$S_D = 2.4$

Consequently

$$E^2_{Y|X} = \frac{303 - 3.98}{476.8} + \frac{3.98}{476.8}$$

$$= .627 + .008$$

It is instructive to examine the linear regression of Y on X with respect to the differences between the regression values and the observed values. Applying equation (20.27) to the data of Table 20.8 yields sample regression coefficients and correlations of

$$b_{Y|X} = -.0539 \qquad b_{X|Y} = -.1552$$

$$r_{XY} = -.0914$$

With a mean $\bar{Y} = 49.2$, the consequent difference values $(\tilde{Y}_i - Y_i)$ are easily obtained. The regression values and difference values are given in Table 20.10. These difference values are plotted as a function of X in Figure 20.9. It is clear from the plot of differences that the difference values are a non-constant function of X. This systematic relationship is an indication that the linear regression function is not an adequate model of the relationship between X and Y. The plot of the differences in the mean regression against X in Figure 20.10 indicates the adequacy of that regression.

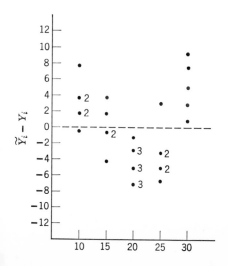

FIGURE 20.9

Scatter plot of differences between observed values Y and linear regression values $\tilde{Y}$ as a function of X.

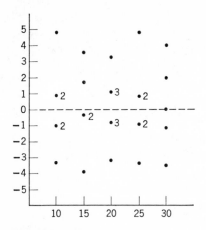

FIGURE 20.10

Scatter plot of differences between observed values Y and mean regression as a function of X.

The statistical and scientific significance of the observed characteristics of the joint distribution, that is, r_{XY}, $b_{Y|X}$, $E^2_{Y|X}$, and so on, can be ascertained under certain conditions to be described in the next section. In the meantime, some comments are in order regarding the adequacy of r_{XY} and $E^2_{Y|X}$ as measures of the relationship of X and Y in a sample of observations on (X, Y). In the extreme case where $N = 2$, it is necessary that the relationship be linear and the $r_{XY} = \pm 1.0$ due to the simple limitation that two separate points can define only a straight line and both points fall on that line (hence, $S^2_{Y|X} = 0$). As N is larger this restriction becomes less important but plays a role in restricting the possible observed values of r_{XY}. With only modest sample sizes this restriction on r_{XY} loses its salience. The problems incurred in r_{XY} as a measure of relationship resulting from the linearity of regression assumption are met head-on in comparison of $E^2_{Y|X}$ and r_{XY}. If the difference between $E^2_{Y|X}$ and r_{XY}^2 is large, r_{XY} cannot be an acceptable measure of the relationship of the observed values.

The effect of sample size on $E^2_{Y|X}$ is more critical than the corresponding effect on r_{XY}. If the number of observed categories on X are equal to the number of observations, $E^2_{Y|X} = 1$ simply because $SS_{\text{residual}} = 0$; that is, there is no variability of Y within the categories defined by the observed X values. If X is measured with a greater degree of accuracy than is reflected in sample size (for example, scores differing by .1 of a unit over the unit interval and sample size no larger than 10), the number of Y values per X value is likely to be very small. Unless the number of values of Y in each interval is reasonably large, the value of $E^2_{Y|X}$ is relatively meaningless.

Sampling Distributions of Statistics for Jointly Distributed Variables

The application of the theory of statistical tests to hypotheses about joint distributions of variables requires a theory of the sampling distribution of statistics from samples of observations on jointly distributed variables. Particularly, we are interested in the sampling distributions of r_{XY} and $E^2_{Y|X}$ in order to construct tests of hypotheses about ρ_{XY} and $\eta^2_{Y|X}$.

The sampling distributions of statistics in samples of observations on jointly distributed random variables depend on the probability functions of the variables involved. The normal distribution plays a heavy role in regression and correlation theory. In the context of interest in the regression of Y on X only, the conditional distributions of Y given X must be normal before certain statistical procedures are applicable. The conditional distributions of X given Y are not involved unless the regression of X on Y is of concern. The same statements hold for $\eta^2_{Y|X}$ and $\eta^2_{X|Y}$. If both regressions, or the Pearson correlation ρ_{XY}, are of interest, then both sets of conditional distributions must be normally distributed before the most widely used statistical procedures are applicable. Actually, a stronger condition is imposed: The joint variables must have a joint normal or bivariate normal distribution.

In the following discussion we limit ourselves to normally distributed conditional variables in order to study the regression of Y on X, and $\eta^2_{Y|X}$ and bivariate normal distributed variables to study ρ_{XY}. Little need be said about the conditional distributions—univariate distribution theory (Chapter 10) covers it completely. Notationally, we specify that the distribution of Y given $X = x$ is normal with expectation $\mu_{Y|x}$ and variance $\sigma^2_{Y|x}$ by Y_x: $N(\mu_{Y|x}, \sigma^2_{Y|x})$. If all variances are equal in a set of conditional distributions, we designate the distributions by Y_x: $N(\mu_{Y|x}, \sigma^2_{Y|X})$.

The bivariate normal distribution is a function of both X and Y, their variances $\sigma_X{}^2$ and $\sigma_Y{}^2$, and the correlation ρ_{XY}. The density function of (X, Y) for a normal distribution is

$$f(x, y) = \frac{1}{2\pi\sigma_X\sigma_Y\sqrt{1 - \rho_{XY}{}^2}}$$

$$\times \exp - \frac{1}{2(1 - \rho_{XY}{}^2)} \left(\frac{(x - \mu_X)^2}{\sigma_X{}^2} + \frac{(y - \mu_Y)^2}{\sigma_Y{}^2} - 2\rho_{XY} \frac{x - \mu_X}{\sigma_X} \frac{y - \mu_Y}{\sigma_Y} \right)$$

where $(\exp - c)$ implies e^{-c}. A schematic picture of the surface of this function is given in Figure 20.11.

By integrating across Y for each value of X, it can be shown that X: $N(\mu_X, \sigma_X{}^2)$. Conversely, if (X, Y) is bivariate normal it follows that Y: $N(\mu_Y, \sigma_Y{}^2)$. If (X, Y) is bivariate normal, each conditional distribution $(Y \mid x)$ and $(X \mid y)$ is normally

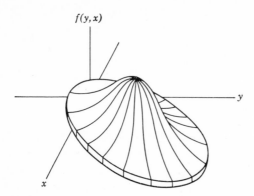

FIGURE 20.11

Schematic representation of the bivariate density function. Adapted with permission from Mood and Graybill, *Introduction to the theory of statistics*, New York: McGraw-Hill, 1963.

distributed with expectations and variances dependent on σ_X^2, σ_Y^2, μ_X, μ_Y, and ρ_{XY}. Thus, $(Y \mid x)$: $N[\mu_Y + (\sigma_Y/\sigma_X)(x - \mu_X)\rho_{XY}, \sigma_Y^2(1 - \rho_{XY}^2)]$. This last expression is related to the fact that if (X, Y) is bivariate normal, the only type of regression of Y on X that is possible is a linear regression. Consequently, the correlation ρ is an important parameter in determining the conditional distributions. One of the implications of this linearity restriction is that regression, correlation, and independence are all tied together in a special way in bivariate normal distributions. In the general bivariate distributions, independence is not insured by $(\rho = 0)$ or by $(\theta_{Y|X} = 0)$ in $\hat{Y} = \mu_Y + \theta_{Y|X}(X - \mu_X)$. However, in bivariate normal distributions $(\rho = 0)$ or $(\theta_{Y|X} = 0)$ insures independence of X

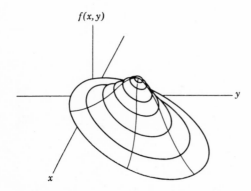

FIGURE 20.12

Schematic representation of the bivariate density function showing equal-density contours.

and Y; that is, $f(X = x, Y = y) = f(X = x)f(Y = y)$. The most important aspect of this is that it is impossible for any other than a linear regression to exist in bivariate normal distributed random variables. If (X, Y) is normal bivariate distributed, inferences about regression and independence can be stated in terms of the parameter ρ.

A useful way of depicting the bivariate normal distribution is in terms of the contour lines corresponding to constant values of $f(x, y)$. These lines, shown in Figure 20.12, correspond to the sets of points of (X, Y) having equal density

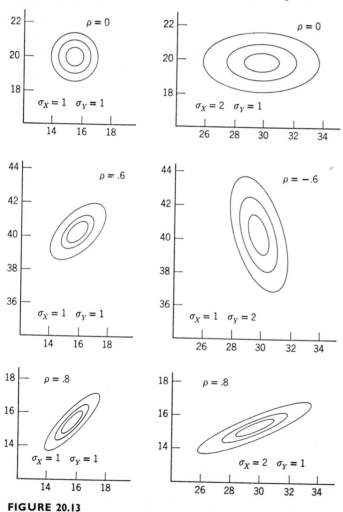

FIGURE 20.13

Illustrative equal-density ellipses in bivariate normal distribution.

function values. Or, if we imagine a slice through the surface of the density function parallel to the (X, Y) plane, the locus of the points of intersection of the slice and the density function surface is a contour. In general, these contour lines are ellipses. Each ellipse has as its center a point corresponding to (μ_X, μ_Y), and the principal axes of all the ellipses are the same (but with different lengths); that is, the ellipses are all centered and oriented in the same way. The slope and orientation of the equal density ellipses are determined by σ_X, σ_Y, and ρ_{XY}. A variety of ellipses are shown in Figure 20.13, illustrating the effects of these parameters on the ellipses.

Regression Problems. When we are interested in the regression of Y on X (or conversely X on Y) the sampling distribution theory needed is that of the conditional distributions of Y given X. This theory is the same as that encountered in the one-way analysis of variance. Each value of X is conceptually the same as a level on a treatment factor having a quantitative meaning. Consequently, the requirement in the analysis of variance that the dependent variable in each treatment be normally distributed is the same as the requirement in regression theory that the conditional distributions be normal.

In developing the statistical model for an analysis of the regression problem, it is convenient to take the analysis-of-variance approach. This is particularly straightforward if we limit our development of regression functions defined by the conditional expectations. Other regression functions can be cast in this same general mold. More advanced texts treat the analysis of specific nonlinear regression functions as "trend" analysis or the analysis of "orthogonal polynomials." The essential feature of this development is an expression of Y as a composite:

$$Y_{ij} = \mu_y + \hat{Y}_j + \varepsilon_{ij} \tag{20.45}$$

where

$$\hat{Y}_j = \theta_{Y|X}(X_j - \mu_x)$$

in which Y_{ij} is the ith score in the jth category of X, with quantity X_j. The quantity ε_{ij} is a random variable (error, residual) associated with the jth category on X and the ith observation in the category. If we let $\tau_j = \theta_{Y|X}(X_j - \mu_x)$, this model clearly is equivalent to the simple fixed effects model of the analysis of variance.

The partition of the variance of Y in the regression model is similar to the corresponding partition in the analysis of variance:

$$(Y_{ij} - \overline{Y}) = (Y_{ij} - \overline{Y}_j) + (\overline{Y}_j - \tilde{Y}_j) + (\tilde{Y}_j - \overline{Y})$$

where $\overline{Y}_j$ is the mean of the Y values in the jth category on X and $\tilde{Y}_j$ is the regression value in the jth category on X. Squaring and summing over j and i gives the sums-of-squares partition:

$$\sum_{j=1}^{M}\sum_{i=1}^{N_j}(Y_{ij} - \bar{Y})^2 = \sum_{j=1}^{M}\sum_{i=1}^{N_j}(Y_{ij} - \bar{Y}_j)^2 + \sum_{j=1}^{M}N_j(\bar{Y}_j - \tilde{Y}_j)^2 + \sum_{j=1}^{M}N_j(\tilde{Y}_j - \bar{Y})^2 \quad (20.46)$$

$$SS_T = SS_{\text{residual}} + SS_{\text{deviations from linear}} + SS_{\text{linear}} \quad (20.47)$$

where M is the number of categories on X. It can easily be shown that

$$SS_{\text{deviations from linear}} + SS_{\text{linear}}$$

is equal to $SS_{\text{treatments}}$ in the one-way analysis of variance. The degrees of freedom associated with each of these sums of squares is

$$df_T = df_{\text{residual}} + df_{\text{deviations from linear}} + df_{\text{linear}} \quad (20.48)$$

$$N_T - 1 = (N_T - M) + (M - 2) + (1) \quad (20.49)$$

The respective mean squares are obtained by dividing the SS's by the appropriate df's. By reexpressing the various components in the sums of squares in terms of the parameters σ_X^2, ρ_{XY}, and $\eta_{Y|X}^2$, it can be shown that, if $\sigma_{\varepsilon_j}^2 = \sigma_\varepsilon^2$ for all $j = 1, \ldots, M$ and that if ε_{ij}'s are independent, then

$$E(MS_{\text{residual}}) = \sigma_\varepsilon^2 \quad (20.50)$$

$$E(MS_{\text{between}}) = \sigma_\varepsilon^2 + \frac{N_T \eta_{Y|X}^2 \sigma_Y^2}{M - 1} \quad (20.51)$$

$$E(MS_{\text{linear}}) = \sigma_\varepsilon^2 + N_T \rho_{XY}^2 \sigma_Y^2 \quad (20.52)$$

$$E(MS_{\text{deviations from linear}}) = \sigma_\varepsilon^2 + \frac{(\eta_{Y|X}^2 - \rho_{XY}^2)N_T \sigma_Y^2}{M - 2} \quad (20.53)$$

The difference $(\eta_{Y|X}^2 - \rho_{XY}^2)$ is a measure of the departure of the regression function of expectations from the linear least squares regression function. The expectations may be reexpressed in terms of the regression coefficient $\theta_{Y|X}$ by noting the following identities:

$$N_T \rho_{XY}^2 \sigma_Y^2 = N_T \sigma_X^2 \theta_{Y|X}^2$$

$$\frac{(\eta_{Y|X}^2 - \rho_{XY}^2)N_T \sigma_Y^2}{M - 2} = \frac{(\sigma_Y^2 \eta_{Y|X}^2 - \sigma_X^2 \theta_{Y|X}^2)N_T}{M - 2}$$

Consequently, hypotheses regarding ρ_{XY} can be replaced by hypotheses regarding $\theta_{Y|X}$ and conversely. We state all hypotheses about $\theta_{Y|X}$ in terms of ρ.

Within the framework of this discussion the distribution theory basis of the tests for the presence of regression is the same as that for the one-way analysis of

TABLE 20.11 Analysis of Variance in Regression

Source	SS	df	MS	E(MS)	F
Between-X Categories (Treatments)	$\sum_{j=1}^{M}\sum_{i=1}^{N_j}(\bar{Y}_j - \bar{Y})^2$	$M-1$	MS_{η^2}	$\sigma_\varepsilon^2 + \dfrac{N_T\eta_{Y|X}^2\sigma_Y^2}{M-1}$	(1) $\dfrac{MS_{\eta^2}}{MS_\varepsilon}$
Linear	$\sum_{j=1}^{M}\sum_{i=1}^{N_j}N_j(\hat{Y}_j - \bar{Y})^2$	1	MS_{ρ^2}	$\sigma_\varepsilon^2 + N_T\rho_{XY}^2\sigma_Y^2$	(2) $\dfrac{MS_{\rho^2}}{MS_\varepsilon}$
Deviation from Linear	$\sum_{j=1}^{M}\sum_{i=1}^{N_j}N_j(\bar{Y}_j - \hat{Y}_j)^2$	$M-2$	$MS_{\eta^2-\rho^2}$	$\sigma_\varepsilon^2 + \dfrac{(\eta_{Y|X}^2 - \rho_{XY}^2)N_T\sigma_Y^2}{M-2}$	(3) $\dfrac{MS_{\eta^2-\rho^2}}{MS_\varepsilon}$
Residual	$\sum_{j=1}^{M}\sum_{i=1}^{N_j}(Y_{ij} - \bar{Y}_j)^2$	$N_T - M$	MS_ε	σ_ε^2	
Total	$\sum_{j=1}^{M}\sum_{i=1}^{N_j}(Y_{ij} - \bar{Y})^2$	$N_T - 1$			

variance. It is assumed that each of the observations is drawn independently from a random variable that is normally distributed $N(\mu_j, \sigma_\varepsilon^2)$ where μ_j depends on the value of X but σ_ε^2 is constant for all observations.

The test hypotheses generally of interest in the context of the regression problem are:

$$H_{T:\eta}: \eta_{Y|X}^2 = 0$$

$$H_{T:\rho}: \rho_{XY} = 0$$

$$H_{T:\eta-\rho}: (\eta_{Y|X}^2 - \rho_{XY}^2) = 0$$

The last hypothesis simply states that there is no departure from linearity in the regression function defined by the conditional expectations. All of these hypotheses can be tested by an analysis of variance in regression based on the sums of squares and mean squares indicated above. Table 20.11 is a summary of the analysis of variance in general. The F ratios defined in this table need some explanation. If the hypothesis regarding $\eta_{Y|X}^2$, that is, (1), is to be tested, then tests of the other two hypotheses, that is, (2) and (3), should be interpreted only as a contingent set of tests under the outcome of (1). That is, F ratios (2) and (3) in the table are redundant with respect to (1). However, tests of hypotheses about ρ_{XY}^2 and $(\eta_{Y|X}^2 - \rho_{XY}^2)$, involving F ratios (2) and (3) are independent.

Applying the analysis of variance in regression to the data of Table 20.10 using SS's already calculated gives the results of Table 20.12. The significance probabilities indicate a very low type I error probability for the rejection of the hypothesis of zero regression, $\eta_{Y|X}^2$. The linear regression component does not approach significance, and the deviation from linearity of regression is clearly the primary source of significance in test (1). The interpretation of these results depends on the character of the data observed. In the prediction context, the significance of (1)

TABLE 20.12 Analysis of Variance in Regression Applied to Data of Table 20.10

Source	SS	df	MS	F
Between-X Categories (Treatments)	303	4	75.75	(1) 10.90
Linear	3.98	1	3.98	(2) .57
Deviation from Linear.	299.02	3	99.67	(3) 14.34
Residual	173.80	25	6.95	
Total	476.80	29		

Significance Probabilities
Test (1): $\Psi < .001$
Test (2): $\Psi > .5$
Test (3): $\Psi < .001$

indicates that, if X were used to predict an unobserved Y in some sample from the same population, the accuracy of prediction would be enhanced in comparison with knowledge only of μ_Y. The lack of significance in the linear component implies that linear regression values are not different (statistically) from μ_Y. Hence, μ_Y is as good a prediction of Y as the prediction based on $Y = \mu_Y + b_{Y|X}(X - \mu_X)$. The score-correcting point of view leads to the inference that Y is not influenced by X in a linear way and the rescaled values of Y given the linear regression residuals will be insignificantly different from the original Y values. On the other hand, if the Y values are rescaled about the nonlinear regression, the reduction in variability in Y would be statistically significant.

It should be clear that the scientific significance of the statistical results of an analysis of variance in regression is dependent on the value of the type II error probability β and the alternative hypothesis. In establishing a decision or rejection rule for H_T: $\eta^2_{Y|X} = 0$, an alternative must be defined. The one-way fixed effects analysis of variance model can be used as a guide. The principles of establishing an (α, β, δ) test are the same in both problems. For linear alternatives to the hypothesis $(H_T$: $\rho_{XY} = 0)$, the differences of conditional expectations for adjacent values on X are constant throughout X. As a consequence, the linear alternative is equivalent to the generalized M-effect alternative in the one-way analysis of variance.

Correlation Problems.   In this section we turn our attention to the correlation coefficient in samples from bivariate normal random variables. Our primary interest is in r_{XY}, either as an estimate of ρ_{XY} or in a test of a hypothesis about ρ_{XY}. The first component in the discussion is the sampling theory of r_{XY} when (X, Y) is normal bivariate.

One property of the sampling distribution of r_{XY} is immediately clear: $-1.0 \le r_{XY} \le 1.0$. Consequently, it is clear that r_{XY} cannot be distributed normally, or as Student's t. However, R. A. Fisher showed that, under certain circumstances, a simple transformation of r_{XY} is approximately normal in distribution. Under another condition a different transformation of r_{XY} can be shown to have a Student's t distribution.

The sampling distribution of r_{XY} depends on sample size and ρ_{XY}. The density function equations are not given here. However, Figure 20.14 shows the density function of r_{XY} for $N = 10$, $N = 50$, $\rho = 0$, $\rho = .5$, and $\rho = .8$. The sampling distribution density function for r_{XY} is oddly behaved when N is very small. For $N = 2$ the function is not defined because $r = \pm1.0$ of necessity. For $N = 4$ the density function is uniform, if $\rho = 0$. The graphs of density functions in Figure 20.14 illustrate how rapidly the functions approach a normal form.

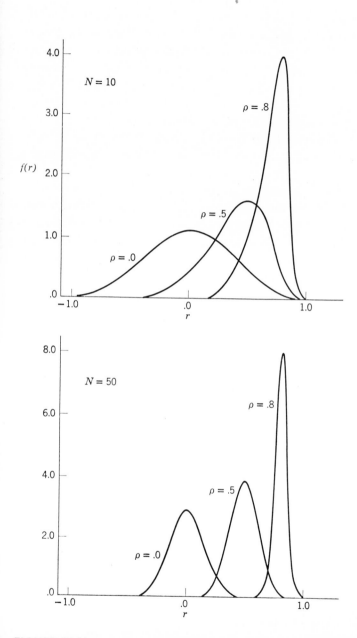

FIGURE 20.14

Density functions for sampling distributions of r_{XY}.

For sample sizes of moderate magnitude (or more), for example, $N > 10$, the Fisher Z_r transformation of r values is the preferred treatment. If (X, Y) is bivariate normal, the sample statistic

$$Z_r = \tfrac{1}{2} \log_e \left(\frac{1 + r_{XY}}{1 - r_{XY}} \right) \tag{20.54}$$

is normally distributed. The expectation of this statistic is approximated by

$$Z_\rho = \tfrac{1}{2} \log_e \left(\frac{1 + \rho_{XY}}{1 - \rho_{XY}} \right) \tag{20.55}$$

The variance of Z_r is approximated by

$$\frac{1}{N - 3}$$

As a consequence, Z_r has a sampling distribution the density function of which is approximately normal, $N[Z_\rho, 1/(N - 3)]$. Figure 20.15 illustrates the Z_r distributions corresponding to the density functions of Figure 20.15. Table S at the back of the book is an extensive table of Z_r values. The Z_r values should be signed the same as r_{XY}.

A special case is referred to in much of the statistics literature widely used in research. If $\rho_{XY} = 0$, the distribution of r_{XY} is transformed easily to the Student's t distribution. That is,

$$t = \frac{r_{XY}\sqrt{N - 2}}{\sqrt{1 - r_{XY}{}^2}} \tag{20.56}$$

will be distributed as Student's t with $(N - 2)$ degrees of freedom. This result is related to the analysis of variance test of the linear component of regression. The relationship of this distribution theory and the analysis of variance is not developed here; it is based on the random effects model of the analysis of variance.

The Z_r and Student transformation of r_{XY} lead immediately to statistical tests. Take the more general result, the Z_r transformation. If a specific value of ρ_{XY} is implied by some theory or model, then the statistical hypothesis

$$H_T\colon \rho_{XY} = \rho_T$$

can be tested by the methods of a single-sample test of a hypothesis about expectations. If (X, Y) is bivariate normal, then the sample statistic Z_r is $N[Z_\rho, 1/(N - 3)]$ under H_T. Hence the test is an example of a Case A one-sample test for expectations. Specifying the alternative values of ρ takes the same steps in this context as

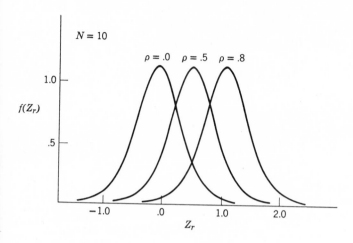

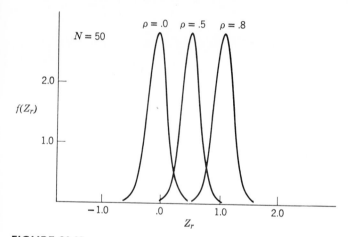

FIGURE 20.15
Density functions for sampling distributions of Z_r.

in the single-sample problem. The possible alternative hypotheses are

$$H_1: \rho_{XY} = \rho_1 < \rho_T$$
$$H_2: \rho_{XY} = \rho_2 > \rho_T$$
$$H_3: \rho_{XY} = \rho_3 \neq \rho_T$$

Imagine that we wished to test H_T against H_1. The single-sample procedures are applied to show that

$$Z = \frac{Z_r - Z_\rho}{\sqrt{1/(N-3)}} \tag{20.57}$$

is $N(0, 1)$. The hypothesis H_T is rejected if Z is smaller than Z_α for a lower tailed test. For example, if $N = 40$, $r_{XY} = .170$, $\alpha = .05$, $\rho_T = .45$, then

$$Z = \frac{.1717 - .4847}{\sqrt{1/37}}$$

$$= \frac{-.313}{.164}$$

$$= -1.91$$

Since $Z_{.05}$ for a lower tailed test is -1.645, the observed r_{XY} is significantly smaller than the value specified in H_T. The significance probability is given by $\Phi(-1.91) = .028$. The value of a specific alternative ρ_A that would satisfy the requirements $\beta = .05$ is given in a straightforward manner. The critical value of the test is such that

$$\frac{Z_c - .4847}{.164} = -1.645$$

Hence

$$Z_c = .4847 - 1.645(.164)$$

$$= .2149$$

For a lower tailed test and $\beta = .05$ the value Z_c must be 1.645 standard deviate units above Z_{ρ_A}. Consequently

$$\frac{.2149 - Z_{\rho_A}}{.164} = 1.645$$

and

$$Z_{\rho_A} = -1.645(.164) + .2149$$

$$= -.0549$$

The value of r_{XY} corresponding to this is $-.055$ approximately. The question now arises whether the difference between the effective alternative ρ_A and ρ_T is scientifically meaningful or interesting. The full development of an (α, β, δ) procedure depends on an understanding of the implications of the magnitude of ρ_{XY}.

Where an alternative ρ_A is naturally defined, the sample size necessary to achieve specific α and β values is easily determined. For example, let $\rho_T = .3$,

$\rho_A = .4$, $\alpha = .01$, and $\beta = .05$. Then

$$\frac{Z_c - .3095}{\sqrt{1/(N-3)}} = 2.326$$

$$\frac{Z_c - .4236}{\sqrt{1/(N-3)}} = -1.645$$

$$Z_c = 2.326 \sqrt{\frac{1}{N-3}} + .3095$$

$$Z_c = -1.645 \sqrt{\frac{1}{N-3}} + .4236$$

By simple algebra, the last two equations are simultaneously true if

$$N = 1216$$

The magnitude of N may seem out of proportion until it is recalled that ρ^2 represents the proportion of variance in Y resulting from regression on X and vice versa. The hypotheses indicate $\rho_T{}^2 = .09$, $\rho_A{}^2 = .16$. Consequently, we are trying to detect a difference equivalent to 7 percent of the variance $\sigma_Y{}^2$ or $\sigma_X{}^2$. The sample size needed for $\rho_T = 0$ and $\rho_A = .9$ with $\alpha = .01$, $\beta = .05$ is approximately 10. The general equation for sample size is

$$N = \frac{|Z_\alpha| + |Z_\beta|}{[Z_{\rho_T} - Z_{\rho_A}]^2} + 3 \qquad (20.58)$$

Where $\rho_T = 0$ the Student t distribution for (20.56) can be used in determining the significance probability of the observed value of r_{XY}. Because of the problems in evaluating the noncentral t distribution when H_T is false, this procedure is not recommended. The calculation is simpler and more accurate than the Fisher Z_r method (Z_r values are only approximate normal) when N is small. However, the advantages of the Z_r method outweigh the disadvantages.

Hypotheses about the correlation coefficients in two bivariate normal distributions can be tested using the Z_r transformation. Imagine a population H and a population G in which a theory specified correlation coefficients ρ_H and ρ_G, respectively; that is, the theory leads to

$$H_T: \rho_H - \rho_G = \Delta_T$$

with one or another of the alternatives

$$H_1: \rho_H - \rho_G = \Delta_1 < \Delta_T$$

$$H_2: \rho_H - \rho_G = \Delta_2 > \Delta_T$$

$$H_3: \rho_H - \rho_G = \Delta_3 \neq \Delta_T$$

A sample correlation, r_H and r_G, for each of the two populations is obtained in the usual way. By transforming r_H, r_G, ρ_H, and ρ_G into the respective Z_r values, the problem is transformed into a Case A two-sample problem in expectations. The statistic

$$Z = \frac{(Z_{r_H} - Z_{r_G}) - (Z_{\rho_H} - Z_{\rho_G})}{\sqrt{\dfrac{1}{N_H - 3} + \dfrac{1}{N_G - 3}}} \tag{20.59}$$

is normally distributed, $N(0, 1)$. The sample sizes N_H and N_G must both be reasonably large (10 or more) and the two samples must be independently drawn from bivariate normal random variables.

Some Special Techniques

Several special techniques are available for the study of the relationship between variables where the variables are not apparently amenable to the techniques already presented. Generally these techniques are applicable where one of two conditions prevails: The data are orderings and not values in the sense of random variables, or the data are categories instead of variables. These same techniques can be used in the analysis of data as transformation methods. That is, even if the observations are on continuous random variables, there may be some reason, such as nonnormality of distributions, for dealing with dichotomized or multiple categories of values of the variable or the rank orders of the values.

The Concordance of Orders. Where the values of the observed variables are simple ranks or orders, a special technique in describing their relationship is useful. This technique capitalizes on the fact that the "scores" assigned to the subjects are the first N integers. Statistical tests of hypotheses about the concordance of the ranks are based on the differences of "randomly" assigned ranks in the two variables. Another use of this technique is to provide a discrete transformation technique for studying the relationship of two variables with magnitudes giving only ordinal representations of characteristics of the observed objects. If magnitude differences in values observed are thought to be badly out of coordination with magnitude differences in the characteristics, this technique permits statements about the monotonicity of the relationship of the two variables. Two statistics are described here, although there are others described in the references given at the end of this chapter.

A caution is timely at this point. The indexes of concordance that are developed in this section are not correlations in the sense of the earlier portions of this chapter. If the conditions for the linear model of correlation are met and the values are transformed to ranks, the technique for rank order concordance (most often referred to as rank order correlation) does give an indication of the relative degree of correlation between the variables. However, the index itself is not an estimate of the parameter ρ. And, in particular, the square of the index is not a measure of the proportion of variance common to the two variables. Substituting the measures of concordance discussed here for correlations (because the conditions for application of correlation techniques are not met) does not solve any problems; it simply introduces a new set of considerations.

The two techniques are Spearman's rank order correlation and Kendall's tau. Both of these apply when there is a sample of N observations made on variables that are order valued. Imagine the following example. Two characteristics of subjects are observable only in a process of paired comparison in which a greater-than or smaller-than decision is the result of the observation procedures. Both variables are observed in this way for a sample of subjects, giving two sets of rankings or orders for the single set of subjects. Another example deals with orders directly. Imagine two athletic events in which the order that the participants completed a course of effort was the variable observed. The participants each finish the courses in some order relative to the other participants. Thus, each participant has an order number on both courses. An example of such a data set is given in Table 20.13, which also indicates the difference between the order values, and the squares of the differences, values used in calculating the rank order correlation. The calculation of the rank order correlation value is also shown in Table 20.13.

The correlation of the ranks, calculated by the equations for the linear correlation coefficient already discussed, can be obtained by an equation introduced by Spearman, who showed that the fact that the first N integers were the data led to a simplification in calculation. The calculation is called the Spearman rank order correlation, r_S. Because the sums and sums of squares of the two variables are fixed for all samples of size N (since they are functions of the first N integers only), the calculation depends only on the squares of the differences in the ranks:

$$r_S = 1 - \frac{6 \sum_{i=1}^{N} D_i^2}{N(N^2 - 1)} \tag{20.60}$$

The derivation of this equation from an expression of the correlation coefficient is not difficult. We do not reproduce it here.

The computation is applied to the data of Table 20.13 and the resulting rank order correlation coefficient reported there. The value can be interpreted as the

TABLE 20.13 Illustrative Data for Concordance of Orders

Athlete	Event 1	Event 2	Difference, D	D^2
1	3	3	−0	0
2	2	4	−2	4
3	5	6	−1	1
4	7	5	2	4
5	8	9	−1	1
6	1	2	−1	1
7	9	8	1	1
8	4	1	3	9
9	6	7	−1	1

$$\sum_{i=1}^{9} D_i^2 = 22$$

$$r_S = 1 - \frac{6\sum_{i=1}^{N} D_i^2}{N(N^2 - 1)}$$

$$= 1 - \frac{6(22)}{9(81 - 1)}$$

$$= 1 - \frac{132}{720}$$

$$= .817$$

correlation between the ranks in the two rank orders of the data. However, the interpretation of the coefficient as a proportion of variance is not warranted, nor is the interpretation of the coefficient as indication of the degree that the conditional distributions are less variable than the nonconditional distributions.

When ties exist in the rank ordering it is not possible to use equation (20.60). However, a simple alternative is to assign tied rank orders the mean of the ranks that would have been assigned if the ties were not present. Thus, if individuals ranked fifth, sixth, seventh, and eighth all have the same tie value they would all be assigned a value of 6.5. The linear correlation coefficient calculation is applied to the ranks just as if the ranks and mean ranks for ties were ordinary score values. The resulting coefficient is not a Spearman rank order correlation coefficient but may be interpreted as a Spearman coefficient corrected for ties. Another procedure is to break ties by a random process, such as looking up numbers in a table of random numbers, and assigning tied subjects different ranks by that process.

For data in which there are no ties, the sampling distribution of the rank order correlation coefficient can be readily calculated. Under the condition that the assignment of ranks is completely an equal probability process, the probability of obtaining a given coefficient, and hence subsets of coefficients, can be calculated. The equations for these calculations are not given here. They are developed in the books by Kendall and by Siegel, referenced at the end of this chapter. Under the condition that the number of observations is reasonably large, say 10 or more, the hypothesis that $r_S = 0$ can be tested by methods similar to those involved in single-sample tests of hypotheses about correlation in the linear case. The Student t distribution is approximated by the distribution of

$$t = \frac{r_S\sqrt{N-2}}{\sqrt{1 - r_S^2}} \qquad (20.61)$$

If the distributions of the variables that are the sources of the ranks are normal, then values of r_S will be good approximations of r_{XY}, particularly if the correlation of the random variables, ρ_{XY}, is equal to zero.

The second measure of the concordance of orders is the Kendall tau coefficient, τ. This coefficient is not a correlation in the sense that we have used the word. It is a transformation technique for asking questions about a degree of concordance, much as the Wilcoxon tests were techniques for asking questions about the distribution of random variables. The basis of the procedure is the rationale that if there is a high degree of concordance, there will be few inversions of order of individual subjects on the two variables, if the individuals are ordered in accordance with their ranks on the two variables. The other side of the coin is the relatively large number of inversions of subjects ordered by their ranks when there is a small degree of concordance of their ranks. An inversion between two individuals occurs when on one ranking one of the individuals is ranked above another individual, but on a second ranking the order of the individuals is reversed: individual A is higher than B on the first variable, and individual B is higher than A on the second variable. When the two rankings are the same, no inversions are observed. At the other extreme, when the rankings are directly in the reverse order, there is an inversion for each pair of individuals, that is, $\binom{N}{2}$ different inversions, one for each of the $\binom{N}{2}$ pairs. By rather simple logic it is apparent that the ratio

$$\frac{2(\text{number of inversions})}{\text{number of pairs}}$$

will be 2.0 for reverse order in the two groups and 0 for no inversions. Subtracting this ratio from 1.0 produces a coefficient with sign coordinated with the usual meaning of a positive and negative relationship, Kendall's τ:

$$\tau = 1 - \frac{2(\text{number of inversions})}{\text{number of pairs}} \qquad (20.62)$$

There is no weighting of inversions in terms of the degree of the inversion, as might be represented by the number of other individuals interposed in the inversion. In rank order correlation the differences are differentially weighted as the squares of the differences. Nevertheless, there is a strong resemblance between the two coefficients and they are highly correlated in sample data. Where the rank order coefficient is probably most frequently interpreted as a substitute for the linear correlation coefficient, the Kendall τ has a straightforward interpretation as the relative frequency of concordance in ranking.

Calculation of τ for a sample is really a problem in counting the number of inversions observed. The denominator is evaluated by methods familiar to the reader from our work on combinatorials. A graphical device serves a useful role in counting the number of inversions. The subjects are arrayed separately in order of their ranking on the two variables. Lines are drawn from one array to another, connecting each subject with himself in the two arrays. This is illustrated, with the data of Table 20.13, in Figure 20.16. The number of inversions is given by the number of pairs of lines that cross. In our example there are seven such crossing points. Hence the value of τ is given by

$$\tau = 1 - \frac{2(7)}{\binom{10}{2}} = 1 - \frac{14}{45}$$

$$= .689$$

Where there are a number of ties in the data, they can be broken by a random device, and the Kendall statistic calculated for the tie-broken data. However, there

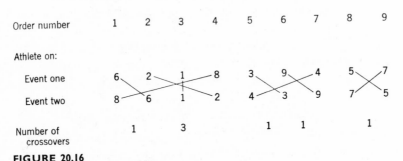

FIGURE 20.16

Crossovers in data of Table 20.13 for calculation of Kendall's tau.

is a computational procedure available for calculating the τ statistic with ties. This procedure is defined in Kendall and also in Siegel.

The distribution of τ is readily defined under the hypothesis of no concordance except as might be expected with a random assortment of ranks in the two variables. This distribution is described in the references already cited. For the hypothesis $H_T: E(\tau) = 0$, a simple approximation is available that enables the use of the normal distribution in a test. For sample sizes of 10 or more the statistic

$$Z = \frac{\tau - 1 \Big/ \binom{N}{2}}{\sqrt{\dfrac{2(2N + 5)}{9N(N - 1)}}} \tag{20.63}$$

is normally distributed, $N(0, 1)$. The $1/\binom{N}{2}$ term is a correction for continuity.

The Relationship of Categorical Variables. Behavioral scientists frequently deal with variables that are dichotomies, that is, two valued or two-state variables. These kinds of variables arise in at least two major ways. First, the phenomena being studied may be two-valued, such as sex, or true versus false. Second, the phenomena being studied may be observed only crudely so that a dichotomous distinction is all that is possible, such as high versus low, short versus long. Statistical techniques have been devised to deal with the problem of describing the relationship of such variables and of applying statistical decision theory to hypotheses about the relationships.

Point Biserial Correlation. When one of two variables is continuous and the other is dichotomous, the correlation between the two variables is referred to as a biserial correlation. To distinguish the situation in which the dichotomous variable is a true dichotomy and the situation in which the dichotomous variable is a dichotomy imposed on a continuous variable (a transformation, perhaps), the former is called a point biserial. We develop the point biserial here, and not the so-called Pearson biserial.

The dichotomous variable is represented by two arbitrary values, a convenient choice being 0 and 1. The subjects observed have two scores, one on the continuous variable and the other a 0 or a 1 on the dichotomous variable. The point biserial coefficient of correlation r_p is calculated by the standard linear correlation formula, using these scores. The interpretation of this coefficient and statistical tests of hypotheses about the expected value of the correlation are the same as in two continuous variables. However, because of the simplicity of evaluation of the mean and variance of the dichotomous variable, the equation for the coefficient is simplified. If Y is the continuous variable, we let $\overline{Y}_1$ be the mean of the Y values for those subjects who have a value of 1 on the dichotomous variable. Likewise let $\overline{Y}_0$ be the mean of the Y values for subjects with the score of 0. If the relative frequency of 1 scores is p and the relative frequency of 0 scores is q then the standard deviation of the number of 1's is $\sqrt{Npq}$, where N is the number of observations. The point biserial coefficient is

$$r_p = \frac{(\overline{Y}_1 - \overline{Y}_0)\sqrt{Npq}}{\sqrt{\sum_{i=1}^N (Y_i - \overline{Y})^2}} \tag{20.64}$$

This is reexpressible in terms of the mean of the Y values for the least frequent value of the dichotomy, for example $\overline{Y}_a$, the standard deviation of Y, and the number of observations N_a in that category, and N_b in the other:

$$r_p = \frac{(\overline{Y}_a - \overline{Y})}{S_Y} \sqrt{\frac{N_a N}{N_b(N-1)}}$$

The square of the point biserial correlation coefficient is equivalent to that portion of the variance of Y that is associated with the dichotomy. The theory of the analysis of variance is directly applicable to the tests of hypotheses about the magnitude of the correlation and regression of Y on the dichotomy. The dichotomy serves as a designator of the two levels in a single-factor random groups design. Also, the difference in the means of the Y values in the groups of subjects defined by the 0 and 1 scores on the dichotomy is directly proportional to the correlation. As a consequence, tests of hypotheses about the point biserial correlation are expressible as tests of hypotheses about the difference in expectations in a two-sample problem. A test of $H_T: E(r_p) = 0$ is equivalent to a t test or a normal test of $H_T: E(Y_1) = E(Y_0)$ in a two-sample problem with the two samples being defined by the dichotomy. All of the statistical theory and probabilistic assumptions needed in the two-sample tests of hypotheses about expectations are relevant in the test of hypotheses about r_p.

A caution is warranted at this point. When one part of the dichotomy is not as frequent as the other, that is, when $p \neq q$, equivalent to the unequal sample size problems in the two-sample problem, the meaningfulness of r_p can be impaired. The greater the disparity in p and q, the more caution the scientist should exercise in interpretation. Most scientists would look askance at a two-sample test about hypotheses with sample size of 10 in one group and 90 in another group. However, it is not infrequent in correlational analysis that one sees dichotomies with such extreme distributions. They may be natural and of great interest, but r_p is afflicted with all of the problems that a two-sample problem would have with such a poor balance in representation of one of the groups.

Fourfold Correlation. When we have two dichotomous variables, a correlation coefficient called the fourfold correlation coefficient, ϕ or phi coefficient, is commonly used in the description of the degree of relationship of the variables and in testing hypotheses about the significance of departure from independence of the two variables. Since both variables are dichotomies, there are exactly four categories for each of the subjects to fall into. It is convenient to arrange these categories into a two-way layout with two cells in each row and two cells in each column, as shown in Figure 20.17, for two variables X and Y. The most common procedure for evaluating the correlation coefficient in these data is to enter the frequency of each kind of subject in the cells of the layout and on the margins of the rows and columns, as shown in Figure 20.17 with the symbols a, b, c, and d.

The General Layout

		Y		
		1	0	X Totals
X	1	a	b	$a + b$
	0	c	d	$c + d$
	Y Totals	$a + c$	$b + d$	N

An Example Data Set

		Y		
		1	0	X Totals
X	1	12	16	28
	0	14	9	23
	Y Totals	26	25	51

$$\phi = \frac{(12)(9) - (16)(14)}{\sqrt{(12 + 16)(14 + 9)(12 + 14)(16 + 9)}}$$

$$= -.179$$

FIGURE 20.17

A two-way layout for fourfold correlation.

It can be shown that the simple product-moment correlation coefficient between the two variables for the sample reduces, with simple algebra, to

$$r_{XY} = \phi = \frac{ad - bc}{\sqrt{(a + b)(c + d)(a + c)(b + d)}} \tag{20.65}$$

The sign of this coefficient is determined by the way the dichotomy is defined. In order for the coefficient to indicate a positive relationship with a positive value, the dichotomy should be established with a and d representing the frequencies of individuals who possess both attributes being related (that is, both present, or both absent; both true or both false; and so on).

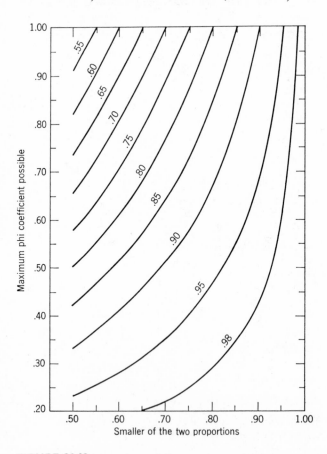

FIGURE 20.18

Maximum phi coefficients. Adapted with permission from Guilford, Fundamental statistics in psychology and education, New York: McGraw-Hill, 1950. Used with permission of McGraw-Hill Book Co.

The interpretation of the fourfold point coefficient is most clearly one of independence of the two dichotomies. If the dichotomies are independent, the interior entries in the fourfold table will be mutually proportional to the marginal entries. In connection with this interpretation, a test of hypotheses about independence of the variables is available. Under the hypothesis that X and Y are independent dichotomous variables, the square of the ϕ coefficient is χ^2/N distributed with one degree of freedom. Consequently, when we wish to determine whether the value of ϕ indicates a departure from independence that is statistically significant, we consult the χ^2 distribution table to determine the significance probability in the χ^2 distribution with one degree of freedom. The table of χ^2 is entered with the value of $N\phi^2$.

The phi coefficient has one very serious fault as a measure of the correlation between variables—its range can be severely restricted if the pattern of frequencies of the dichotomy on both variables is not the same. Let p_X be the frequency of 1's on X, and $q_X = 1 - p_X$. Let p_Y be the frequency of 1's on Y, and $q_Y = 1 - p_Y$. The maximum value of ϕ for any combination of values of p and q is given by

$$\phi_{\max} = \sqrt{\left(\frac{p_X}{q_X}\right)\left(\frac{q_Y}{p_Y}\right)} \tag{20.66}$$

A graph of these maximum values for selected values of p is presented in Figure 20.18. In the graph, the smaller of the two proportions p_X and p_Y is given at the base (if $p < q$ use q); the other proportion is represented on each of the curved lines of the chart.

Some Special Considerations in Regression and Correlation

The distinction between regression and correlation studies is accompanied by preferences for the use of regression over correlation and vice versa. In choosing the statistical techniques to be used in a study of some phenomenon, the properties of the techniques should determine the choice. Any absolute preference for correlation or for regression techniques is not appropriate. For many purposes, the use of correlation is preferred, and for other purposes the use of regression is preferred. The basic statistics are the same in both techniques. When the purpose of a study is to determine the degree and type of dependency of a dependent variable on an independent variable, regression techniques are generally the more appropriate. The analysis of variance model of regression analysis is applicable to these situations without ever considering correlation. On the other hand, many studies are concerned about the joint dependency of two variables, without distinction between dependent and independent variables. Correlation techniques have been criticized because the magnitude of r is a function of sample size. This criticism is no more pertinent in dealing with correlation than for any other statistic. When proper controls over statistical decision making are established, the magnitude and variation of the sample correlation coefficient are placed in proper context. Although the regression coefficient is not as directly a function of sample size as is the correlation coefficient, the considerations of statistical decision theory and the principles of interpretation of descriptive statistics apply to both techniques.

Restrictions in the range of a variable when it is observed will give a poor measure of the correlation or regression of that variable with another variable. Two consequences of the restriction are important to distinguish: The population is not what it is supposed to be, and numerical properties affect the values obtained. In the first place, any inferences made from the statistics calculated (correlation or regression) will not be relevant to the population the sample was intended to represent. If the population is truncated at the extremes, the actual population relevant to the statistic is the truncated population, not the population with full range of variation. In the second place, the regression or correlation coefficient is vitally affected by the extreme scores on the two variables. If these extreme scores are not represented in the sample, the relationship will generally be less apparent. Figure 20.19 illustrates the effect of truncating the sample by restriction of scores in the population to a narrow class interval on each of the two variables. An example

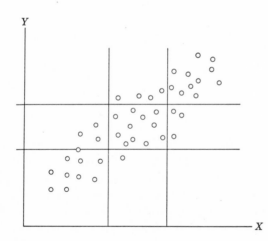

FIGURE 20.19

Scatter graph illustrating a strong positive correlation over full range, and low correlation over restricted ranges (marked off by lines in the scatter).

of this sort of truncation is the use of graduate students in a study of the relationship between intelligence and performance. Since the extreme of intelligence is presumably represented in the subsample of the general population of ordinary citizens that we have in such a study, the relationship between the sample of observations on the variables cannot be taken as a representation of the relationship of the variables in general. Also, in such studies the relationship between variables is likely to be insignificant because within the range of the variables the substantive individual differences are small in comparison with the "random" variation.

Where several distinct populations of subjects are sampled simultaneously in a correlation and regression study, the effect may be a serious misrepresentation of the relationship between the variables. Figure 20.20 illustrates three types of mixtures of relationships that lead to incorrect statistical representation of the relationship. The correlation or regression coefficients may be correct representations of the relationships of the variables studied for the population as defined by the mix in the sample. However, within the separate populations the relationship may be utterly different. This effect is not always predictable, as illustrated in Figure 20.20. A large correlation or regression effect can be generated by combining samples from populations in which the relationship of the variables is nil (or negative), providing that the mean values of the variables in the populations are correlated as in the top graph in Figure 20.20. On the other hand, if there are significant differences in mean values on one variable in the different populations mixed, but not in the other variable, a strong correlation and regression in the populations can become lost in the mix, as illustrated in the middle graph of the figure. If the direction of the regression is reversed in the populations mixed, as in the third graph of the figure, the regression and correlation of the variables can be hidden in the sample.

Two relatively common practices produce misleading results in correlation studies. Imagine several variables measured on a sample of subjects. These variables may be used to define other variables, such as the sum of some of the variables (like items on an intelligence test being counted, or summed, to provide a total score). If the individual variables are correlated with the sum, part of the correlation is generated by the fact that the variables are included in the sum. Thus the magnitude of the correlation is not representative of a relationship between sources of variation outside of the artifact of the mechanical process of summation of the variables. The expected correlation between the summation and one of the variables in the summation is a simple function of the number of variables in the summation. If there are n variables in the summation, the expected correlation of a variable and the summation owing to the presence of the one variable in the summation is

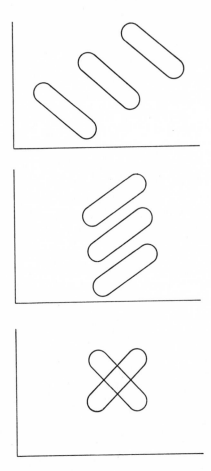

FIGURE 20.20

Illustration of effects of mixing samples from different populations.

$1/\sqrt{n}$. In other forms of the composite of variables, the correlation of a variable in the composite and the composite itself is a function of the fact that the variable is a part of the composite. Frequently the composite is a ratio of the original variables. In this case the correlation between the variable and the composite is called an index correlation.

Correlation coefficients are affected by unreliability in measurement of the variables being correlated. If a variable is a function of two sources of variation and another variable is a function of one of these two sources of variation plus another source of variation, the correlation will be a function of the proportion of the total variation associated with the common source of variation. Since the correlation coefficient is a measure of the proportion of variance in that sense, any randomness in the variables (unreliability) will result in a reduction in the magnitude of the correlation. Thus, if two variables are measuring a single source of variation, they would be correlated perfectly if there were no random error in the measurements. However, if the randomness or unreliability of measurement is uncorrelated in the observations of the two variables, the correlation will be depressed in magnitude as a function of the proportion of the variation associated with the unreliability. The student is referred to texts on psychometric methods for detailed discussions of these issues in correlation and regression.

Multivariate Regression and Correlation

This chapter began with a general statement of the correlation and regression problem with many variables. The main development in the chapter, however, centers on regression and correlation problems of bivariate situations only. When multiple, more than two, variables are involved, the basic pairwise relationship between two variables exists for each pair of variables in the set. The idea of linear and nonlinear regression of one variable on another can be extended directly to the linear and nonlinear regression of one variable on several other variables by way of the parallel between the multivariate case and the multiple-factor case in the analysis of variance. The complexity of the regression is rather greater than that of the analysis of variance, because of the possibility that the independent variables may be correlated themselves, whereas in the analysis of variance the independent variables (factors) are designed to be independent (orthogonal). In many applications the investigator is not interested in a single dependent variable and a collection of independent variables, but rather in the entire collection of variables and in their mutual interrelationships. In this section we focus our attention on the former problem, where one variable is the dependent variable and the others are independent variables. We shall not address the question of investigations of the meaning and interpretation of the mutual intercorrelations of the variables. The latter problem is dealt with in books on multivariate analysis, such as the books by Anderson and by Morrison listed at the end of this chapter.

Regression of Y on X can be generalized to the regression of Y on $X_1, X_2, \ldots, X_k$. The analysis of the regression in this generalization is referred to as multiple regression. In addition, a single coefficient expressing the degree to which the variance of Y is represented in the collection of independent variables is available, and referred to as a multiple correlation coefficient. The linear case, the general nonlinear case, and the coefficients involved are expressed in this section without detailed development, and without developing the calculational aspects. Computer methods of evaluation of the regression coefficients and correlation coefficients are the methods of choice in multivariate analysis, and development of those methods here is beyond the scope of this text. Sources of information about computer programs to perform multivariate analyses are listed in the Lehman and Bailey book included in the suggested reading section of this chapter.

In the general case of regression, the sample space is determined by the value set or the categories in the value set of the independent variable X. Proceeding in the same manner when there are multiple independent variables would segment the sample space into all of the combinations of the values of the variables (or value categories), similar to a multifactor analysis of variance design. Let a given sector in this k-variate layout of the independent variables be defined by $(X_1 = x_1)$, $(X_2 = x_2), \ldots, (X_k = x_k)$. The expected value of Y within the sector is expressed by $\mu_{Y|x_1 x_2 \cdots x_k}$, and the variance within the sector is expressed by $\sigma_{Y|x_1 x_2 \cdots x_k}$. All of the sectors defined by combinations of the X variables are treated in this way, and the sum of squares within the sectors, about the sector expectations, is obtained to estimate the variance of Y within sectors for the entire set of variables X. The within variance is expressed by $\sigma^2_{Y|X_1 X_2 \cdots X_k}$. A measure of the degree to which the distribution of variable Y is affected by the k-way distribution of the X variables is the ratio

$$\zeta^2_{Y|X_1 X_2 \cdots X_k} = \frac{\sigma_Y{}^2 - \sigma^2_{Y|X_1 X_2 \cdots X_k}}{\sigma_Y{}^2}$$

This measure of the relationship of Y and the X variables is a reasonable representation of the degree to which values of Y can be assorted into more-or-less homogeneous distributions when the combination of X values are known, relative to the variance of Y. The relationships among the X variables is largely irrelevant unless there is a desire to reduce the number of X variables and still retain the same reduction of variation in Y given the X's. If there are highly correlated X variables, the assortment of the elements in the sample space by the two correlated X variables will be quite similar, and there is no advantage to having both of them in the set. In samples of multivariate observations, the procedures and interpretation of results are parallel to the procedures and interpretation of results in the random variable theory. We would use the analysis of variance procedures for regression analysis, treating each of the sectors as a level on a factor, or a combination of levels on several factors. In order to test hypotheses about ζ^2 with sample values, assumptions must be made about the conditional distributions of Y within the sectors in the k-way distribution. The assumptions here are identical to the assumptions of the analysis of variance. However, unless the X variables are independent, we cannot partition the variance of Y into parts associated with individual X variables or combinations of X's.

Imposition of a linear model on the regression problem involves developments that we can present only in the sketchiest detail here. In general, the simple algorithmic approach to the general regression problem just described is not possible with a linear model, with a least squares solution to the coefficients in the model. The linear regression estimate of the variable Y is given by

$$\hat{Y} = \xi + \theta_1 X_1 + \theta_2 X_2 + \cdots + \theta_k X_k$$

where the values of ξ and the θ's are obtained so as to minimize the expected square of the differences between the Y values and the corresponding $\hat{Y}$ values. The regression coefficients, the θ's, are not necessarily unique—they will take on a variety of values depending on how they are calculated. The values of the regression coefficients will be configured in such a way that the total effect is constant, but because of the correlation between the X variables, the impact of any given variable in the set is relative to the impact on any other given variable and the correlation between the variables. Imagine that we evaluated the coefficients one at a time, each step holding the previous X variables constant or correcting the regression coefficients to reflect impact of the correlation of variables in previous steps with the successive variable on the regression. The successive distillation of dependency of Y on the independent variables successively considered in this step-wise procedure will become smaller and smaller if the variables are highly correlated. However, should we evaluate the independent variables in another order, the impact of the successively considered variables will appear to be different.

The reader is urged to extend this discussion by reference to one or more of the texts listed in the suggested readings for this chapter. A very useful and practical guide to the use of multivariate correlational techniques is given in the book by Guilford.

SUGGESTED READING

Anderson, T. W. (1956), *An introduction to multivariate statistical analysis*, Wiley, New York.

Cramér, Harold. (1946), *Mathmeatical methods of statistics*, Princeton University Press, New Jersey.

Deming, W. E. (1964), *Statistical adjustment of data*, Dover, New York.

Guenther, W. C. (1965), *Concepts of statistical inference*, McGraw-Hill, New York.

Guilford, J. P. (1950), *Fundamental statistics in psychology and education*, McGraw-Hill, New York.

Hays, W. (1963), *Statistics*, Holt, Rinehart and Winston, New York.

Kendall, M. G. (1955), *Rank correlation methods* (2d ed.), Griffin, London.

Lehman, R. S., and Bailey, D. E. (1967), *Digital computing*, Wiley, New York.

McNemar, Q. (1969), *Psychological statistics* (4th ed.), Wiley, New York.

Morrison, D. F. (1967), *Multivariate statistical methods*, McGraw-Hill, New York.

Siegel, S. (1956), *Nonparametric methods for the behavioral sciences*, McGraw-Hill, New York.

GLOSSARY

Joint Distribution. Two random variables defined on a single sample space are said to be jointly distributed, or simply bivariate distributed. Where multiple variables are defined on a single sample space they are said to be multivariate distributed.

$$(20.1) \qquad P(X = x) = \sum_{y \in Y} P(X = x, Y = y)$$

Independent Random Variables. Two jointly distributed random variables are said to be independent if the joint probability distribution is equal to the product of the separate probability distributions.

$$(20.2) \qquad P(X = x \mid Y = y) = \frac{P(X = x, Y = y)}{P(Y = y)}$$

$$(20.3) \qquad P(X = x \mid Y = y) = P(X = x)$$

$$(20.4) \qquad P(X = x, Y = y) = P(X = x)P(Y = y)$$

Conditional Distribution. A conditional distribution of a variable Y is the distribution of the random variable Y for those events in the sample space defined by a specific value or subset of values from the value set of the conditioning variable. The probabilities of the distribution are conditional probabilities, given the conditioning variable value or subset of values.

Conditional Expectation. A conditional expectation of a variable Y is the expectation of the random variable Y for those events in the sample space defined by a specific value or subset of values from the value set of the conditioning variable and the conditional probabilities of the conditional distribution.

Conditional Variance. A conditional variance of a variable Y is the variance of the random variable Y for those events in the sample space defined by a specific value or subset

of values from the value set of the conditioning variable and the conditional probabilities
of the conditional ditributions.

(20.5)
$$\mu_{Y|x} = E(Y|x) = \sum_{y} P(y|x)y$$

(20.6)
$$\sigma^2_{Y|x} = \text{var}(Y|x) = \sum_{y} P(y|x)(y - \mu_{Y|x})^2$$

Generalized Regression Function. A function of the conditioning variable X in the range
of the value set of the conditioned variable Y is a generalized regression function of Y
on X.

(20.7)
$$g(x) = E(Y|X = x) = \mu_{Y|x}$$

(20.8)
$$g(x) = \xi + \theta x$$

(20.9)
$$\sigma^2_{y|h(X)} < \sigma^2_{y|g(X)}$$

(20.10)
$$\sigma^2_{Y|g(x)} = \sum_{x} P(X = x)E[Y - g(x)]^2$$

(20.11)
$$\sum_{x} P(X = x)E[Y - (\xi + \theta x)]^2$$

(20.12)
$$\theta = \frac{E(XY) - E(X)E(Y)}{E(X^2) - E(X)^2}$$

(20.13)
$$\xi = E(Y) - \theta E(X)$$

(20.14)
$$\hat{Y} = \xi + \theta_1 X + \theta_2 X^2 + \cdots + \theta_n X^n$$

(20.15)
$$\sum_{x} P(X = x)E(Y - \mu_{Y|x})^2 = \sum_{x} P(X = x)E[Y - f(x)]^2$$

(20.16)
$$\sigma^2_{Y|g(x)} = \sum_{Y} P(Y|x)(Y - \lambda_x - \mu_{Y|x})^2 \qquad \text{for each } x \text{ in } X$$

$$= \sigma^2_{Y|X} + \sigma_\lambda^2$$

(20.17)
$$\sigma_Y^2 = \sigma_{\hat{Y}}^2 + \sigma^2_{Y|g(X)}$$

Correlation Ratio. The correlation ratio is the ratio of the difference of the variance of
the dependent variable, Y, and the pooled variance of Y about the conditional expectation
of Y given the independent variable X, to the variance of Y.

(20.18)
$$\eta^2_{Y|X} = \frac{\sigma_Y^2 - \sigma^2_{Y|X}}{\sigma_Y^2}$$

Covariance. The covariance of two variables is a measure of the relationship of two
variables in linear regression, in which case it is a measure of the joint variance of the two
variables.

(20.19)
$$\sigma_{XY} = E(XY) - E(X)E(Y)$$

$$(20.20) \qquad \sigma_{X^*Y^*} = E(X^*Y^*) - E(X^*)E(Y^*)$$

$$= \frac{\sigma_{XY}}{\sigma_X \sigma_Y}$$

$$(20.21) \qquad \rho_{XY} = \sigma_{X^*Y^*}$$

Pearson Product-Moment Correlation. The covariance of two standardized random variables is the Pearson product-moment correlation coefficient.

$$(20.22) \qquad \theta_{Y|X} = \rho_{XY} \frac{\sigma_Y}{\sigma_X}$$

$$(20.23) \qquad \pm\sqrt{\rho_{XY}{}^2} = \pm\sqrt{\theta_{X|Y}\theta_{Y|X}}$$

$$(20.24) \qquad \sigma_{Y|X}^2 = \sigma_Y{}^2(1 - \rho_{XY}{}^2)$$

$$(20.25) \qquad \sigma_Y{}^2 = \sigma_{Y|X}^2 + \sigma_Y{}^2\rho_{XY}$$

$$= \sigma_Y{}^2(1 - \rho_{XY}{}^2) + \sigma_Y{}^2\rho_{XY}{}^2$$

$$(20.26) \qquad \tilde{Y} = \bar{Y} + b_{Y|X}(X_i - \bar{X})$$

$$(20.27) \qquad b_{Y|X} = \frac{\frac{1}{N}\sum_{i=1}^{N} X_i Y_i - \bar{X}\bar{Y}}{\frac{1}{N}\sum_{i=1}^{N} X_i{}^2 - \bar{X}^2}$$

$$= \frac{N\sum_{i=1}^{N} X_i Y_i - (\sum_{i=1}^{N} X_i)(\sum_{i=1}^{N} Y_i)}{N\sum_{i=1}^{N} X_i{}^2 - (\sum_{i=1}^{N} X_i)^2}$$

$$(20.28) \qquad D_i = \tilde{Y}_i - Y_i$$

$$(20.29) \qquad S_Y{}^2 = S_D{}^2 + S_{\tilde{Y}}{}^2$$

$$(20.30) \qquad r_{XY}{}^2 = \frac{S_Y{}^2 - S_D{}^2}{S_Y{}^2}$$

$$(20.31) \qquad S_Y{}^2 = S_Y{}^2 r_{XY}{}^2 + S_Y{}^2(1 - r_{XY}{}^2)$$

$$(20.32) \qquad b_{X|Y} = \frac{\frac{1}{N}\sum_{i=1}^{N} X_i Y_i - \bar{X}\bar{Y}}{\frac{1}{N}\sum_{i=1}^{N} Y_i{}^2 - \bar{Y}^2}$$

$$(20.33) \qquad \pm\sqrt{r_{XY}} = \pm\sqrt{b_{Y|X}b_{X|Y}}$$

$$(20.34) \qquad r_{XY} = \frac{1}{N}\sum_{i=1}^{N} X_i^* Y_i^*$$

$$(20.35) \quad r_{XY} = \frac{N \sum_{i=1}^{N} X_i Y_i - (\sum_{i=1}^{N} X_i)(\sum_{i=1}^{N} Y_i)}{\sqrt{[N \sum_{i=1}^{N} X_i^2 - (\sum_{i=1}^{N} X_i)^2][N \sum_{i=1}^{N} Y_i^2 - (\sum_{i=1}^{N} Y_i)^2]}}$$

$$(20.36) \quad Z = \frac{\overline{Y}^* - \mu_T}{S_{Y^*}/\sqrt{N}} = \frac{\overline{Y} - \mu_T}{S_D/\sqrt{N}}$$

$$(20.37) \quad t = \frac{\overline{Y}^* - \mu_T}{S_{Y^*}/\sqrt{N-1}} = \frac{\overline{Y} - \mu_T}{S_D/\sqrt{N-1}}$$

$$(20.38) \quad Z = \frac{\overline{Y} - \mu_T}{S_Y \sqrt{1-r^2}/\sqrt{N}}$$

$$(20.39) \quad t = \frac{\overline{Y} - \mu_T}{S_Y \sqrt{1-r^2}/\sqrt{N-1}}$$

$$(20.40) \quad E_{Y|X}^2 = \frac{SS_{\text{between}}}{SS_{\text{total}}}$$

$$(20.41) \quad SS_{\text{total}} = SS_{\text{residual}} + SS_{\text{nonlinear}} + SS_{\text{linear}}$$

$$(20.42) \quad SS_{\text{between}} = SS_{\text{linear}} + SS_{\text{nonlinear}}$$

$$(20.43) \quad SS_{\text{linear}} = \frac{N[\sum_{i=1}^{N} X_i Y_i - (\sum_{i=1}^{N} X_i)(\sum_{i=1}^{N} Y_i)/N]^2}{N(\sum_{i=1}^{N} X_i^2) - (\sum_{i=1}^{N} X_i)^2}$$

$$(20.44) \quad E_{Y|X}^2 = \frac{SS_{\text{between}}}{SS_{\text{total}}} = \frac{SS_{\text{linear}} + SS_{\text{nonlinear}}}{SS_{\text{total}}}$$

$$= \frac{SS_{\text{linear}}}{SS_{\text{total}}} + \frac{SS_{\text{nonlinear}}}{SS_{\text{total}}}$$

$$(20.45) \quad Y_{ij} = \mu_y + \hat{Y}_j + \varepsilon_{ij}$$

$$(20.46) \quad \sum_{j=1}^{M} \sum_{i=1}^{Nj} (Y_{ij} - \overline{Y})^2 = \sum_{j=1}^{M} \sum_{i=1}^{Nj} (Y_{ij} - \overline{Y}_j)^2$$

$$+ \sum_{j=1}^{M} N_j(\overline{Y}_j - \tilde{Y}_j)^2 + \sum_{j=1}^{M} N_j(\tilde{Y}_j - \overline{Y})^2$$

$$(20.47) \quad SS_T = SS_{\text{residual}} + SS_{\text{deviations from linear}} + SS_{\text{linear}}$$

$$(20.48) \quad df_T = df_{\text{residual}} + df_{\text{deviations from linear}} + df_{\text{linear}}$$

$$(20.49) \quad N_T - 1 = (N_T - M) + (M - 2) + (1)$$

$$(20.50) \quad E(MS_{\text{residual}}) = \sigma_\varepsilon^2$$

$$(20.51) \quad E(MS_{\text{between}}) = \sigma_\varepsilon^2 + \frac{N_T \eta_{Y|X}^2 \sigma_Y^2}{M-1}$$

$$(20.52) \quad E(MS_{\text{linear}}) = \sigma_\varepsilon^2 + N_T \rho_{XY}^2 \sigma_Y^2$$

$$(20.53) \qquad E(MS_{\text{deviations from linear}}) = \sigma_{\varepsilon}^2 + \frac{(\eta_{Y|X}^2 - \rho_{XY}^2)N_T\sigma_Y^2}{M-2}$$

Fisher Z_r Transformation. The Fisher Z_r transformation transforms the Pearson product-moment correlation coefficient into a variable that is approximately normal in distribution.

$$(20.54) \qquad Z_r = \tfrac{1}{2}\log_e \left(\frac{1 + r_{XY}}{1 - r_{XY}}\right)$$

$$(20.55) \qquad Z_\rho = \tfrac{1}{2}\log_e \left(\frac{1 + \rho_{XY}}{1 - \rho_{XY}}\right)$$

$$(20.56) \qquad t = \frac{r_{XY}\sqrt{N-2}}{\sqrt{1 - r_{XY}^2}}$$

$$(20.57) \qquad Z = \frac{Z_r - Z_\rho}{\sqrt{1/(N-3)}}$$

$$(20.58) \qquad N = \left[\frac{|Z_\alpha| + |Z_\beta|}{Z_{\rho_T} - Z_{\rho_A}}\right]^2 + 3$$

$$(20.59) \qquad Z = \frac{(Z_{r_H} - Z_{r_G}) - (Z_{\rho_H} - Z_{\rho_G})}{\sqrt{\dfrac{1}{N_H - 3} + \dfrac{1}{N_G - 3}}}$$

Kendall's Tau. Kendall's tau coefficient is a measure of concordance of ordered observations, depending on the number of inversions of order in the jointly distributed orders.

Spearman Rank Order Correlation. The Pearson product-moment correlation of ranks is called the Spearman rank order correlation of the ranks. A special calculational form is defined by the sums and sums of squares of integers.

$$(20.60) \qquad r_S = 1 - \frac{6\sum_{i=1}^{N} D_i^2}{N(N^2 - 1)}$$

$$(20.61) \qquad t = \frac{r_S\sqrt{N-2}}{\sqrt{1 - r_S^2}}$$

$$(20.62) \qquad \tau = 1 - \frac{2(\text{number of inversions})}{\text{number of pairs}}$$

$$(20.63) \qquad Z = \frac{\tau - 1 \Big/ \dbinom{N}{2}}{\sqrt{\dfrac{2(2N + 5)}{9N(N - 1)}}}$$

Point Biserial Correlation. The correlation between a dichotomous variable and a continuous variable is called the point biserial correlation.

$$(20.64) \qquad r_p = \frac{(\bar{Y}_1 - \bar{Y}_0)\sqrt{Npq}}{\sqrt{\sum_{i=1}^{N}(Y_i - \bar{Y})^2}}$$

Fourfold Correlation, Phi. The correlation between a pair of dichotomous variables is called the fourfold correlation, or phi coefficient.

$$(20.65) \qquad r_{XY} = \phi = \frac{ad - bc}{\sqrt{(a + b)(c + d)(a + c)(b + d)}}$$

$$(20.66) \qquad \phi_{\max} = \sqrt{\left(\frac{p_X}{q_X}\right)\left(\frac{q_Y}{p_Y}\right)}$$

PROBLEMS

1. Define four random variables on the experiment "toss a coin four times, noting the sequence of heads and tails": (a) the number of heads on the first two tosses, (b) the number of heads on the last two tosses, (c) the number of heads on the first three tosses, and (d) the number of heads on the last three tosses. Develop the bivariate distribution of the following pairs of variables: (a, b), (a, c), (a, d), and (b, c). Show the conditional expectations, conditional variances, and probabilities expected under a hypothesis of independence. Calculate the coefficients of correlation η^2 and ρ for the four pairs of variables, commenting on the relative values of the coefficients for the four sets in terms of the overlap in the definition of the variables.

2. Perform an experiment to sample from the random variables defined in problem 1. That is, actually toss a coin four times, noting the sequence of heads and tails, repeating the four-toss procedures N times (select a practical value of N, say 20 or 25). Record values from the replications of the experiment for all four random variables. Plot the joint frequency of the variables, taken pairwise. Compare the relative frequencies in the joint distribution of the obtained data with the joint probability distribution of the random variables. Calculate estimates of the correlation coefficients η^2 and ρ from the obtained data, that is, E^2 and r. Test the significance of the departure of r from the expected value ρ. Test the significance of departure of r and E^2 from 0.

3. Substitute the values $+1.0$ and -1.0 for ρ in equations (20.24) and (20.25), explaining the implication of the consequences of these substitutions.

4. Select a sample of N standard unit normal random numbers, simulating a three-variable problem, calling the variables X, Y, and Z respectively. These three variables

are theoretically independent and uncorrelated. Calculate the statistics and produce the graphs for these data, similar to those that were calculated and drawn for the data of Table 20.6. Modify two of the variables, say X and Y, in the following way. Sample N values from a table of standard unit random numbers, say W, and form two new variables, $U = X + 5W$ and $V = Y + 10W$. Calculate the correlation coefficients for all of the variables involved, X, Y, Z, U, V, and W. Make a logical analysis of the coefficients, illustrating the operation of the presence of the implicit variable W in the more or less variable form (variance contribution of W to U is 25 and to V is 100). In particular, calculate the regression of U and V on X and Y respectively, and calculate the regression of U and V on X and Y respectively, and calculate the residual values of U and V about the regression lines. Compare the variability of these residual values as a function of the contribution of W to U and V.

5. Using data selected for problem 4, for the variables X, Y, and Z, perform statistical tests regarding the presence of significant correlations, both linear and nonlinear. Use the analysis of variance procedures analysis, selecting an appropriate number of class intervals on the independent variable, in order to test hypotheses about the regression of X on Y and of Z on Y.

6. By reexpressing the sums and sums of squares in the correlation coefficient, equation (20.35), as the sums and the sums of squares of the first N integers, and substituting these quantities into equation (20.35), the equivalence between (20.35) and (20.60) can be shown. Carry out this proof by noting that the sum of the first N integers is $N(N + 1)/2$ and that the sum of squares of the first N integers is $N(N + 1)(2N + 1)/6$.

7. Place the data listed below into rank orders, each variable separately ranked. Calculate the Spearman rank order correlation on the ranked data and compare the resultant value with the correlation calculated with the Pearson product-moment equation. The difference in the two procedures is in the transformation to ranks.

Replication	X Variable	Y Variable
1	26	14
2	32	19
3	30	16
4	28	17
5	38	18
6	31	20
7	36	21
8	24	15

8. For the data of problem 7, test the hypothesis that the Pearson product-moment correlation is significantly different from zero. Apply the test of the hypothesis of zero Spearman rank order correlation given in equation (20.61) to the same data. Comment on the differences in the significance probabilities of the two tests.

9. For the data of problem 7, calculate the Kendall tau coefficient and use this coefficient in the hypothesis that the expected value of tau is zero.

10. Using data of Table 20.6, dichotomize the Y variable using the median as the point of dichotomy, assigning the value 1 to large values and 0 to values less than the median. On the basis of this dichotomy calculate the point biserial correlation coefficient. Draw the scatter plot of the X variable as a function of the Y variable. Calculate the t test of the hypothesis that the difference between the mean values of X in each of the dichotomized-Y groups is zero. This is a test of the hypothesis of no correlation. Compare the significance probability from the test of the hypothesis of zero correlation based on the Pearson product-moment correlation.

11. Dichotomize the X variable as well as the Y variable from Table 20.6 and calculate the fourfold correlation coefficient. Perform a test of the hypothesis of a zero correlation using the χ^2 test indicated in the text.

12. Write an essay discussing the basis of the different results one obtains in the various tests of hypotheses about correlation, for problem 11.

CHAPTER 21

Pearson Chi Square

In Chapter 12 the sampling distribution of sums of squares of unit normal random deviates was introduced as the chi square distribution. That particular development is generally a departure from what most textbooks on applied statistics present as "chi square." Ordinarily, references to χ^2 distributions are in connection with so-called tests of goodness of fit and tests of independence of jointly distributed qualitative or categorical observations. In this chapter we discuss the use of an approximation of the χ^2 distribution presented in Chapter 12 in tests of hypotheses about the goodness of fit and independence. Karl Pearson showed that under certain circumstances, the χ^2 distribution (as in Chapter 12) was an approximation to certain distributions arising from calculations comparing theoretical and observed frequencies. Use of these results have long been called χ^2 tests, a slight misnomer—they properly should be called Pearson χ^2 tests. We shall call them Pearson χ^2 tests. Applications of the Pearson χ^2 to several research problems are presented in this chapter. Derivations of the results presented here are not presented. The derivations are dependent on mathematical sophistication that is beyond the level assumed in this book. We deal with a "heuristic" derivation and leave out the detail.

Three kinds of applications of Pearson χ^2 are developed in this chapter. The first is a test of the hypothesis that obtained frequencies for a set of categories, or a discrete valued random variable, are consistent (in a statistical sampling sense) with the frequencies expected on the basis of some theory or model. That is, we have k simple events in a model of an experiment with probabilities $p_1, p_2, \ldots, p_k$. The experiment is repeated n times, giving rise to expected frequencies of np_1, $np_2, \ldots, np_k$. An empirical experiment gives rise to observation of each of the simple results, corresponding to the simple events in the model, with frequencies $n_1, n_2, \ldots, n_k$. Pearson χ^2 provides a way of testing the hypothesis that the expected and the observed frequencies are the same, comparing the expected with the obtained in each category for all categories simultaneously. The second application is a test of hypotheses about the joint distribution of discrete valued random variables or jointly distributed nominal sample spaces (categories). This application can be a simple test of the hypothesis of expected frequencies, as in the first application just described. Or the application can be framed in such a way that it is a test of the independence of the jointly distributed variables or events. The formal properties of this application are similar to the first described application, with the exception that the k categories are cells in the joint sample space of the experiment. The third application is to obtain a measure of the degree of association between jointly distributed sets of categories, that is, in pairs of nominal categories in a sample space. This application is related to the fourfold correlation coefficient ϕ presented in Chapter 20. However, it is extended to cover experiments with r by c jointly distributed categories.

The General Case and Its Theoretical Basis

Although there are other techniques for deriving the Pearson χ^2, the most readily understood is the argument from the multinomial and binomial random variables. For the binomial, in which each event observed is independent of the others, with constant probability p of being in category S, and q of being in category F, the probability of s S's out of n trials is the binomial probability. The variable S has an expectation np and a variance npq. In Chapter 10 we pointed out that the standardized binomial probability distribution was approximately normal in form when n was very large. Consequently $S^* = (S - np)/\sqrt{npq}$ is unit standard normal, $N(0, 1)$, in the limit on n. Also, it follows that the square of S^* is distributed as a χ^2 variable with one degree of freedom:

$$Z_1^2 = \frac{(S - np)^2}{npq} \text{ is distributed as } \chi_1^2 \tag{21.1}$$

The right-hand term in equation (21.1) can be reexpressed by simple algebra, for a particular value of s:

$$\frac{(s - np)^2}{npq} = \frac{(s - np)^2}{np} + \frac{(n - s - nq)^2}{nq} \tag{21.2}$$

Note that s is the observed frequency of events in category S, say O_S, and np is the expected frequency of events in category S, say E_S; likewise, $(n - s)$ and nq are the observed and expected frequencies of events in category F, say O_F and E_F. The upshot of this is the standard statistics textbook expression for Pearson χ^2 in a two-category distribution with expected frequencies E_S and E_F in the two categories:

$$Z_1^2 = \frac{(O_S - E_S)^2}{E_S} + \frac{(O_F - E_F)^2}{E_F} \tag{21.3}$$

Under the assumptions of the binomial and with unlimited sample size, (21.3) is distributed as a χ^2 variable with one degree of freedom. We must know the value of p in order to calculate the expected frequencies. The degree of goodness of the approximation of the χ^2 distribution to the statistic (21.3) depends on the sample size, and the satisfaction of the independence rule in the observations. In order to get a fair idea of the degree of approximation in terms of sample size, recall the illustrations of the fit of a normal density function to the binomial probability distribution in Chapter 10. In addition, the value of p plays a role in the degree of goodness of the approximation. Extreme values of p give less satisfactory degrees of approximation. In Pearson χ^2 applications this is usually translated into statements about the minimal value of np, a function of sample size as well as p. If p is extreme (small), then n would have to be large to produce a satisfactory approximation. Hence, for small values of p if there are enough trials, observations, to make np equal to or greater than some minimum, the approximation is judged to be satisfactory. The usual rule is that in a two-category experiment that $np \geq 10$ is sufficient. If p is greater than .5, the rule is applied to $(1 - p = q)$, so that $nq \geq 10$ is the standard.

In the two-category experiment the use of the above result would probably be second choice in tests of hypotheses about p. The theory of the binomial random variable, normal approximations to it, and the appropriate tables are accessible enough to make the χ^2 approach unattractive because of the extra calculational work and the fact that it is an approximation. However, this is not true when we have a multiple-category experiment. Imagine an experiment with event categories $e_1, e_2, \ldots, e_k$, with event probabilities $p_1, p_2, \ldots, p_k$. If the experiment is repeated n times, each event in the event set will occur, respectively, $n_1, n_2, \ldots, n_k$ times, where $n_1 + n_2 + \cdots + n_k = n$. The probability of observing a specific combination of the number of events in the k categories is given by the multinomial probability function

$$P\left(\bigcap_{i=1}^{k} n_i\right) = \frac{n!}{n_1!\, n_2! \cdots n_k!} p_1^2 p_2^2 p_3^2 \cdots p_k^2 \tag{21.4}$$

If the number n is unlimitedly large, and if the k categories are mutually exclusive, and if the trials of the experiment are strictly independent, then, by numerical approximation it can be shown that

$$Z = \frac{n_i - np_i}{\sqrt{np_i}}$$

is approximately unit standard normal, $N(0, 1)$. Squaring this equation and summing over the k categories gives

$$\frac{(n_1 - np_1)^2}{np_1} + \frac{(n_2 - np_2)^2}{np_2} + \cdots + \frac{(n_k - np_k)^2}{np_k} \tag{21.5}$$

This expression is much like the right-hand portion of expression (21.2) which was χ^2 distributed with $(k - 1)$ degrees of freedom, $k = 2$ in the special case. It can be shown that the expression (21.5) can be expressed as a Pearson χ^2 distribution with $(k - 1)$ degrees of freedom, Z_{k-1}^2, in accordance with our notation of Chapter 10. The usual notation for this is

$$Z_{k-1}^2 = \frac{(O_1 - E_1)^2}{E_1} + \frac{(O_2 - E_2)^2}{E_2} + \cdots + \frac{(O_k - E_k)^2}{E_k} \tag{21.6}$$

In general, for k categories defined as above, the probability distribution of the quantity in equation (21.6) is approximately χ^2 with $(k - 1)$ degrees of freedom. In a later section of this chapter we shall introduce the notion that a set of jointly distributed categories can be treated essentially the same way with the same result. However, under certain circumstances, the parameters in the cells of the jointly distributed categories may be estimated by the marginal cells, and this introduces restrictions that in turn show up in a reduction of the number of degrees of freedom in the accumulation similar to equation (21.5). Wherever the parameters p_i in expression (21.5) are provided by some theory or hypothesis (such as a model of the experiment), the resulting Pearson χ^2 distributed variable has $(k - 1)$ degrees of freedom. By using portions of the data in an actual experiment to estimate the parameters, restrictions on the deviations of the observed frequencies and the expected frequencies, np_i as contrasted with n_i, are introduced with a consequent loss of degrees of freedom in the distribution.

The number of categories in the multicategory experiment has influenced practice regarding the number of expected events in each category needed before the approximations are satisfactory. A conservative stance requires ten or more in each category, that is, $np_i \geq 10$. However, a more liberal rule, requiring an expected frequency of 5 or more in each category, is often cited in the literature, particularly in connection with relatively numerous categories (k of ten or more).

Pearson χ^2 is not precisely χ^2 as introduced in Chapter 10. However, the approximation is satisfactory if the assumptions underlying the Pearson derivations are met. In the first place, the categories of the experiment, the events of the sample space of the experiment, must be mutually exclusive. Second, the observations of the experiment must be independent. Third, the number of observations must be sufficiently large to assure that the numerical approximations involved in using the χ^2 distribution are sufficiently good. This last requirement is generally expressed as a minimum on the value of the expected frequency in the categories of the experiment.

The use of the Pearson χ^2 in tests of hypotheses depends on the hypotheses and the manner in which the parameters are specified. In the examples used above, the values of the probabilities of the categories of observation were assumed to be given by some model by hypothesis. There are two primary ways of determining the parameters. The first is from a statement of a hypothesis specifying the probability distribution on the sample space of the experiment. For example, an experiment based on the hypothesis that a variable is uniformly distributed, discrete valued, random variable. If the hypothesis of uniformity is true, the probabilities associated with each of the values in the random variable are $1/k$, where there are k values in the value set of the random variable. If we are interested in testing the independence of jointly distributed discrete random variables or nominal sets of categories that are jointly defined on a sample space, the marginal relative frequencies can be used as estimates of the probabilities in the separate variables or sets of categories. The product rule for independence permits specification of the joint probabilities as a product of the estimates of the marginal probabilities. The estimates of the joint probabilities are then used to calculate the expected frequencies.

Hypotheses about Goodness of Fit

Goodness of fit is a phrase used to describe the degree to which a theoretical probability distribution of some variable, set of categories, or jointly distributed categories (or variables) is the same as or different from the corresponding obtained relative frequencies. Three special forms of the goodness-of-fit procedures are given here.

Imagine a simple, discrete valued random variable specified by some theory of behavior, for example, the random variable T in Table 21.1. The random variable might be specified as the number of trials required to complete some puzzle and the probabilities determined from a theoretical model of problem solving. In order to test the hypothesis that the distribution of trials to solution is empirically the distribution specified in the model, we use the Pearson χ^2 test of goodness of fit. Table 21.1 also includes the expected frequencies for a sample of 100 observations and the obtained frequencies in the 100 observations. The probabilities $P(T = t)$ are provided entirely by the theory of problem solving. The expected frequencies are simply the number of observations to be made in the experiment times the probabilities of the various categories. The table also lists the frequencies that might have been observed in an actual experiment. Evaluating equation (21.6) for the data and model of Table 21.1 gives

$$Z_{5-1}^2 = \frac{(18 - 20)^2}{20} + \frac{(25 - 30)^2}{30} + \frac{(32 - 25)^2}{25}$$
$$+ \frac{(12 - 15)^2}{15} + \frac{(13 - 10)^2}{10}$$
$$= \frac{4}{20} + \frac{25}{30} + \frac{49}{25} + \frac{9}{15} + \frac{9}{10}$$
$$= 4.493$$

Looking up the χ^2 values for 4 degrees of freedom in the χ^2 table, we find the significance probability of such a value is between .5 and .3. Consequently, we would not reject the hypothesis that the observed frequencies could come from an experiment sampling the random variable with the specified probability distribution.

In general, when some theory specifies the probabilities in a discrete valued variable, as in our example of Table 21.1, there will be k categories and $(k - 1)$ degrees of freedom for a test of the hypothesis that the probabilities are as specified

TABLE 21.1 Example Data Used in a Pearson χ^2 Goodness-of-Fit Test

Trials to Solution, T	1	2	3	4	5
$P(T = t)$	.20	.30	.25	.15	.10
Expected Frequency: $nP(T = t)$	20	30	25	15	10
Observed Frequency	18	25	32	12	13

by the theory. The test hypothesis is given in definite terms by the statement of the probabilities. The test of the hypothesis is by way of the Pearson χ^2 statistic of equation (21.5) or (21.6). The test hypothesis is rejected if the χ^2 distribution indicates a significance probability smaller than the level of significance selected by the investigator prior to the test. Notice that the sample size does not play a direct role in the degrees of freedom, only in determining the expected frequencies in the calculation.

In another application, we test the hypothesis that a continuous random variable has some specified density function. In continuous functions, if the measurement or observational process were sufficiently accurate, the expected frequency of any given value in the value set is zero. Consequently we must aggregate values in the value set of the random variable and deal with the aggregates as categories or values in a discrete random variable. The trick is to select the aggregates in an appropriate way. Imagine that considerations of constant and function indicated that if our theory is correct, the density function of a certain random variable X is the normal density function. The theory must be more specific if we are to proceed. The expectation and variance of the distribution must be specified before we can deal with the appropriate density function. Imagine a density function of a variable with an expectation of 50 and a standard deviation of 10. The first step is to define several class intervals on the value set of the random variable, the categories to be used in the evaluation of (21.6) for the Pearson χ^2. Two options are immediately apparent—equal class intervals with unequal probabilities, and unequal class intervals with equal probabilities. The latter option is the more attractive of the two. In the former option we find categories, class intervals, that have very tiny probabilities. If we ignore these categories, we shall not have an exhaustive set of categories. If we do not ignore these categories, we shall never be able to obtain a large enough sample size to be able to have the expected frequency in those intervals large enough to satisfy our basic assumptions. The choice of intervals on the equal probability criterion permits us to include the extreme values in the two most extreme class intervals and still meet the equal probability criterion. The number of class intervals can be selected to balance with the number of observations that will be made. Under any circumstances, the number of intervals should be no more than the smallest number giving an expected frequency of 5 or more in each interval. Thus, if there were to be 100 observations made, the number of intervals at maximum would be 20. Table 21.2 presents an example in which a normal random variable with expectation of 50 and standard deviation 10 is partitioned into eight segments, each with probability $\frac{1}{8}$. Also, in Table 21.2, the expected frequency and illustrative "empirical" or obtained frequencies from an experiment with $n = 400$ are listed.

TABLE 21.2 Equal Probability Class Intervals on a Normally Distributed Random Variable with Expectation of 50 and Standard Deviation of 10, with Illustrative Data

Class Interval	Probability	Expected Frequency	Obtained Frequency
61.5 and above	$\frac{1}{8}$	50	75
56.8 to 61.5	$\frac{1}{8}$	50	70
53.2 to 56.8	$\frac{1}{8}$	50	58
50 to 53.2	$\frac{1}{8}$	50	46
46.8 to 50	$\frac{1}{8}$	50	51
43.2 to 46.8	$\frac{1}{8}$	50	40
38.5 to 43.2	$\frac{1}{8}$	50	35
Below 38.5	$\frac{1}{8}$	50	25

The Pearson χ^2 calculation is based on ($k = 8$) categories with no special restrictions from the data, and hence the number of degrees of freedom is $k - 1$, or 7. The value obtained from the calculation is

$$Z_7^2 = \frac{(75 - 50)^2}{50} + \frac{(70 - 50)^2}{50} + \frac{(58 - 50)^2}{50} + \frac{(46 - 50)^2}{50}$$

$$+ \frac{(51 - 50)^2}{50} + \frac{(40 - 50)^2}{50} + \frac{(35 - 50)^2}{50} + \frac{(25 - 50)^2}{50}$$

$$= 41.12$$

This statistic indicates that a highly significant departure from normality is present in the obtained frequencies, the .005 level of significance having a χ^2 value of 20.3.

The class intervals used in this example were selected quite arbitrarily. Other class intervals are just as reasonable. In an experiment with as large a sample size as the one just reported, the number of intervals could be as large as 80 and still preserve the satisfactory approximation in the Pearson χ^2. Although the data reported in the experiment leave little doubt that the frequency distribution is nonnormal in form, there might be less assurance if there had been a less extreme test statistic. In that case, we might have learned more about the actual distribution by having a more finely segmented value set. In particular, when there are relatively rapid changes in the probability distribution specified by the model, and the class intervals are defined so broadly that one interval covers the subset of values in which the rapid change occurs, the test will be insensitive to departures from the shape in that interval. In general, we need to know what portions of the value set of the variable are of most interest. These portions of the value set are then represented with appropriately narrow class intervals in order to ascertain the degree of fit in those intervals with the desired precision. In portions of the value set where there is little need for precision, the class intervals can be relatively broad. However, it should be pointed out that when the class intervals are all of the same probability, the Pearson χ^2 statistic is more nearly approximated by the χ^2 distribution.

In the example above, we assumed that the expectation and the variance of the random variable were 50 and 10, respectively. In most applications this information is unknown. However, it is necessary to know these parameters, or to have some estimate of them, before the class intervals can be defined. If the values that are used to calculate the class intervals are not actually the appropriate values, then the observed values will not be placed in appropriate class intervals with respect to the expected frequencies. Imagine that the expectation of the variable were 80 and the class intervals were calculated on the assumption that the expectation was 50. The actually observed values would almost all be above the highest class interval lower limit of 61.5 in the example of Table 21.2. A similar result would occur if the standard deviation were 20 instead of 10 as is assumed in the expected frequency calculations. Specifically, the probability of observing a score above the score value of 61.5 is $\frac{1}{8}$ in a distribution with expectation of 50 and standard deviation of 10— whereas if the expectation is 80 and the standard deviation is 10 the probability is approximately .968, and if the expectation is 50 and the standard deviation is 20 the probability is approximately .283.

In order to avoid comparing the wrong expected and observed frequency distributions, where there is some doubt about the expectation and standard deviation of the random variable, the sample mean and unbiased estimate of the variance from the sample are used to determine the class intervals. The values in the standard unit normal random variable satisfying the probabilities in the class interval are obtained from a table of that distribution. These then are converted to the value set of the nonstandardized random variable. Thus, in the data of Table 21.2 the lower limit of the upper interval is the value corresponding to the value of x in $P(X > x) = .125$. Consulting a table of the standard unit normal distribution, we find that ($x = 1.15$) fulfills the requirement. The calculation to get back to the location and dispersion of the nonstandard variable is performed simply by multiplying the unit standard normal deviate by the desired standard deviation and adding the desired expectations; $10(1.15) + 50 = 61.5$. All of the other values of the limits of the class intervals are found in a similar fashion. Frequently, the easiest way to proceed is to standardize the observed values to correspond to a variable with expectation of 0 and a standard deviation of 1. When either of these procedures is followed, restrictions are placed on the way the observed values will be assorted into the class intervals. No discrepancy between the observed and expected frequencies can be due to a discrepancy in the mean and the expectation or to a discrepancy between the standard deviations. In the initial procedure, where assumed parameters were used in determining the expected frequencies, any discrepancy between the assumed parameters and sample values entered into the Pearson χ^2 statistic. The restriction imposed on the Pearson χ^2 statistic can be shown to involve 2 degrees of freedom; 2 degrees of freedom are lost. Thus, the degrees of freedom for the test in which the mean and sample standard deviation are used in finding the class intervals is $(k - 1 - 2 = k - 3)$.

The third application of the Pearson χ^2 to a goodness-of-fit problem involves jointly distributed category sets or jointly distributed random variables. If continuous random variables are involved, definition of class intervals provides the mutually exclusive and discrete categories in the joint distribution much in the same way as in the single-variable example above. Table 21.3 illustrates

TABLE 21.3 Jointly Distributed Categories Indicating Theoretical Joint Distribution Probabilities and Illustrative Data

Basic layout	Freshman	Sophomore	Junior	Senior
Female	p_{11}	p_{12}	p_{13}	p_{14}
Male	p_{21}	p_{22}	p_{23}	p_{24}

Hypothesized Probabilities	Freshman	Sophomore	Junior	Senior
Female	.05	.10	.15	.20
Male	.20	.15	.10	.05

Data—Expected and Obtained Frequencies, n = 100	Freshman	Sophomore	Junior	Senior
Female	$E = 5$ $O = 8$	$E = 10$ $O = 16$	$E = 15$ $O = 10$	$E = 20$ $O = 16$
Male	$E = 20$ $O = 25$	$E = 15$ $O = 12$	$E = 10$ $O = 10$	$E = 5$ $O = 3$

the problem with two sets of categories that are jointly distributed: sex and class level in college.

Except for the fact that the categories are arranged along two dimensions, the structure of this experiment is the same as in the others we have just dealt with. The number of categories in each of the two sets of categories are not really involved in the test. The hypothesis is that the experiment is a sampling of events defined by the intersection of two elements from a more finely differentiated set of observations than for any one of the two sets of categories taken separately. We could readily conceptualize the joint categories as defining a set of mutually exclusive and exhaustive categories in the following way: female freshman, male freshman, female sophomore, . . . , male senior.

The Pearson χ^2 for the illustrative data of Table 21.3 is 10.52. The degrees of freedom for the test of the goodness of fit of the obtained frequencies and the expected frequencies is given by the number of categories minus 1, that is, $8 - 1 = 7$. A χ^2 distributed variable has probabilities of .25 of being greater than 9.04 and .1 of being greater than 12.0. Consequently, the significance probability of the Pearson χ^2 for the illustrative data of Table 21.3 is something less than .25 but somewhat greater than .1.

The reader should be alert to the difference in this last example and the examples in the next section on testing hypotheses about independence of jointly distributed categories and variables. In this example, we are testing the goodness of fit of two joint distributions, the expected and the obtained, with no implications whatsoever about dependence and independence.

Hypotheses about Independence and Association

The last example in the previous section illustrated the use of a Pearson χ^2 test of a joint distribution. A special use of this is encountered when the theoretical joint probabilities are the probabilities required by the product rule of independence of the jointly distributed categories or variables. Recall that independent jointly distributed variables are characterized by joint probabilities that are the products of the marginal probabilities. The same theorem applies to jointly distributed categorical, qualitative, events. If the marginal probabilities are known for two jointly distributed variables or categories, the joint probabilities can be calculated by multiplying the appropriate marginal probabilities. For example, Table 21.4 shows two sets of categories that are jointly distributed with known marginal probabilities. Also shown are the joint probabilities calculated by the product rule

TABLE 21.4 Illustration of Pearson χ^2 Procedures for Tests of Independence with Known Marginal Probabilities

Known Marginals

			A		
		a_1	a_2	a_3	Marginal on B
B	b_1				.4
	b_2				.6
Marginal on A		.2	.3	.5	

Implied Joint Probabilities for Independence

		A		
		a_1	a_2	a_3
B	b_1	$(.4)(.2) = .08$	$(.4)(.3) = .12$	$(.4)(.5) = .20$
	b_2	$(.6)(.2) = .12$	$(.6)(.3) = .18$	$(.6)(.5) = .30$

Expected and Obtained Frequencies with ($n = 200$) Observations

		A		
		a_1	a_2	a_3
B	b_1	$E = 16$ $O = 28$	$E = 24$ $O = 32$	$E = 40$ $O = 34$
	b_2	$E = 24$ $O = 31$	$E = 36$ $O = 36$	$E = 60$ $O = 39$

and a set of expected and observed frequencies obtained from ($n = 200$) observations. The Pearson χ^2 statistic is obtained in the usual manner:

$$Z_{6-1}^2 = \frac{(28 - 16)^2}{16} + \frac{(32 - 24)^2}{24} + \frac{(34 - 40)^2}{40}$$
$$+ \frac{(31 - 24)^2}{24} + \frac{(36 - 36)^2}{36} + \frac{(39 - 60)^2}{60}$$
$$= 21.96$$

The degrees of freedom for this application is determined by the number of categories in the joint distribution. Where there are c columns and r rows in the joint layout there are $(rc = k)$ categories in the joint distribution. Since no data were used to determine the joint probabilities, the number of degrees of freedom is 1 fewer than the number of categories in the sum, $(k - 1)$ or $[rc - 1 = (2)(3) - 1 = 6 - 1 = 5]$ in our example. The χ^2 value for level of significance of .005 is 16.7. The data indicate a significant departure from independence.

A more realistic application is the situation where the marginal probabilities are not known and must be estimated from the data. The best estimates that can be made are the relative frequency of observations for the marginal categories. These values are calculated by accumulating the number of observations obtained in each cell of the joint distribution for each column and for each row. The row totals and the column totals are divided by the total number of observations to obtain the relative frequency of observations in the row and column marginals. These relative frequencies are treated as estimates of the probabilities on the marginal distributions. The estimates of the marginal probabilities are then used to calculate the joint distribution expected under the hypothesis of independence. From that point, the calculation is precisely as in the first example. However, we have used the data to determine (estimate) r parameters for the rows and c parameters for the columns. Two of the $(r + c)$ parameters are fixed by the fact that the marginal probability estimates must sum to 1.0. In all, some $[(r - 1) + (c - 1)]$ parameters are estimated. For the $(k = rc)$ categories there are $(k - 1)$ degrees of freedom in the Pearson χ^2 without additional restrictions. However, there are $[(r - 1) + (c - 1)]$ restrictions associated with the parameters that are estimated. Consequently there are $[rc - 1 - (r - 1) - (c - 1) = (r - 1)(c - 1)]$ degrees of freedom for the test.

This application is illustrated with the data of Table 21.4. The obtained marginal frequencies are derived from the obtained joint frequencies. From these marginal frequencies, the relative frequencies and the estimates of the joint distribution probabilities are calculated, as illustrated in Table 21.5. The number of degrees of freedom is $[(2 - 1)(3 - 1) = 2]$, compared with 5 in the example where the marginal probabilities were known (not estimated). The Pearson χ^2 is calculated from the expected and obtained frequencies in the familiar way from the data in the lower portion of Table 21.5. The value of the statistic is .0003, too small to permit rejection of the hypothesis of independence.

In applications of this test to 2 by 2, or fourfold tables, that is, where the categories are both dichotomous, a special result is obtained. Consider the illustration of the fourfold point correlation table of Figure 20.17, and equation (20.65) for the ϕ coefficient. It can be shown algebraically that the Pearson χ^2 statistic for the test of independence, using the observed marginal relative frequencies to calculate the joint probabilities, is equivalent to $n\phi^2$. The Pearson χ^2 has 1 degree

TABLE 21.5 Illustration of the Use of Marginal Obtained Relative Frequencies in Pearson χ^2 Tests of Hypotheses of Independence

The Observed Data and Marginal Relative Frequencies with Product Rule Joint Probabilities

		a_1	a_2	a_3	Marginal Observed Values	
B	b_1	(.470)(.295) = .13865	(.470)(.340) = .15980	(.470)(.365) = .17155	94	.470
	b_2	(.530)(.295) = .15635	(.530)(.340) = .18020	(.530)(.365) = .19345	106	.530
Marginal Observed Values		59	68	73	200	
		.295	.340	.365		

(header: *A*)

Expected and Observed with ($n = 200$) Observations

A

		a_1	a_2	a_3
B	b_1	$E = 27.7$ $O = 28$	$E = 32.0$ $O = 32$	$E = 34.3$ $O = 34$
	b_2	$E = 31.3$ $O = 31$	$E = 36.0$ $O = 36$	$E = 38.7$ $O = 39$

of freedom in this special case: $(2 - 1)(2 - 1) = 1$. It can be shown that the χ^2 approximation is given by

$$Z_1^2 = \frac{n(ad - bc)^2}{(a + b)(c + d)(a + c)(b + d)} \tag{21.7}$$

A slightly better approximation is achieved if the Yates correction for continuity is used:

$$Z_1^2 = \frac{n(|ad - bc| - n/2)^2}{(a + b)(c + d)(a + c)(b + d)} \tag{21.8}$$

It should be noted that only in the 2 by 2 design should the correction for continuity be used. The Pearson χ^2 is thus a test for the departure of the fourfold point correlation coefficient from 0.

The relationship between ϕ and Pearson χ^2 can be generalized to obtain a measure of strength of association from jointly distributed categories. Reversing the roles of Z^2 and ϕ, we have $\phi^2 = Z^2/n$. In an r by c joint distribution where the obtained marginals are used to calculate the expected frequencies by the product rule for independence, the Pearson χ^2 statistic has $[d = (r-1)(c-1)]$ degrees of freedom: $Z_d{}^2$. Cramér defined a statistic with range between 0 and 1 that is sometimes interpreted as a measure of the degree of association of the two sets of categories:

$$\phi' = \sqrt{\frac{\phi^2}{S-1}} = \sqrt{\frac{Z_d{}^2}{n(S-1)}} \qquad (21.9)$$

where S is the smaller of r and c.

The meaning of this statistic is not intuitively clear. However, it is clear that it is not a measure of shared variance or predictability in the sense that the product-moment correlation coefficient is.

Some Calculational Conveniences

The equation most well known in the calculation of Pearson χ^2 statistics is

$$\sum_{i=1}^{k} \frac{(O_i - E_i)^2}{E_i}$$

This equation can be considerably simplified by some algebra. The numerator is squared out and the summation distributed through the simplified result. Noting that

$$\sum_{i=1}^{k} O_i = \sum_{i=1}^{k} E_i = n$$

and making the corresponding substitutions, we obtain the result

$$\sum_{i=1}^{k} \frac{(O_i - E_i)^2}{E_i} = \sum_{i=1}^{k} \frac{O_i{}^2}{E_i} - n \qquad (21.10)$$

Where the data are expressed as proportions for observed values, and probabilities for expected values, formula (21.10) is algebraically equivalent to

$$n\left(\sum_{i=1}^{k} \frac{f_i{}^2}{p_i} - 1\right) \qquad (21.11)$$

where f_i is the proportion, the relative frequency, of observations in the ith category and p_i is the probability of an observation in that category.

The calculational forms of equations (21.10) and (21.11) are often of greater convenience than the more conventional form. The algebraic equivalence of the conventional form and the forms given in this section permit the use of the form most convenient for a given application, with the same value being obtained regardless of the calculational form used.

SUGGESTED READING

Maxwell, A. E. (1961), *Analysing qualitative data*, Wiley, New York.

GLOSSARY

Goodness of Fit Test. A test of the hypothesis that obtained frequencies for a set of categories, or a discrete valued random variable, are consistent within sampling limits with the frequencies expected on the basis of some theory or model, usually a probability distribution.

Test of Independence. A test of the hypothesis that obtained joint category or discrete valued random variables are consistent within sampling limits with the frequencies expected on the basis of the hypothesis of independence of the categories or variables.

Pearson Chi Square. Under certain circumstances the statistic calculated by equations (21.3), (21.6), and others in this chapter are approximately chi square distributed and are called Pearson chi square test statistics.

(21.1) $$Z_1{}^2 = \frac{(S - np)^2}{npq} \text{ is distributed as } \chi_1{}^2$$

(21.2) $$\frac{(s - np)^2}{npq} = \frac{(s - np)^2}{np} + \frac{(n - s - nq)^2}{nq}$$

(21.3) $$Z_1{}^2 = \frac{(O_S - E_S)^2}{E_S} + \frac{(O_F - E_F)^2}{E_F}$$

(21.4) $$P\left(\bigcap_{i=1}^{k} n_i\right) = \frac{n!}{n_1!\, n_2! \cdots n_k!} p_1{}^2 p_2{}^2 p_3{}^2 \cdots p_k{}^2$$

(21.5)
$$\frac{(n_1 - np_1)^2}{np_1} + \frac{(n_2 - np_2)^2}{np_2} + \cdots + \frac{(n_k - np_k)^2}{np_k}$$

(21.6)
$$Z_{k-1}^2 = \frac{(O_1 - E_1)^2}{E_1} + \frac{(O_2 - E_2)^2}{E_2} + \cdots + \frac{(O_k - E_k)^2}{E_k}$$

(21.7)
$$Z_1^2 = \frac{n(ad - bc)^2}{(a + b)(c + d)(a + c)(b + d)}$$

(21.8)
$$Z_1^2 = \frac{n(|ad - bc| - n/2)^2}{(a + b)(c + d)(a + c)(b + d)}$$

(21.9)
$$\phi' = \sqrt{\frac{\phi^2}{S - 1}} = \sqrt{\frac{Z_d^2}{n(S - 1)}}$$

(21.10)
$$\sum_{i=1}^{k} \frac{(O_i - E_i)^2}{E_i} = \sum_{i=1}^{k} \frac{O_i^2}{E_i} - n$$

(21.11)
$$n\left(\sum_{i=1}^{k} \frac{f_i^2}{p_i} - 1\right)$$

PROBLEMS

1. Carry out the algebra expressed in equation (21.2).

2. Toss a coin four times and record the number of heads. This experiment is modeled by a binomial with $n = 4$ and $p = .5$. No comparison of the single result against the binomial probability distribution is possible with the single repetition of the experiment. Determine the number of repetitions that are necessary for a test of the hypothesis that the distribution is binomial. Perform the experiment the specified number of times and test the hypothesis with a goodness-of-fit χ^2 test.

3. Using the data actually gathered in the experiment of problem 2, test the goodness of fit of the normal distribution to the obtained data using $N(np, npq)$ as the normal distribution. Write down a justification for the number of class intervals used in the normal model.

4. Take 100 numbers from the table of the uniformly distributed random numbers at the back of the book. Test the hypothesis that the population from which these numbers were drawn is uniformly distributed. Defend your selection of the number of class intervals for the test. What is the expected frequency with which the hypothesis will be rejected in this exercise?

5. A model specifies that two dichotomous categorical events are jointly distributed with probabilities indicated in the left-hand table of the two tables below and the obtained frequencies in the right-hand table. Test the hypothesis that the data fit the joint distribution.

	A_1	A_2		A_1	A_2
B_1	.1	.4	B_1	18	79
B_2	.2	.3	B_2	40	63

6. Using the obtained data from problem 5, test the hypothesis that A and B are independent. If the obtained frequencies fit the joint probability distribution of problem 5, is the test of independence appropriate as compared with an application to the model of the rule for independent outcomes?

7. The following data are collected in an experiment in which three B categories and four A categories are defined, each observation entering into one or another of the 12 cells of the layout. Test the hypothesis that the joint probability distribution of A and B is the uniform distribution. Also, test the hypothesis that the two category sets are independent. Discuss the differences of these two tests.

	A_1	A_2	A_3	A_4
B_1	10	14	22	17
B_2	12	16	20	24
B_3	14	17	24	26

Appendix: Tables and Charts

ACKNOWLEDGMENTS

Acknowledgment is made to the authors, editors, and publishers whose materials are used in this appendix; I wish to thank them for permission to reproduce the materials. The tables and charts were reproduced from the following publications.

Beyer, W. H. (1966), *Handbook of tables for probability and statistics*, The Chemical Rubber Co., Cleveland, Ohio.

Davies, O. L. (1956), *Design and analysis of industrial experiments*, Vol. 1, Oliver and Boyd, London.

Eisenhart, C., Hastay, M. W., and Wallis, W. A. (1947), *Selected techniques of statistical analysis*, McGraw-Hill, New York.

Fisher, R. A., and Yates, F. (1938), *Statistical tables for biological, agricultural and medical research*, Oliver and Boyd, London.

Guenther, W. C. (1965), *Concepts of statistical inference*, McGraw-Hill, New York.

Hodges, J. L., Jr., and Lehmann, E. L. (1964), *Basic concepts of probability and statistics*, Holden-Day, San Francisco.

Owen, D. B. (1962), *Handbook of statistical tables*, John Wiley and Sons, Inc., New York.

Pearson, E. S., and Hartley, H. O. (1962), *Biometrika tables for statisticians*, Vol. 1, Biometrika Trustees, London.

Pearson, E. S., and Hartley, H. O. (1951), Power Functions of the Analysis-of-Variance Tests, *Biometrika*, Vol. *38*, 112–130.

TABLE A.I Uniformly Distributed Random Numbers

Line/Col.	(1)	(2)	(3)	(4)	(5)	(6)	(7)	(8)	(9)	(10)	(11)	(12)	(13)	(14)
1	10480	15011	01536	02011	81647	91646	69179	14194	62590	36207	20969	99570	91291	90700
2	22368	46573	25595	85393	30995	89198	27982	53402	93965	34095	52666	19174	39615	99505
3	24130	48360	22527	97265	76393	64809	15179	24830	49340	32081	30680	19655	63348	58629
4	42167	93093	06243	61680	07856	16376	39440	53537	71341	57004	00849	74917	97758	16379
5	37570	39975	81837	16656	06121	91782	60468	81305	49684	60672	14110	06927	01263	54613
6	77921	06907	11008	42751	27756	53498	18602	70659	90655	15053	21916	81825	44394	42880
7	99562	72905	56420	69994	98872	31016	71194	18738	44013	48840	63213	21069	10634	12952
8	96301	91977	05463	07972	18876	20922	94595	56869	69014	60045	18425	84903	42508	32307
9	89579	14342	63661	10281	17453	18103	57740	84378	25331	12566	58678	44947	05585	56941
10	85475	36857	43342	53988	53060	59533	38867	62300	08158	17983	16439	11458	18593	64952
11	28918	69578	88231	33276	70997	79936	56865	05859	90106	31595	01547	85590	91610	78188
12	63553	40961	48235	03427	49626	69445	18663	72695	52180	20847	12234	90511	33703	90322
13	09429	93969	52636	92737	88974	33488	36320	17617	30015	08272	84115	27156	30613	74952
14	10365	61129	87529	85689	48237	52267	67689	93394	01511	26358	85104	20285	29975	89868
15	07119	97336	71048	08178	77233	13916	47564	81056	97735	85977	29372	74461	28551	90707
16	51085	12765	51821	51259	77452	16308	60756	92144	49442	53900	70960	63990	75601	40719
17	02368	21382	52404	60268	89368	19885	55322	44819	01188	65255	64835	44919	05944	55157
18	01011	54092	33362	94904	31273	04146	18594	29852	71585	85030	51132	01915	92747	64951
19	52162	53916	46369	58586	23216	14513	83149	98736	23495	64350	94738	17752	35156	35749
20	07056	97628	33787	09998	42698	06691	76988	13602	51851	46104	88916	19509	25625	58104
21	48663	91245	85828	14346	09172	30168	90229	04734	59193	22178	30421	61666	99904	32812
22	54164	58492	22421	74103	47070	25306	76468	26384	58151	06646	21524	15227	96909	44592
23	32639	32363	05597	24200	13363	38005	94342	28728	35806	06912	17012	64161	18296	22851
24	29334	27001	87637	87308	58731	00256	45834	15398	46557	41135	10367	07684	36188	18510
25	02488	33062	28834	07351	19731	92420	60952	61280	50001	67658	32586	86679	50720	94953
26	81525	72295	04839	96423	24878	82651	66566	14778	76797	14780	13300	87074	79666	95725
27	29676	20591	68086	26432	46901	20849	89768	81536	86645	12659	92259	57102	80428	25280
28	00742	57392	39064	66432	84673	40027	32832	61362	98947	96067	64760	64584	96096	98253
29	05366	04213	25669	26422	44407	44048	37937	63904	45766	66134	75470	66520	34693	90449
30	91921	26418	64117	94305	26766	25940	39972	22209	71500	64568	91402	42416	07844	69618
31	00582	04711	87917	77341	42206	35126	74087	99547	81817	42607	43808	76655	62028	76630
32	00725	69884	62797	56170	86324	88072	76222	36086	84637	93161	76038	65855	77919	88006
33	69011	65797	95876	55293	18988	27354	26575	08625	40801	59920	29841	80150	12777	48501
34	25976	57948	29888	88604	67917	48708	18912	82271	65424	69774	33611	54262	85963	03547
35	09763	83473	73577	12908	30883	18317	28290	35797	05998	41688	34952	37888	38917	88050
36	91567	42595	27958	30134	04024	86385	29880	99730	55536	84855	29080	09250	79656	73211
37	17955	56349	90999	49127	20044	59931	06115	20542	18059	02008	73708	83517	36103	42791
38	46503	18584	18845	49618	02304	51038	20655	58727	28168	15475	56942	53389	20562	87338
39	92157	89634	94824	78171	84610	82834	09922	25417	44137	48413	25555	21246	35509	20468
40	14577	62765	35605	81263	39667	47358	56873	56307	61607	49518	89656	20103	77490	18062
41	98427	07523	33362	64270	01638	92477	66969	98420	04880	45585	46565	04102	46880	45709
42	34914	63976	88720	82765	34476	17032	87589	40836	32427	70002	70663	88863	77775	69348
43	70060	28277	39475	46473	23219	53416	94970	25832	69975	94884	19661	72828	00102	66794
44	53976	54914	06990	67245	68350	82948	11398	42878	80287	88267	47363	46634	06541	97809
45	76072	29515	40980	07391	58745	25774	22987	80059	39911	96189	41151	14222	60697	59583
46	90725	52210	83974	29992	65831	38857	50490	83765	55657	14361	31720	57375	56228	41546
47	64364	67412	33339	31926	14883	24413	59744	92351	97473	89286	35931	04110	23726	51900
48	08962	00358	31662	25388	61642	34072	81249	35648	56891	69352	48373	45578	78547	81788
49	95012	68379	93526	70765	10593	04542	76463	54328	02349	17247	28865	14777	62730	92277
50	15664	10493	20492	38391	91132	21999	59516	81652	27195	48223	46751	22923	32261	85653

Source. Tables A.1, A.2, B, and C are reprinted, with permission, from the *Handbook of tables for probability and statistics*, 2nd edition, 1968, The Chemical Co., Cleveland, Ohio.

TABLE A.I (continued)

Line/Col.	(1)	(2)	(3)	(4)	(5)	(6)	(7)	(8)	(9)	(10)	(11)	(12)	(13)	(14)
51	16408	81899	04153	53381	79401	21438	83035	92350	36693	31238	59649	91754	72772	02338
52	18629	81953	05520	91962	04739	13092	97662	24822	94730	06496	35090	04822	86772	98299
53	73115	35101	47498	87637	99016	71060	88824	71013	18735	20286	23153	72924	35165	43040
54	57491	16703	23167	49323	45021	33132	12544	41035	80780	45393	44812	12515	98931	91202
55	30405	83946	23792	14422	15059	45799	22716	19792	09983	74353	68668	30429	70735	25499
56	16631	35006	85900	98275	32388	52390	16815	69298	82732	38480	73817	32523	41961	44437
57	96773	20206	42559	78985	05300	22164	24369	54224	35083	19687	11052	91491	60383	19746
58	38935	64202	14349	82674	66523	44133	00697	35552	35970	19124	63318	29686	03387	59846
59	31624	76384	17403	53363	44167	64486	64758	75366	76554	31601	12614	33072	60332	92325
60	78919	19474	23632	27889	47914	02584	37680	20801	72152	39339	34806	08930	85001	87820
61	03931	33309	57047	74211	63445	17361	62825	39908	05607	91284	68833	25570	38818	46920
62	74426	33278	43972	10119	89917	15665	52872	73823	73144	88662	88970	74492	51805	99378
63	09066	00903	20795	95452	92648	45454	09552	88815	16553	51125	79375	97596	16296	66092
64	42238	12426	87025	14267	20979	04508	64535	31355	86064	29472	47689	05974	52468	16834
65	16153	08002	26504	41744	81959	65642	74240	56302	00033	67107	77510	70625	28725	34191
66	21457	40742	29820	96783	29400	21840	15035	34537	33310	06116	95240	15957	16572	06004
67	21581	57802	02050	89728	17937	37621	47075	42080	96403	48626	68995	43805	33386	21597
68	55612	78095	83197	33732	05810	24813	86902	60397	16489	03264	88525	42786	05269	92532
69	44657	66999	99324	51281	84463	60563	79312	93454	68876	25471	93911	25650	12682	73572
70	91340	84979	46949	81973	37949	61023	43997	15263	80644	43942	89203	71795	99533	50501
71	91227	21199	31935	27022	84067	05462	35216	14486	29891	68607	41867	14951	91696	85065
72	50001	38140	66321	19924	72163	09538	12151	06878	91903	18749	34405	56087	82790	70925
73	65390	05224	72958	28609	81406	39147	25549	48542	42627	45233	57202	94617	23772	07896
74	27504	96131	83944	41575	10573	08619	64482	73923	36152	05184	94142	25299	84387	34925
75	37169	94851	39117	89632	00959	16487	65536	49071	39782	17095	02330	74301	00275	48280
76	11508	70225	51111	38351	19444	66499	71945	05422	13442	78675	84081	66938	93654	59894
77	37449	30362	06694	54690	04052	53115	62757	95348	78662	11163	81651	50245	34971	52924
78	46515	70331	85922	38329	57015	15765	97161	17869	45349	61796	66345	81073	49106	79860
79	30986	81223	42416	58353	21532	30502	32305	86482	05174	07901	54339	58861	74818	46942
80	63798	64995	46583	09765	44160	78128	83991	42865	92520	83531	80377	35909	81250	54238
81	82486	84846	99254	67632	43218	50076	21361	64816	51202	88124	41870	52689	51275	83556
82	21885	32906	92431	09060	64297	51674	64126	62570	26123	05155	59194	52799	28225	85762
83	60336	98782	07408	53458	13564	59089	26445	29789	85205	41001	12535	12133	14645	23541
84	43937	46891	24010	25560	86355	33941	25786	54990	71899	15475	95434	98227	21824	19585
85	97656	63175	89303	16275	07100	92063	21942	18611	47348	20203	18534	03862	78095	50136
86	03299	01221	05418	38982	55758	92237	26759	86367	21216	98442	08303	56613	91511	75928
87	79626	06486	03574	17668	07785	76020	79924	25651	83325	88428	85076	72811	22717	50585
88	85636	68335	47539	03129	65651	11977	02510	26113	99447	68645	34327	15152	55230	93448
89	18039	14367	61337	06177	12143	46609	32989	74014	64708	00533	35398	58408	13261	47908
90	08362	15656	60627	36478	65648	16764	53412	09013	07832	41574	17639	82163	60859	75567
91	79556	29068	04142	16268	15387	12856	66227	38358	22478	73373	88732	09443	82558	05250
92	92608	82674	27072	32534	17075	27698	98204	63863	11951	34648	88022	56148	34925	57031
93	23982	25835	40055	67006	12293	02753	14827	22235	35071	99704	37543	11601	35503	85171
94	09915	96306	05908	97901	28395	14186	00821	80703	70426	75647	76310	88717	37890	40129
95	50937	33300	26695	62247	69927	76123	50842	43834	86654	70959	79725	93872	28117	19233
96	42488	78077	69882	61657	34136	79180	97526	43092	04098	73571	80799	76536	71255	64239
97	46764	86273	63003	93017	31204	36692	40202	35275	57306	55543	53203	18098	47625	88684
98	03237	45430	55417	63282	90816	17349	88298	90183	36600	78406	06216	95787	42579	90730
99	86591	81482	52667	61583	14972	90053	89534	76036	49199	43716	97548	04379	46370	28672
100	38534	01715	94964	87288	65680	43772	39560	12918	86537	62738	19636	51132	25739	56947

TABLE A.2 Normal Random Deviates from $N(0, 1)$

01	02	03	04	05	06	07	08	09	10
.464	.137	2.455	−.323	−.068	.296	−.288	1.298	.241	−.957
.060	−2.526	−.531	−.194	.543	−1.558	.187	−1.190	.022	.525
1.486	−.354	−.634	.697	.926	1.375	.785	−.963	−.853	−1.865
1.022	−.472	1.279	3.521	.571	−1.851	.194	1.192	−.501	−.273
1.394	−.555	.046	.321	2.945	1.974	−.258	.412	.439	−.035
.906	−.513	−.525	.595	.881	−.934	1.579	.161	−1.885	.371
1.179	−1.055	.007	.769	.971	.712	1.090	−.631	−.255	−.702
−1.501	−.488	−.162	−.136	1.033	.203	.448	.748	−.423	−.432
−.690	.756	−1.618	−.345	−.511	−2.051	−.457	−.218	.857	−.465
1.372	.225	.378	.761	.181	−.736	.960	−1.530	−.260	.120
−.482	1.678	−.057	−1.229	−.486	.856	−.491	−1.983	−2.830	−.238
−1.376	−.150	1.356	−.561	−.256	−.212	.219	.779	.953	−.869
−1.010	.598	−.918	1.598	.065	.415	−.169	.313	−.973	−1.016
−.005	−.899	.012	−.725	1.147	−.121	1.096	.481	−1.691	.417
1.393	−1.163	−.911	1.231	−.199	−.246	1.239	−2.574	−.588	.056
−1.787	−.261	1.237	1.046	−.508	−1.630	−.146	−.392	−.627	.561
−.105	−.357	−1.384	.360	−.992	−.116	−1.698	−2.832	−1.108	−2.357
−1.339	1.827	−.959	.424	.969	−1.141	−1.041	.362	−1.726	1.956
1.041	.535	.731	1.377	.983	−1.330	1.620	−1.040	.524	−.281
.279	−2.056	.717	−.873	−1.096	−1.396	1.047	.089	−.573	.932
−1.805	−2.008	−1.633	.542	.250	−.166	.032	.079	.471	−1.029
−1.186	1.180	1.114	.882	1.265	−.202	.151	−.376	−.310	.479
.658	−1.141	1.151	−1.210	−.927	.425	.290	−.902	.610	2.709
−.439	.358	−1.939	.891	−.227	.602	.873	−.437	−.220	−.057
−1.399	−.230	.385	−.649	−.577	.237	−.289	.513	.738	−.300
.199	.208	−1.083	−.219	−.291	1.221	1.119	.004	−2.015	−.594
.159	.272	−.313	.084	−2.828	−.439	−.792	−1.275	−.623	−1.047
2.273	.606	.606	−.747	.247	1.291	.063	−1.793	−.699	−1.347
.041	−.307	.121	.790	−.584	.541	.484	−.986	.481	.996
−1.132	−2.098	.921	.145	.446	−1.661	1.045	−1.363	−.586	−1.023
.768	.079	−1.473	.034	−2.127	.665	.084	−.880	−.579	.551
.375	−1.658	−.851	.234	−.656	.340	−.086	−.158	−.120	.418
−.513	−.344	.210	−.735	1.041	.008	.427	−.831	.191	.074
.292	−.521	1.266	−1.206	−.899	.110	−.528	−.813	.071	.524
1.026	2.990	−.574	−.491	−1.114	1.297	−1.433	−1.345	−3.001	.479
−1.334	1.278	−.568	−.109	−.515	−.566	2.923	.500	.359	.326
−.287	−.144	−.254	.574	−.451	−1.181	−1.190	−.318	−.094	1.114
.161	−.886	−.921	−.509	1.410	−.518	.192	−.432	1.501	1.068
−1.346	.193	−1.202	.394	−1.045	.843	.942	1.045	.031	.772
1.250	−.199	−.288	1.810	1.378	.584	1.216	.733	.402	.226
.630	−.537	.782	.060	.499	−.431	1.705	1.164	.884	−.298
.375	−1.941	.247	−.491	−.665	−.135	−.145	−.498	.457	1.064
−1.420	.489	−1.711	−1.186	.754	−.732	−.066	1.006	−.798	.162
−.151	−.243	−.430	−.762	.298	1.049	1.810	2.885	−.768	−.129
−.309	.531	.416	−1.541	1.456	2.040	−.124	.196	.023	−1.204
.424	−.444	.593	.993	−.106	.116	.484	−1.272	1.066	1.097
.593	.658	−1.127	−1.407	−1.579	−1.616	1.458	1.262	.736	−.916
.862	−.885	−.142	−.504	.532	1.381	.022	−.281	−.342	1.222
.235	−.628	−.023	−.463	−.899	−.394	−.538	1.707	−.188	−1.153
−.853	.402	.777	.833	.410	−.349	−1.094	.580	1.395	1.298

11	12	13	14	15	16	17	18	19	20
−1.329	−.238	−.838	−.988	−.445	.964	−.266	−.322	−1.726	2.252
1.284	−.229	1.058	.090	.050	.523	.016	.277	1.639	.554
.619	.628	.005	.973	−.058	.150	−.635	−.917	.313	−1.203
.699	−.269	.722	−.994	−.807	−1.203	1.163	1.244	1.306	−1.210
.101	.202	−.150	.731	.420	.116	−.496	−.037	−2.466	.794
−1.381	.301	.522	.233	.791	−1.017	−.182	.926	−1.096	1.001
−.574	1.366	−1.843	.746	.890	.824	−1.249	−.806	−.240	.217
.096	.210	1.091	.990	.900	−.837	−1.097	−1.238	.030	−.311
1.389	−.236	.094	3.282	.295	−.416	.313	.720	.007	.354
1.249	.706	1.453	.366	−2.654	−1.400	.212	.307	−1.145	.639
.756	−.397	−1.772	−.257	1.120	1.188	−.527	.709	.479	.317
−.860	.412	−.327	.178	.524	−.672	−.831	.758	.131	.771
−.778	−.979	.236	−1.033	1.497	−.661	.906	1.169	−1.582	1.303
.037	.062	.426	1.220	.471	.784	−.719	.465	1.559	−1.326
2.619	−.440	.477	1.063	.320	1.406	−.701	−.128	.518	−.676
−.420	−.287	−.050	−.481	1.521	−1.367	.609	.292	.048	.592
1.048	.220	1.121	−1.789	−1.211	−.871	−.740	.513	−.558	−.395
1.000	−.638	1.261	.510	−.150	.034	.054	−.055	.639	−.825
.170	−1.131	−.985	.102	−.939	−1.457	1.766	1.087	−1.275	2.362
.389	−.435	.171	.891	1.158	1.041	1.048	−.324	−.404	1.060
−.305	.838	−2.019	−.540	.905	1.195	−1.190	.106	.571	.298
−.321	−.039	1.799	−1.032	−2.225	−.148	.758	−.862	.158	−.726
1.900	1.572	−.244	−1.721	1.130	.495	−.484	.014	−.778	−1.483
−.778	−.288	−.224	−1.324	−.072	.890	−.410	.752	.376	−.224
.617	−1.718	−.183	−.110	1.719	.696	−1.339	−.614	1.071	−.386
−1.430	−.953	.770	−.007	−1.872	1.075	−.913	−1.168	1.775	.238
.267	−.048	.972	.734	−1.408	−1.955	−.848	2.002	.232	−1.273
.978	−.520	−.368	1.690	−1.479	.985	1.475	−.098	−1.633	2.399
−1.235	−1.168	.325	1.421	2.652	−.486	−1.253	.270	−1.103	.118
−.258	.638	2.309	.741	−.161	−.679	.336	1.973	.370	−2.277
.243	.629	−1.516	−.157	.693	1.710	.800	−.265	1.218	.655
−.292	−1.455	−1.451	1.492	−.713	.821	−.031	−.780	1.330	.977
−.505	.389	.544	−.042	1.615	−1.440	−.989	−.580	.156	.052
.397	−.287	1.712	.289	−.904	.259	−.600	−1.635	−.009	−.799
−.605	−.470	.007	.721	−1.117	.635	.592	−1.362	−1.441	.672
1.360	.182	−1.476	−.599	−.875	.292	−.700	.058	−.340	−.639
.480	−.699	1.615	−.225	1.014	−1.370	−1.097	.294	.309	−1.389
−.027	−.487	−1.000	−.015	.119	−1.990	−.687	−1.964	−.366	1.759
−1.482	−.815	−.121	1.884	−.185	.601	.793	.430	−1.181	.426
−1.256	−.567	−.994	1.011	−1.071	−.623	−.420	−.309	1.362	.863
−1.132	2.039	1.934	−.222	.386	1.100	.284	1.597	−1.718	−.560
−.780	−.239	−.497	−.434	−.284	−.241	−.333	1.348	−.478	−.169
−.859	−.215	.241	1.471	.389	−.952	.245	.781	1.093	−.240
.447	1.479	.067	.426	−.370	−.675	−.972	.225	.815	.389
.269	.735	−.066	−.271	−1.439	1.036	−.306	−1.439	−.122	−.336
.097	−1.883	−.218	.202	−.357	.019	1.631	1.400	.223	−.793
−.686	1.596	−.286	.722	.655	−.275	1.245	−1.504	.066	−1.280
.957	.057	−1.153	.701	−.280	1.747	−.745	1.338	−1.421	.386
−.976	−1.789	−.696	−1.799	−.354	.071	2.355	.135	−.598	1.883
.274	.226	−.909	−.572	.181	1.115	.406	.453	−1.218	−.115

TABLE A.2 (continued)

21	22	23	24	25	26	27	28	29	30
−1.752	−.329	−1.256	.318	1.531	.349	−.958	−.059	.415	−1.084
−.291	.085	1.701	−1.087	−.443	−.292	.248	−.539	−1.382	.318
−.933	.130	.634	.899	1.409	−.883	−.095	.229	.129	.367
−.450	−.244	.072	1.028	1.730	−.056	−1.488	−.078	−2.361	−.992
.512	−.882	.490	−1.304	−.266	.757	−.361	.194	−1.078	.529
−.702	.472	.429	−.664	−.592	1.442	−1.515	−1.209	−1.043	.278
.284	.039	−.518	1.351	1.473	.889	.300	.339	−.206	1.392
−.509	1.420	−.782	−.429	−1.266	.627	−1.165	.819	−.261	.409
−1.776	−1.033	1.977	.014	.702	−.435	−.816	1.131	.656	.061
−.044	1.807	.342	−2.510	1.071	−1.220	−.060	−.764	.079	−.964
.263	−.578	1.612	−.148	−.383	−1.007	−.414	.638	−.186	.507
.986	.439	−.192	−.132	.167	.883	−.400	−1.440	−.385	−1.414
−.441	−.852	−1.446	−.605	−.348	1.018	.963	−.004	2.504	−.847
−.866	.489	.097	.379	.192	−.842	.065	1.420	.426	−1.191
−1.215	.675	1.621	.394	−1.447	2.199	−.321	−.540	−.037	.185
−.475	−1.210	.183	.526	.495	1.297	−1.613	1.241	−1.016	−.090
1.200	.131	2.502	.344	−1.060	−.909	−1.695	−.666	−.838	−.866
−.498	−1.202	−.057	−1.354	−1.441	−1.590	.987	.441	.637	−1.116
−.743	.894	−.028	1.119	−.598	.279	2.241	.830	.267	−.156
.779	−.780	−.954	.705	−.361	−.734	1.365	1.297	−.142	−1.387
−.206	−.195	1.017	−1.167	−.079	−.452	.058	−1.068	−.394	−.406
−.092	−.927	−.439	.256	.503	.338	1.511	−.465	−.118	−.454
−1.222	−1.582	1.786	−.517	−1.080	−.409	−.474	−1.890	.247	.575
.068	.075	−1.383	−.084	.159	1.276	1.141	.186	−.973	−.266
.183	1.600	−.335	1.553	.889	.896	−.035	.461	.486	1.246
−.811	−2.904	.618	.588	.533	.803	−.696	.690	.820	.557
−1.010	1.149	1.033	.336	1.306	.835	1.523	.296	−.426	.004
1.453	1.210	−.043	.220	−.256	−1.161	−2.030	−.046	.243	1.082
.759	−.838	−.877	−.177	1.183	−.218	−3.154	−.963	−.822	−1.114
.287	.278	−.454	.897	−.122	.013	.346	.921	.238	−.586
−.669	.035	−2.007	1.077	.525	−.154	−1.036	.015	−.220	.882
.392	.106	−1.430	−.204	−.326	.825	−.432	−.094	−1.566	.679
−.337	.199	−.160	.625	−.891	1.464	−.318	1.297	.932	−.032
.369	−1.990	1.190	.666	−1.614	.082	.922	−.139	−.833	.091
−1.694	.710	−.655	−.546	1.654	.134	.466	.033	−.039	.838
.985	.340	.276	.911	−.170	−.551	1.000	−.838	.275	−.304
−1.063	−.594	1.526	−.787	.873	−.405	1.324	.162	−.163	−2.716
.033	1.527	1.422	.308	.854	.151	.741	.064	1.212	.823
.597	.362	3.760	1.159	.874	−.794	−.915	1.215	1.627	−1.248
−1.601	−.570	.133	−.660	1.485	.682	−.898	.686	.658	.346
−.266	−1.309	.597	.989	.934	1.079	−.656	−.999	−.036	−.537
.901	1.531	−.889	−1.019	.084	1.531	−.144	−1.920	.678	−.402
−1.433	−1.008	−.990	.090	.940	.207	−.745	.638	1.469	1.214
1.327	.763	−1.724	−.709	−1.100	−1.346	−.946	−.157	.522	−1.264
−.248	.788	−.577	.122	−.536	.293	1.207	−2.243	1.642	1.353
−.401	−.679	.921	.476	1.121	−.864	.128	−.551	−.872	1.511
.344	−.324	.686	−1.487	−.126	.803	−.961	.183	−.358	−.184
.441	−.372	1.336	.062	1.506	−.315	−.112	−.452	1.594	−.264
.824	.040	−1.734	.251	.054	−.379	1.298	−.126	.104	−.529
1.385	1.320	−.509	−.381	−1.671	−.524	−.805	1.348	.676	.799

TABLE A.2 (continued)

31	32	33	34	35	36	37	38	39	40
1.556	.119	−.078	.164	−.455	.077	−.043	−.299	.249	−.182
.647	1.029	1.186	.887	1.204	−.657	.644	−.410	−.652	−.165
.329	.407	1.169	−2.072	1.661	.891	.233	−1.628	−.762	−.717
−1.188	1.171	−1.170	−.291	.863	−.045	−.205	.574	−.926	1.407
−.917	−.616	−1.589	1.184	.266	.559	−1.833	−.572	−.648	−1.090
.414	.469	−.182	.397	1.649	1.198	.067	−1.526	−.081	−.192
.107	−.187	1.343	.472	−.112	1.182	.548	2.748	.249	.154
−.497	1.907	.191	.136	−.475	.458	.183	−1.640	−.058	1.278
.501	.083	−.321	1.133	1.126	−.299	1.299	1.617	1.581	2.455
−1.382	−.738	1.225	1.564	−.363	−.548	1.070	.390	−1.398	.524
−.590	.699	−.162	−.011	1.049	−.689	1.225	.339	−.539	−.445
−1.125	1.111	−1.065	.534	.102	.425	−1.026	.695	−.057	.795
.849	.169	−.351	.584	2.177	.009	−.696	−.426	−.692	−1.638
−1.233	−.585	.306	.773	1.304	−1.304	.282	−1.705	.187	−.880
.104	−.468	.185	.498	−.624	−.322	−.875	1.478	−.691	−.281
.261	−1.883	−.181	1.675	−.324	−1.029	−.185	.004	−.101	−1.187
−.007	1.280	.568	−1.270	1.405	1.731	2.072	1.686	.728	−.417
.794	−.111	.040	−.536	−.976	2.192	1.609	−.190	−.279	−1.611
.431	−2.300	−1.081	−1.370	2.943	.653	−2.523	.756	.886	−.983
−.149	1.294	−.580	.482	−1.449	−1.067	1.996	−.274	.721	.490
−.216	−1.647	1.043	.481	−.011	−.587	−.916	−1.016	−1.040	−1.117
1.604	−.851	−.317	−.686	−.008	1.939	.078	−.465	.533	.652
−.212	.005	.535	.837	.362	1.103	.219	.488	1.332	−.200
.007	−.076	1.484	.455	−.207	−.554	1.120	.913	−.681	1.751
−.217	.937	.860	.323	1.321	−.492	−1.386	−.003	−.230	.539
−.649	.300	−.698	.900	.569	.842	.804	1.025	.603	−1.546
−1.541	.193	2.047	−.552	1.190	−.087	2.062	−2.173	−.791	−.520
.274	−.530	.112	.385	.656	.436	.882	.312	−2.265	−.218
.876	−1.498	−.128	−.387	−1.259	−.856	−.353	.714	.863	1.169
−.859	−1.083	1.288	−.078	−.081	.210	.572	1.194	−1.118	−1.543
−.015	−.567	.113	2.127	−.719	3.256	−.721	−.663	−.779	−.930
−1.529	−.231	1.223	.300	−.995	−.651	.505	.138	−.064	1.341
.278	−.058	−2.740	−.296	−1.180	.574	1.452	.846	−.243	−1.208
1.428	.322	2.302	−.852	.782	−1.322	−.092	−.546	.560	−1.430
.770	−1.874	.347	.994	−.485	−1.179	.048	−1.324	1.061	.449
−.303	−.629	.764	.013	−1.192	−.475	−1.085	−.880	1.738	−1.225
−.263	−2.105	.509	−.645	1.362	.504	−.755	1.274	1.448	.604
.997	−1.187	−.242	.121	2.510	−1.935	.350	.073	.458	−.446
−.063	−.475	−1.802	−.476	.193	−1.199	.339	.364	−.684	1.353
−.168	1.904	−.485	−.032	−.554	.056	−.710	−.778	.722	−.024
.366	−.491	.301	−.008	−.894	−.945	.384	−1.748	−1.118	.394
.436	−.464	.539	.942	−.458	.445	−1.883	1.228	1.113	−.218
.597	−1.471	−.434	.705	−.788	.575	.086	.504	1.445	−.513
−.805	−.624	1.344	.649	−1.124	.680	−.986	1.845	−1.152	−.393
1.681	−1.910	.440	.067	−1.502	−.755	−.989	−.054	−2.320	.474
−.007	−.459	1.940	.220	−1.259	−1.729	.137	−.520	−.412	2.847
.209	−.633	.299	.174	1.975	−.271	.119	−.199	.007	2.315
1.254	1.672	−1.186	−1.310	.474	.878	−.725	−.191	.642	−1.212
−1.016	−.697	.017	−.263	−.047	−1.294	−.339	2.257	−.078	−.049
−1.169	−.355	1.086	−.199	.031	.396	−.143	1.572	.276	.027

41	42	43	44	45	46	47	48	49	50
−.856	−.063	.787	−2.052	−1.192	−.831	1.623	1.135	.759	−.189
−.276	−1.110	.752	−1.378	−.583	.360	.365	1.587	.621	1.344
.379	−.440	.858	1.453	−1.356	.503	−1.134	1.950	−1.816	−.283
1.468	.131	.047	.355	.162	−1.491	−.739	−1.182	−.533	−.497
−1.805	−.772	1.286	−.636	−1.312	−1.045	1.559	−.871	−.102	−.123
2.285	.554	.418	−.577	−1.489	−1.255	.092	−.597	−1.051	− 980
−.602	.399	1.121	−1.026	.087	1.018	−1.437	.661	.091	−.637
.229	−.584	.705	.124	.341	1.320	−.824	−1.541	−.163	2.329
1.382	−1.454	1.537	−1.299	.363	−.356	−.025	.294	2.194	−.395
.978	.109	1.434	−1.094	−.265	−.857	−1.421	−1.733	.570	−.053
−.678	−2.335	1.202	−1.697	.547	−.201	−.373	−1.363	−.081	.958
−.366	−1.084	−.626	.798	1.706	−1.160	−.838	1.462	.636	.570
−1.074	−1.379	.086	−.331	−.288	−.309	−1.527	−.408	.183	.856
−.600	−.096	.696	.446	1.417	−2.140	.599	−.157	1.485	1.387
.918	1.163	−1.445	.759	.878	−1.781	−.056	−2.141	−.234	.975
−.791	−.528	.946	1.673	−.680	−.784	1.494	−.086	−1.071	−1.196
.598	−.352	.719	−.341	.056	−1.041	1.429	.235	.314	−1.693
.567	−1.156	−.125	−.534	.711	−.511	.187	−.644	−1.090	−1.281
.963	.052	.037	.637	−1.335	.055	.010	−.860	−.621	.713
.489	−.209	1.659	.054	1.635	.169	.794	−1.550	1.845	−.388
−1.627	−.017	.699	.661	−.073	.188	1.183	−1.054	−1.615	−.765
−1.096	1.215	.320	.738	−1.865	−1.169	−.667	−.674	−.062	1.378
−2.532	1.031	−.799	1.665	−2.756	−.151	−.704	.602	−.672	1.264
.024	−1.183	−.927	−.629	.204	−.825	.496	2.543	.262	−.785
.192	.125	.373	−.931	−.079	.186	−.306	.621	−.292	1.131
−1.324	−1.229	−.648	−.430	.811	.868	.787	1.845	−.374	−.651
−.726	−.746	1.572	−1.420	1.509	−.361	−.310	−3.117	1.637	.642
−1.618	1.082	−.319	.300	1.524	−.418	−1.712	.358	−1.032	.537
1.695	.843	2.049	.388	−.297	1.077	−.462	.655	.940	−.354
.790	.605	−3.077	1.009	−.906	−1.004	.693	−1.098	1.300	.549
1.792	−.895	−.136	−1.765	1.077	.418	−.150	.808	.697	.435
.771	−.741	−.492	−.770	−.458	−.021	1.385	−1.225	−.066	−1.471
−1.438	.423	−1.211	.723	−.731	.883	−2.109	−2.455	−.210	1.644
−.294	1.266	−1.994	−.730	.545	.397	1.069	−.383	−.097	−.985
−1.966	.909	.400	.685	−.800	1.759	.268	1.387	−.414	1.615
.999	1.587	1.423	.937	−.943	.090	1.185	−1.204	.300	−1.354
.581	.481	−2.400	.000	.231	.079	−2.842	−.846	−.508	−.516
.370	−1.452	−.580	−1.462	−.972	1.116	−.994	.374	−3.336	−.058
.834	−1.227	−.709	−1.039	−.014	−.383	−.512	−.347	.881	−.638
−.376	−.813	.660	−1.029	−.137	.371	.376	.968	1.338	−.786
−1.621	.815	−.544	−.376	−.852	.436	1.562	.815	−1.048	.188
.163	−.161	2.501	−.265	−.285	1.934	1.070	.215	−.876	.073
1.786	−.538	−.437	.324	.105	−.421	−.410	−.947	.700	−1.006
2.140	1.218	−.351	−.068	.254	.448	−1.461	.784	.317	1.013
.064	.410	.368	.419	−.982	1.371	.100	−.505	.856	.890
.789	−.131	1.330	.506	−.645	−1.414	2.426	1.389	−.169	−.194
−.011	−.372	−.699	2.382	−1.395	−.467	1.256	−.585	−1.359	−1.804
−.463	.003	−1.470	1.493	.960	.364	−1.267	−.007	.616	.624
−1.210	−.669	.009	1.284	−.617	.355	−.589	−.243	−.015	−.712
−1.157	.481	.560	1.287	1.129	−.126	.006	1.532	1.328	.980

TABLE B Binomial Probability Function

n	s	.05	.10	.15	.20	.25	p .30	.35	.40	.45	.50
1	0	.9500	.9000	.8500	.8000	.7500	.7000	.6500	.6000	.5500	.5000
	1	.0500	.1000	.1500	.2000	.2500	.3000	.3500	.4000	.4500	.5000
2	0	.9025	.8100	.7225	.6400	.5625	.4900	.4225	.3600	.3025	.2500
	1	.0950	.1800	.2550	.3200	.3750	.4200	.4550	.4800	.4950	.5000
	2	.0025	.0100	.0225	.0400	.0625	.0900	.1225	.1600	.2025	.2500
3	0	.8574	.7290	.6141	.5120	.4219	.3430	.2746	.2160	.1664	.1250
	1	.1354	.2430	.3251	.3840	.4219	.4410	.4436	.4320	.4084	.3750
	2	.0071	.0270	.0574	.0960	.1406	.1890	.2389	.2880	.3341	.3750
	3	.0001	.0010	.0034	.0080	.0156	.0270	.0429	.0640	.0911	.1250
4	0	.8145	.6561	.5220	.4096	.3164	.2401	.1785	.1296	.0915	.0625
	1	.1715	.2916	.3685	.4096	.4219	.4116	.3845	.3456	.2995	.2500
	2	.0135	.0486	.0975	.1536	.2109	.2646	.3105	.3456	.3675	.3750
	3	.0005	.0036	.0115	.0256	.0469	.0756	.1115	.1536	.2005	.2500
	4	.0000	.0001	.0005	.0016	.0039	.0081	.0150	.0256	.0410	.0625
5	0	.7738	.5905	.4437	.3277	.2373	.1681	.1160	.0778	.0503	.0312
	1	.2036	.3280	.3915	.4096	.3955	.3602	.3124	.2592	.2059	.1562
	2	.0214	.0729	.1382	.2048	.2637	.3087	.3364	.3456	.3369	.3125
	3	.0011	.0081	.0244	.0512	.0879	.1323	.1811	.2304	.2757	.3125
	4	.0000	.0004	.0022	.0064	.0146	.0284	.0488	.0768	.1128	.1562
	5	.0000	.0000	.0001	.0003	.0010	.0024	.0053	.0102	.0185	.0312
6	0	.7351	.5314	.3771	.2621	.1780	.1176	.0754	.0467	.0277	.0156
	1	.2321	.3543	.3993	.3932	.3560	.3025	.2437	.1866	.1359	.0938
	2	.0305	.0984	.1762	.2458	.2966	.3241	.3280	.3170	.2780	.2344
	3	.0021	.0146	.0415	.0819	.1318	.1852	.2355	.2765	.3032	.3125
	4	.0001	.0012	.0055	.0154	.0330	.0595	.0951	.1382	.1861	.2344
	5	.0000	.0001	.0004	.0015	.0044	.0102	.0205	.0369	.0609	.0938
	6	.0000	.0000	.0000	.0001	.0002	.0007	.0018	.0041	.0083	.0156
7	0	.6983	.4783	.3206	.2097	.1335	.0824	.0490	.0280	.0152	.0078
	1	.2573	.3720	.3960	.3670	.3115	.2471	.1848	.1306	.0872	.0547
	2	.0406	.1240	.2097	.2753	.3115	.3177	.2985	.2613	.2140	.1641
	3	.0036	.0230	.0617	.1147	.1730	.2269	.2679	.2903	.2918	.2734
	4	.0002	.0026	.0109	.0287	.0577	.0972	.1442	.1935	.2388	.2734
	5	.0000	.0002	.0012	.0043	.0115	.0250	.0466	.0774	.1172	.1641
	6	.0000	.0000	.0001	.0004	.0013	.0036	.0084	.0172	.0320	.0547
	7	.0000	.0000	.0000	.0000	.0001	.0002	.0006	.0016	.0037	.0078

Linear interpolations with respect to p will in general be accurate at most to two decimal places.

TABLE B (continued)

n	s	.05	.10	.15	.20	.25	p .30	.35	.40	.45	.50
8	0	.6634	.4305	.2725	.1678	.1001	.0576	.0319	.0168	.0084	.0039
	1	.2793	.3826	.3847	.3355	.2670	.1977	.1373	.0896	.0548	.0312
	2	.0515	.1488	.2376	.2936	.3115	.2965	.2587	.2090	.1569	.1094
	3	.0054	.0331	.0839	.1469	.2076	.2541	.2786	.2787	.2568	.2188
	4	.0004	.0046	.0185	.0459	.0865	.1361	.1875	.2322	.2627	.2734
	5	.0000	.0004	.0026	.0092	.0231	.0467	.0808	.1239	.1719	.2188
	6	.0000	.0000	.0002	.0011	.0038	.0100	.0217	.0413	.0703	.1094
	7	.0000	.0000	.0000	.0001	.0004	.0012	.0033	.0079	.0164	.0312
	8	.0000	.0000	.0000	.0000	.0000	.0001	.0002	.0007	.0017	.0039
9	0	.6302	.3874	.2316	.1342	.0751	.0404	.0207	.0101	.0046	.0020
	1	.2985	.3874	.3679	.3020	.2253	.1556	.1004	.0605	.0339	.0176
	2	.0629	.1722	.2597	.3020	.3003	.2668	.2162	.1612	.1110	.0703
	3	.0077	.0446	.1069	.1762	.2336	.2668	.2716	.2508	.2119	.1641
	4	.0006	.0074	.0283	.0661	.1168	.1715	.2194	.2508	.2600	.2461
	5	.0000	.0008	.0050	.0165	.0389	.0735	.1181	.1672	.2128	.2461
	6	.0000	.0001	.0006	.0028	.0087	.0210	.0424	.0743	.1160	.1641
	7	.0000	.0000	.0000	.0003	.0012	.0039	.0098	.0212	.0407	.0703
	8	.0000	.0000	.0000	.0000	.0001	.0004	.0013	.0035	.0083	.0176
	9	.0000	.0000	.0000	.0000	.0000	.0000	.0001	.0003	.0008	.0020
10	0	.5987	.3487	.1969	.1074	.0563	.0282	.0135	.0060	.0025	.0010
	1	.3151	.3874	.3474	.2684	.1877	.1211	.0725	.0403	.0207	.0098
	2	.0746	.1937	.2759	.3020	.2816	.2335	.1757	.1209	.0763	.0439
	3	.0105	.0574	.1298	.2013	.2503	.2668	.2522	.2150	.1665	.1172
	4	.0010	.0112	.0401	.0881	.1460	.2001	.2377	.2508	.2384	.2051
	5	.0001	.0015	.0085	.0264	.0584	.1029	.1536	.2007	.2340	.2461
	6	.0000	.0001	.0012	.0055	.0162	.0368	.0689	.1115	.1596	.2051
	7	.0000	.0000	.0001	.0008	.0031	.0090	.0212	.0425	.0746	.1172
	8	.0000	.0000	.0000	.0001	.0004	.0014	.0043	.0106	.0229	.0439
	9	.0000	.0000	.0000	.0000	.0000	.0001	.0005	.0016	.0042	.0098
	10	.0000	.0000	.0000	.0000	.0000	.0000	.0000	.0001	.0003	.0010
11	0	.5688	.3138	.1673	.0859	.0422	.0198	.0088	.0036	.0014	.0004
	1	.3293	.3835	.3248	.2362	.1549	.0932	.0518	.0266	.0125	.0055
	2	.0867	.2131	.2866	.2953	.2581	.1998	.1395	.0887	.0513	.0269
	3	.0137	.0710	.1517	.2215	.2581	.2568	.2254	.1774	.1259	.0806
	4	.0014	.0158	.0536	.1107	.1721	.2201	.2428	.2365	.2060	.1611
	5	.0001	.0025	.0132	.0388	.0803	.1321	.1830	.2207	.2360	.2256
	6	.0000	.0003	.0023	.0097	.0268	.0566	.0985	.1471	.1931	.2256
	7	.0000	.0000	.0003	.0017	.0064	.0173	.0379	.0701	.1128	.1611
	8	.0000	.0000	.0000	.0002	.0011	.0037	.0102	.0234	.0462	.0806

n	s	.05	.10	.15	.20	.25	.30	.35	.40	.45	.50
	9	.0000	.0000	.0000	.0000	.0001	.0005	.0018	.0052	.0126	.0269
	10	.0000	.0000	.0000	.0000	.0000	.0000	.0002	.0007	.0021	.0054
	11	.0000	.0000	.0000	.0000	.0000	.0000	.0000	.0000	.0002	.0005
12	0	.5404	.2824	.1422	.0687	.0317	.0138	.0057	.0022	.0008	.0002
	1	.3413	.3766	.3012	.2062	.1267	.0712	.0368	.0174	.0075	.0029
	2	.0988	.2301	.2924	.2835	.2323	.1678	.1088	.0639	.0339	.0161
	3	.0173	.0852	.1720	.2362	.2581	.2397	.1954	.1419	.0923	.0537
	4	.0021	.0213	.0683	.1329	.1936	.2311	.2367	.2128	.1700	.1208
	5	.0002	.0038	.0193	.0532	.1032	.1585	.2039	.2270	.2225	.1934
	6	.0000	.0005	.0040	.0155	.0401	.0792	.1281	.1766	.2124	.2256
	7	.0000	.0000	.0006	.0033	.0115	.0291	.0591	.1009	.1489	.1934
	8	.0000	.0000	.0001	.0005	.0024	.0078	.0199	.0420	.0762	.1208
	9	.0000	.0000	.0000	.0001	.0004	.0015	.0048	.0125	.0277	.0537
	10	.0000	.0000	.0000	.0000	.0000	.0002	.0008	.0025	.0068	.0161
	11	.0000	.0000	.0000	.0000	.0000	.0000	.0001	.0003	.0010	.0029
	12	.0000	.0000	.0000	.0000	.0000	.0000	.0000	.0000	.0001	.0002
13	0	.5133	.2542	.1209	.0550	.0238	.0097	.0037	.0013	.0004	.0001
	1	.3512	.3672	.2774	.1787	.1029	.0540	.0259	.0113	.0045	.0016
	2	.1109	.2448	.2937	.2680	.2059	.1388	.0836	.0453	.0220	.0095
	3	.0214	.0997	.1900	.2457	.2517	.2181	.1651	.1107	.0660	.0349
	4	.0028	.0277	.0838	.1535	.2097	.2337	.2222	.1845	.1350	.0873
	5	.0003	.0055	.0266	.0691	.1258	.1803	.2154	.2214	.1989	.1571
	6	.0000	.0008	.0063	.0230	.0559	.1030	.1546	.1968	.2169	.2095
	7	.0000	.0001	.0011	.0058	.0186	.0442	.0833	.1312	.1775	.2095
	8	.0000	.0000	.0001	.0011	.0047	.0142	.0336	.0656	.1089	.1571
	9	.0000	.0000	.0000	.0001	.0009	.0034	.0101	.0243	.0495	.0873
	10	.0000	.0000	.0000	.0000	.0001	.0006	.0022	.0065	.0162	.0349
	11	.0000	.0000	.0000	.0000	.0000	.0001	.0003	.0012	.0036	.0095
	12	.0000	.0000	.0000	.0000	.0000	.0000	.0000	.0001	.0005	.0016
	13	.0000	.0000	.0000	.0000	.0000	.0000	.0000	.0000	.0000	.0001
14	0	.4877	.2288	.1028	.0440	.0178	.0068	.0024	.0008	.0002	.0001
	1	.3593	.3559	.2539	.1539	.0832	.0407	.0181	.0073	.0027	.0009
	2	.1229	.2570	.2912	.2501	.1802	.1134	.0634	.0317	.0141	.0056
	3	.0259	.1142	.2056	.2501	.2402	.1943	.1366	.0845	.0462	.0222
	4	.0037	.0349	.0998	.1720	.2202	.2290	.2022	.1549	.1040	.0611
	5	.0004	.0078	.0352	.0860	.1468	.1963	.2178	.2066	.1701	.1222
	6	.0000	.0013	.0093	.0322	.0734	.1262	.1759	.2066	.2088	.1833
	7	.0000	.0002	.0019	.0092	.0280	.0618	.1082	.1574	.1952	.2095

n	s	.05	.10	.15	.20	.25	p .30	.35	.40	.45	.50
	8	.0000	.0000	.0003	.0020	.0082	.0232	.0510	.0918	.1398	.1833
	9	.0000	.0000	.0000	.0003	.0018	.0066	.0183	.0408	.0762	.1222
	10	.0000	.0000	.0000	.0000	.0003	.0014	.0049	.0136	.0312	.0611
	11	.0000	.0000	.0000	.0000	.0000	.0002	.0010	.0033	.0093	.0222
	12	.0000	.0000	.0000	.0000	.0000	.0000	.0001	.0005	.0019	.0056
	13	.0000	.0000	.0000	.0000	.0000	.0000	.0000	.0001	.0002	.0009
	14	.0000	.0000	.0000	.0000	.0000	.0000	.0000	.0000	.0000	.0001
15	0	.4633	.2059	.0874	.0352	.0134	.0047	.0016	.0005	.0001	.0000
	1	.3658	.3432	.2312	.1319	.0668	.0305	.0126	.0047	.0016	.0005
	2	.1348	.2669	.2856	.2309	.1559	.0916	.0476	.0219	.0090	.0032
	3	.0307	.1285	.2184	.2501	.2252	.1700	.1110	.0634	.0318	.0139
	4	.0049	.0428	.1156	.1876	.2252	.2186	.1792	.1268	.0780	.0417
	5	.0006	.0105	.0449	.1032	.1651	.2061	.2123	.1859	.1404	.0916
	6	.0000	.0019	.0132	.0430	.0917	.1472	.1906	.2066	.1914	.1527
	7	.0000	.0003	.0030	.0138	.0393	.0811	.1319	.1771	.2013	.1964
	8	.0000	.0000	.0005	.0035	.0131	.0348	.0710	.1181	.1647	.1964
	9	.0000	.0000	.0001	.0007	.0034	.0116	.0298	.0612	.1048	.1527
	10	.0000	.0000	.0000	.0001	.0007	.0030	.0096	.0245	.0515	.0916
	11	.0000	.0000	.0000	.0000	.0001	.0006	.0024	.0074	.0191	.0417
	12	.0000	.0000	.0000	.0000	.0000	.0001	.0004	.0016	.0052	.0139
	13	.0000	.0000	.0000	.0000	.0000	.0000	.0001	.0003	.0010	.0032
	14	.0000	.0000	.0000	.0000	.0000	.0000	.0000	.0000	.0001	.0005
	15	.0000	.0000	.0000	.0000	.0000	.0000	.0000	.0000	.0000	.0000
16	0	.4401	.1853	.0743	.0281	.0100	.0033	.0010	.0003	.0001	.0000
	1	.3706	.3294	.2097	.1126	.0535	.0228	.0087	.0030	.0009	.0002
	2	.1463	.2745	.2775	.2111	.1336	.0732	.0353	.0150	.0056	.0018
	3	.0359	.1423	.2285	.2463	.2079	.1465	.0888	.0468	.0215	.0085
	4	.0061	.0514	.1311	.2001	.2252	.2040	.1553	.1014	.0572	.0278
	5	.0008	.0137	.0555	.1201	.1802	.2099	.2008	.1623	.1123	.0667
	6	.0001	.0028	.0180	.0550	.1101	.1649	.1982	.1983	.1684	.1222
	7	.0000	.0004	.0045	.0197	.0524	.1010	.1524	.1889	.1969	.1746
	8	.0000	.0001	.0009	.0055	.0197	.0487	.0923	.1417	.1812	.1964
	9	.0000	.0000	.0001	.0012	.0058	.0185	.0442	.0840	.1318	.1746
	10	.0000	.0000	.0000	.0002	.0014	.0056	.0167	.0392	.0755	.1222
	11	.0000	.0000	.0000	.0000	.0002	.0013	.0049	.0142	.0337	.0667
	12	.0000	.0000	.0000	.0000	.0000	.0002	.0011	.0040	.0115	.0278
	13	.0000	.0000	.0000	.0000	.0000	.0000	.0002	.0008	.0029	.0085

n	s	.05	.10	.15	.20	.25	p .30	.35	.40	.45	.50
	14	.0000	.0000	.0000	.0000	.0000	.0000	.0000	.0001	.0005	.0018
	15	.0000	.0000	.0000	.0000	.0000	.0000	.0000	.0000	.0001	.0002
	16	.0000	.0000	.0000	.0000	.0000	.0000	.0000	.0000	.0000	.0000
17	0	.4181	.1668	.0631	.0225	.0075	.0023	.0007	.0002	.0000	.0000
	1	.3741	.3150	.1893	.0957	.0426	.0169	.0060	.0019	.0005	.0001
	2	.1575	.2800	.2673	.1914	.1136	.0581	.0260	.0102	.0035	.0010
	3	.0415	.1556	.2359	.2393	.1893	.1245	.0701	.0341	.0144	.0052
	4	.9076	.0605	.1457	.2093	.2209	.1868	.1320	.0796	.0411	.0182
	5	.0010	.0175	.0668	.1361	.1914	.2081	.1849	.1379	.0875	.0472
	6	.0001	.0039	.0236	.0680	.1276	.1784	.1991	.1839	.1432	.0944
	7	.0000	.0007	.0065	.0267	.0668	.1201	.1685	.1927	.1841	.1484
	8	.0000	.0001	.0014	.0084	.0279	.0644	.1134	.1606	.1883	.1855
	9	.0000	.0000	.0003	.0021	.0093	.0276	.0611	.1070	.1540	.1855
	10	.0000	.0000	.0000	.0004	.0025	.0095	.0263	.0571	.1008	.1484
	11	.0000	.0000	.0000	.0001	.0005	.0026	.0090	.0242	.0525	.0944
	12	.0000	.0000	.0000	.0000	.0001	.0006	.0024	.0081	.0215	.0472
	13	.0000	.0000	.0000	.0000	.0000	.0001	.0005	.0021	.0068	.0182
	14	.0000	.0000	.0000	.0000	.0000	.0000	.0001	.0004	.0016	.0052
	15	.0000	.0000	.0000	.0000	.0000	.0000	.0000	.0001	.0003	.0010
	16	.0000	.0000	.0000	.0000	.0000	.0000	.0000	.0000	.0000	.0001
	17	.0000	.0000	.0000	.0000	.0000	.0000	.0000	.0000	.0000	.0000
18	0	.3972	.1501	.0536	.0180	.0056	.0016	.0004	.0001	.0000	.0000
	1	.3763	.3002	.1704	.0811	.0338	.0126	.0042	.0012	.0003	.0001
	2	.1683	.2835	.2556	.1723	.0958	.0458	.0190	.0069	.0022	.0006
	3	.0473	.1680	.2406	.2297	.1704	.1046	.0547	.0246	.0095	.0031
	4	.0093	.0700	.1592	.2153	.2130	.1681	.1104	.0614	.0291	.0117
	5	.0014	.0218	.0787	.1507	.1988	.2017	.1664	.1146	.0666	.0327
	6	.0002	.0052	.0301	.0816	.1436	.1873	.1941	.1655	.1181	.0708
	7	.0000	.0010	.0091	.0350	.0820	.1376	.1792	.1892	.1657	.1214
	8	.0000	.0002	.0022	.0120	.0376	.0811	.1327	.1734	.1864	.1669
	9	.0000	.0000	.0004	.0033	.0139	.0386	.0794	.1284	.1694	.1855
	10	.0000	.0000	.0001	.0008	.0042	.0149	.0385	.0771	.1248	.1669
	11	.0000	.0000	.0000	.0001	.0010	.0046	.0151	.0374	.0742	.1214
	12	.0000	.0000	.0000	.0000	.0002	.0012	.0047	.0145	.0354	.0708
	13	.0000	.0000	.0000	.0000	.0000	.0002	.0012	.0045	.0134	.0327
	14	.0000	.0000	.0000	.0000	.0000	.0000	.0002	.0011	.0039	.0117
	15	.0000	.0000	.0000	.0000	.0000	.0000	.0000	.0002	.0009	.0031
	16	.0000	.0000	.0000	.0000	.0000	.0000	.0000	.0000	.0001	.0006
	17	.0000	.0000	.0000	.0000	.0000	.0000	.0000	.0000	.0000	.0001
	18	.0000	.0000	.0000	.0000	.0000	.0000	.0000	.0000	.0000	.0000

TABLE B (continued)

n	s	.05	.10	.15	.20	.25 p	.30	.35	.40	.45	.50
19	0	.3774	.1351	.0456	.0144	.0042	.0011	.0003	.0001	.0000	.0000
	1	.3774	.2852	.1529	.0685	.0268	.0093	.0029	.0008	.0002	.0000
	2	.1787	.2852	.2428	.1540	.0803	.0358	.0138	.0046	.0013	.0003
	3	.0533	.1796	.2428	.2182	.1517	.0869	.0422	.0175	.0062	.0018
	4	.0112	.0798	.1714	.2182	.2023	.1491	.0909	.0467	.0203	.0074
	5	.0018	.0266	.0907	.1636	.2023	.1916	.1468	.0933	.0497	.0222
	6	.0002	.0069	.0374	.0955	.1574	.1916	.1844	.1451	.0949	.0518
	7	.0000	.0014	.0122	.0443	.0974	.1525	.1844	.1797	.1443	.0961
	8	.0000	.0002	.0032	.0166	.0487	.0981	.1489	.1797	.1771	.1442
	9	.0000	.0000	.0007	.0051	.0198	.0514	.0980	.1464	.1771	.1762
	10	.0000	.0000	.0001	.0013	.0066	.0220	.0528	.0976	.1449	.1762
	11	.0000	.0000	.0000	.0003	.0018	.0077	.0233	.0532	.0970	.1442
	12	.0000	.0000	.0000	.0000	.0004	.0022	.0083	.0237	.0529	.0961
	13	.0000	.0000	.0000	.0000	.0001	.0005	.0024	.0085	.0233	.0518
	14	.0000	.0000	.0000	.0000	.0000	.0001	.0006	.0024	.0082	.0222
	15	.0000	.0000	.0000	.0000	.0000	.0000	.0001	.0005	.0022	.0074
	16	.0000	.0000	.0000	.0000	.0000	.0000	.0000	.0001	.0005	.0018
	17	.0000	.0000	.0000	.0000	.0000	.0000	.0000	.0000	.0001	.0003
	18	.0000	.0000	.0000	.0000	.0000	.0000	.0000	.0000	.0000	.0000
	19	.0000	.0000	.0000	.0000	.0000	.0000	.0000	.0000	.0000	.0000
20	0	.3585	.1216	.0388	.0115	.0032	.0008	.0002	.0000	.0000	.0000
	1	.3774	.2702	.1368	.0576	.0211	.0068	.0020	.0005	.0001	.0000
	2	.1887	.2852	.2293	.1369	.0669	.0278	.0100	.0031	.0008	.0002
	3	.0596	.1901	.2428	.2054	.1339	.0716	.0323	.0123	.0040	.0011
	4	.0133	.0898	.1821	.2182	.1897	.1304	.0738	.0350	.0139	.0046
	5	.0022	.0319	.1028	.1746	.2023	.1789	.1272	.0746	.0365	.0148
	6	.0003	.0089	.0454	.1091	.1686	.1916	.1712	.1244	.0746	.0370
	7	.0000	.0020	.0160	.0545	.1124	.1643	.1844	.1659	.1221	.0739
	8	.0000	.0004	.0046	.0222	.0609	.1144	.1614	.1797	.1623	.1201
	9	.0000	.0001	.0011	.0074	.0271	.0654	.1158	.1597	.1771	.1602
	10	.0000	.0000	.0002	.0020	.0099	.0308	.0686	.1171	.1593	.1762
	11	.0000	.0000	.0000	.0005	.0030	.0120	.0336	.0710	.1185	.1602
	12	.0000	.0000	.0000	.0001	.0008	.0039	.0136	.0355	.0727	.1201
	13	.0000	.0000	.0000	.0000	.0002	.0010	.0045	.0146	.0366	.0739
	14	.0000	.0000	.0000	.0000	.0000	.0002	.0012	.0049	.0150	.0370
	15	.0000	.0000	.0000	.0000	.0000	.0000	.0003	.0013	.0049	.0148
	16	.0000	.0000	.0000	.0000	.0000	.0000	.0000	.0003	.0013	.0046
	17	.0000	.0000	.0000	.0000	.0000	.0000	.0000	.0000	.0002	.0011
	18	.0000	.0000	.0000	.0000	.0000	.0000	.0000	.0000	.0000	.0002
	19	.0000	.0000	.0000	.0000	.0000	.0000	.0000	.0000	.0000	.0000
	20	.0000	.0000	.0000	.0000	.0000	.0000	.0000	.0000	.0000	.0000

TABLE C Normal Distribution and Related Functions

x	$F(x)$ Φ	$1 - F(x)$	$f(x)$ ϕ	x	$F(x)$ Φ	$1 - F(x)$	$f(x)$ ϕ
.00	.5000	.5000	.3989	.35	.6368	.3632	.3752
.01	.5040	.4960	.3989	.36	.6406	.3594	.3739
.02	.5080	.4920	.3989	.37	.6443	.3557	.3726
.03	.5120	.4880	.3988	.38	.6480	.3520	.3712
.04	.5160	.4840	.3986	.39	.6517	.3483	.3697
.05	.5199	.4801	.3984	.40	.6554	.3446	.3683
.06	.5239	.4761	.3982	.41	.6591	.3409	.3668
.07	.5279	.4721	.3980	.42	.6628	.3372	.3653
.08	.5319	.4681	.3977	.43	.6664	.3336	.3637
.09	.5359	.4641	.3973	.44	.6700	.3300	.3621
.10	.5398	.4602	.3970	.45	.6736	.3264	.3605
.11	.5438	.4562	.3965	.46	.6772	.3228	.3589
.12	.5478	.4522	.3961	.47	.6808	.3192	.3572
.13	.5517	.4483	.3956	.48	.6844	.3156	.3555
.14	.5557	.4443	.3951	.49	.6879	.3121	.3538
.15	.5596	.4404	.3945	.50	.6915	.3085	.3521
.16	.5636	.4364	.3939	.51	.6950	.3050	.3503
.17	.5675	.4325	.3932	.52	.6985	.3015	.3485
.18	.5714	.4286	.3925	.53	.7019	.2981	.3467
.19	.5753	.4247	.3918	.54	.7054	.2946	.3448
.20	.5793	.4207	.3910	.55	.7088	.2912	.3429
.21	.5832	.4168	.3902	.56	.7123	.2877	.3410
.22	.5871	.4129	.3894	.57	.7157	.2843	.3391
.23	.5910	.4090	.3885	.58	.7190	.2810	.3372
.24	.5948	.4052	.3876	.59	.7224	.2776	.3352
.25	.5987	.4013	.3867	.60	.7257	.2743	.3332
.26	.6026	.3974	.3857	.61	.7291	.2709	.3312
.27	.6064	.3936	.3847	.62	.7324	.2676	.3292
.28	.6103	.3897	.3836	.63	.7357	.2643	.3271
.29	.6141	.3859	.3825	.64	.7389	.2611	.3251
.30	.6179	.3821	.3814	.65	.7422	.2578	.3230
.31	.6217	.3783	.3802	.66	.7454	.2546	.3209
.32	.6255	.3745	.3790	.67	.7486	.2514	.3187
.33	.6293	.3707	.3778	.68	.7517	.2483	.3166
.34	.6331	.3669	.3765	.69	.7549	.2451	.3144

TABLE C (continued)

x	$F(x)$ Φ	$1 - F(x)$	$f(x)$ ϕ	x	$F(x)$ Φ	$1 - F(x)$	$f(x)$ ϕ
.70	.7580	.2420	.3123	1.05	.8531	.1469	.2299
.71	.7611	.2389	.3101	1.06	.8554	.1446	.2275
.72	.7642	.2358	.3079	1.07	.8577	.1423	.2251
.73	.7673	.2327	.3056	1.08	.8599	.1401	.2227
.74	.7704	.2296	.3034	1.09	.8621	.1379	.2203
.75	.7734	.2266	.3011	1.10	.8643	.1357	.2179
.76	.7764	.2236	.2989	1.11	.8665	.1335	.2155
.77	.7794	.2206	.2966	1.12	.8686	.1314	.2131
.78	.7823	.2177	.2943	1.13	.8708	.1292	.2107
.79	.7852	.2148	.2920	1.14	.8729	.1271	.2083
.80	.7881	.2119	.2897	1.15	.8749	.1251	.2059
.81	.7910	.2090	.2874	1.16	.8770	.1230	.2036
.82	.7939	.2061	.2850	1.17	.8790	.1210	.2012
.83	.7967	.2033	.2827	1.18	.8810	.1190	.1989
.84	.7995	.2005	.2803	1.19	.8830	.1170	.1965
.85	.8023	.1977	.2780	1.20	.8849	.1151	.1942
.86	.8051	.1949	.2756	1.21	.8869	.1131	.1919
.87	.8079	.1921	.2732	1.22	.8888	.1112	.1895
.88	.8106	.1894	.2709	1.23	.8907	.1093	.1872
.89	.8133	.1867	.2685	1.24	.8925	.1075	.1849
.90	.8159	.1841	.2661	1.25	.8944	.1056	.1826
.91	.8186	.1814	.2637	1.26	.8962	.1038	.1804
.92	.8212	.1788	.2613	1.27	.8980	.1020	.1781
.93	.8238	.1762	.2589	1.28	.8997	.1003	.1758
.94	.8264	.1736	.2565	1.29	.9015	.0985	.1736
.95	.8289	.1711	.2541	1.30	.9032	.0968	.1714
.96	.8315	.1685	.2516	1.31	.9049	.0951	.1691
.97	.8340	.1660	.2492	1.32	.9066	.0934	.1669
.98	.8365	.1635	.2468	1.33	.9082	.0918	.1647
.99	.8389	.1611	.2444	1.34	.9099	.0901	.1626
1.00	.8413	.1587	.2420	1.35	.9115	.0885	.1604
1.01	.8438	.1562	.2396	1.36	.9131	.0869	.1582
1.02	.8461	.1539	.2371	1.37	.9147	.8053	.1561
1.03	.8485	.1515	.2347	1.38	.9162	.0838	.1539
1.04	.8508	.1492	.2323	1.39	.9177	.0823	.1518

TABLE C (continued)

x	$F(x)$ Φ	$1 - F(x)$	$f(x)$ ϕ	x	$F(x)$ Φ	$1 - F(x)$	$f(x)$ ϕ
1.40	.9192	.0808	.1497	1.75	.9599	.0401	.0863
1.41	.9207	.0793	.1476	1.76	.9608	.0392	.0848
1.42	.9222	.0778	.1456	1.77	.0616	.0384	.0833
1.43	.9236	.0764	.1435	1.78	.9625	.0375	.0818
1.44	.9251	.0749	.1415	1.78	.9633	.0367	.0804
1.45	.9265	.0735	.1394	1.80	.9641	.0359	.0790
1.46	.9279	.0721	.1374	1.81	.9649	.0351	.0775
1.47	.9292	.0708	.1354	1.82	.9656	.0344	.0761
1.48	.9306	.0694	.1334	1.83	.9664	.0336	.0748
1.49	.9319	.0681	.1315	1.84	.9671	.0329	.0734
1.50	.9332	.0668	.1295	1.85	.9678	.0322	.0721
1.51	.9345	.0655	.1276	1.86	.9686	.0314	.0707
1.52	.9357	.0643	.1257	1.87	.9693	.0307	.0694
1.53	.9370	.0630	.1238	1.88	.9699	.0301	.0681
1.54	.9382	.0618	.1219	1.89	.9706	.0294	.0669
1.55	.9394	.0606	.1200	1.90	.9713	.0287	.0656
1.56	.9406	.0594	.1182	1.91	.9719	.0281	.0644
1.57	.9418	.0582	.1163	1.92	.9276	.0274	.0632
1.58	.9429	.0571	.1145	1.93	.9732	.0268	.0620
1.59	.9441	.0559	.1127	1.94	.9738	.0262	.0608
1.60	.9452	.0548	.1109	1.95	.9744	.0256	.0596
1.61	.9463	.0537	.1092	1.96	.9750	.0250	.0584
1.62	.9474	.0526	.1074	1.97	.9756	.0244	.0573
1.63	.9484	.0516	.1057	1.98	.9761	.0239	.0562
1.64	.9495	.0505	.1040	1.99	.9767	.0233	.0551
1.65	.9505	.0495	.1023	2.00	.9773	.0227	.0540
1.66	.9515	.0485	.1006	2.01	.9778	.0222	.0529
1.67	.9525	.0475	.0989	2.02	.9783	.0217	.0519
1.68	.9535	.0465	.0973	2.03	.9788	.0212	.0508
1.69	.9545	.0455	.0957	2.04	.9793	.0207	.0498
1.70	.9554	.0446	.0940	2.05	.9798	.0202	.0488
1.71	.9564	.0436	.0925	2.06	.9803	.0197	.0478
1.72	.9573	.0427	.0909	2.07	.9808	.0192	.0468
1.73	.9582	.0418	.0893	2.08	.9812	.0188	.0459
1.74	.9591	.0409	.0878	2.09	.9817	.0183	.0449

TABLE C (continued)

x	$F(x)$ Φ	$1 - F(x)$	$f(x)$ ϕ	x	$F(x)$ Φ	$1 - F(x)$	$f(x)$ ϕ
2.10	.9821	.0179	.0440	2.45	.9929	.0071	.0198
2.11	.9826	.0174	.0431	2.46	.9931	.0069	.0194
2.12	.9830	.0170	.0422	2.47	.9932	.0068	.0189
2.13	.9834	.0166	.0413	2.48	.9934	.0066	.0184
2.14	.9838	.0162	.0404	2.49	.9936	.0064	.0180
2.15	.9842	.0158	.0396	2.50	.9938	.0062	.0175
2.16	.9846	.0154	.0387	2.51	.9940	.0060	.0171
2.17	.9850	.0150	.0379	2.52	.9941	.0059	.0167
2.18	.9854	.0146	.0371	2.53	.9943	.0057	.0163
2.19	.9857	.0143	.0363	2.54	.9945	.0055	.0158
2.20	.9861	.0139	.0355	2.55	.9946	.0054	.0155
2.21	.9864	.0136	.0347	2.56	.9948	.0052	.0151
2.22	.9868	.0132	.0339	2.57	.9949	.0051	.0147
2.23	.9871	.0129	.0332	2.58	.9951	.0049	.0143
2.24	.9875	.0125	.0325	2.59	.9952	.0048	.0139
2.25	.9878	.0122	.0317	2.60	.9953	.0047	.0136
2.26	.9881	.0119	.0310	2.61	.9955	.0045	.0132
2.27	.9884	.0116	.0303	2.62	.9956	.0044	.0129
2.28	.9887	.0113	.0297	2.63	.9957	.0043	.0126
2.29	.9890	.0110	.0290	2.64	.9959	.0041	.0122
2.30	.9893	.0107	.0283	2.65	.9960	.0040	.0119
2.31	.9896	.0104	.0277	2.66	.9961	.0039	.0116
2.32	.9898	.0102	.0270	2.67	.9962	.0038	.0113
2.33	.9901	.0099	.0264	2.68	.9963	.0037	.0110
2.34	.9904	.0096	.0258	2.69	.9964	.0036	.0107
2.35	.9906	.0094	.0252	2.70	.9965	.0035	.0104
2.36	.9909	.0091	.0246	2.71	.9966	.0034	.0101
2.37	.9911	.0089	.0241	2.72	.9967	.0033	.0099
2.38	.9913	.0087	.0235	2.73	.9968	.0032	.0096
2.39	.9916	.0084	.0229	2.74	.9969	.0031	.0093
2.40	.9918	.0082	.0224	2.75	.9970	.0030	.0091
2.41	.9920	.0080	.0219	2.76	.9971	.0029	.0088
2.42	.9922	.0078	.0213	2.77	.9972	.0028	.0086
2.43	.9925	.0075	.0208	2.78	.9973	.0027	.0084
2.44	.9927	.0073	.0203	2.79	.9974	.0026	.0081

TABLE C (continued)

x	$F(x)$ Φ	$1 - F(x)$	$f(x)$ ϕ	x	$F(x)$ Φ	$1 - F(x)$	$f(x)$ ϕ
2.80	.9974	.0026	.0079	3.15	.9992	.0008	.0028
2.81	.9975	.0025	.0077	3.16	.9992	.0008	.0027
2.82	.9976	.0024	.0075	3.17	.9992	.0008	.0026
2.83	.9977	.0023	.0073	3.18	.9993	.0007	.0025
2.84	.9977	.0023	.0071	3.19	.9993	.0007	.0025
2.85	.9978	.0022	.0069	3.20	.9993	.0007	.0024
2.86	.9979	.0021	.0067	3.21	.9993	.0007	.0023
2.87	.9979	.0021	.0065	3.22	.9994	.0006	.0022
2.88	.9980	.0020	.0063	3.23	.9994	.0006	.0022
2.89	.9981	.0019	.0061	3.24	.9994	.0006	.0021
2.90	.9981	.0019	.0060	3.25	.9994	.0006	.0020
2.91	.9982	.0018	.0058	3.26	.9994	.0006	.0020
2.92	.9983	.0017	.0056	3.27	.9995	.0005	.0019
2.93	.9983	.0017	.0055	3.28	.9995	.0005	.0018
2.94	.9984	.0016	.0053	3.29	.9995	.0005	.0018
2.95	.9984	.0016	.0051	3.30	.9995	.0005	.0017
2.96	.9985	.0015	.0050	3.31	.9995	.0005	.0017
2.97	.9985	.0015	.0048	3.32	.9996	.0004	.0016
2.98	.9986	.0014	.0047	3.33	.9996	.0004	.0016
2.99	.9986	.0014	.0046	3.34	.9996	.0004	.0015
3.00	.9987	.0013	.0044	3.35	.9996	.0004	.0015
3.01	.9987	.0013	.0043	3.36	.9996	.0004	.0014
3.02	.9987	.0013	.0042	3.37	.9996	.0004	.0014
3.03	.9988	.0012	.0040	3.38	.9996	.0004	.0013
3.04	.9988	.0012	.0039	3.39	.9997	.0003	.0013
3.05	.9989	.0011	.0038	3.40	.9997	.0003	.0012
3.06	.9989	.0011	.0037	3.41	.9997	.0003	.0012
3.07	.9989	.0011	.0036	3.42	.9997	.0003	.0012
3.08	.9990	.0010	.0035	3.43	.9997	.0003	.0011
3.09	.9990	.0010	.0034	3.44	.9997	.0003	.0011
3.10	.9990	.0010	.0033	3.45	.9997	.0003	.0010
3.11	.9991	.0009	.0032	3.46	.9997	.0003	.0010
3.12	.9991	.0009	.0031	3.47	.9997	.0003	.0010
3.13	.9991	.0009	.0030	3.48	.9997	.0003	.0009
3.14	.9992	.0008	.0029	3.49	.9998	.0002	.0009

TABLE C (continued)

x	$F(x)$ Φ	$1 - F(x)$	$f(x)$ ϕ	x	$F(x)$ Φ	$1 - F(x)$	$f(x)$ ϕ
3.50	.9998	.0002	.0009	3.75	.9999	.0001	.0004
3.51	.9998	.0002	.0008	3.76	.9999	.0001	.0003
3.52	.9998	.0002	.0008	3.77	.9999	.0001	.0003
3.53	.9998	.0002	.0008	3.78	.9999	.0001	.0003
3.54	.9998	.0002	.0008	3.79	.9999	.0001	.0003
3.55	.9998	.0002	.0007	3.80	.9999	.0001	.0003
3.56	.9998	.0002	.0007	3.81	.9999	.0001	.0003
3.57	.9998	.0002	.0007	3.82	.9999	.0001	.0003
3.58	.9998	.0002	.0007	3.83	.9999	.0001	.0003
3.59	.9998	.0002	.0006	3.84	.9999	.0001	.0003
3.60	.9998	.0002	.0006	3.85	.9999	.0001	.0002
3.61	.9998	.0002	.0006	3.86	.9999	.0001	.0002
3.62	.9999	.0001	.0006	3.87	.9999	.0001	.0002
3.63	.9999	.0001	.0005	3.88	.9999	.0001	.0002
3.64	.9999	.0001	.0005	3.89	1.0000	.0000	.0002
3.65	.9999	.0001	.0005	3.90	1.0000	.0000	.0002
3.66	.9999	.0001	.0005	3.91	1.0000	.0000	.0002
3.67	.9999	.0001	.0005	3.92	1.0000	.0000	.0002
3.68	.9999	.0001	.0004	3.93	1.0000	.0000	.0002
3.69	.9999	.0001	.0004	3.94	1.0000	.0000	.0002
3.70	.9999	.0001	.0004	3.95	1.0000	.0000	.0002
3.71	.9999	.0001	.0004	3.96	1.0000	.0000	.0002
3.72	.9999	.0001	.0004	3.97	1.0000	.0000	.0002
3.73	.9999	.0001	.0004	3.98	1.0000	.0000	.0001
3.74	.9999	.0001	.0004	3.99	1.0000	.0000	.0001

x	1.282	1.645	1.960	2.326	2.576	3.090
$F(x)$	.90	.95	.975	.99	.995	.999
$2[1 - F(x)]$	.20	.10	.05	.02	.01	.002

TABLE D Chi Square Distribution Function

F df	.995	.990	.975	.950	.900	.750	.500	.250	.100	.050	.025	.010	.005
1	7.88	6.63	5.02	3.84	2.71	1.32	.455	.102	.0158	.00393	.000982	.000157	.0000393
2	10.6	9.21	7.38	5.99	4.61	2.77	1.39	.575	.211	.103	.0506	.0201	.0100
3	12.8	11.3	9.35	7.81	6.25	4.11	2.37	1.21	.584	.352	.216	.115	.0717
4	14.9	13.3	11.1	9.49	7.78	5.39	3.36	1.92	1.06	.711	.484	.297	.207
5	16.7	15.1	12.8	11.1	9.24	6.63	4.35	2.67	1.61	1.15	.831	.554	.412
6	18.5	16.8	14.4	12.6	10.6	7.84	5.35	3.45	2.20	1.64	1.24	.872	.676
7	20.3	18.5	16.0	14.1	12.0	9.04	6.35	4.25	2.83	2.17	1.69	1.24	.989
8	22.0	20.1	17.5	15.5	13.4	10.2	7.34	5.07	3.49	2.73	2.18	1.65	1.34
9	23.6	21.7	19.0	16.9	14.7	11.4	8.34	5.90	4.17	3.33	2.70	2.09	1.73
10	25.2	23.2	20.5	18.3	16.0	12.5	9.34	6.74	4.87	3.94	3.25	2.56	2.16
11	26.8	24.7	21.9	19.7	17.3	13.7	10.3	7.58	5.58	4.57	3.82	3.05	2.60
12	28.3	26.2	23.3	21.0	18.5	14.8	11.3	8.44	6.30	5.23	4.40	3.57	3.07
13	29.8	27.7	24.7	22.4	19.8	16.0	12.3	9.30	7.04	5.89	5.01	4.11	3.57
14	31.3	29.1	26.1	23.7	21.1	17.1	13.3	10.2	7.79	6.57	5.63	4.66	4.07
15	32.8	30.6	27.5	25.0	22.3	18.2	14.3	11.0	8.55	7.26	6.26	5.23	4.60
16	34.3	32.0	28.8	26.3	23.5	19.4	15.3	11.9	9.31	7.96	6.91	5.81	5.14
17	35.7	33.4	30.2	27.6	24.8	20.5	16.3	12.8	10.1	8.67	7.56	6.41	5.70
18	37.2	34.8	31.5	28.9	26.0	21.6	17.3	13.7	10.9	9.39	8.23	7.01	6.26
19	38.6	36.2	32.9	30.1	27.2	22.7	18.3	14.6	11.7	10.1	8.91	7.63	6.84
20	40.0	37.6	34.2	31.4	28.4	23.8	19.3	15.5	12.4	10.9	9.59	8.26	7.43
21	41.4	38.9	35.5	32.7	29.6	24.9	20.3	16.3	13.2	11.6	10.3	8.90	8.03
22	42.8	40.3	36.8	33.9	30.8	26.0	21.3	17.2	14.0	12.3	11.0	9.54	8.64
23	44.2	41.6	38.1	35.2	32.0	27.1	22.3	18.1	14.8	13.1	11.7	10.2	9.26
24	45.6	43.0	39.4	36.4	33.2	28.2	23.3	19.0	15.7	13.8	12.4	10.9	9.89
25	46.9	44.3	40.6	37.7	34.4	29.3	24.3	19.9	16.5	14.6	13.1	11.5	10.5
26	48.3	45.6	41.9	38.9	35.6	30.4	25.3	20.8	17.3	15.4	13.8	12.2	11.2
27	49.6	47.0	43.2	40.1	36.7	31.5	26.3	21.7	18.1	16.2	14.6	12.9	11.8
28	51.0	48.3	44.5	41.3	37.9	32.6	27.3	22.7	18.9	16.9	15.3	13.6	12.5
29	52.3	49.6	45.7	42.6	39.1	33.7	28.3	23.6	19.8	17.7	16.0	14.3	13.1
30	53.7	50.9	47.0	43.8	40.3	34.8	29.3	24.5	20.6	18.5	16.8	15.0	13.8

Source. Reprinted, with permission, from Pearson, E. S., and Hartley, H. O. (1962), *Biometrika tables for statisticians*, Vol. 1, Biometrika Trustees, London.

TABLE E Student's *t* Distribution Function

df \ F	.60	.75	.90	.95	.975	.99	.995	.9995
1	.325	1.000	3.078	6.314	12.706	31.821	63.657	636.619
2	.289	.816	1.886	2.920	4.303	6.965	9.925	31.598
3	.277	.765	1.638	2.353	3.182	4.541	5.841	12.941
4	.271	.741	1.533	2.132	2.776	3.747	4.604	8.610
5	.267	.727	1.476	2.015	2.571	3.365	4.032	6.859
6	.265	.718	1.440	1.943	2.447	3.143	3.707	5.959
7	.263	.711	1.415	1.895	2.365	2.998	3.499	5.405
8	.262	.706	2.397	1.860	2.306	2.896	3.355	5.041
9	.261	.703	1.383	1.833	2.262	2.821	3.250	4.781
10	.260	.700	1.372	1.812	2.228	2.764	3.169	4.587
11	.260	.697	1.363	1.796	2.201	2.718	3.106	4.437
12	.259	.695	1.356	1.782	2.179	2.681	3.055	4.318
13	.259	.694	1.350	1.771	2.160	2.650	3.012	4.221
14	.258	.692	1.345	1.761	2.145	2.624	2.977	4.140
15	.258	.691	1.341	1.753	2.131	2.602	2.947	4.073
16	.258	.690	1.337	1.746	2.120	2.583	2.921	4.015
17	.257	.689	1.333	1.740	2.110	2.567	2.898	3.965
18	.257	.688	1.330	1.734	2.101	2.552	2.878	3.922
19	.257	.688	1.328	1.729	2.093	2.539	2.861	3.883
20	.257	.687	1.325	1.725	2.086	2.528	2.845	3.850
21	.257	.686	1.323	1.721	2.080	2.518	2.831	3.819
22	.256	.686	1.321	1.717	2.074	2.508	2.819	3.792
23	.256	.685	1.319	1.714	2.069	2.500	2.807	3.767
24	.256	.685	1.318	1.711	2.064	2.492	2.797	3.745
25	.256	.684	1.316	1.708	2.060	2.485	2.787	3.725
26	.256	.684	1.315	1.706	2.056	2.479	2.779	3.707
27	.256	.684	1.314	1.703	2.052	2.473	2.771	3.690
28	.256	.683	1.313	1.701	2.048	2.467	2.763	3.674
29	.256	.683	1.311	1.699	2.045	2.462	2.756	3.659
30	.256	.683	1.310	1.697	2.042	2.457	2.750	3.646
40	.255	.681	1.303	1.684	2.021	2.423	2.704	3.551
60	.254	.679	1.296	1.671	2.000	2.390	2.660	3.460
120	.254	.677	1.289	1.658	1.980	2.358	2.617	3.373
∞	.253	.674	1.282	1.645	1.960	2.326	2.576	3.291

Source. Table E is taken from Fisher and Yates: *Statistical tables for biological, agricultural and medical research*, published by Oliver and Boyd, Edinburgh, and by permission of the authors and publishers.

TABLE F Number of Observations for t-Test of Expectation, Single Samples

Level of t-Test

β =	α = .005 / α = .01					α = .01 / α = .02					α = .025 / α = .05					α = .05 / α = .1				
(Single-Sided / Double-Sided)	.01	.05	.1	.2	.5	.01	.05	.1	.2	.5	.01	.05	.1	.2	.5	.01	.05	.1	.2	.5
.05																				
.10																				
.15																				122
.20										139					99					70
.25					110					90				128	64			139	101	45
.30				134	78				115	63			119	90	45		122	97	71	32
.35			125	99	58			109	85	47		109	88	67	34		90	72	52	24
.40		115	97	77	45		101	85	66	37	117	84	68	51	26	101	70	55	40	19
.45		92	77	62	37	110	81	68	53	30	93	67	54	41	21	80	55	44	33	15
.50	100	75	63	51	30	90	66	55	43	25	76	54	44	34	18	65	45	36	27	13
.55	83	63	53	42	26	75	55	46	36	21	63	45	37	28	15	54	38	30	22	11
.60	71	53	45	36	22	63	47	39	31	18	53	38	32	24	13	46	32	26	19	9
.65	61	46	39	31	20	55	41	34	27	16	46	33	27	21	12	39	28	22	17	8
.70	53	40	34	28	17	47	35	30	24	14	40	29	24	19	10	34	24	19	15	8
.75	47	36	30	25	16	42	31	27	21	13	35	26	21	16	9	30	21	17	13	7
.80	41	32	27	22	14	37	28	24	19	12	31	22	19	15	9	27	19	15	12	6
.85	37	29	24	20	13	33	25	21	17	11	28	21	17	13	8	24	17	14	11	6
.90	34	26	22	18	12	29	23	19	16	10	25	19	16	12	7	21	15	13	10	5
.95	31	24	20	17	11	27	21	18	14	9	23	17	14	11	7	19	14	11	9	5
1.00	28	22	19	16	10	25	19	16	13	9	21	16	13	10	6	18	13	11	8	5
1.1	24	19	16	14	9	21	16	14	12	8	18	13	11	9	6	15	11	9	7	
1.2	21	16	14	12	8	18	14	12	10	7	15	12	10	8	5	13	10	8	6	
1.3	18	15	13	11	8	16	13	11	9	6	14	10	9	7		11	8	7	6	
1.4	16	13	12	10	7	14	11	10	9	6	12	9	8	7		10	8	7	5	
1.5	15	12	11	9	7	13	10	9	8	6	11	8	7	6		9	7	6	5	

$$\frac{d}{|\mu_1 - \mu_2|} = \varrho$$

TABLE F (continued)

Level of t-Test

| Single-Sided Test | $\alpha = .005$ | | | | | $\alpha = .01$ | | | | | $\alpha = .025$ | | | | | $\alpha = .05$ | | | | |
| Double-Sided Test | $\alpha = .01$ | | | | | $\alpha = .02$ | | | | | $\alpha = .05$ | | | | | $\alpha = .1$ | | | | |
$\beta =$	.01	.05	.1	.2	.5	.01	.05	.1	.2	.5	.01	.05	.1	.2	.5	.01	.05	.1	.2	.5
1.6	13	11	10	8	6	12	10	9	7	5	10	8	7	6	6	8	6	6		
1.7	12	10	9	8	6	11	9	8	7		9	7	6	5		8	6	5		
1.8	12	10	9	8	6	10	8	7	7		8	7	6			7	6			
1.9	11	9	8	7	6	10	8	7	6		8	6	6			7	5			
2.0	10	8	8	7	5	9	7	7	6		7	6	5			6				
2.1	10	8	7	7		8	7	6	6		7	6				6				
2.2	9	8	7	6		8	7	6	5		7	6				6				
2.3	9	7	7	6		8	6	6			6	5				5				
2.4	8	7	7	6		7	6	6			6									
2.5	8	7	6	6		7	6	6			6									
3.0	7	6	6	5		6	5	5			5									
3.5	6	5	5			5														
4.0	6																			

$$\frac{D}{|\mu_A - \mu_L|} = \rho$$

Source. Reprinted, with permission, from Davies, O. L. (1956), *Design and analysis of industrial experiments*, Vol. 1, Oliver and Boyd, Edinburgh.

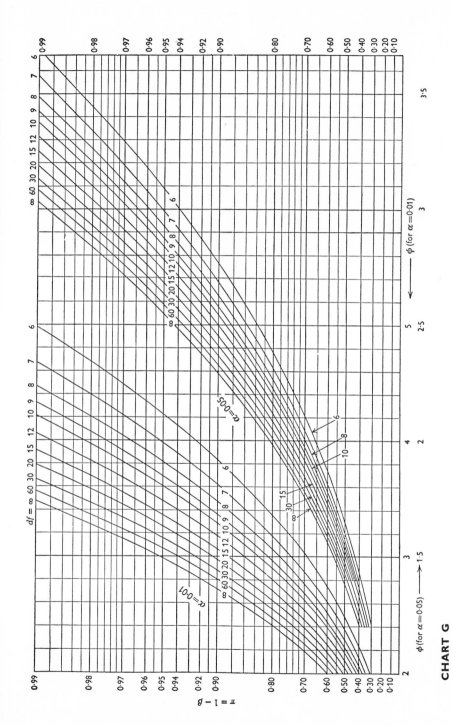

CHART G

Power function of the t-test.

Source. Reprinted with permission from Pearson, E. S. and Hartley, H. O. (1962), *Biometrika tables for statisticians*, Vol. 1, Biometrika Trustees, London.

TABLE H Sample Size for Single-Sample Tests of Hypotheses about Variances

| | α = .01 | | | | α = .05 | | | |
df	β = .01	β = .05	β = .1	β = .5	β = .01	β = .05	β = .01	β = .5
1	42,240	1,687	420.2	14.58	25,450	977.0	243.3	8.444
2	458.2	89.78	43.71	6.644	298.1	58.40	28.43	4.322
3	98.79	32.24	19.41	4.795	68.05	22.21	13.37	3.303
4	44.69	18.68	12.48	3.955	31.93	13.35	8.920	2.826
5	27.22	13.17	9.369	3.467	19.97	9.665	6.875	2.544
6	19.28	10.28	7.628	3.144	14.44	7.699	5.713	2.354
7	14.91	8.524	6.521	2.911	11.35	6.491	4.965	2.217
8	12.20	7.352	5.757	2.736	9.418	5.675	4.444	2.112
9	10.38	6.516	5.198	2.597	8.103	5.088	4.059	2.028
10	9.072	5.890	4.770	2.484	7.156	4.646	3.763	1.960
12	7.343	5.017	4.159	2.312	5.889	4.023	3.335	1.854
15	5.847	4.211	3.578	2.132	4.780	3.442	2.925	1.743
20	4.548	3.462	3.019	1.943	3.802	2.895	2.524	1.624
24	3.959	3.104	2.745	1.842	3.354	2.630	2.326	1.560
30	3.403	2.752	2.471	1.735	2.927	2.367	2.125	1.492
40	2.874	2.403	2.192	1.619	2.516	2.103	1.919	1.418
60	2.358	2.046	1.902	1.490	2.110	1.831	1.702	1.333
120	1.829	1.661	1.580	1.332	1.686	1.532	1.457	1.228
∞	1.000	1.000	1.000	1.000	1.000	1.000	1.000	1.000

Table entries are $\delta^2 = \sigma^2/\sigma_T^2$ for $H_T: \sigma^2 = \sigma_T^2$ against $H_A: \sigma^2 > \sigma_T^2$ for an $(\alpha, \beta, \delta^2)$ test with df degrees of freedom. For $H_A: \sigma^2 < \sigma_T^2$, exchange α and β values and use $1/\delta^2$.

Source. From *Selected techniques of statistical analysis* by Eisenhart, C., Hastay, M. W., and Wallis, W. Copyright 1947 by McGraw-Hill, Inc. Used by permission of McGraw-Hill Book Company.

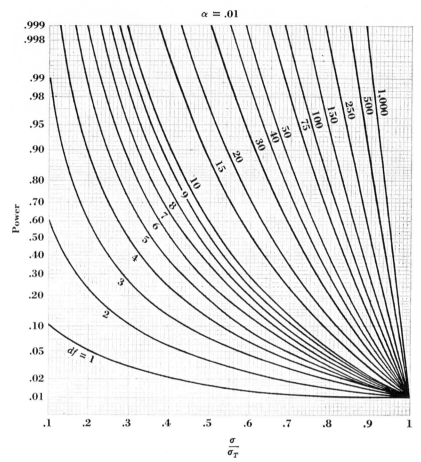

$\alpha = .01$

CHART I

Power curves for testing H_T: $\sigma^2 = \sigma_T{}^2$ against H_A: $\sigma^2 < \sigma_T{}^2$.

Source. From *Concepts of statistical inference* by Guentner, W. C. Copyright 1965 by McGraw-Hill, Inc. Used with permission of McGraw-Hill Book Co.

651

$\alpha = .025$

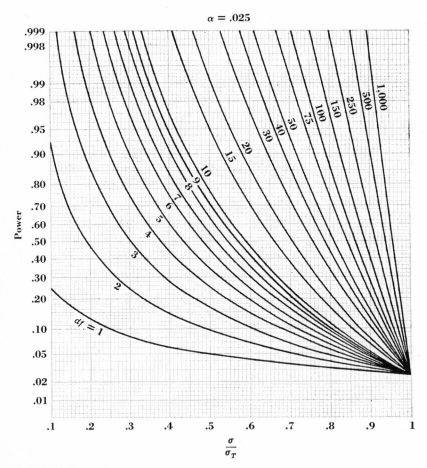

CHART I (continued)

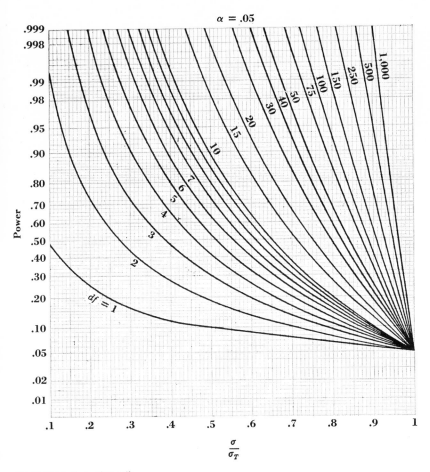

$\alpha = .05$

Power

$\dfrac{\sigma}{\sigma_T}$

CHART I (continued)

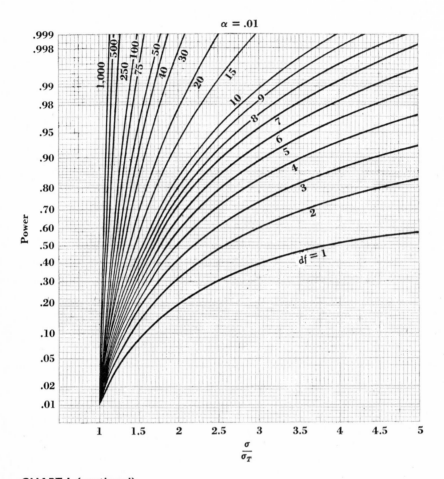

$$\alpha = .01$$

Power

$$\frac{\sigma}{\sigma_T}$$

CHART I (continued)

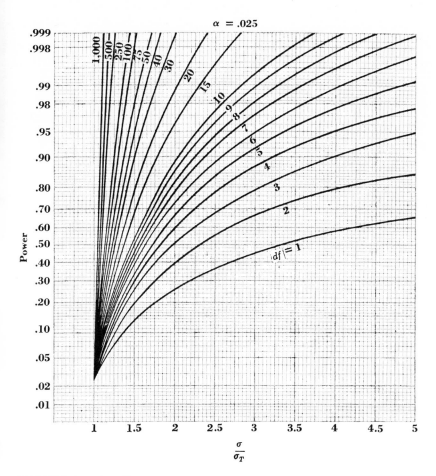

$\alpha = .025$

CHART I (continued)

α = .05

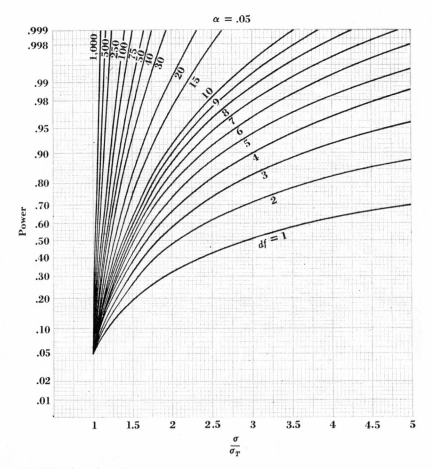

CHART I (continued)

TABLE J Sample Size for t-Test of Difference Between Two Expectations

Level of t-Test

Single-Sided Test: $\alpha = .005$, $\alpha = .01$, $\alpha = .025$, $\alpha = .05$
Double-Sided Test: $\alpha = .01$, $\alpha = .02$, $\alpha = .05$, $\alpha = .1$

Row variable: $\rho = \dfrac{D}{|\mu_1 - \mu_2|}$

ρ	$\alpha=.005 / .01$ .01	.05	.1	.2	.5	$\alpha=.01 / .02$ .01	.05	.1	.2	.5	$\alpha=.025 / .05$ .01	.05	.1	.2	.5	$\alpha=.05 / .1$ .01	.05	.1	.2	.5
.05																				
.10																				
.15																				
.20																				137
.25															124					88
.30										123					87					61
.35					110					90					64				102	45
.40					85				101	70				100	50			108	78	35
.45				118	68				82	55			105	79	39		108	86	62	28
.50				96	55			106	68	45		106	86	64	32		88	70	51	23
.55			101	79	46		106	88	58	38		87	71	53	27	112	73	58	42	19
.60		101	85	67	39		90	74	48	32	104	74	60	45	23	89	61	49	36	16
.65		87	73	57	34	104	77	64	43	27	88	63	51	39	20	76	52	42	30	14
.70	100	75	63	50	29	90	66	55	38	24	76	55	44	34	17	66	45	36	26	12
.75	88	66	55	44	26	79	58	48	34	21	67	48	39	29	15	57	40	32	23	11
.80	77	58	49	39	23	70	51	43	33	19	59	42	34	26	14	50	35	28	21	10
.85	69	51	43	35	21	62	46	38	30	17	52	37	31	23	12	45	31	25	18	9
.90	62	46	39	31	19	55	41	34	27	15	47	34	27	21	11	40	28	22	16	8
.95	55	42	35	28	17	50	37	31	24	14	42	30	25	19	10	36	25	20	15	7
1.00	50	38	32	26	15	45	33	28	22	13	38	27	23	17	9	33	23	18	14	7
1.1	42	32	27	22	13	38	28	23	19	11	32	23	19	14	8	27	19	15	12	6
1.2	36	27	23	18	11	32	24	20	16	9	27	20	16	12	7	23	16	13	10	5
1.3	31	23	20	16	10	28	21	17	14	8	23	17	14	11	6	20	14	11	9	5
1.4	27	20	17	14	9	24	18	15	12	8	20	15	12	10	6	17	12	10	8	4
1.5	24	18	15	13	8	21	16	14	11	7	18	13	11	9	5	15	11	9	7	4

The five sub-columns under each α heading are values of $\beta = .01, .05, .1, .2, .5$.

TABLE J (continued)

Level of t-Test

Single-Sided Test	α = .005					α = .01					α = .025					α = .05				
Double-Sided Test	α = .01					α = .02					α = .05					α = .1				
β =	.01	.05	.1	.2	.5	.01	.05	.1	.2	.5	.01	.05	.1	.2	.5	.01	.05	.1	.2	.5
1.6	21	16	14	11	7	19	14	12	10	6	16	12	10	8	5	14	10	8	6	4
1.7	19	15	13	10	7	17	13	11	9	6	14	11	9	7	4	12	9	7	6	3
1.8	17	13	11	10	6	15	12	10	8	5	13	10	8	6	4	11	8	7	5	
1.9	16	12	11	9	6	14	11	9	8	5	12	9	7	6	4	10	7	6	5	
2.0	14	11	10	8	6	13	10	9	7	5	11	8	7	6	4	9	7	6	4	
2.1	13	10	9	8	5	12	9	8	7	5	10	8	6	5	3	8	6	5	4	
2.2	12	10	8	7	5	11	9	7	6	4	9	7	6	5		8	6	5	4	
2.3	11	9	8	7	5	10	8	7	6	4	9	7	6	5		7	5	5	4	
2.4	11	9	8	6	5	10	8	7	6	4	8	6	5	4		7	5	4	4	
2.5	10	8	7	6	4	9	7	6	5	4	8	6	5	4		6	5	4	3	
3.0	8	6	6	5	4	7	6	5	4	3	6	5	4	4		5	4	3		
3.5	6	5	5	4	3	6	5	4	4		5	4	4	3		4	3			
4.0	6	5	4	4		5	4	4	3		4	3				4				

$$\varrho = \frac{\sigma}{|\mu_1 - \mu_2|}$$

Source. Reprinted, with permission, from Davies, O. L. (1956), *Design and analysis of industrial experiments*, Vol. 1, Oliver and Boyd, London.

TABLE K Wilcoxon Distribution for Two Independent Samples

Header spanning columns 0–20: $w - ($ Minimum Value of $w)$

N_S	N_L	$\binom{N}{N_S}$	0	1	2	3	4	5	6	7	8	9	10	11	12	13	14	15	16	17	18	19	20
3	3	20	1	2	4	7	10	13	16	18	19	20											
3	4	35	1	2	4	7	11	15	20	24	28	31	33	34	35								
4	4	70	1	2	4	7	12	17	24	31	39	46	53	58	63	66	68	69	70				
3	5	56	1	2	4	7	11	16	22	28	34	40	45	49	52	54	55	56					
4	5	126	1	2	4	7	12	18	26	35	46	57	69	80	91	100	108	114	119	122	124	125	126
5	5	252	1	2	4	7	12	19	28	39	53	69	87	106	126	146	165	183	199	213	224	233	240
6	6	84	1	2	4	7	11	16	23	30	38	46	54	61	68	73	77	80	82	83	84		
4	6	210	1	2	4	7	12	18	27	37	50	64	80	96	114	130	146	160	173	183	192	198	203
5	6	462	1	2	4	7	12	19	29	41	57	76	99	124	153	183	215	247	279	309	338	363	386
6	6	924	1	2	4	7	12	19	30	43	61	83	111	143	182	224	272	323	378	433	491	546	601
3	7	120	1	2	4	7	11	16	23	31	40	50	60	70	80	89	97	104	109	113	116	118	119
4	7	330	1	2	4	7	12	18	27	38	52	68	87	107	130	153	177	200	223	243	262	278	292
5	7	792	1	2	4	7	12	19	29	42	59	80	106	136	171	210	253	299	347	396	445	493	539
6	7	1716	1	2	4	7	12	19	30	44	63	87	118	155	201	253	314	382	458	539	627	717	811
7	7	3432	1	2	4	7	12	19	30	45	65	91	125	167	220	283	358	445	545	657	782	918	1064
3	8	165	1	2	4	7	11	16	23	31	41	52	64	76	89	101	113	124	134	142	149	154	158
4	8	495	1	2	4	7	12	18	27	38	53	70	91	114	141	169	200	231	264	295	326	354	381
5	8	1287	1	2	4	7	12	19	29	42	60	82	110	143	183	228	280	337	400	466	536	607	680
6	8	3003	1	2	4	7	12	19	30	44	64	89	122	162	213	272	343	424	518	621	737	860	994
7	8	6435	1	2	4	7	12	19	30	45	66	93	129	174	232	302	388	489	609	746	904	1080	1277
8	8	12870	1	2	4	7	12	19	30	45	67	95	133	181	244	321	418	534	675	839	1033	1254	1509

Source. Tables K, L, M, and N reprinted from *Basic concepts of probability and statistics* by J. L. Hodges, Jr. and E. Lehman, published by Holden-Day, Inc.

TABLE L Auxiliary Table for the Wilcoxon Repeated Observations Test

N	2^N	N	2^N	$E(W)$	$SD(W)$	N	$E(W)$	$SD(W)$
1	2	11	2,048	33.0	11.25	21	115.5	28.77
2	4	12	4,096	39.0	12.75	22	126.5	30.80
3	8	13	8,192	45.0	14.31	23	138.0	32.88
4	16	14	16,384	52.0	15.93	24	150.0	35.00
5	32	15	32,768	60.0	17.61	25	162.5	37.17
6	64	16	65,536	68.0	19.34	26	175.5	39.37
7	128	17	131,072	76.5	21.12	27	189.0	41.62
8	256	18	262,144	85.5	22.96	28	203.0	43.91
9	512	19	524,288	95.0	24.85	29	217.5	46.25
10	1,024	20	1,048,576	105.0	26.79	30	232.5	48.62

TABLE M $N(W \leq w)$ for $w \leq N$ in the Wilcoxon Repeated Observations Test

w	$N(W \leq w)$	w	$N(W \leq w)$	w	$N(W \leq w)$	w	$N(W \leq w)$
0	1	6	14	11	55	16	169
1	2	7	19	12	70	17	207
2	3	8	25	13	88	18	253
3	5	9	33	14	110	19	307
4	7	10	43	15	137	20	371
5	10						

TABLE N $N(W \leq w)$ for $w \geq N$ in the Wilcoxon Repeated Observations Test

$w - N$	N3	4	5	6	7	8	9	10	11	12	13	14	15	16	17	18	19	20
1	6	9	13	18	24	32	42	54	69	87	109	136	168	206	252	306	370	446
2	7	11	16	22	30	40	52	67	85	107	134	166	204	250	304	368	444	533
3	8	13	19	27	37	49	64	82	104	131	163	201	247	301	365	441	530	634
4		14	22	32	44	59	77	99	126	158	196	242	296	360	436	525	629	751
5		15	25	37	52	70	92	119	151	189	235	289	353	429	518	622	744	886
6		16	27	42	60	82	109	141	179	225	279	343	419	508	612	734	876	1,041
7			29	46	68	95	127	165	211	265	329	405	494	598	720	862	1,027	1,219
8			30	50	76	108	146	192	246	310	386	475	579	701	843	1,008	1,200	1,422
9			31	54	84	121	167	221	285	361	450	554	676	818	983	1,175	1,397	1,653
10			32	57	91	135	188	252	328	417	521	643	785	950	1,142	1,374	1,620	1,916
11				59	98	148	210	285	374	478	600	742	907	1,099	1,321	1,577	1,873	2,213
12				61	104	161	233	320	423	545	687	852	1,044	1,266	1,522	1,818	2,158	2,548
13				62	109	174	256	356	476	617	782	974	1,196	1,452	1,748	2,088	2,478	2,926
14				63	114	186	279	394	532	695	886	1,108	1,364	1,660	2,000	2,390	2,838	3,350
15				64	118	197	302	433	591	779	999	1,254	1,550	1,890	2,280	2,728	3,240	3,825

TABLE N (continued)

w − N	N3	4	5	6	7	8	9	10	11	12	13	14	15	16	17	18	19	20
16					121	207	324	472	653	868	1,120	1,414	1,753	2,143	2,591	3,103	3,688	4,356
17					123	216	345	512	717	962	1,251	1,587	1,975	2,422	2,934	3,519	4,187	4,947
18					125	224	366	552	783	1,062	1,391	1,774	2,218	2,728	3,312	3,980	4,740	5,604
19					126	231	385	591	851	1,166	1,539	1,976	2,481	3,062	3,728	4,487	5,351	6,333
20					127	237	403	630	920	1,274	1,697	2,192	2,766	3,427	4,183	5,045	6,026	7,139
21					128	242	420	668	989	1,387	1,863	2,423	3,074	3,823	4,680	5,658	6,769	8,028
22						246	435	704	1,059	1,502	2,037	2,669	3,404	4,251	5,222	6,328	7,584	9,008
23						249	448	739	1,128	1,620	2,219	2,929	3,757	4,714	5,810	7,059	8,478	10,084
24						251	460	772	1,197	1,741	2,408	3,203	4,135	5,212	6,447	7,856	9,455	11,264
25						253	470	803	1,265	1,863	2,603	3,492	4,536	5,746	7,136	8,721	10,520	12,557
26						254	479	832	1,331	1,986	2,805	3,794	4,961	6,318	7,878	9,658	11,681	13,968
27						255	487	859	1,395	2,110	3,012	4,109	5,411	6,928	8,675	10,673	12,941	15,506
28						256	493	883	1,457	2,233	3,223	4,437	5,884	7,576	9,531	11,766	14,306	17,180
29							498	905	1,516	2,355	3,438	4,776	6,380	8,265	10,445	12,942	15,783	18,997
30							502	925	1,572	2,476	3,656	5,126	6,901	8,993	11,420	14,206	17,377	20,966

TABLE O Distribution Function of Fisher's F

x, for F(x) = .90

Denominator df	Num. df 1	2	3	4	5	6	7	8	9	10	12	15	20	24	30	40	60	120	∞
1	39.86	49.50	53.59	55.83	57.24	58.20	58.91	59.44	59.86	60.19	60.71	61.22	61.74	62.00	62.26	62.53	62.79	63.06	63.33
2	8.53	9.00	9.16	9.24	9.29	9.33	9.35	9.37	9.38	9.39	9.41	9.42	9.44	9.45	9.46	9.47	9.47	9.48	9.49
3	5.54	5.46	5.39	5.34	5.31	5.28	5.27	5.25	5.24	5.23	5.22	5.20	5.18	5.18	5.17	5.16	5.15	5.14	5.13
4	4.54	4.32	4.19	4.11	4.05	4.01	3.98	3.95	3.94	3.92	3.90	3.87	3.84	3.83	3.82	3.80	3.79	3.78	3.76
5	4.06	3.78	3.62	3.52	3.45	3.40	3.37	3.34	3.32	3.30	3.27	3.24	3.21	3.19	3.17	3.16	3.14	3.12	3.10
6	3.78	3.46	3.29	3.18	3.11	3.05	3.01	2.98	2.96	2.94	2.90	2.87	2.84	2.82	2.80	2.78	2.76	2.74	2.72
7	3.59	3.26	3.07	2.96	2.88	2.83	2.78	2.75	2.72	2.70	2.67	2.63	2.59	2.58	2.56	2.54	2.51	2.49	2.47
8	3.46	3.11	2.92	2.81	2.73	2.67	2.62	2.59	2.56	2.54	2.50	2.46	2.42	2.40	2.38	2.36	2.34	2.32	2.29
9	3.36	3.01	2.81	2.69	2.61	2.55	2.51	2.47	2.44	2.42	2.38	2.34	2.30	2.28	2.25	2.23	2.21	2.18	2.16
10	3.29	2.92	2.73	2.61	2.52	2.46	2.41	2.38	2.35	2.32	2.28	2.24	2.20	2.18	2.16	2.13	2.11	2.08	2.06
11	3.23	2.86	2.66	2.54	2.45	2.39	2.34	2.30	2.27	2.25	2.21	2.17	2.12	2.10	2.08	2.05	2.03	2.00	1.97
12	3.18	2.81	2.61	2.48	2.39	2.33	2.28	2.24	2.21	2.19	2.15	2.10	2.06	2.04	2.01	1.99	1.96	1.93	1.90
13	3.14	2.76	2.56	2.43	2.35	2.28	2.23	2.20	2.16	2.14	2.10	2.05	2.01	1.98	1.96	1.93	1.90	1.88	1.85
14	3.10	2.73	2.52	2.39	2.31	2.24	2.19	2.15	2.12	2.10	2.05	2.01	1.96	1.94	1.91	1.89	1.86	1.83	1.80
15	3.07	2.70	2.49	2.36	2.27	2.21	2.16	2.12	2.09	2.06	2.02	1.97	1.92	1.90	1.87	1.85	1.82	1.79	1.76
16	3.05	2.67	2.46	2.33	2.24	2.18	2.13	2.09	2.06	2.03	1.99	1.94	1.89	1.87	1.84	1.81	1.78	1.75	1.72
17	3.03	2.64	2.44	2.31	2.22	2.15	2.10	2.06	2.03	2.00	1.96	1.91	1.86	1.84	1.81	1.78	1.75	1.72	1.69
18	3.01	2.62	2.42	2.29	2.20	2.13	2.08	2.04	2.00	1.98	1.93	1.89	1.84	1.81	1.78	1.75	1.72	1.69	1.66
19	2.99	2.61	2.40	2.27	2.18	2.11	2.06	2.02	1.98	1.96	1.91	1.86	1.81	1.79	1.76	1.73	1.70	1.67	1.63
20	2.97	2.59	2.38	2.25	2.16	2.09	2.04	2.00	1.96	1.94	1.89	1.84	1.79	1.77	1.74	1.71	1.68	1.64	1.61
21	2.96	2.57	2.36	2.23	2.14	2.08	2.02	1.98	1.95	1.92	1.87	1.83	1.78	1.75	1.72	1.69	1.66	1.62	1.59
22	2.95	2.56	2.35	2.22	2.13	2.06	2.01	1.97	1.93	1.90	1.86	1.81	1.76	1.73	1.70	1.67	1.64	1.60	1.57
23	2.94	2.55	2.34	2.21	2.11	2.05	1.99	1.95	1.92	1.89	1.84	1.80	1.74	1.72	1.69	1.66	1.62	1.59	1.55
24	2.93	2.54	2.33	2.19	2.10	2.04	1.98	1.94	1.91	1.88	1.83	1.78	1.73	1.70	1.67	1.64	1.61	1.57	1.53
25	2.92	2.53	2.32	2.18	2.09	2.02	1.97	1.93	1.89	1.87	1.82	1.77	1.72	1.69	1.66	1.63	1.59	1.56	1.52
26	2.91	2.52	2.31	2.17	2.08	2.01	1.96	1.92	1.88	1.86	1.81	1.76	1.71	1.68	1.65	1.61	1.58	1.54	1.50
27	2.90	2.51	2.30	2.17	2.07	2.00	1.95	1.91	1.87	1.85	1.80	1.75	1.70	1.67	1.64	1.60	1.57	1.53	1.49
28	2.89	2.50	2.29	2.16	2.06	2.00	1.94	1.90	1.87	1.84	1.79	1.74	1.69	1.66	1.63	1.59	1.56	1.52	1.48
29	2.89	2.50	2.28	2.15	2.06	1.99	1.93	1.89	1.86	1.83	1.78	1.73	1.68	1.65	1.62	1.58	1.55	1.51	1.47
30	2.88	2.49	2.28	2.14	2.05	1.98	1.93	1.88	1.85	1.82	1.77	1.72	1.67	1.64	1.61	1.57	1.54	1.50	1.46
40	2.84	2.44	2.23	2.09	2.00	1.93	1.87	1.83	1.79	1.76	1.71	1.66	1.61	1.57	1.54	1.51	1.47	1.42	1.38
60	2.79	2.39	2.18	2.04	1.95	1.87	1.82	1.77	1.74	1.71	1.66	1.60	1.54	1.51	1.48	1.44	1.40	1.35	1.29
120	2.75	2.35	2.13	1.99	1.90	1.82	1.77	1.72	1.68	1.65	1.60	1.55	1.48	1.45	1.41	1.37	1.32	1.26	1.19
∞	2.71	2.30	2.08	1.94	1.85	1.77	1.72	1.67	1.63	1.60	1.55	1.49	1.42	1.38	1.34	1.30	1.24	1.17	1.00

Source. Reprinted, with permission, from Pearson, E. S. and Hartley, H. O. (1962), Biometrika tables for statisticians, Vol. 1, Biometrika Trustees, London.

TABLE O (continued)

x, for F(x) = .95

Num. df	1	2	3	4	5	6	7	8	9	10	12	15	20	24	30	40	60	120	∞
1	161.4	199.5	215.7	224.6	230.2	234.0	236.8	238.9	240.5	241.9	243.9	245.9	248.0	249.1	250.1	251.1	252.2	253.3	254.3
2	18.51	19.00	19.16	19.25	19.30	19.33	19.35	19.37	19.38	19.40	19.41	19.43	19.45	19.45	19.46	19.47	19.48	19.49	19.50
3	10.13	9.55	9.28	9.12	9.01	8.94	8.89	8.85	8.81	8.79	8.74	8.70	8.66	8.64	8.62	8.59	8.57	8.55	8.53
4	7.71	6.94	6.59	6.39	6.26	6.16	6.09	6.04	6.00	5.96	5.91	5.86	5.80	5.77	5.75	5.72	5.69	5.66	5.63
5	6.61	5.79	5.41	5.19	5.05	4.95	4.88	4.82	4.77	4.74	4.68	4.62	4.56	4.53	4.50	4.46	4.43	4.40	4.36
6	5.99	5.14	4.76	4.53	4.39	4.28	4.21	4.15	4.10	4.06	4.00	3.94	3.87	3.84	3.81	3.77	3.74	3.70	3.67
7	5.59	4.74	4.35	4.12	3.97	3.87	3.79	3.73	3.68	3.64	3.57	3.51	3.44	3.41	3.38	3.34	3.30	3.27	3.23
8	5.32	4.46	4.07	3.84	3.69	3.58	3.50	3.44	3.39	3.35	3.28	3.22	3.15	3.12	3.08	3.04	3.01	2.97	2.93
9	5.12	4.26	3.86	3.63	3.48	3.37	3.29	3.23	3.18	3.14	3.07	3.01	2.94	2.90	2.86	2.83	2.79	2.75	2.71
10	4.96	4.10	3.71	3.48	3.33	3.22	3.14	3.07	3.02	2.98	2.91	2.85	2.77	2.74	2.70	2.66	2.62	2.58	2.54
11	4.84	3.98	3.59	3.36	3.20	3.09	3.01	2.95	2.90	2.85	2.79	2.72	2.65	2.61	2.57	2.53	2.49	2.45	2.40
12	4.75	3.89	3.49	3.26	3.11	3.00	2.91	2.85	2.80	2.75	2.69	2.62	2.54	2.51	2.47	2.43	2.38	2.34	2.30
13	4.67	3.81	3.41	3.18	3.03	2.92	2.83	2.77	2.71	2.67	2.60	2.53	2.46	2.42	2.38	2.34	2.30	2.25	2.21
14	4.60	3.74	3.34	3.11	2.96	2.85	2.76	2.70	2.65	2.60	2.53	2.46	2.39	2.35	2.31	2.27	2.22	2.18	2.13
15	4.54	3.68	3.29	3.06	2.90	2.79	2.71	2.64	2.59	2.54	2.48	2.40	2.33	2.29	2.25	2.20	2.16	2.11	2.07
16	4.49	3.63	3.24	3.01	2.85	2.74	2.66	2.59	2.54	2.49	2.42	2.35	2.28	2.24	2.19	2.15	2.11	2.06	2.01
17	4.45	3.59	3.20	2.96	2.81	2.70	2.61	2.55	2.49	2.45	2.38	2.31	2.23	2.19	2.15	2.10	2.06	2.01	1.96
18	4.41	3.55	3.16	2.93	2.77	2.66	2.58	2.51	2.46	2.41	2.34	2.27	2.19	2.15	2.11	2.06	2.02	1.97	1.92
19	4.38	3.52	3.13	2.90	2.74	2.63	2.54	2.48	2.42	2.38	2.31	2.23	2.16	2.11	2.07	2.03	1.98	1.93	1.88
20	4.35	3.49	3.10	2.87	2.71	2.60	2.51	2.45	2.39	2.35	2.28	2.20	2.12	2.08	2.04	1.99	1.95	1.90	1.84
21	4.32	3.47	3.07	2.84	2.68	2.57	2.49	2.42	2.37	2.32	2.25	2.18	2.10	2.05	2.01	1.96	1.92	1.87	1.81
22	4.30	3.44	3.05	2.82	2.66	2.55	2.46	2.40	2.34	2.30	2.23	2.15	2.07	2.03	1.98	1.94	1.89	1.84	1.78
23	4.28	3.42	3.03	2.80	2.64	2.53	2.44	2.37	2.32	2.27	2.20	2.13	2.05	2.01	1.96	1.91	1.86	1.81	1.76
24	4.26	3.40	3.01	2.78	2.62	2.51	2.42	2.36	2.30	2.25	2.18	2.11	2.03	1.98	1.94	1.89	1.84	1.79	1.73
25	4.24	3.39	2.99	2.76	2.60	2.49	2.40	2.34	2.28	2.24	2.16	2.09	2.01	1.96	1.92	1.87	1.82	1.77	1.71
26	4.23	3.37	2.98	2.74	2.59	2.47	2.39	2.32	2.27	2.22	2.15	2.07	1.99	1.95	1.90	1.85	1.80	1.75	1.69
27	4.21	3.35	2.96	2.73	2.57	2.46	2.37	2.31	2.25	2.20	2.13	2.06	1.97	1.93	1.88	1.84	1.79	1.73	1.67
28	4.20	3.34	2.95	2.71	2.56	2.45	2.36	2.29	2.24	2.19	2.12	2.04	1.96	1.91	1.87	1.82	1.77	1.71	1.65
29	4.18	3.33	2.93	2.70	2.55	2.43	2.35	2.28	2.22	2.18	2.10	2.03	1.94	1.90	1.85	1.81	1.75	1.70	1.64
30	4.17	3.32	2.92	2.69	2.53	2.42	2.33	2.27	2.21	2.16	2.09	2.01	1.93	1.89	1.84	1.79	1.74	1.68	1.62
40	4.08	3.23	2.84	2.61	2.45	2.34	2.25	2.18	2.12	2.08	2.00	1.92	1.84	1.79	1.74	1.69	1.64	1.58	1.51
60	4.00	3.15	2.76	2.53	2.37	2.25	2.17	2.10	2.04	1.99	1.92	1.84	1.75	1.70	1.65	1.59	1.53	1.47	1.39
120	3.92	3.07	2.68	2.45	2.29	2.17	2.09	2.02	1.96	1.91	1.83	1.75	1.66	1.61	1.55	1.50	1.43	1.35	1.25
∞	3.84	3.00	2.60	2.37	2.21	2.10	2.01	1.94	1.88	1.83	1.75	1.67	1.57	1.52	1.46	1.39	1.32	1.22	1.00

Denominator df

Num. df	1	2	3	4	5	6	7	8	9	10	12	15	20	24	30	40	60	120	∞
1	647.8	799.5	864.2	899.6	921.8	937.1	948.2	956.7	963.3	968.6	976.7	984.9	993.1	997.2	1001	1006	1010	1014	1018
2	38.51	39.00	39.17	39.25	39.30	39.33	39.36	39.37	39.39	39.40	39.41	39.43	39.45	39.46	39.46	39.47	39.48	39.49	39.50
3	17.44	16.04	15.44	15.10	14.88	14.73	14.62	14.54	14.47	14.42	14.34	14.25	14.17	14.12	14.08	14.04	13.99	13.95	13.90
4	12.22	10.65	9.98	9.60	9.36	9.20	9.07	8.98	8.90	8.84	8.75	8.66	8.56	8.51	8.46	8.41	8.36	8.31	8.26
5	10.01	8.43	7.76	7.39	7.15	6.98	6.85	6.76	6.68	6.62	6.52	6.43	6.33	6.28	6.23	6.18	6.12	6.07	6.02
6	8.81	7.26	6.60	6.23	5.99	5.82	5.70	5.60	5.52	5.46	5.37	5.27	5.17	5.12	5.07	5.01	4.96	4.90	4.85
7	8.07	6.54	5.89	5.52	5.29	5.12	4.99	4.90	4.82	4.76	4.67	4.57	4.47	4.42	4.36	4.31	4.25	4.20	4.14
8	7.57	6.06	5.42	5.05	4.82	4.65	4.53	4.43	4.36	4.30	4.20	4.10	4.00	3.95	3.89	3.84	3.78	3.73	3.67
9	7.21	5.71	5.08	4.72	4.48	4.32	4.20	4.10	4.03	3.96	3.87	3.77	3.67	3.61	3.56	3.51	3.45	3.39	3.33
10	6.94	5.46	4.83	4.47	4.24	4.07	3.95	3.85	3.78	3.72	3.62	3.52	3.42	3.37	3.31	3.26	3.20	3.14	3.08
11	6.72	5.26	4.63	4.28	4.04	3.88	3.76	3.66	3.59	3.53	3.43	3.33	3.23	3.17	3.12	3.06	3.00	2.94	2.88
12	6.55	5.10	4.47	4.12	3.89	3.73	3.61	3.51	3.44	3.37	3.28	3.18	3.07	3.02	2.96	2.91	2.85	2.79	2.72
13	6.41	4.97	4.35	4.00	3.77	3.60	3.48	3.39	3.31	3.25	3.15	3.05	2.95	2.89	2.84	2.78	2.72	2.66	2.60
14	6.30	4.86	4.24	3.89	3.66	3.50	3.38	3.29	3.21	3.15	3.05	2.95	2.84	2.79	2.73	2.67	2.61	2.55	2.49
15	6.20	4.77	4.15	3.80	3.58	3.41	3.29	3.20	3.12	3.06	2.96	2.86	2.76	2.70	2.64	2.59	2.52	2.46	2.40
16	6.12	4.69	4.08	3.73	3.50	3.34	3.22	3.12	3.05	2.99	2.89	2.79	2.68	2.63	2.57	2.51	2.45	2.38	2.32
17	6.04	4.62	4.01	3.66	3.44	3.28	3.16	3.06	2.98	2.92	2.82	2.72	2.62	2.56	2.50	2.44	2.38	2.32	2.25
18	5.98	4.56	3.95	3.61	3.38	3.22	3.10	3.01	2.93	2.87	2.77	2.67	2.56	2.50	2.44	2.38	2.32	2.26	2.19
19	5.92	4.51	3.90	3.56	3.33	3.17	3.05	2.96	2.88	2.82	2.72	2.62	2.51	2.45	2.39	2.33	2.27	2.20	2.13
20	5.87	4.46	3.86	3.51	3.29	3.13	3.01	2.91	2.84	2.77	2.68	2.57	2.46	2.41	2.35	2.29	2.22	2.16	2.09
21	5.83	4.42	3.82	3.48	3.25	3.09	2.97	2.87	2.80	2.73	2.64	2.53	2.42	2.37	2.31	2.25	2.18	2.11	2.04
22	5.79	4.38	3.78	3.44	3.22	3.05	2.93	2.84	2.76	2.70	2.60	2.50	2.39	2.33	2.27	2.21	2.14	2.08	2.00
23	5.75	4.35	3.75	3.41	3.18	3.02	2.90	2.81	2.73	2.67	2.57	2.47	2.36	2.30	2.24	2.18	2.11	2.04	1.97
24	5.72	4.32	3.72	3.38	3.15	2.99	2.87	2.78	2.70	2.64	2.54	2.44	2.33	2.27	2.21	2.15	2.08	2.01	1.94
25	5.69	4.29	3.69	3.35	3.13	2.97	2.85	2.75	2.68	2.61	2.51	2.41	2.30	2.24	2.18	2.12	2.05	1.98	1.91
26	5.66	4.27	3.67	3.33	3.10	2.94	2.82	2.73	2.65	2.59	2.49	2.39	2.28	2.22	2.16	2.09	2.03	1.95	1.88
27	5.63	4.24	3.65	3.31	3.08	2.92	2.80	2.71	2.63	2.57	2.47	2.36	2.25	2.19	2.13	2.07	2.00	1.93	1.85
28	5.61	4.22	3.63	3.29	3.06	2.90	2.78	2.69	2.61	2.55	2.45	2.34	2.23	2.17	2.11	2.05	1.98	1.91	1.83
29	5.59	4.20	3.61	3.27	3.04	2.88	2.76	2.67	2.59	2.53	2.43	2.32	2.21	2.15	2.09	2.03	1.96	1.89	1.81
30	5.57	4.18	3.59	3.25	3.03	2.87	2.75	2.65	2.57	2.51	2.41	2.31	2.20	2.14	2.07	2.01	1.94	1.87	1.79
40	5.42	4.05	3.46	3.13	2.90	2.74	2.62	2.53	2.45	2.39	2.29	2.18	2.07	2.01	1.94	1.88	1.80	1.72	1.64
60	5.29	3.93	3.34	3.01	2.79	2.63	2.51	2.41	2.33	2.27	2.17	2.06	1.94	1.88	1.82	1.74	1.67	1.58	1.48
120	5.15	3.80	3.23	2.89	2.67	2.52	2.39	2.30	2.22	2.16	2.05	1.94	1.82	1.76	1.69	1.61	1.53	1.43	1.31
∞	5.02	3.69	3.12	2.79	2.57	2.41	2.29	2.19	2.11	2.05	1.94	1.83	1.71	1.64	1.57	1.48	1.39	1.27	1.00

Denominator df

TABLE O (continued)

x, for F(x) = .99

Denominator df \ Num. df	1	2	3	4	5	6	7	8	9	10	12	15	20	24	30	40	60	120	∞
1	4052	4999.5	5403	5625	5764	5859	5928	5982	6022	6056	6106	6157	6209	6235	6261	6287	6313	6339	6366
2	98.50	99.00	99.17	99.25	99.30	99.33	99.36	99.37	99.39	99.40	99.42	99.43	99.45	99.46	99.47	99.47	99.48	99.49	99.50
3	34.12	30.82	29.46	28.71	28.24	27.91	27.67	27.49	27.35	27.23	27.05	26.87	26.69	26.60	26.50	26.41	26.32	26.22	26.13
4	21.20	18.00	16.69	15.98	15.52	15.21	14.98	14.80	14.66	14.55	14.37	14.20	14.02	13.93	13.84	13.75	13.65	13.56	13.46
5	16.26	13.27	12.06	11.39	10.97	10.67	10.46	10.29	10.16	10.05	9.89	9.72	9.55	9.47	9.38	9.29	9.20	9.11	9.02
6	13.75	10.92	9.78	9.15	8.75	8.47	8.26	8.10	7.98	7.87	7.72	7.56	7.40	7.31	7.23	7.14	7.06	6.97	6.88
7	12.25	9.55	8.45	7.85	7.46	7.19	6.99	6.84	6.72	6.62	6.47	6.31	6.16	6.07	5.99	5.91	5.82	5.74	5.65
8	11.26	8.65	7.59	7.01	6.63	6.37	6.18	6.03	5.91	5.81	5.67	5.52	5.36	5.28	5.20	5.12	5.03	4.95	4.86
9	10.56	8.02	6.99	6.42	6.06	5.80	5.61	5.47	5.35	5.26	5.11	4.96	4.81	4.73	4.65	4.57	4.48	4.40	4.31
10	10.04	7.56	6.55	5.99	5.64	5.39	5.20	5.06	4.94	4.85	4.71	4.56	4.41	4.33	4.25	4.17	4.08	4.00	3.91
11	9.65	7.21	6.22	5.67	5.32	5.07	4.89	4.74	4.63	4.54	4.40	4.25	4.10	4.02	3.94	3.86	3.78	3.69	3.60
12	9.33	6.93	5.95	5.41	5.06	4.82	4.64	4.50	4.39	4.30	4.16	4.01	3.86	3.78	3.70	3.62	3.54	3.45	3.36
13	9.07	6.70	5.74	5.21	4.86	4.62	4.44	4.30	4.19	4.10	3.96	3.82	3.66	3.59	3.51	3.43	3.34	3.25	3.17
14	8.86	6.51	5.56	5.04	4.69	4.46	4.28	4.14	4.03	3.94	3.80	3.66	3.51	3.43	3.35	3.27	3.18	3.09	3.00
15	8.68	6.36	5.42	4.89	4.56	4.32	4.14	4.00	3.89	3.80	3.67	3.52	3.37	3.29	3.21	3.13	3.05	2.96	2.87
16	8.53	6.23	5.29	4.77	4.44	4.20	4.03	3.89	3.78	3.69	3.55	3.41	3.26	3.18	3.10	3.02	2.93	2.84	2.75
17	8.40	6.11	5.18	4.67	4.34	4.10	3.93	3.79	3.68	3.59	3.46	3.31	3.16	3.08	3.00	2.92	2.83	2.75	2.65
18	8.29	6.01	5.09	4.58	4.25	4.01	3.84	3.71	3.60	3.51	3.37	3.23	3.08	3.00	2.92	2.84	2.75	2.66	2.57
19	8.18	5.93	5.01	4.50	4.17	3.94	3.77	3.63	3.52	3.43	3.30	3.15	3.00	2.92	2.84	2.76	2.67	2.58	2.49
20	8.10	5.85	4.94	4.43	4.10	3.87	3.70	3.56	3.46	3.37	3.23	3.09	2.94	2.86	2.78	2.69	2.61	2.52	2.42
21	8.02	5.78	4.87	4.37	4.04	3.81	3.64	3.51	3.40	3.31	3.17	3.03	2.88	2.80	2.72	2.64	2.55	2.46	2.36
22	7.95	5.72	4.82	4.31	3.99	3.76	3.59	3.45	3.35	3.26	3.12	2.98	2.83	2.75	2.67	2.58	2.50	2.40	2.31
23	7.88	5.66	4.76	4.26	3.94	3.71	3.54	3.41	3.30	3.21	3.07	2.93	2.78	2.70	2.62	2.54	2.45	2.35	2.26
24	7.82	5.61	4.72	4.22	3.90	3.67	3.50	3.36	3.26	3.17	3.03	2.89	2.74	2.66	2.58	2.49	2.40	2.31	2.21
25	7.77	5.57	4.68	4.18	3.85	3.63	3.46	3.32	3.22	3.13	2.99	2.85	2.70	2.62	2.54	2.45	2.36	2.27	2.17
26	7.72	5.53	4.64	4.14	3.82	3.59	3.42	3.29	3.18	3.09	2.96	2.81	2.66	2.58	2.50	2.42	2.33	2.23	2.13
27	7.68	5.49	4.60	4.11	3.78	3.56	3.39	3.26	3.15	3.06	2.93	2.78	2.63	2.55	2.47	2.38	2.29	2.20	2.10
28	7.64	5.45	4.57	4.07	3.75	3.53	3.36	3.23	3.12	3.03	2.90	2.75	2.60	2.52	2.44	2.35	2.26	2.17	2.06
29	7.60	5.42	4.54	4.04	3.73	3.50	3.33	3.20	3.09	3.00	2.87	2.73	2.57	2.49	2.41	2.33	2.23	2.14	2.03
30	7.56	5.39	4.51	4.02	3.70	3.47	3.30	3.17	3.07	2.98	2.84	2.70	2.55	2.47	2.39	2.30	2.21	2.11	2.01
40	7.31	5.18	4.31	3.83	3.51	3.29	3.12	2.99	2.89	2.80	2.66	2.52	2.37	2.29	2.20	2.11	2.02	1.92	1.80
60	7.08	4.98	4.13	3.65	3.34	3.12	2.95	2.82	2.72	2.63	2.50	2.35	2.20	2.12	2.03	1.94	1.84	1.73	1.60
120	6.85	4.79	3.95	3.48	3.17	2.96	2.79	2.66	2.56	2.47	2.34	2.19	2.03	1.95	1.86	1.76	1.66	1.53	1.38
∞	6.63	4.61	3.78	3.32	3.02	2.80	2.64	2.51	2.41	2.32	2.18	2.04	1.88	1.79	1.70	1.59	1.47	1.32	1.00

x, for F(x) = .995

Num. df

Denominator df	∞	120	60	40	30	24	20	15	12	10	9	8	7	6	5	4	3	2	1
1	25465	25359	25253	25148	25044	24940	24836	24630	24426	24224	24091	23925	23715	23437	23056	22500	21615	20000	16211
2	199.5	199.5	199.5	199.5	199.5	199.5	199.4	199.4	199.4	199.4	199.4	199.4	199.4	199.3	199.3	199.2	199.2	199.0	198.5
3	41.83	41.99	42.15	42.31	42.47	42.62	42.78	43.08	43.39	43.69	43.88	44.13	44.43	44.84	45.39	46.19	47.47	49.80	55.55
4	19.32	19.47	19.61	19.75	19.89	20.03	20.17	20.44	20.70	20.97	21.14	21.35	21.62	21.97	22.46	23.15	24.26	26.28	31.33
5	12.14	12.27	12.40	12.53	12.66	12.78	12.90	13.15	13.38	13.62	13.77	13.96	14.20	14.51	14.94	15.56	16.53	18.31	22.78
6	8.88	9.00	9.12	9.24	9.36	9.47	9.59	9.81	10.03	10.25	10.39	10.57	10.79	11.07	11.46	12.03	12.92	14.54	18.63
7	7.08	7.19	7.31	7.42	7.53	7.65	7.75	7.97	8.18	8.38	8.51	8.68	8.89	9.16	9.52	10.05	10.88	12.40	16.24
8	5.95	6.06	6.18	6.29	6.40	6.50	6.61	6.81	7.01	7.21	7.34	7.50	7.69	7.95	8.30	8.81	9.60	11.04	14.69
9	5.19	5.30	5.41	5.52	5.62	5.73	5.83	6.03	6.23	6.42	6.54	6.69	6.88	7.13	7.47	7.96	8.72	10.11	13.61
10	4.64	4.75	4.86	4.97	5.07	5.17	5.27	5.47	5.66	5.85	5.97	6.12	6.30	6.54	6.87	7.34	8.08	9.43	12.83
11	4.23	4.34	4.44	4.55	4.65	4.76	4.86	5.05	5.24	5.42	5.54	5.68	5.86	6.10	6.42	6.88	7.60	8.91	12.23
12	3.90	4.01	4.12	4.23	4.33	4.43	4.53	4.72	4.91	5.09	5.20	5.35	5.52	5.76	6.07	6.52	7.23	8.51	11.75
13	3.65	3.76	3.87	3.97	4.07	4.17	4.27	4.46	4.64	4.82	4.94	5.08	5.25	5.48	5.79	6.23	6.93	8.19	11.37
14	3.44	3.55	3.66	3.76	3.86	3.96	4.06	4.25	4.43	4.60	4.72	4.86	5.03	5.26	5.56	6.00	6.68	7.92	11.06
15	3.26	3.37	3.48	3.58	3.69	3.79	3.88	4.07	4.25	4.42	4.54	4.67	4.85	5.07	5.37	5.80	6.48	7.70	10.80
16	3.11	3.22	3.33	3.44	3.54	3.64	3.73	3.92	4.10	4.27	4.38	4.52	4.69	4.91	5.21	5.64	6.30	7.51	10.58
17	2.98	3.10	3.21	3.31	3.41	3.51	3.61	3.79	3.97	4.14	4.25	4.39	4.56	4.78	5.07	5.50	6.16	7.35	10.38
18	2.87	2.99	3.10	3.20	3.30	3.40	3.50	3.68	3.86	4.03	4.14	4.28	4.44	4.66	4.96	5.37	6.03	7.21	10.22
19	2.78	2.89	3.00	3.11	3.21	3.31	3.40	3.59	3.76	3.93	4.04	4.18	4.34	4.56	4.85	5.27	5.92	7.09	10.07
20	2.69	2.81	2.92	3.02	3.12	3.22	3.32	3.50	3.68	3.85	3.96	4.09	4.26	4.47	4.76	5.17	5.82	6.99	9.94
21	2.61	2.73	2.84	2.95	3.05	3.15	3.24	3.43	3.60	3.77	3.88	4.01	4.18	4.39	4.68	5.09	5.73	6.89	9.83
22	2.55	2.66	2.77	2.88	2.98	3.08	3.18	3.36	3.54	3.70	3.81	3.94	4.11	4.32	4.61	5.02	5.65	6.81	9.73
23	2.48	2.60	2.71	2.82	2.92	3.02	3.12	3.30	3.47	3.64	3.75	3.88	4.05	4.26	4.54	4.95	5.58	6.73	9.63
24	2.43	2.55	2.66	2.77	2.87	2.97	3.06	3.25	3.42	3.59	3.69	3.83	3.99	4.20	4.49	4.89	5.52	6.66	9.55
25	2.38	2.50	2.61	2.72	2.82	2.92	3.01	3.20	3.37	3.54	3.64	3.78	3.94	4.15	4.43	4.84	5.46	6.60	9.48
26	2.33	2.45	2.56	2.67	2.77	2.87	2.97	3.15	3.33	3.49	3.60	3.73	3.89	4.10	4.38	4.79	5.41	6.54	9.41
27	2.29	2.41	2.52	2.63	2.73	2.83	2.93	3.11	3.28	3.45	3.56	3.69	3.85	4.06	4.34	4.74	5.36	6.49	9.34
28	2.25	2.37	2.48	2.59	2.69	2.79	2.89	3.07	3.25	3.41	3.52	3.65	3.81	4.02	4.30	4.70	5.32	6.44	9.28
29	2.24	2.33	2.45	2.56	2.66	2.76	2.86	3.04	3.21	3.38	3.48	3.61	3.77	3.98	4.26	4.66	5.28	6.40	9.23
30	2.18	2.30	2.42	2.52	2.63	2.73	2.82	3.01	3.18	3.34	3.45	3.58	3.74	3.95	4.23	4.62	5.24	6.35	9.18
40	1.93	2.06	2.18	2.30	2.40	2.50	2.60	2.78	2.95	3.12	3.22	3.35	3.51	3.71	3.99	4.37	4.98	6.07	8.83
60	1.69	1.83	1.96	2.08	2.19	2.29	2.39	2.57	2.74	2.90	3.01	3.13	3.29	3.49	3.76	4.14	4.73	5.79	8.49
120	1.43	1.61	1.75	1.87	1.98	2.09	2.19	2.37	2.54	2.71	2.81	2.93	3.09	3.28	3.55	3.92	4.50	5.54	8.18
∞	1.00	1.36	1.53	1.67	1.79	1.90	2.00	2.19	2.36	2.52	2.62	2.74	2.90	3.09	3.35	3.72	4.28	5.30	7.88

Denominator d

TABLE O (continued)

x, for F(x) = .999

Denominator df

Num. df	1	2	3	4	5	6	7	8	9	10	12	15	20	24	30	40	60	120	∞
1	4053ᵃ	5000ᵃ	5404ᵃ	5625ᵃ	5764ᵃ	5859ᵃ	5929ᵃ	5981ᵃ	6023ᵃ	6056ᵃ	6107ᵃ	6158ᵃ	6209ᵃ	6235ᵃ	6261ᵃ	6287ᵃ	6313ᵃ	6340ᵃ	6366ᵃ
2	998.5	999.0	999.2	999.2	999.3	999.3	999.4	999.4	999.4	999.4	999.4	999.4	999.4	999.5	999.5	999.5	999.5	999.5	999.5
3	167.0	148.5	141.1	137.1	134.6	132.8	131.6	130.6	129.9	129.2	128.3	127.4	126.4	125.9	125.4	125.0	124.5	124.0	123.5
4	74.14	61.25	56.18	53.44	51.71	50.53	49.66	49.00	48.47	48.05	47.41	46.76	46.10	45.77	45.43	45.09	44.75	44.40	44.05
5	47.18	37.12	33.20	31.09	29.75	28.84	28.16	27.64	27.24	26.92	26.42	25.91	25.39	25.14	24.87	24.60	24.33	24.06	23.79
6	35.51	27.00	23.70	21.92	20.81	20.03	19.46	19.03	18.69	18.41	17.99	17.56	17.12	16.89	16.67	16.44	16.21	15.99	15.75
7	29.25	21.69	18.77	17.19	16.21	15.52	15.02	14.63	14.33	14.08	13.71	13.32	12.93	12.73	12.53	12.33	12.12	11.91	11.70
8	25.42	18.49	15.83	14.39	13.49	12.86	12.40	12.04	11.77	11.54	11.19	10.84	10.48	10.30	10.11	9.92	9.73	9.53	9.33
9	22.86	16.39	13.90	12.56	11.71	11.13	10.70	10.37	10.11	9.89	9.57	9.24	8.90	8.72	8.55	8.37	8.19	8.00	7.81
10	21.04	14.91	12.55	11.28	10.48	9.92	9.52	9.20	8.96	8.75	8.45	8.13	7.80	7.64	7.47	7.30	7.12	6.94	6.76
11	19.69	13.81	11.56	10.35	9.58	9.05	8.66	8.35	8.12	7.92	7.63	7.32	7.01	6.85	6.68	6.52	6.35	6.17	6.00
12	18.64	12.97	10.80	9.63	8.89	8.38	8.00	7.71	7.48	7.29	7.00	6.71	6.40	6.25	6.09	5.93	5.76	5.59	5.42
13	17.81	12.31	10.21	9.07	8.35	7.86	7.49	7.21	6.98	6.80	6.52	6.23	5.93	5.78	5.63	5.47	5.30	5.14	4.97
14	17.14	11.78	9.73	8.62	7.92	7.43	7.08	6.80	6.58	6.40	6.13	5.85	5.56	5.41	5.25	5.10	4.94	4.77	4.60
15	16.59	11.34	9.34	8.25	7.57	7.09	6.74	6.47	6.26	6.08	5.81	5.54	5.25	5.10	4.95	4.80	4.64	4.47	4.31
16	16.12	10.97	9.00	7.94	7.27	6.81	6.46	6.19	5.98	5.81	5.55	5.27	4.99	4.85	4.70	4.54	4.39	4.23	4.06
17	15.72	10.66	8.73	7.68	7.02	6.56	6.22	5.96	5.75	5.58	5.32	5.05	4.78	4.63	4.48	4.33	4.18	4.02	3.85
18	15.38	10.39	8.49	7.46	6.81	6.35	6.02	5.76	5.56	5.39	5.13	4.87	4.59	4.45	4.30	4.15	4.00	3.84	3.67
19	15.08	10.16	8.28	7.26	6.62	6.18	5.85	5.59	5.39	5.22	4.97	4.70	4.43	4.29	4.14	3.99	3.84	3.68	3.51
20	14.82	9.95	8.10	7.10	6.46	6.02	5.69	5.44	5.24	5.08	4.82	4.56	4.29	4.15	4.00	3.86	3.70	3.54	3.38
21	14.59	9.77	7.94	6.95	6.32	5.88	5.56	5.31	5.11	4.95	4.70	4.44	4.17	4.03	3.88	3.74	3.58	3.42	3.26
22	14.38	9.61	7.80	6.81	6.19	5.76	5.44	5.19	4.99	4.83	4.58	4.33	4.06	3.92	3.78	3.63	3.48	3.32	3.15
23	14.19	9.47	7.67	6.69	6.08	5.65	5.33	5.09	4.89	4.73	4.48	4.23	3.96	3.82	3.68	3.53	3.38	3.22	3.05
24	14.03	9.34	7.55	6.59	5.98	5.55	5.23	4.99	4.80	4.64	4.39	4.14	3.87	3.74	3.59	3.45	3.29	3.14	2.97
25	13.88	9.22	7.45	6.49	5.88	5.46	5.15	4.91	4.71	4.56	4.31	4.06	3.79	3.66	3.52	3.37	3.22	3.06	2.89
26	13.74	9.12	7.36	6.41	5.80	5.38	5.07	4.83	4.64	4.48	4.24	3.99	3.72	3.59	3.44	3.30	3.15	2.99	2.82
27	13.61	9.02	7.27	6.33	5.73	5.31	5.00	4.76	4.57	4.41	4.17	3.92	3.66	3.52	3.38	3.23	3.08	2.92	2.75
28	13.50	8.93	7.19	6.25	5.66	5.24	4.93	4.69	4.50	4.35	4.11	3.86	3.60	3.46	3.32	3.18	3.02	2.86	2.69
29	13.39	8.85	7.12	6.19	5.59	5.18	4.87	4.64	4.45	4.29	4.05	3.80	3.54	3.41	3.27	3.12	2.97	2.81	2.64
30	13.29	8.77	7.05	6.12	5.53	5.12	4.82	4.58	4.39	4.24	4.00	3.75	3.49	3.36	3.22	3.07	2.92	2.76	2.59
40	12.61	8.25	6.60	5.70	5.13	4.73	4.44	4.21	4.02	3.87	3.64	3.40	3.15	3.01	2.87	2.73	2.57	2.41	2.23
60	11.97	7.76	6.17	5.31	4.76	4.37	4.09	3.87	3.69	3.54	3.31	3.08	2.83	2.69	2.55	2.41	2.25	2.08	1.89
120	11.38	7.32	5.79	4.95	4.42	4.04	3.77	3.55	3.38	3.24	3.02	2.78	2.53	2.40	2.26	2.11	1.95	1.76	1.54
∞	10.83	6.91	5.42	4.62	4.10	3.74	3.47	3.27	3.10	2.96	2.74	2.51	2.27	2.13	1.99	1.84	1.66	1.45	1.00

ᵃMultiply these entries by 100.

TABLE P Number of Degrees of Freedom Required for Comparison of Two Variances, Equal Sample Sizes

df	α = 0.01				α = 0.05				α = 0.5			
	β = 0.01	β = 0.05	β = 0.1	β = 0.5	β = 0.01	β = 0.05	β = 0.1	β = 0.5	β = 0.01	β = 0.05	β = 0.1	β = 0.5
1	16,420,000	654,200	161,500	4052	654,200	26,070	6,436	161.5	4,052	161.5	39.85	1.000
2	9,000	1,881	891.0	99.00	1,881	361.0	171.0	19.00	99.00	19.00	9.000	1.000
3	867.7	273.3	158.8	29.46	273.3	86.06	50.01	9.277	29.46	9.277	5.391	1.000
4	255.3	102.1	65.62	15.98	102.1	40.81	26.24	6.388	15.98	6.388	4.108	1.000
5	120.3	55.39	37.87	10.97	55.39	25.51	17.44	5.050	10.97	5.050	3.453	1.000
6	71.67	36.27	25.86	8.466	36.27	18.35	13.09	4.284	8.466	4.284	3.056	1.000
7	48.90	26.48	19.47	6.993	26.48	14.34	10.55	3.787	6.993	3.787	2.786	1.000
8	36.35	20.73	15.61	6.029	20.73	11.82	8.902	3.438	6.029	3.438	2.589	1.000
9	28.63	17.01	13.06	5.351	17.01	10.11	7.757	3.179	5.351	3.179	2.440	1.000
10	23.51	14.44	11.26	4.849	14.44	8.870	6.917	2.978	4.849	2.978	2.323	1.000
12	17.27	11.16	8.923	4.155	11.16	7.218	5.769	2.687	4.155	2.687	2.147	1.000
15	12.41	8.466	6.946	3.522	8.466	5.777	4.740	2.404	3.522	2.404	1.972	1.000
20	8.630	6.240	5.270	2.938	6.240	4.512	3.810	2.124	2.938	2.124	1.794	1.000
24	7.071	5.275	4.526	2.659	5.275	3.935	3.376	1.984	2.659	1.984	1.702	1.000
30	5.693	4.392	3.833	2.386	4.392	3.389	2.957	1.841	2.386	1.841	1.606	1.000
40	4.470	3.579	3.183	2.114	3.579	2.866	2.549	1.693	2.114	1.693	1.506	1.000
60	3.372	2.817	2.562	1.836	2.817	2.354	2.141	1.534	1.836	1.534	1.396	1.000
120	2.350	2.072	1.939	1.533	2.072	1.828	1.710	1.352	1.533	1.352	1.265	1.000
∞	1.000	1.000	1.000	1.000	1.000	1.000	1.000	1.000	1.000	1.000	1.000	1.000

$\delta^2 = \sigma_1^2/\sigma_2^2$ under H_A for an (α, β, δ), where $H_T: \sigma_1^2 = \sigma_2^2$ against $H_A: \sigma_1^2 > \sigma_2^2$.

Source. From *Selected techniques of statistical analysis* by Eisenhart, C., Hastey, M. W., and Wallis, W. A. Copyright 1947 by McGraw-Hill, Inc. Used with permission of McGraw-Hill Book Company.

$$\alpha = .01$$

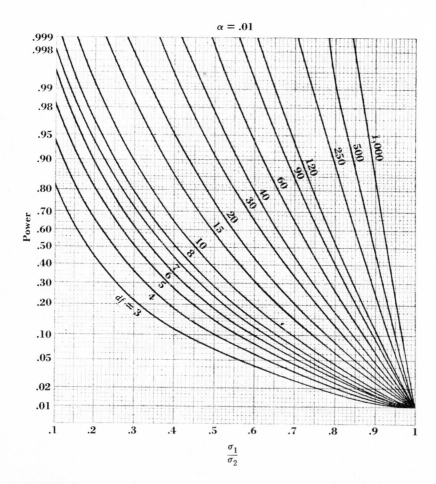

CHART Q

Power functions for two-sample tests of hypotheses about variances, equal sample size.

Source. From *Concepts of statistical interference* by Guenther, W. C. Copyright 1965 by McGraw-Hill Inc. Used with permission of McGraw-Hill Book Company.

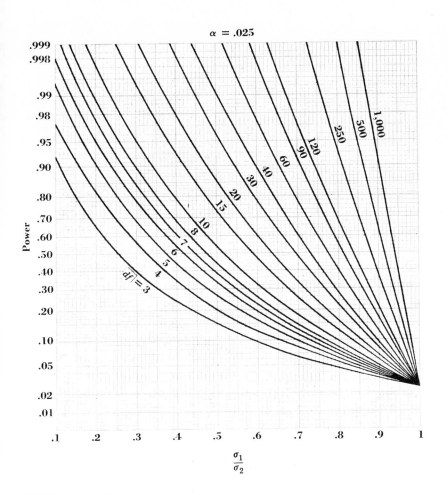

α = .025

Power

.999
.998
.99
.98
.95
.90
.80
.70
.60
.50
.40
.30
.20
.10
.05
.02
.01

.1   .2   .3   .4   .5   .6   .7   .8   .9   1

$\frac{\sigma_1}{\sigma_2}$

df = 3
4
5
6
7
8
10
15
20
30
40
60
90
120
250
500
1,000

CHART Q (continued)

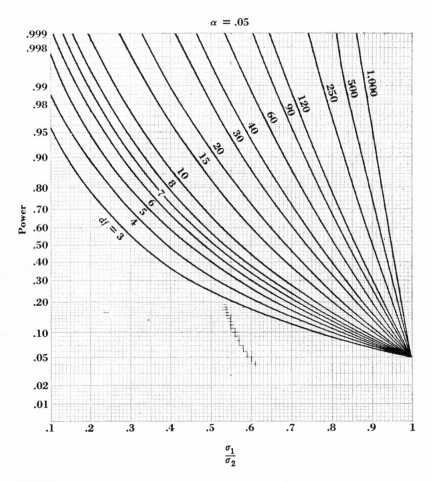

$\alpha = .05$

CHART Q (continued)

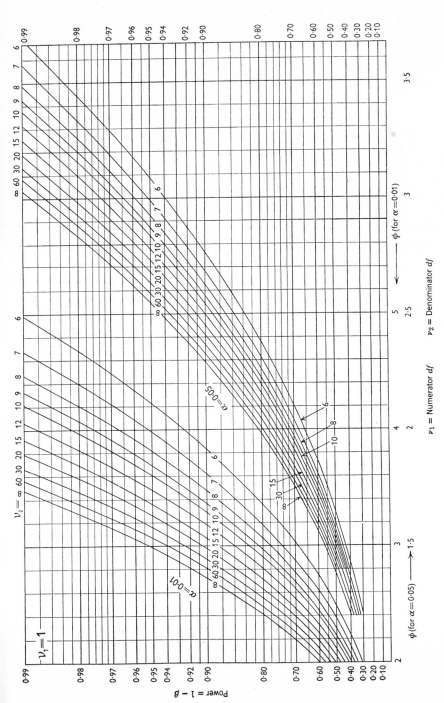

CHART R

Power functions for F-tests in the analysis of variance.

Source. Reprinted, with permission, from Vol. 38 of *Biometrika*.

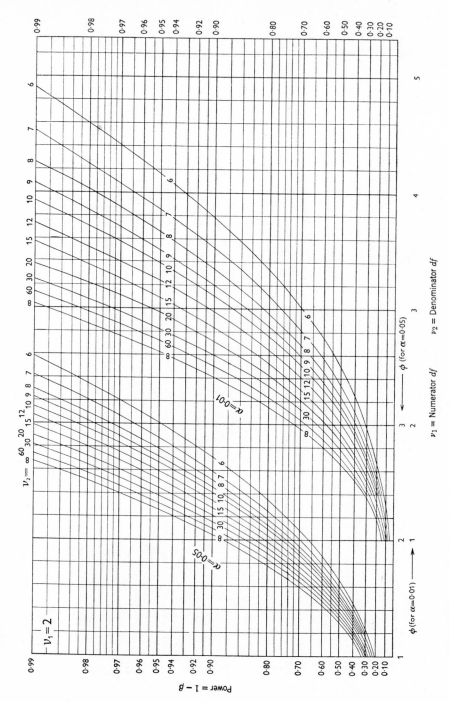

CHART R (continued)

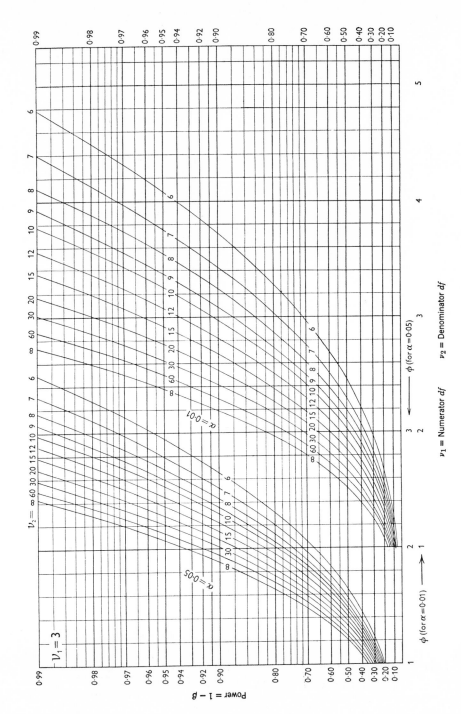

$\nu_1 = 3$

$\nu_2 = \infty$ 60 30 20 15 12 10 9 8 7 6

Power = $1 - \beta$

ϕ (for $\alpha = 0.01$) ⟶

$\alpha = 0.05$

$\alpha = 0.01$

ϕ (for $\alpha = 0.05$) ⟶

ν_1 = Numerator df ν_2 = Denominator df

CHART R (continued)

675

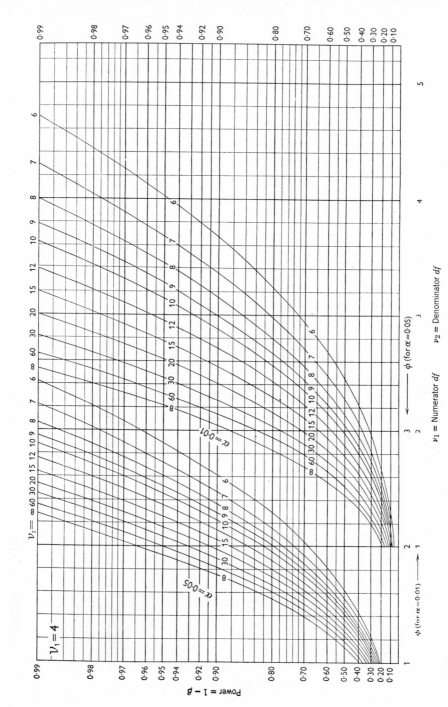

$\nu_1 = 4$

Power $= 1 - \beta$

ϕ (for $\alpha = 0.01$)

ϕ (for $\alpha = 0.05$)

$\nu_1 =$ Numerator df $\nu_2 =$ Denominator df

CHART R (continued)

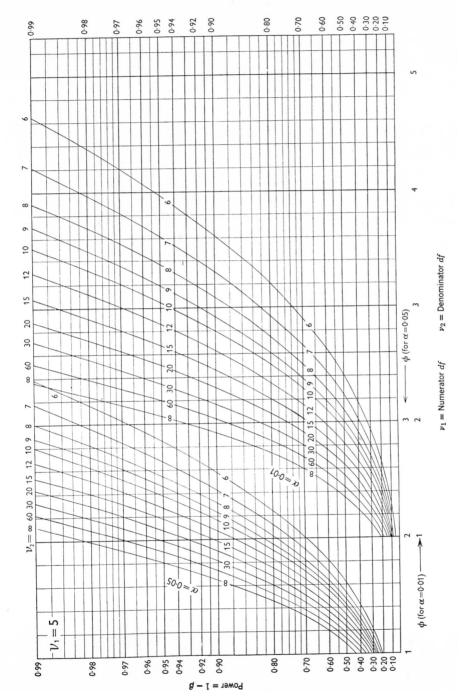

$\nu_1 = $ Numerator df $\nu_2 = $ Denominator df

CHART R (continued)

677

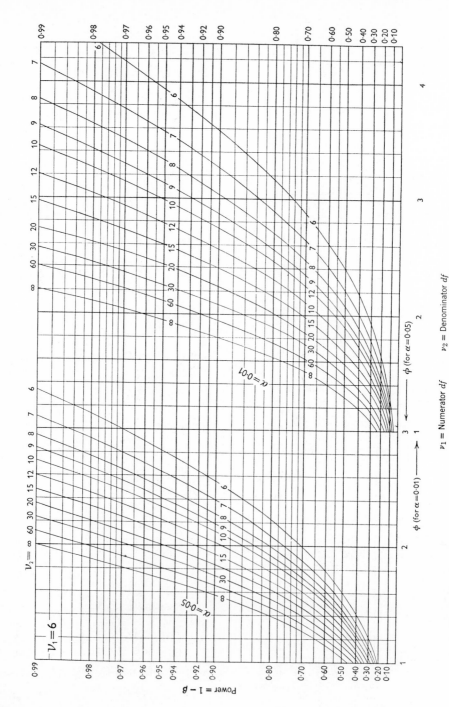

ν_1 = Numerator df ν_2 = Denominator df

CHART R (continued)

678

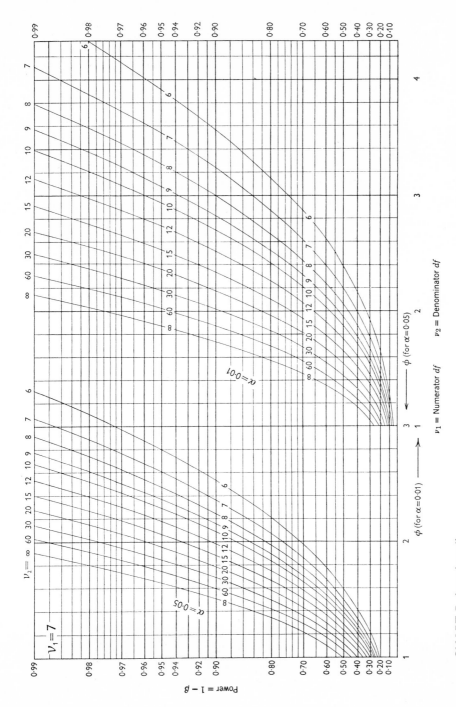

$\nu_1 = 7$

Power $= 1 - \beta$

ϕ (for $\alpha = 0.01$) ⟶

ϕ (for $\alpha = 0.05$) ⟶

$\alpha = 0.05$

$\alpha = 0.01$

$\nu_1 = $ Numerator df $\nu_2 = $ Denominator df

CHART R (continued)

CHART R (continued)

TABLE S Fisher's r to Z_r

r or ρ	z_r or z_ρ	r or ρ	z_r or z_ρ	r or ρ	z_r or z_ρ
.00	.0000	.35	.3654	.70	.8673
.01	.0100	.36	.3769	.71	.8872
.02	.0200	.37	.3884	.72	.9076
.03	.0300	.38	.4001	.73	.9287
.04	.0400	.39	.4110	.74	.9505
.05	.0500	.40	.4236	.75	.9730
.06	.0601	.41	.4356	.76	.9962
.07	.0701	.42	.4477	.77	1.0203
.08	.0802	.43	.4599	.78	1.0454
.09	.0902	.44	.4722	.79	1.0714
.10	.1003	.45	.4847	.80	1.0986
.11	.1104	.46	.4973	.81	1.1270
.12	.1206	.47	.5101	.82	1.1568
.13	.1307	.48	.5230	.83	1.1881
.14	.1409	.49	.5361	.84	1.2212
.15	.1511	.50	.5493	.85	1.2562
.16	.1614	.51	.5627	.86	1.2933
.17	.1717	.52	.5763	.87	1.3331
.18	.1820	.53	.5901	.88	1.3758
.19	.1923	.54	.6042	.89	1.4219
.20	.2027	.55	.6184	.90	1.4722
.21	.2132	.56	.6328	.91	1.5275
.22	.2237	.57	.6475	.92	1.5890
.23	.2342	.58	.6625	.93	1.6584
.24	.2448	.59	.6777	.94	1.7380
.25	.2554	.60	.6931	.95	1.8318
.26	.2661	.61	.7089	.96	1.9459
.27	.2769	.62	.7250	.961	1.9588
.28	.2877	.63	.7414	.962	1.9721
.29	.2986	.64	.7582	.963	1.9857
.30	.3095	.65	.7753	.964	1.9996
.31	.3205	.66	.7928	.965	2.0139
.32	.3316	.67	.8107	.966	2.0287
.33	.3428	.68	.8291	.967	2.0439
.34	.3541	.69	.8480	.968	2.0595

r or ρ	z_r or z_ρ	r or ρ	z_r or z_ρ	r or ρ	z_r or z_ρ
.969	2.0756	.979	2.2729	.989	2.5987
.970	2.0923	.980	2.2976	.990	2.6467
.971	2.1095	.981	2.3235	.991	2.6996
.972	2.1273	.982	2.3507	.992	2.7587
.973	2.1457	.983	2.3796	.993	2.8257
.974	2.1649	.984	2.4104	.994	2.9031
.975	2.1847	.985	2.4427	.995	2.9945
.976	2.2054	.986	2.4774	.996	3.1063
.977	2.2269	.987	2.5147	.997	3.2504
.978	2.2494	.988	2.5550	.998	3.4534

Source. Reprinted, with permission, from Owen, D. B. (1962), *Handbook of statistical tables*, John Wiley and Sons, Inc., New York.

Index